PROTEOMICS TODAY

WILEY-INTERSCIENCE SERIES IN MASS SPECTROMETRY

Series Editors

John R. de Laeter • *Applications of Inorganic Mass Spectrometry*
Michael Kinter and Nicholas E. Sherman • *Protein Sequencing and Identification Using Tandem Mass Spectrometry*
Chhabil Dass, *Principles and Practice of Biological Mass Spectrometry*
Mike S. Lee • *LC/MS Applications in Drug Development*
Jerzy Silberring and Rolf Eckman • *Mass Spectrometry and Hyphenated Techniques in Neuropeptide Research*
J. Wayne Rabalais • *Principles and Applications of Ion Scattering Spectrometry: Surface Chemical and Structural Analysis*
Mahmoud Hamdan and Pier Giorgio Righetti • *Proteomics Today: Protein Assessment and Biomarkers Using Mass Spectrometry, 2D Electrophoresis, and Microarray Technology*
Igor A. Kaltashov and Stephen J. Eyles • *Mass Spectrometry in Biophysics: Confirmation and Dynamics of Biomolecules*

PROTEOMICS TODAY

Protein Assessment and Biomarkers Using Mass Spectrometry, 2D Electrophoresis, and Microarray Technology

MAHMOUD HAMDAN
GlaxoSmithKline

PIER GIORGIO RIGHETTI
University of Verona, Italy

A JOHN WILEY & SONS, INC., PUBLICATION

Published by John Wiley & Sons, Inc., Hoboken, New Jersey.
Published simultaneously in Canada.

Library of Congress Cataloging-in-Publication Data

Proteomics today: Protein assessment and biomarkers using mass spectrometry, 2D electrophoresis, and microarray technology / Mahmoud Hamdan, Pier Giorgio Righetti.

p. cm.

Includes bibliographical references and index.

ISBN 0-471-64817-5 (cloth)

1. Proteins–Spectra. 2. Mass spectrometry. 3. Proteomics. I. Hamdan, Mahmoud, 1947-II. Righetti, P. G.

QP551.M315 2005
572′.6–dc22

2004050921

Printed in the United States of America

10 9 8 7 6 5 4 3 2 1

To Fatima, Giovanna, Mohammed, Jamilah
To my nephews Mohsin, Hamdan, Mahmoud and Suad

M.H.

CONTENTS

PREFACE TO PART I

Prior to starting Part I of this book, and while searching existing literature on current proteomic activities, I came across a number of statements that caught my attention and in a way have influenced my choice of the material in Part I. Therefore, it is reasonable to list them in here: First, mass spectrometry is a central component in modern proteomic research; second, the ability to determine statistically significant alterations in protein expression that might be provoked by disease, environmental, pharmacological, or genetic factors is a central component in current proteomic research; third, at the biochemical level, proteins rarely act alone, rather they interact with other proteins to perform a given cellular task. Although data obtained by various expression proteomics strategies have functional relevance by detecting changes in protein abundance, such measurements offer only an indirect readout of dynamics in protein activity. This means that numerous post-translational forms of protein regulation, including those governed by protein–protein interactions, remain undetected; and forth, biomarkers are not necessarily proteins. They can be DNA, RNA, or metabolites that can be associated with a measurable change with a given disease. Having stated that, protein-based analyses have two attractive features: First, proteins can be found regularly in blood, urine, and other biological fluids which make such approaches rather noninvasive; and second, proteins are the real executioners of various biological functions which make them key players in many diseases. To resolve the various issues contained within these statements, the last decade has witnessed an unprecedented use of a wide range of technologies and the fall of conventional barriers among the various disciplines. Within the wide host of technologies employed in the area of proteomic research, mass spectrometry emerges as a central component. The first three chapters of this book are an attempt to capture the recent contribution of this technology and its interaction with other technologies to tackle various proteomic challenges. The organization of Part I begins with a chapter dedicated to the major components of current mass spectrometers with a particular emphasis on the developments which have influenced sensitivity, resolution and the capability to perform various tandem mass spectrometry functions. Chapter 2 deals with the various approaches including mass spctrometry in the field of disease biomarkers. To provide a broader perspective of the role of mass spectrometry in this particular area, some material has been included that is not circumscribed by mass spectrometry, per se. I have dedicated Chapter 3 to the argument of protein quantification including the emerging strategies for the quantification of phosphorylated and glycosylated proteins.

Of course regardless of the size of any given book and the good intentions of the author, it would be a pure scientific arrogance to pretend that such book would provide a comprehensive coverage of the arguments raised within. On the other hand, I would like to think that this text will be looked upon by prospective readers as a contribution to a vast and continuously evolving debate, where single contributions are required to enrich and possibly stimulate such debate.

I cannot end this Preface without acknowledging that the years I spent at the University of Wales, Swansea working with Professor J. H. Beynon (founding editor-in-chief of the journal, *Rapid Communications of Mass Spectrometry*) had an immense impact on my appreciation of mass spectrometry as a tool that can be applied in a wide range of applications including present day proteomics.

MAHMOUD HAMDAN

Verona, Italy
November 2004

ACKNOWLEDGMENTS

The author of Part I of this book would like to acknowledge the scientific inspiration and the lasting friendship of professors, Keith Birkinshaw, Pietro Traldi, and John H. Beynon. The author is also grateful for the support and patience of all members of the Mass Spectrometry & Separation Technologies group at GlaxoSmithKline Research Centre in Verona, Italy.

M.H.

The striking developments in Part II of this book, reported as part of the research from my group in the field of immobilized pH gradients, would have been impossible without the heroic efforts of a group of close collaborators, among whom I would like to mention Drs. E. Gianazza, C. Gelfi, and M. Chiari. For our own developments in the field of capillary isoelectric focusing and the use of isoelectric buffers, I would like to express my appreciation of the work of Drs. C. Gelfi, A. Bossi, E. Olivieri, L. Castelletti, and B. Verzola. Our recent work in two-dimensional map analysis would have been impossible without the help and close collaboration of Drs. B. Herbert, A. Castagna, F. Antonucci, D. Cecconi, N. Campostrini, C. Rustichelli, and P. Antonioli. The research from my own group reported here has been supported over the years by grants from MURST (Ministero Università e Ricerca Scientifica e Technologica) in 1999 for Protein Folding and in 2000 for New Techniques in Proteome Analysis; by FIRB (Fondo Investimenti per Ricerca di Base) in 2001, Grant number RBNE01KJHT; and by PRIN (Progetti di Ricerca di Rilevante Interesse Nazionale) in 2003. Finally, I would like to thank the colleagues who have supplied me with original photographs of their work.

P.G.R

I

INTRODUCTION TO PART I

It can be argued that there are few if any analytical instruments that surpass the mass spectrometer in the diversity of its application in both basic and applied research. To appreciate such diversity, it is sufficient to point out that the same mass spectrometer can be used for applications extending from the characterization of electronic excited states and vibrational levels of simple molecules to the construction of protein interaction maps in multicellular organisms.

Although mass spectrometry (MS) was first conceived and applied almost a century ago by the celebrated physicist J. J. Thomson (1913), its application on a large scale had to wait for almost half a century. The application of the same technique to tackle biochemical problems in general and proteins in particular had to wait for almost another half a century. Limiting ourselves to the applications of MS in present-day proteomics, we find it relevant to point out some of the elements which directly or indirectly have influenced the current success of MS-based approaches in proteomics. Of course, the choice of these elements is meant to reflect the opinion of the present authors, which is not necessarily in full agreement with those of others; however, we retain that the reader is entitled to access different opinions regarding a fairly complex and continuously evolving subject.

1. *Ionizing Biomolecules.* For over 80 years ionization methods have been limited to thermally stable, nonpolar, and volatile molecules. These limitations on the ionization side have excluded large biomolecules, including proteins. In the early and late 1980s three ionization methods, fast atom/ion bombardment (FAB) (Barber et al.,

Proteomics Today. By Mahmoud Hamdan and Pier Giorgio Righetti
ISBN 0-471-64817-5

1981), electrospray ionization (ESI) (Yamashita and Fenn, 1984; Meng et al., 1988), and matrix-assisted laser desorption ionization (MALDI) (Tanaka et al., 1987, 1988; Karas and Hillenkamp, 1988), have become commercially available. These methods and in particular ESI and MALDI have given the user the right tools to generate biomolecular ions over a wide range of molecular masses. These simple, versatile and sensitive ionization methods, were soon coupled to a variety of analyzers such as triple quadrupoles, three-dimensional (3D) ion trap, time of flight (TOF), including its orthogonal version, which allowed the coupling of TOF to both pulsed (MALDI) and continuous (ESI) ionization. A further impetus was given to the process of ion analysis through the commercialization of a wide variety of hybrid configurations. These include TOF–TOF, ion mobility–TOF, ion trap–Fourier transform (FT)–ion cyclotron resonance (ICR), quadrupole–TOF, ion trap–TOF, and others. These variations had a direct impact on the sensitivity and resolution of the sequencing information, which can be obtained by performing various tandem MS analyses. Furthermore, some of these hybrid configurations could perform additional functions. For example, as well as for ion analysis, the configuration linear ion trap–FT–ICR is used for an enhanced external ion accumulation, which had a positive impact on the sensitivity and overall performance of this type of instrument, particularly when coupled to ESI and MALDI. The combination ion mobility–TOF is another example on how these hybrid analyzers can perform other functions. As well as ion analysis, this combination adds a further dimension to the separation process in the analysis of complex mixtures.

2. *Definition of Proteome.* The word *proteome*, first coined in 1994 at the Siena 2D Electrophoresis meeting, had a direct impact on the issues, which had to be addressed by MS-based approaches. In other words, it gave a reasonable indication on what type of information was needed and the type of technology to provide such information. The initial definition of this term was simply understood as "the identification and characterization of all proteins expressed by an organism or tissue." Deeper examination of the same term by those working in proteomics resulted in a more comprehensive definition. For example, it became evident that the word *characterization* also encompasses the determination of the function(s), localization, posttranslational modifications, and concentration in time of all expressed proteins. Subsequent studies have demonstrated that even this definition was not comprehensive enough. Within such a definition we should be able to correlate these elements to factors controlling them, such as environmental, pharmacological, genetic, and pathological circumstances. This extended definition of the word "proteome" has proved to be a valuable indicator to both instrument manufacturers and those using such instruments to perform a given task.

3. *Integrating MS with Other Methods.* One other element which contributed to the present-day role of proteomics is the realization that most MS-based approaches have to integrate other analytical methods. Protein quantification including posttranslational modifications and the construction of protein interaction maps are two representative examples where MS has to work in harmony with other components, such as separation methods, labeling, derivatization, and enrichment protocols. It is becoming more evident that, for MS to deliver its full promise in proteomic analyses,

MS not only has to keep pace with the emerging needs but also has to interact in harmony with other protocols, which have origins in both chemistry and biology. The concerted effort to optimize multidimentional chromatography is a good example of such integration.

4. *Bioinformatics.* The role of bioinformatics in current proteomic activities certainly needs more than one dedicated book to be described; therefore it is wise to limit ourselves to database-searching tools and how their combination with MS has become a central tool in any meaningful proteomic analysis. At the present time there are three main approaches for identifying proteins using MS data: (i) the top-down approach, which identifies proteins by matching the measured intact molecular masses against a protein sequence database; (ii) the bottom-up approach, which identifies proteins by matching masses of peptides and possibly fragmentation patterns of proteins against a protein sequence database; and (iii) de novo sequencing, which builds a protein/peptide sequence based on acquired peaks within a tandem mass spectrum. The bottom-up approach currently represents the preferred method for protein identification. It typically involves protein digestion, chromatographic separation, and MS–MS analysis. Two types of data can be used by this method: peptide mass fingerprinting (PMF) and MS–MS. The first type of data is commonly generated by MALDI–MS analysis of intact tryptic peptides, while the latter type of data is generally provided by collision-induced dissociation (CID) of a mass-to-charge (m/z) ratio associated with a given peptide. Currently there are a number of tools which can be used for high-throughput peptide identification, including SEQUEST, Mascot, SONAR, and others, which are described in the present text. Understandably, none of these tools is perfect and there are continuous attempts to improve their performance, in particular the reliability of assignment. Recently, statistical models have been developed to back these search tools. For example, a statistical approach was reported (Keller et al, 2002a) to assign reliabilities to peptide hits using a database consisting of 18 protein sequences derived from *Drosophila* with possible human contaminants (Keller et al., 2002b). This statistical model was designed to filter out a large number of database search results with predictable false identification rates. In another recent study, a support vector machine (SVM) technique was applied to separate correct SEQUEST identifications from incorrect ones (Anderson et al., 2003). This approach is a binary classifier that learns to distinguish between correctly and incorrectly identified peptides by using a vector of parameters describing each peptide identification. In a recent article Razumovskaya et al. (2004) have described a model combining a statistical decision-making procedure with a neural network. This model was used for assessing reliability of peptide identifications made by SEQUEST. What these examples and others in the recent literature indicate is a continuous and determined effort for improving the existing tools for MS-based protein identification.

At this point it is appropriate to ask what are the arguments which are capturing the imagination of the scientists working in the proteomic field? A partial answer to this question can be found within the material of this text; however we choose to underline some of them right at the onset of this text: (i) Having identified suitable strategies for high-throughput protein interaction networks, two challenging problems associated with this type of analysis are attracting attention: first, how to extend these strategies to more complex organisms and, second, how to separate credible interactions from artifacts and background noise. (ii) Accumulation of large-scale protein interaction networks are likely to enhance our understanding of various living organisms. This understanding in turn raises a question of extreme importance: How do these large-scale interactions reflect the functional properties of the various cellular compartments? This question is attracting various efforts based on the construction of protein interaction databases and the functional annotation of the various proteins. (iii) Protein tyrosine phosphorylation is another argument which is capturing an immense attention in current proteomic activities. This process controls diverse signaling pathways, and disregulated tryrosine kinase activity plays a direct role in human diseases, including various forms of cancer. Because activated kinases exert their effects by phosphorylating multiple substrate proteins, it is difficult or almost impossible to assess experimentally the contribution of a specific substrate to a cellular response or activity. Recently there have been some attempts (Sharma et al., 2003) to direct a modified tyrosine kinase to specifically phosphorylate a single substrate of choice in vivo, an approch which is termed *functional interaction trap*. (iv) Another emerging concept is tissue imaging by MS (Stoeckli et al., 2001), where tissue sections are directly analyzed by MALDI–TOF MS. The method, while currently not providing protein quantification, has already provided the proof-of-concept that clinically diagnostic protein patterns can be generated. The use of surface-enhanced laser desorption (SELDI) (Hutchens and Yip, 1993) as a tool for high-throughput biomarker discovery in a variety of biological and clinical samples is growing rapidly and will no doubt contribute to an area in current proteomic research aimed at translating basic research into clinical practice. (v) Protein quantification in general and the quantification of Posttranslationally modified proteins in particular remain the focus of vast proteomic activities. Although there has been some progress regarding the design of strategies for the quantification of phosphorylation and glycation, other forms of modifications are still difficult to capture. Ubiquitination is a good example of such a challenging task. There are rare examples on the use of multidimensional chromatography coupled to tandem mass spectrometry to assess such modification. There is no doubt that the coming years will see more effort along these lines to characterize and possibly quantify this biologically significant modification.

In Part I, the authors tried to capture the various proteomic efforts to tackle some of the above arguments. They also attempted to highlight the most recent development in MS instrumentation and their likely impact on the sensitivity, mass resolution, and sequencing information. Some of these recent development are associated with newly emerging hybrid instruments such as MALDI–TOF–TOF and linear Ion trap–FT–ICR, where the first can provide truly high-energy collision-induced dissociation, while the second can provide enhanced ion intensity combined with extremely high mass resolution, which will prove to be a valuable tool for mass tag measurements in high-throughput protein identification.

REFERENCES

Anderson, D. C., Li, W., Payan, D., Noble, W. S. (2003) *J. Proteome Res.* **2,** 137.

Barber, M., Bordoli, R. S., Sedgwick, R. D., Tetler, L. W. (1981) *Organic Mass Spectrom.* **16,** 256.

Hutchens, T. W., Yip, T. T. (1993) *Rapid Commun. Mass Spectrom.* **7,** 576.

Karas, M., Hillenkamp, F. (1988) *Anal. Chem.* **60,** 2299.

Keller, A., Nesvizhskii, A., Kolker, E., Aebersold, R. (2002a) *Anal. Chem.* **74,** 5383.

Keller, A., Purvine, S., Nesvizhskii, A., Stolyar, S. et al. (2002b) *OMICS.* **6,** 207.

Meng, C. K., Mann, M., Fenn, J. B., Phys. Z., D. (1988) **10,** 361.

Razumovskaya, J., Olman, V., Xu, D., Uberbacher, E. C., VerBerkmoes, N. C., Hettich, R. L., Xu, Y. (2004) *Proteomics.* **4,** 961.

Sharma, A., Antoku, S., Fujiwara, K., Mayer, B. (2003) *Mol. Cell. Proteomics.* **2.11,** 1217.

Stoeckli, M., Chaurand, P., Hallamann, D., Caprioli, R. M. (2001) *Nature Medicine.* **7,** 493.

Tanaka, K., Ido, Y., Akita, S., Yoshida, T. 2nd Japan-China Symposium on Mass Spectrom., Osaka,1987.

Tanaka, K., Ido, Y., Akita, S., Yoshida, Y., Yoshida, T. (1988) *Rapid Commun. Mass Spectrom.* **2,** 151.

Thomson, J. J., "Rays of positive electricity and their application to chemical analyses", Longmans Green, London, 1913.

Yamashita, M., Fenn, J. B. (1984) *J. Phys. Chem.* **88,** 4451.

1

INSTRUMENTATION AND DEVELOPMENTS

1.1. INTRODUCTION

The analysis of complex biochemical systems comprised of interacting proteins, peptides, and other components starts with the elucidation of the molecular weights of the various components followed by the identification of their amino acid sequences. A higher level in such analysis involves the identification of protein modifications, which affect protein localization and protein function. The last 30 years have witnessed the use of several complementary techniques to tackle such a task. Two-dimensional gel electrophoresis (2DE) is one of these techniques, which can provide excellent visualization of the biological complexity of the sample, rough quantitative estimate of the individual proteins, their approximate molecular weights, and the detection of certain types of posttranslational modifications. This powerful technique, however, has a number of limitations, including relatively long analysis time, limited possibility of automation, and poor results at low and very high molecular weights and at the extremes of working pH. Another technique is Edman degradation, which has been always considered the method of choice for amino acid sequencing. In this approach chemical reagents are used to remove one amino acid at a time from the unblocked terminus of intact protein or peptide. The resulting amino acid derivative is purified and identification is obtained by comparing the liquid chromatography (LC) retention time of the sample to those of standard amino acid derivatives. As in the case of 2DE there are a number of limitations associated with this approach: The technique requires an unmodified amino terminus and pure sample quantities in the range of 10 pmol. Furthermore, posttranslational modifications can result in anomalous retention times

Proteomics Today. By Mahmoud Hamdan and Pier Giorgio Righetti
ISBN 0-471-64817-5

with the direct result of wrong identification. Analysis times for a peptide containing 30 amino acids can be as long as 15 h.

Mass spectrometry has a number of intrinsic characteristics, which made it an attractive choice for addressing most of the limitations associated with 2DE and Edman degradation techniques. Mass spectrometry has high sensitivity, is highly amenable to automation, and requires low- to mid-femtomole sample quantities to yield faster and more reliable molecular weight and sequencing information. For example, the recent increase in the use of TOF and FT–ICR instruments allows the determination of protein molecular weights with almost two orders of magnitude improvement in mass accuracy over 2DE. Various tandem mass spectrometry functions in combination with powerful algorithms and database search allow sequencing and in some cases the detection of postranslational modifications in times as short as 1 min. Because of these and other capabilities, MS has developed into an essential tool for biochemical and biological research. The range of problems that MS is currently being applied to include the analysis of posttranslational modifications, protein–protein interactions, protein–deoxyribonucleic acid (DNA) and protein–ribonucleic acid (RNA) interactions, protein quantification, and proteins involved in signal transduction pathways.

Many of us within the MS community tend to link modern proteomics with the introduction of two ionization techniques: ESI and MALDI. In reality proteomic analysis (analysis of proteins and peptides) started a decade earlier with the introduction of two soft ionization techniques: ^{252}Cf plasma desorption (^{252}Cf PD), first introduced by Macfarlane and Torgerson (1976a), and FAB, first introduced by Barber's group (1981). It can be argued that the introduction of ESI and MALDI represented the most significant step toward the achievement of what today is considered modern proteomics. Having said that, the contribution of FAB and ^{252}Cf PD to our understanding of desorption ionization mechanism(s), the understanding and implementation of various scan modes, and how to use them to obtain sequencing information are good reasons to give these ionization methods certain attention, which will certainly help the reader to appreciate their contribution to what we call today proteome analysis.

1.2. IONIZATION TECHNIQUES FOR MACROMOLECULES

The introduction of plasma desorption ionization (Macfarlane and Torgerson, 1976a) can be considered the first tangible step toward the use of MS to investigate large biomolecules. In this technique ^{252}Cf gives two fission fragments, one of which is used to ionize the sample while the other travels in the opposite direction to be detected and used to establish the zero-time marker. This inherent characteristic of the ionization process allows fairly accurate *m*/*z* measurements of molecules having molecular masses in the kilodalton range. Despite a number of attractive features of this technique, the use of a radioactive element hindered its dissemination and limited its use to highly specialized research groups.

The second major development in biological MS came in the early 1980s when the late Barber and his group (1981) introduced FAB capable of ionizing nonvolatile and thermally labile large molecules. In this technique the analyte is dissolved in an excess of a suitable matrix and the resulting solution is bombarded by a kiloelectron-volt inert gas atomic beam or a Cs^+ ion beam. In a relatively short period of time this technique became a household name which could be coupled to a wide range of MS configurations.

It is no exaggeration to state that the introduction of FAB was one of the major steps to pave the way for the latter introduction of MALDI. For example, many of the ideas on which today's understanding of MALDI were proposed and discussed during the period in which FAB has dominated the scene. Some concepts such as the explosive expansion of high-pressure gas into vacuum, including the entertainment of analyte molecules, the role of the matrix in isolating the analyte molecules to reduce aggregation, and the role of preformed ions and ion–molecule reactions are all concepts which have been pointed out in the 1980s to explain the ionization mechanism and ion formation in FAB (Sunner, 1993).

The dominance of this technique in the field of peptide/protein analysis lasted all through the 1980s until the birth of commercial ESI (Yamashita and Fenn,1984) and MALDI ion sources (Tanaka et al., 1987, 1988; Karas and Hillenkamp, 1988). There is no doubt that both techniques have pushed MS-based approaches to the forefront of problem resolving in the areas of proteins, biochemistry, and biomedical research.

By definition, a mass spectrometer consists of an ion source, an ion analyzer, and a detector. Over the last 20 years all three elements have witnessed major developments, which have been reflected in various applications, particularly those relevant to proteomic analyses.

In today's world of MS there are two ionization methods, ESI and MALDI, which represent the first and possibly the only choice for the analysis of macromolecules. Prior to these methods the same analysis would have been conducted by the use of ^{252}Cf plasma desorption or FAB. Although both ionization methods have been surpassed by more powerful ionization methods, we believe there are two good reasons for including them in the present chapter. First, FAB and ^{252}Cf PD were the first ionization methods to bring MS to the arena of macromolecules. Second, the principles on which these methods are based and the experience gained through their application have paved the way to the birth of current ionization methods, in particular MALDI–TOF.

1.2.1. ^{252}Cf Plasma Desorption Ionization

This method was first developed by Macfarlane and Torgerson (1976a) and was the first to be used for the ionization of nonvolatile and thermally unstable molecules, including proteins and peptides. In this approach the radioactive element releases fission fragments in the mass range 80–160 Da which possess kinetic energy up to 200 MeV. In the case of ^{252}Cf, a typical fission fragment pair is $^{142}Ba^{18+}$ and $^{106}Tc^{22+}$, having kinetic energies of approximately 79 and 104 MeV, respectively. When either of these fission fragments hits a given sample, it dissipates its high kinetic energy within 10^{-12} s, resulting in a highly localized "hot spot" leading to both volatilization and ionization.

The first mass spectra obtained by ^{252}Cf ion source were reported by Macfarlane and Torgeson (1976b; Macfarlane, 1983). In such measurements, the sample is dissolved in an appropriate solvent and deposited on a thin (~1-μm) nickel foil. This sample foil is then precisely aligned very close to the ^{252}Cf source. One of the fission fragments penetrates the sample, and ions emerge from the opposite side and are accelerated into a TOF analyzer. Both positive and negative ions are generated, and masses have been measured with accuracy within 5×10^{-4} Da. This impressive accuracy is made possible because one of the fission fragments travels in the opposite direction to its companion, which penetrates the sample and therefore can be detected and used to establish a precise zero time.

Following the initial work by Macfarlane and Torgeson (1976a, b), the technique has been adapted by other research groups and its application to protein and peptide analysis has been consolidated to the extent that a commercial instrument housing this type of source was introduced in 1983 by a Swedish company, Bio-Ion, Nordic. Chait et al. (1984) used this technique to generate mass spectra of polyether oligomers having nominal molecular weights of 750–2000 Da. The use of ^{252}Cf MS in the investigation of proteins and peptides has been demonstrated by various groups; a summary of these works can be found in an extensive chapter by Macfarlane et al. (1991).

Although ^{252}Cf MS has amply demonstrated its capability for the analysis of proteins and peptides, the diffusion of this highly promising technique remained very slow and could not attract the required investment to facilitate its dissemination. The failure of this technique to attract enough users can be attributed to a number of reasons: First, the presence of a radioactive element was automatically associated with a hazardous working environment. Here we should point out that ^{252}Cf has a short half-life and its fission fragments have a very short range in solids, and therefore a well-shielded ion source should not represent a hazardous working environment. A more serious limitation was associated with the availability (rate of production) of ^{252}Cf, which could not be guaranteed easily for potential users of the technique. Second, at the time of its introduction in the mid-1970s the electronics serving TOF instruments together with the software backup were not as powerful as we know them today. The third and possibly main reason for the eclipse of ^{252}Cf ionization technique is the contemporary introduction of two simpler and more efficient ionization techniques, ESI and MALDI.

1.2.2. Fast Atom/Ion Bombardment

The FAB ionization method introduced by Barber's group (1981) was a natural extension to the earlier secondary ion mass spectrometry (SIMS) (Honig, 1958). In fact, the subsequent introduction of the Cs^+ion beam to replace Ar or Xe neutral beams for ion generation in FAB and the more frequent use of liquid matrices in SIMS made the differentiation between FAB and SIMS rather difficult. Very frequently, the terms FAB and liquid SIMS are interchanged. It is relevant to point out that the replacement of a neutral Ar or Xe beam with a Cs^+ ion beam has improved the sensitivity of the technique and its overall performance. This is because generation of the neutral primary beam was usually produced through resonant charge transfer in a collision cell operated at a relatively high pressure. This meant that acceleration above 8 keV resulted in voltage breakdown within the ion source and other vacuum-related problems. In the case of the Cs ion gun the pressure within the ion source was kept low and the desorbed ions could be accelerated at voltages as high as 30 kV. Developments prior to the commercialization of the modern FAB ion sources also included an interesting work by Devienne and Grandclement (1966). These authors reported ion sputtering from metallic surfaces bombarded with a kiloelectron-volt Ar beam. The approach was named molecular beam solid analysis (MBSA).

Two commercial versions of the FAB technique have been in use, static (Barber et al., 1981) and dynamic FAB (Ito et al., 1985; Caprioli et al., 1986). Both versions use the same type of ion gun but differ in the modality of sample delivery to the ion source. In the static mode the sample is loaded onto a probe tip in the form of a solution in a macroscopic droplet of the matrix (e.g., glycerol). This sample is then bombarded with a primary kiloelectron-volt beam of atoms or ions to produce secondary ions of both the analyte and the matrix. The dynamic FAB was devised to eliminate or at least diminish a number of negative aspects of its static counterpart. These drawbacks include very intense background chemical noise as a result of radiation damage and the presence of aggregates of matrix molecules, which reduced the eventual detection of the desired analyte ions. In the initial manifestation of the dynamic FAB (Ito et al., 1985), a dilute acqeous solution (95% water–5% glycerol) was forced through a porous frit to provide a thin film of the analyte–glycerol solution over a large surface area which is bombarded by the primary beam. A more popular version of the dynamic FAB, called continuous-flow (CF-FAB) (Caprioli et al., 1986), uses a microbore capillary that supplies a continuous flow (5–10 μL/min) of a dilute (5%) aqueous solution of glycerol to the bombardment area. Continuous-flow FAB was developed to accommodate certain LC/MS needs. In this type of arrangement a fused silica capillary [75 μm inside diameter (ID)] transfers the eluate from the LC column to the FAB bombardment target, commonly a stainless steel direct-insertion probe tip. Flow rates of 2–5 μL/min are nominally used with mild heating of the target, to about 50°C, to aid evaporation of the solvent. A wick made of porous cellulose fiber is also used to absorb excess liquid. The continuos flow of the solvent provides a cleansing action of the surface, thus minimizing memory effect. Continuous-flow FAB has also been coupled to capillary electrophoresis and was used by various groups for on-line separation and analysis of mixtures (Minard et al., 1988; Caprioli et al., 1989; Reinhoud et al., 1989)

Needless to say, at the present time the role of FAB in proteome research is almost nonexistent, and therefore we will not enter into further experimental details regarding this technique. Instead we will attempt to highlight some aspects, which no doubt had a profound influence on the current approaches, which paradoxically are responsible for the demise of the FAB technique. The emergence of ESI and MALDI and their combination with nonscanning analyzers such as the TOF put in focus the shortcomings of the FAB ionization method in handling high-throughput analysis of complex protein mixtures. However, before giving specific examples on such contribution the following general considerations are of interest: (i) A good part of our current knowledge of MALDI can be traced back to previous studies and developments of FAB ionization. This is not surprising in view of the fact that ^{252}Cf plasma desorption, FAB, and MALDI can be collectively described as desorption ionization methods; the first two use energetic particles, while the latter uses photons. Setting aside quantum conservation rules, which differ from particles to photons, we can reasonably assume that the initial step of energy deposition is a common feature of all three methods, and that is why the resulting mass spectra bear a close resemblance. (ii) The use of an energy-absorbing matrix is an essential element in the successful operation of all three methods; the matrix may vary, but its role is almost universally accepted. (iii) Recent developments in orthogonal TOF technology have removed the last barrier in ion analysis, which separates pulsed ionization methods (^{252}Cf, MALDI) and a continuos ion formation method such as FAB and ESI. (iv) For over a decade FAB has been the method of choice for coupling with different types of analyzers which allowed the development and refinement of various ion scan modes, nomenclature of peptide fragmentation, database searching, and other associated software development. Most of these developments have found their way to present-day proteomic approaches, in which FAB has been replaced by ESI and MALDI. To support the above considerations, we have chosen a number of aspects together with some examples which have matured due to the FAB technique.

1.2.3. Type of Fragment Ions and Nomenclature

Fragmentation of peptide ions generated by FAB was first discussed by Williams et al. (1981). This study together with several others which followed (Dass and Desiderio, 1987; Seki et al., 1985) have established that the fragmentation of a charged peptide to its specific charged fragments was the result of cleavage involving three types of bonds in the backbone of the peptide: the bond between the carbonyl group and the α-carbon (–CHR–CO–), the peptide amide bond (–CO–NH–), and the amino–alkyl bond (–NH–CHR–), with charge retention by either N- or C-terminus fragments. In addition to these major fragmentation channels, ions due to internal fragmentation formed by contemporary cleavage of two bonds in the peptide chain and certain immonium ions are also commonly observed. Furthermore, cleavage of the side chains of certain amino acids can also yield sequence-specific ions. Of course, these fragmentation pathways are based on experimental observations, which means that certain peptides may not display all expected sequence ion series. The nomenclature described by Roepstorf and Fohlman (1984) and the latter additions by Biemann (1988) are summarized in Scheme 1.1. It is worth noting here that d_n, w_n, and v_n fragment ions are only formed at relatively high (>500 eV) collision energy. Of these, the d_n and w_n ions allow the differentiation of leucine and isoleucine because they retain the β-carbon atom of the amino acid. These fragmentation rules and associated nomenclature have been based on FAB analysis and are used in today's MS–MS data interpretation.

1.3. EXAMPLES ON ANALYTICAL SOLUTIONS BASED ON FAB–MS

The versatility of a given analytical technique can be assessed by the diversity of the problems which such a technique can successfully tackle. For over 10 years FAB–MS has been the method of choice for peptide/protein analyses. During the same period it had been coupled to a wide range of analyzers and used to effect various scanning

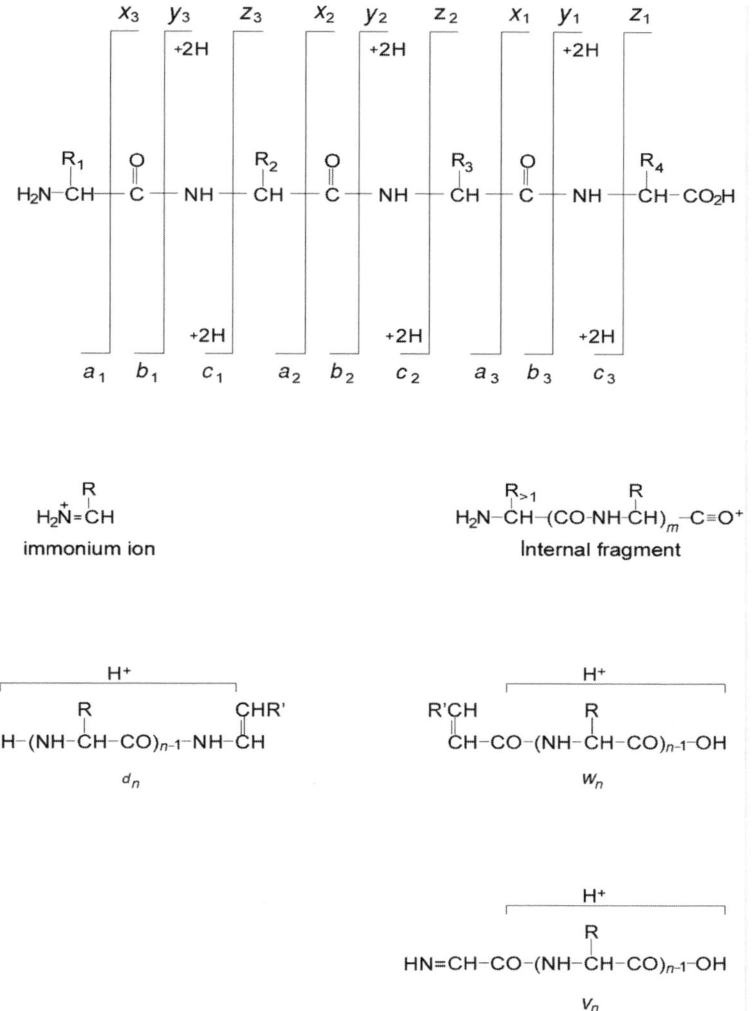

SCHEME 1.1. Nomenclature of possible sequence ions of peptides proposed by Roepstorf and Fohlman (1984) and extended by Biemann (1988).

functions both in the MS and MS/MS modes. To underline such a contribution, we have chosen two representative examples which can be easily described as the contribution of FAB to what is today called proteomic analyses.

1.3.1. Detection of Abnormalities in Hemoglobin

Abnormalities in hemoglobin, known as hemoglobinopathies, can be provoked by mutations in the DNA sequences responsible for the structure and expression of the globin chains. It is now commonly acknowledged that a single base change in a globin gene can change the codon for one amino acid to that of another. Most often, the result is a single-amino-acid substitution in the sequence of the synthesized protein. In fact, single-amino-acid replacement accounts for almost 95% of all known hemoglobin variants. Although the substitution of one amino acid out of 141 (α-chain) or 146 (β-chain) may sound like a small difference, it can have substantial influence on the function and stability of the entire molecule. For example, sickle cell anemia (Ingram, 1959) is caused by a β-chain variant which has a Glu replaced by Val in position 6. Other mutations can have more dramatic effects on the sequence of the chain. For instance, if the mutation occurs in one of the bases for the stop codon, which signals the termination of the polypeptide chain, it can be converted to an amino acid codon. In such a case the synthesized protein will have a C-terminal extension, as in the case with hemoglobin constant spring variant (Milner et al., 1971) and several other highly unstable hemoglobin variants (Huisman, 1989).

The capability of FAB-MS to provide specific information on the amino acid substitution in such variants can be appreciated by considering the two FAB mass spectra in Figures 1.1*a*, *b*. The two mass spectra were obtained by direct infusion of their respective digests into a FAB ion source mounted on a sector machine. The spectrum in Figure 1.1*a* refers to a normal β-chain, while the spectrum in Figure 1.1*b* is that of its (Hb Köln) variant. When comparing the two mass spectra, it is not difficult to observe that m/z 1127 is only observed in Figure 1.1*a*, while m/z 1159 is only present in spectrum Figure 1.1*b*. The difference of m/z 32 between the two peaks was correctly assumed to be due to the substitution of a single amino acid (Prome et al., 1987). However, such substitution can involve three different amino acids: Pro > Glu, Asp > Phe, or Val > Met. The authors rightly excluded the first two possibilities on the bases that both would require more than one base change in the DNA sequence and that both would have different electrophoretic mobilities compared to normal hemoglobin. In fact, such substitution was attributed to Val > Met.

1.3.2. Glycoprotein Structure Determination

Some 26 years ago, the first glycoprotein primary structure was determined successfully by an MS-based approach (Morris et al., 1978). These early studies of so-called antifreeze proteins led to the discovery of multiple disaccharides O-linked through threonines on a heterogeneous polypeptide backbone sequence. The ionization methods used for these early studies were electron impact (EI) and chemical ionization (CI), both of which required derivatization to confer volatility on the investigated sample.

The first technique to allow the analysis of intact polymer units, whether oligosaccharide, polypeptide, or glycopeptide, was FAB–MS. Of course, the introduction of ESI and MALDI in the late 1980s has dramatically reduced the role of FAB–MS and MS–MS in the characterization of glycoproteins. Having said that, FAB–MS–MS was commonly conducted on sector machines which allowed the use of relatively high collision energy together with mass accuracy and resolving power.

Without going into too much detail regarding specific examples of this particular argument, it is reasonable to state that high-mass FAB–MS and high-collision-energy

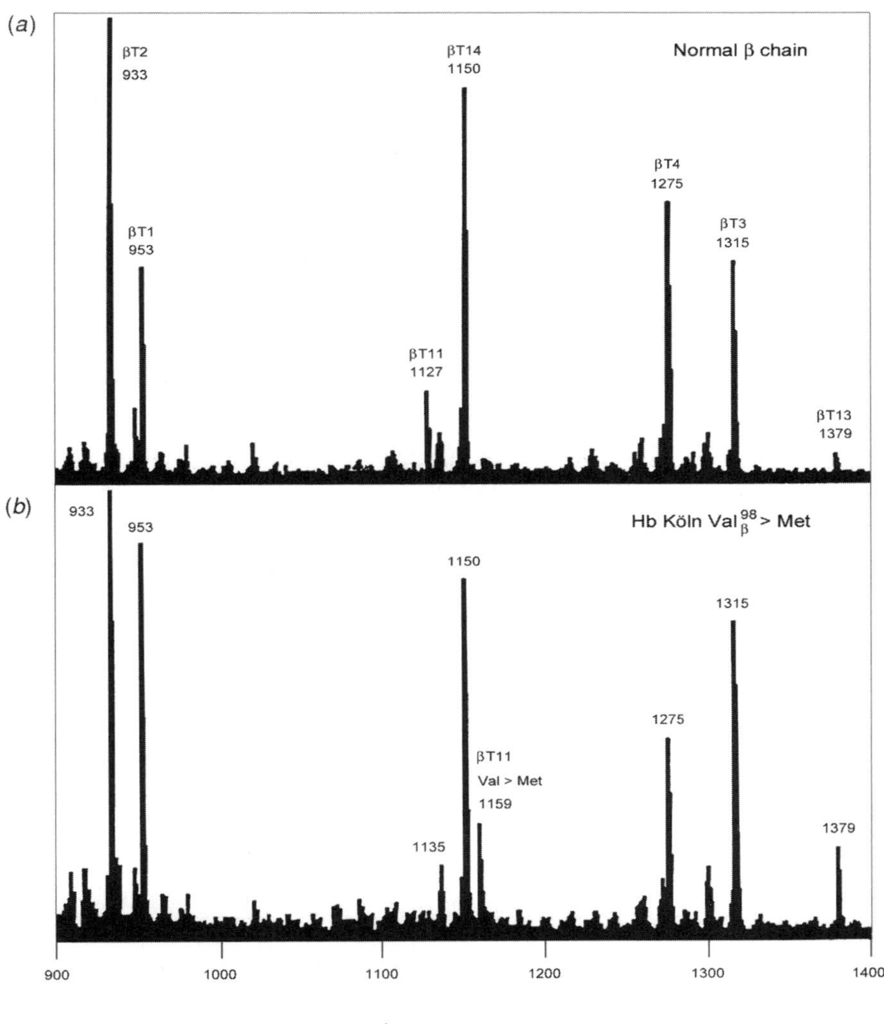

FIGURE 1.1. Positive-ion FAB mass spectra of (*a*) normal and (*b*) abnormal (Hb Köln) β-chain tryptic digest. (From Prome et al., 1987, with permission.)

FAB–CID have given substantial contribution to a wide range of modern protein and glycoprotein structural problems. This MS approach was commonly used as a complementary approach to Edman chemistry; however, the use of FAB analyses gave the user an additional ability to observe and assign blocking groups at the N-terminus or heterogeneity at the C-terminus. In the analysis of both recombinant and native proteins and glycoproteins, FAB mapping procedures were used to provide sequence information; detect errors of translation, mutation, insertion, and deletion; and of course identify the site of glycosylation.

1.3.3. Early Scanning Functions on Sector Machines

As we have stated earlier, the FAB source has been coupled to almost all types of analyzers both hybrid and none. In present day proteomics, the term *tandem mass spectrometry* (MS–MS) is automatically associated with the amino acid sequencing of m/z selected peptide. In this mode of operation, a precursor ion is selected by the first stage of the instrument and injected into a collision cell containing inert gas, while the second stage of the instrument is used to monitor the collision-induced fragmentation. In present-day proteomics sector machines are hardly mentioned because they have been replaced by TOF and ion-trapping instruments. The sector machines have been extensively used for FAB–MS and FAB–MS–MS, the latter usually conducted through the use of two scan functions, the first a daughter ion (B/E) scan, and the second a parent ion (B^2/E) scan. These scan functions are hardly mentioned in today's tandem mass spectrometry, yet during the 1980s these scan functions together with FAB were the main source of sequencing information for peptides and proteins.

The B/E scan is an MS/MS measurement commonly conducted on a double-focusing sector instrument composed of a magnet (B) and an electrostatic analyzer (E). The fragmentation takes place in the first field free region (between the ion source and the first analyzer). Such fragmentation is a combination of unimolecular fragmentation of metastable ions and collision-induced dissociation aided by the introduction of a collision gas. In this scan mode the parent ion is selected by the magnet, while scanning both the electrostatic analyzer and the magnet in a linear mode such that B/E is constant would yield B/E spectra associated with the selected parent ion. A critical factor in this type of measurement is the calibration of the desired scan range of the magnet, which can be established through the use of a simple formula: $M* =$ (lowest daughter ion mass)^2/ (parent ion mass). This means that to observe a daughter ion having 200 Da derived from a parent ion of 2000 Da, the magnet has to be scanned down to 20 Da ($M*$)

The B^2/E scan can identify the parent(s) ions that have fragmented in the first free region to give various daughter ions. The lowest mass in the calibration range of the magnet is established by the formula $M* =$ (daughter ion mass)^2 / (highest parent ion mass). This means that, to detect all the parents of m/z 200 daughter and assuming that the highest parent mass is 2500 Da, the magnet has to be calibrated down to 16 Da ($M*$). To know more about these scan functions, the reader is referred to Jennings and Mason (1983).

The rest of this chapter will be dedicated to the various components of a mass spectrometer, ion sources (limited to ESI and MALDI), analyzers, and ion detectors.

1.4. ELECTROSPRAY IONIZATION

In the late 1960s Dole etal. (1968) demonstrated that electrospraying a liquid containing analyte molecules could generate sufficient ions for subsequent analysis. This approach was revisited by Yamashita and Fenn (1984), who demonstrated the capability of electrospray for small molecules, while Meng et al. (1988) demonstrated the same potential for proteins. It should also be pointed out that, independently and at about the same time, Russian groups have described similar results on small molecules (Alexandrov et al., 1984). Although it was its capability to analyze intact proteins which sparked widespread interest in electrospray mass spectrometry, the technique is now an indispensable tool in a wide range of applications for both small and large molecules.

The simplicity of ESI can be appreciated by the fact that the original version of such a source proposed by Yamashita and Fenn in 1984 is almost identical to current-day ion sources. Basically, a simple mobile-phase (organic solvent–water) together with a given analyte are introduced at few microliters per minute flow rate into a stainless steel hypodermic needle (0.1 mm i.d.) chamfered at the end to form a sharp-edged conical tip which is maintained at 1–5 kV relative to ground. This tip is positioned at a few millimeters distances from a sampling cone (orifice), which separates the atmospheric pressure region from an intermediate-pressure region and the rest of the mass spectrometer. Surrounding the stainless steel capillary there is a concentric tube which carries N_2 gas to the tip of the capillary for coaxial nebulization of the exiting liquid. Of course, there are variations on this basic configurations introduced by various manufacturers regarding dimension of the source, distances, flow rates, direction of spray, operating temperature of the source, and others.

In the mid-1990s a new version of the ESI ion source operated at much lower flow rate (NanoESI) was introduced (Wilm and Mann, 1996). This version is based on a theoretical model for the electrostatic dispersion of liquid in the electrospray which treats the spraying process in analogy with liquid metal ion sources (Wilm and Mann, 1994). The low flow (20–40 nL/min) in this version was realized with gold-coated capillaries drawn with a short taper to a fine tip of $\sim$1 μm i.d. About 1 μL of liquid can be directly loaded into the capillary, and at such flow rates it can last for an hour. In the early days of its introduction NanoES was exclusively used in the direct-infusion mode; however, subsequent development and improvement in chromatography, in particular in the area of microcapillary LC, have made NanoES the ion source of choice for LC/MS and LC/MS–MS analysis of complex protein mixtures.

Current literature cites two models for ion formation from charged droplets: the charge–residue mechanism first proposed by Dole et al. (1968) and the ion evaporation model proposed by Thomson and Iribarne (1979). The first model assumes that following substantial solvent evaporation from a charged droplet the charge density on the surface would increase, reaching the Rayleigh limit, at which the Coulombic repulsive forces between the surface charges would exceed the surface tension, leading to a subdivision of the parent droplet. If this division process continued far enough and provided the original solution was dilute enough, a state would be reached in which each droplet would contain a single molecule, which would retain part of the original charge.

In the ion evaporation model it is assumed that the initial droplet size distribution is in the range 1–3 μm with a charge approaching the Rayleigh limit. Due to evaporation, the droplets will shrink so that charge must be reduced. For large droplets ($>$1 μm), droplet breakup is generally asymmetric and occurs at some fraction of the actual Rayleigh limit. Such droplets have been observed to form a protrusion from which ions

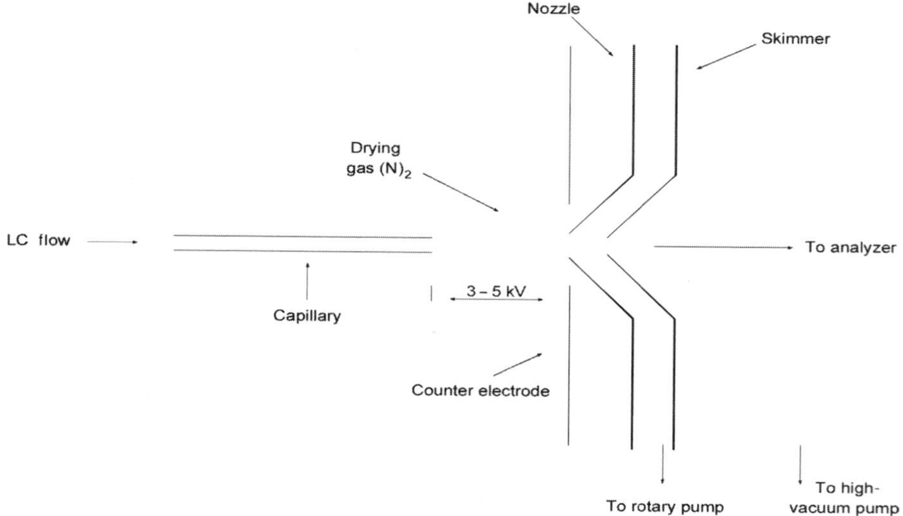

FIGURE 1.2. Schematic representation of main components of electrospray ion source.

and smaller charged droplets are emitted, typically shedding more than 10% of their charge, and about one order of magnitude less mass (Taflin et al., 1989). Furthermore, it has been postulated that droplets undergo several such fission processes, yielding smaller and smaller droplets until the electric field at the droplet surface is sufficient for ion evaporation.

Without going into too many theoretical details, it is not difficult to observe that neither mechanism can give a full account of gas-phase ion formation from a liquid droplet. In fact, clear differences between the charge–residue mechanism and the ion evaporation mechanism are difficult to identify. The first mechanism proposes the formation of droplets which may not be much greater in size than the molecular ion itself, while the other mechanism predicts field-induced evaporation of a substantially solvated molecular ion. In both mechanisms, however, the most speculative step in the ESI process is the mechanism by which a macromolecular ion is transferred to the gas phase. To describe such a transformation as the result of Coulombic forces in combination with evaporation is an oversimplification of a fairly complex sequence of events; however, in the absence of a precise model for such a process, such a description should do for now. The basic components in a typical ESI ion source are schematically given in Figure 1.2.

1.5. MATRIX-ASSISTED LASER DESORPTION IONIZATION

Matrix-assisted laser desorption ionization was first described by Tanaka et al. (1987, 1988) and Karas and Hillenkamp (1988). Together with ESI, MALDI is now among the most important methods for the ionization of nonvolatile, high-molecular-weight compounds, in particular peptides, proteins, oligonucleotides, oligosaccharides, and

TABLE 1.1. Laser Type, Associated Wavelength, Energy, and Pulse Width

Laser Type	Wavelength	Photon Energy (eV)	Pulse Width
Nitrogen	337 nm	3.68	<1–few ns
Nd: YAGx3	355 nm	3.49	Up to 5 ns
Nd: YAGx4	266 nm	4.66	Up to 5 ns
Excimer (XeCl)	308 nm	4.02	Up to 25 ns
Excimer (KrF)	248 nm	5.0	Up to 25 ns
Excimer (ArF)	193 nm	6.42	Typically 15 ns
Er: YAG	2.94 μm	0.42	85 ns
Carbon dioxide	10.6 μm	0.12	100 ns + 1-μs tail

synthetic polymers of high molecular weights. In this approach the sample to be analyzed is dissolved in an appropriate solvent at a concentration of a few picomoles per microliter; 1–2 μL of this solution is mixed with an equal volume of a freshly prepared matrix solution. As the matrix solution is generally at a concentration of ~10 mg/L, this results in an excess of matrix compared to the sample. An amount (1–2 μL) of the final solution is applied to the target plate, which is then left to dry at room temperature. At this point the target plate can be introduced into the ion source of the mass spectrometer. Currently, all the steps described above can be effected automatically through the use of robotic and other sample preparation devices. Although both ultravilot (UV) and infrared (IR) laser beams can be used in MALDI, literature over the last 15 years has shown that UV of 337 nm wavelength derived from N_2 laser is the most popular commercial laser source. Other forms of laser lines suitable for similar analysis are summarized in Table 1.1. The pulsed nature of the ion generation in MALDI has made TOF analyzers the ideal choice for coupling with this type of source, which is schematically represented in Figure 1.3.

Attempts to understand the mechanism(s) of ion formation in MALDI is still attracting various theoretical and experimental efforts. So far a number of chemical and physical pathways have been suggested for ion formation under MALDI conditions.

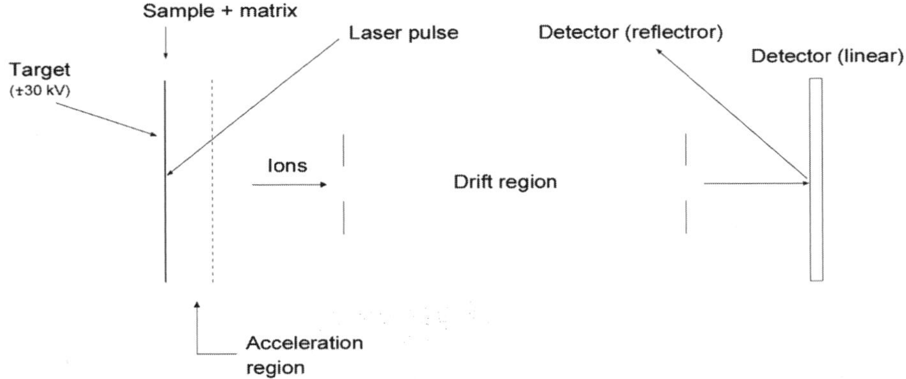

FIGURE 1.3. Simplified representation of MALDI–TOF arrangement.

TABLE 1.2. Commercial MALDI Matrices and Likely Application Areas

Matrix	Application
3,5-Dimethoxy-4-hydroxycinnamic acid (sinapinic acid)	Proteins, peptides, polymers
α-Cyano-4-hydroxycinnamic acid (α-cyano)	Peptides, polymers
2,5-Dihydroxybenzoic acid (DHB)	Sugars, peptides, nucleotides, polymers
2,4,6-Trihydroxyacetophenone (THAP)	Oligonucleotides
Hydroxypicolinic acid (HPA)	Oligonucleotides, peptides, glycoproteins
β-Indole acrylic acid (IAA)	Polymethyl methacylates
2-(4-Hydroxyphenylazo)benzoic acid (HABA)	Glycolipids, peptides, proteins
Dithranol	Polymers
5-Methoxysaliclic acid + 2,5-dihydroxybenzoic acid (1 : 10)	Peptides, proteins
α-Cyano-4-hydroxycinnamic acid + 2,5-dihydroxybenzoic acid (3 : 5)	Polymers

These include gas-phase photoionization, ion–molecule reactions, excited-state proton transfer, energy pooling, thermal ionization, and desorption of preformed ions. These pathways and others have been reviewed in a number of articles (Zenobi and Knochenmuss, 1998; Ehring et al., 1992; Liao and Allison, 1995).

The enormous MALDI data generated over the last 15 years have shown that the ultimate performance of a MALDI source can be influenced by a number of parameters, including the matrix and whether or not delayed extraction (DE) is applied. these two parameters together with other variations on conventional MALDI are briefly considered in the following section.

One of the main characteristics of the matrix is to absorb at the wavelength of the laser. A range of MALDI matrices which are suitable for use with N_2 laser at $\lambda = 337$ nm are listed in Table 1.2, together with the general application areas. It can be seen that there is usually more than one matrix suitable for the same type of analysis, and therefore the user has to make an initial investigation to identify the more suitable matrix for a given application. The choice of the matrix is not only crucial for ion formation itself but it can also influence the extent of fragmentation of the same ions. A limited number of matrices have been studied regarding this particular aspect. Karas et al. (1995) have reported that ion fragmentation outside the source (postsource decay) decreased in the order sinapinic acid > 2,5-dihydroxy benzoic acid (DHB) > hydroxypicolinic acid. Similar results were reported by Spengler et al. (1992), who described sinapinic acid as being a hotter matrix compared to DHB. The latter has also been found to induce more fragmentation than 3-hydroxypicolinic acid (Zhu et al., 1995). In another study, to gain more information on the effect of a matrix on the mass accuracy and the resolution of a given measurement, Juhasz et al. (1997) have measured the initial velocity of ions generated by MALDI in different matrices. In their study the authors conducted a systematic determination of the initial velocity and its dependence on the identity of the matrix, molecular weight of the analyte, ion polarity, and wavelength of irradiation. The authors concluded that the matrix material was the most influential parameter on the initial velocity of the formed ions. For example, sinapinic acid and α-cyano-4-hydroxycinnamic acid matrices ejected slower peptide and protein ions compared to 2,5-dihydroxybenzoic acid and 3-hydroxypicolinic acid, $\sim$300 versus $\sim$550 m/s.

1.5.1. MALDI at High Pressure

In proteomic analyses, ESI and MALDI are considered as two complementary techniques. In an attempt to increase the overlap between the two ionization methods, there have been some attempts to operate MALDI source at subatmospheric and atmospheric pressures. Burlingam's group (Laiko et al., 2000) have introduced what they termed atmospheric pressure MALDI. In such an arrangement ions were formed at atmospheric pressure and a stream of N_2 was used to pneumatically transfer them to the high-vacuum region of the MALDI instrument. Considering the initial data published by these authors and the experimental observations by the same authors regarding the characteristics of this source, it is difficult to identify relevant advantages compared to the conventional vacuum MALDI. It is certainly a useful exercise as far as ion transfer is concerned, but we do not see any additional advantage regarding sensitivity and sample consumption, two elements which current ptoteomic analyses are looking for. On the contrary, it was reported by the same authors that MALDI under vacuum required less sample.

Another variation on the conventional MALDI has been reported by Gillig et al., (2000). The authors described the coupling of high-pressure MALDI with ion mobility orthogonal TOF. This system consists of two sections, the ion mobility chamber and the TOF analyzer separated by a 200-μm-diameter aperture. The ion mobility chamber contains the MALDI ion source and mobility drift cell operated at 1–10 Torr He. The TOF section contains various optics together with the TOF analyzer and maintained at $\sim10^{-6}$ Torr. Ions are formed in the drift cell by irradiating the sample deposited on a direct-insertion probe with 337 nm from an N_2 laser. After exiting the 200-μm aperture separating the two sections, ions are focused into the extraction region with a high-voltage pulsar. This system has been tested for simple mixtures of standard peptides; however, no data on cell lysate or other complex samples have been presented. It is worth noting here that the idea behind such experimental arrangement is to simulate 2DE separation combined with MS detection. The ion mobility side is used as a low-resolution mass spectrometer, while the TOF analyzer and the detector are used to identify the separated components. Atmospheric pressure MALDI sources are currently available on a number of commercial instruments.

1.5.2. Desorption Ionization on Silicon (DIOS)

Over a number of years the data generated by both FAB and MALDI ionization have amply demonstrated the indispensable role of the matrix for getting meaningful mass spectra of macromolecules. The same data have also provided evidence that the presence of the same matrix can influence the complexity and the sensitivity of the acquired mass spectra. This latter feature has encouraged various research groups to attempt generating MALDI mass spectra without the use of a matrix. Direct-desorption ionization without a matrix has been attempted on a variety of surfaces (Zenobi, 1997; Zhan et al., 1997; Wang et al., 1996) but has not achieved much success because of the rapid molecular degradation that is usually observed upon direct exposure to laser radiation.

A more successful attempt to generate meaningful matrix-free MALDI analysis has been reported by Wei et al. (1999). In this matrix-free desorption approach a porous silicon surface is used to trap the analyte molecules and, because of its high absorptivity in the UV, acts as an energy receptacle for the laser radiation. The experimental protocol involves the generation of porous silicon from flat crystalline silicon by using a simple galvanostatic etching procedure (Cullis et al., 1997). A micrometer-thick porous layer with a nanocrystalline structure is produced that exhibits bright photoluminescence upon exposure to UV light. The thickness, morphology, porosity, resistivity, and other characteristics of the material are readily modulated through the choice of silicon-wafer precursor and etching conditions. Freshly etched porous silicon surfaces are hydrophobic owing to the metastable, silicon hydride termination, but through Lewis acid–mediated (Buriak and Allen, 1998) or light-promoted hydrosilylation reactions (Stewart and Buriak, 1998), the surface can be easily stabilized and functionalized as required. Furthermore, and because of the high stability of the hydrophobic, hydrosilylated surfaces to aqueous media, these surfaces can be used with little degradation. To gain further information on the best experimental conditions, these authors have tested four types of porous silicon surfaces, each containing a specific surface modification, including hydrogen, dodecyl, ethylphenyl, and oxide. Of the surfaces tested, all could effectively generate intact ions. However, the more hydrophobic surfaces and in particular ethylphenyl-terminated surfaces gave better signal for the same quantity of analyte. The same authors noted that dissolving the sample in methanol–water at 1 : 1 v/v ratio provided stronger signal than 100% aqueous solutions. This may suggest that that the acqueous analytes are carried less effectively into the pores compared to the less polar methanol.

In the original work by Wei et al. (1999) the authors have used 5×5 well plates in 1.1 cm^2 area on n-type silicon, allowing for the serial analysis of 25 samples. In their preliminary measurements the authors demonstrated the validity of the approach for the analysis of various analytes having molecular masses in the range 150–12,000 Da. These included carbohydrates, peptides, glycolipids, and natural products. Over the last year or so a commercial version of DIOS plates became available, consisting of 96 spot arrays (2 mm for each spot) including some wells to be used for calibration for accurate mass measurements. These target plates are suitable for use with traditional MALDI instrumentation with no need for hardware modifications (apart from the holder of the sample plate).

Porous silicon substrates were also fabricated using nanoelectrochemical H_2O_2–metal–HF etching (Li and Bohn, 2000) and were used to investigate experimental parameters which may influence ion generation in DIOS (Kruse et al., 2001). These parameters included morphology and physical properties of porous silicon, laser wavelength, mode of ion detection, pH, and solvents. The authors reported a strong correlation between DIOS performance and both pore size and overall porosity of the silicon surface. Comparison of porous silicon to that of other porous substrates with similar pore morphologies but different thermal or optical properties showed a unique effectiveness of porous silicon as a substrate for DIOS. This observation implies interplay involving surface morphology, optical absorption strength, and thermal conductivity.

The utility of DIOS–MS for the analysis of protein digests and peptides including proteolytic digests of β-lactoglobulin, bovine serum albumin, and flock house virus capsid proteins has been reported in a number of studies (Lewis et al., 2003). Peptide mixtures were also examined under different experimental conditions (Kruse et al., 2001).

In a recent article Prenni et al. (2003) have described the use of off-line LC separation in conjunction with DIOS. Electrospray deposition was used to generate a spatially preserved linear track of the separated components on a specially designed DIOS chip. According to the authors, a comparison between off-line LC/DIOS and traditional ESI–LC/MS–MS revealed that the first yielded better sequence coverage. We must point out that such comparison was based on two model proteins; whether the deduction by the authors can be extended to other and more challenging protein mixtures remains to be demonstrated.

Based on the limited number of publications describing DIOS, the following general observations can be made: First, optimal performance of DIOS–MS is typically observed for molecules less than 3000 Da, and the capability of the technique to analyze intact proteins is still to be demonstrated. Second, the application of DIOS–MS in the analysis of biological molecules is highly compatible with silicon-based microfluidics and lab-on-chip technology, providing valid motivation for further development and refinement. Third, it is still unclear through what mechanism the analyte ions are formed and subsequently released from the porous surface. A tentative explanation suggests that the energy for such release is transferred from the silicon to the traped analytes through vibrational pathways or as a result of the rapid heating of porous silicon producing H_2 that releases the analyte (Ogata et al., 1998). Alternatively, as porous silicon absorbs hydrocarbons from air or while under vacuum, rapid heating/vaporization of trapped hydrocarbon contaminants or solvent molecules may contribute to the vaporization and ionization of the analyte embedded in the porous silicon (Canham, 1997). Fourth, there is an initial indication that the potential application of DIOS can be broad; however, such potential is strictly related to finding new surfaces which can be easily and efficiently derivatized to trap a wide range of analytes. Currently such a task is not easy to realize, and we believe that it will take a while before matrix-free MALDI becomes a viable alternative to conventional MALDI.

1.5.3. Delayed Extraction

Delayed extraction and reflectron are two elements which brought MALDI–TOF–MS to be one of the two most popular methods in modern proteomic analysis. Prior to the introduction of both elements the technique was considered low resolution, which could not compete with techniques which rely on the use of quadrupoles and trapping techniques. The rationale behind DE originated from the early work of Wiley and McLaren (1955). They devised a simple technique which, by means of pulsed extraction in split-acceleration field geometry, corrects ion flight times for the distributions of either initial ion velocity or ion position but not for the combined effect of the two parameters. It has been known for a while that in MALDI ionization particles are ejected from the solid sample in a jetlike manner (Quist et al., 1994; Beavis and Chait, 1991; Spengler and Bökelmann, 1993), forming a forward peaking cone of both neutrals and ions, which expand with a common velocity distribution. In this almost uniform velocity plume, kinetic energies are roughly proportional to the particle's mass. When ions contained in such a plume are subjected to a strong electrostatic field, they undergo multiple collisions with the neutral particles and, hence, will acquire further kinetic energy dispersion and a net energy deficit with respect to the nominal accelerating voltage. At the same time, the loss of strict correlation between ion position, ion velocity, and ion energy would weaken the possibility for time focusing by the classical reflectron option. This type of scenario is commonly invoked to explain the strong dependence of peak widths on laser irradiance, where the slightest increase above the threshold needed for ion formation is accompanied by deterioration in mass resolution. In modern MADI instruments most of these undesired effects are either prevented or corrected through the application of a split-acceleration stage in which the acceleration voltage is turned on with a time delay (0.1–0.5 μs) after the laser pulse.

At the practical level the application of delayed extraction (time lag focusing) significantly improves the mass resolution and consequently mass accuracy in both the linear and reflectron mode. Usually there are differences in the application of this mode of operation by the various manufacturers of MALDI instruments. However, all of them are based on the same principle, where ions in the source are initially allowed to spread out in a near-zero electric field and therefore the distance the ions travel is almost dependent on their initial velocities. The faster ions will cover a longer distance compared to their low-velocity counterparts. Upon applying the high electric field, the energy acquired by these ions would depend on their position within the ion source, in other words, on the distance they covered during the zero-field interval. This kinetic energy compensation would result in velocity focusing, allowing ions having the same mass to reach the detector at the same time regardless of their initial velocity at the time of their formation.

1.6. ION DETECTION

Current literature leaves no doubt that detection is the component which has not kept pace with the development associated with new ionization techniques and new generation of analyzers. Microchannel plates (MCPs) are still the routine ion detectors of choice for TOF–MS. These detectors are fast and easy to operate. Before pointing out the advantages and drawbacks of these commonly used detectors, we ought to have a brief look at their basic structure and the operation principle on which they are based.

1.6.1. Microchannel Plates (MCPs)

Essentially, an MCP is an array of 10^4–10^7 capillary-type electron multipliers oriented parallel to each other and having channel diameters in the range of 10–100 μm. Figure 1.4*a* shows a schematic view of a disc-shaped MCP; this disc is fabricated from lead glass capillaries and is heated to render the channel walls semiconducting and to optimize the emission characteristics of each channel, while Figure 1.4*b* gives a schematic diagram of a single channel. The main characteristics of MCP devices have been reviewed by Wiza (1979) together with a description of various configurations that have been fabricated for use as ion detectors in MS.

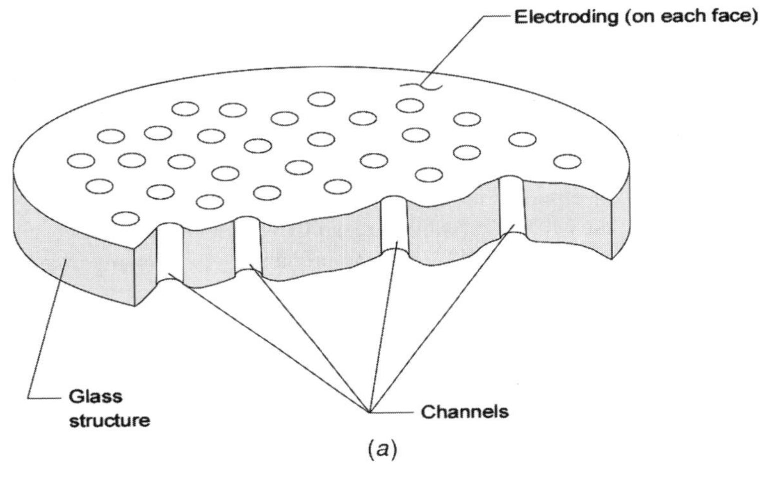

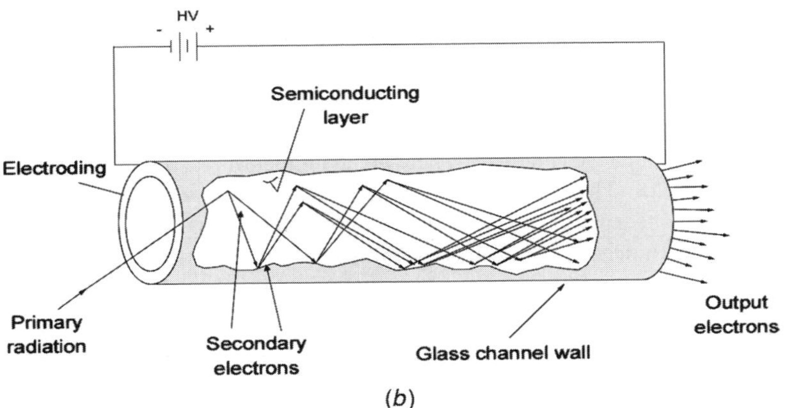

FIGURE 1.4. (*a*) Cutaway view of microchannel plate and (*b*) schematic view of single channel of microchannel plate (based on schemes published in various catalogs by Galileo Electro-Optics).

The salient characteristics of these devices include the following: (a) ultrahigh time resolution (<100 ps); (b) spatial resolution is limited by channel dimension and channels spacing; (c) at 1000 V gains of 10^3–10^4 are typical for straight channels, while curved channels can provide up to 10^6 gain and tend to suppress backward ion feedback; and (d) detection efficiency for pulsed positive ions is about 60%. The main advantages of these devices include high gain, fast response, simple geometry, low applied voltage, and relative insensitivity to low magnetic fields. Having said that, the same devices have a number of drawbacks which encouraged various research groups to look for alternative devices. These drawbacks include the following: The gain of MCPs has been observed to increase with an increase in the charge state of the incident ions regardless of the ionic species (Takagi et al., 1983). When these detectors are used with MALDI ionization, which generate mainly singly charged ions, such limitation is not that serious. A more serious limitation is associated with the fact that MCPs rely on secondary electron emission, which is inversely proportional to the velocity of the impacting ion, rendering this type of detector inefficient for high molecular masses. A third limitation is associated with the recovery time, which is in the range of 10^{-2}–10^{-3} s for each channel. For intense beams such recovery time is too long and can cause poor time resolution; however, for the detection of very low ion beam currents, the MCP can function reasonably as a fast counter because of the very large number of channels (10^6), and the probability of ions impacting the same channel within 10^{-3} s is likely to be very low.

1.6.2. Cryogenic Detectors

Recent proteomic research in general and functional proteomics in particular indicate that increasing the available sensitivity and dynamic range of current MS instrumentation remains a focus of interest. It is becoming increasingly evident that we need to carry out proteomics analyses on ever smaller amounts of material. Such needs are encouraging the search for other types of detectors which can deliver higher sensitivity and possibly address certain drawbacks associated with MCPs.

Over the last decade there has been increasing interest in cryogenic detectors: however, we are still waiting to see their use on commercial instruments. Cryogenic detectors have been developed since the early 1980s for applications in particle physics, astrophysics, and material analysis. There are a number of reasons behind their use in these diverse fields. These include high resolving power, low energy threshold, and large, sensitive absorbing areas. The ability of these detectors to detect slow-moving massive ions with near 100% efficiency and the additional information gained from measuring the energy of the same ions have attracted a number of attempts to use them in TOF–MS. This type of detector has a number of characteristecs, which make its use in TOF–MS highly attractive: On paper cryogenic detectors should not exhibit any decrease in sensitivity for large masses. Any impacting particle will create a signal as long as the energy deposited exceeds the low noise level of the detector, typically a small fraction of electron volts. Cryogenic detectors are therefore expected to detect large slow-moving ions or neutrals with 100% efficiency. The energy resolution provided by these detectors may also be used to differentiate between charge states. This is because a doubly charged ion would have twice the kinetic energy of its singly charged counterpart and thus would result in double the pulse height. Such a feature can be used, for example, to distinguish between a doubly charged dimer and its singly charged monomer, even though they both have the same flight time. The same characteristic can be used to discriminate between fragments and their precursors. Further details on the characteristics and operation principles of cryogenic detectors have been given in a number of recent works (Frank et al., 1999; Kraus, 2002; Twerenbold, 1996; Twerenbold et al., 2001). Two types of cryogenic detectors, superconducting tunnel junctions (STJ) and low-temperature calorimeters, are briefly described here.

1.6.2.1. Superconducting Tunnel Junction This type of detector consists of two thin films of superconductors separated by a thin insulating barrier through which quantum mechanical tunneling of electrons can occur. The operating principle of this type of detector is schematically represented in Figure 1.5*a*. It operates best at a temperature which has to be well below the critical temperature of the superconducting films. The tunnel current through the barrier is proportional to the number of quasi-particles in the STJ. When a particle such as an energetic biomolecular ion strikes the surface of an STJ detector, the kinetic energy of the ion creates nonthermal phonons with a wide distribution of energies, including those with energy greater than the binding energy (2Δ) of the weakly bound electron pairs (Cooper pairs) within the superconducting layers, Δ being the energy gap of the superconducting layer and is typically in the order of 1 meV or less. Phonons whose energy is greater than 2Δ can subsequently be absorbed by Cooper pairs. The resulting electron and holelike excitations (quasi-particles) can quantum mechanically tunnel through the barrier, creating a measurable current signals.

1.6.2.2. Thermal Detectors The other type of cryogenic detector, the thermal detector (bolometer and calorimeter), consists of an absorber in which the energy of an incident ion is converted into an increase in the temperature of the absorbing layer. Such a temperature increase can then be measured by a temperature transducer. Among the prominent examples of this type of detector is the one known as normal conductor–insulator–superconductor (NIS). In this type of sensor, electrons injected from the superconductor through the insulating tunnel barrier into the normal conductor heat the electrons therein. This increase in the electron temperature can be measured as a change in current through the NIS junction. The principal components of a thermal detector are shown in Figure 1.5*b*. These include an absorber with heat capacity C that absorbs particles or radiation, a temperature transducer which converts temperature changes into electrical signals, and the thermal link (g) to a reservoir of constant temperature.

Almost 10 years ago Twerenbold (1996) demonstrated the use of cryogenic detectors in TOF–MS; proof of concept studies have also shown that cryogenic detectors can indeed detect single, large biomolecular ions. Furthermore, comparison of these

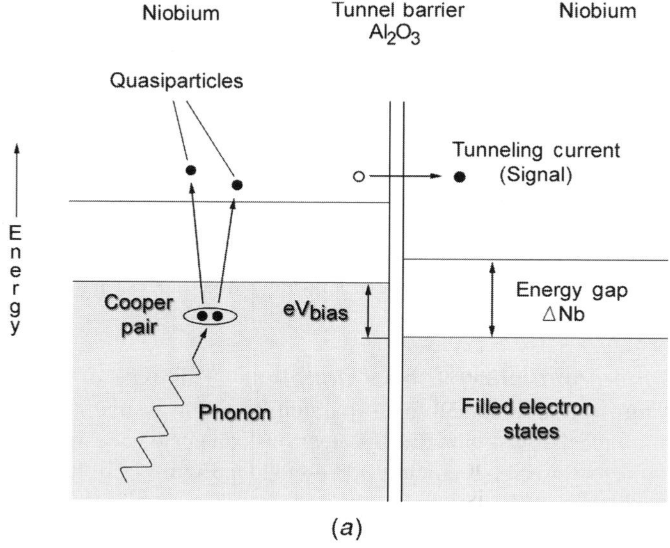

(*a*)

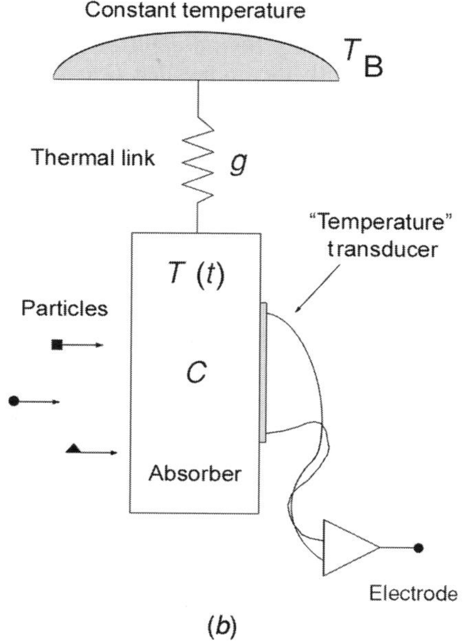

(*b*)

FIGURE 1.5. (*a*) Schematic operating principle of STJ (from Frank et al., 1999, with permission). (*b*) Principal components of thermal detector (calorimeter or bolometer). (From Kraus, 2002, with permission.)

detectors with the well-established MCPs has revealed that the first have much higher sensitivity per area at high molecular masses. This particular characteristic is highly attractive in proteomic studies in which the sensitivity is a key factor which directly influences the success or failure of a given measurement. Having said that, it is natural to ask what are the obstacles to a more diffused use of cryogenic detectors in TOF–MS. Such a question can be partially addressed by the following considerations: First, cryogenic detectors operate at a very low temperature, typically 1 K or below; such a characteristic is a hindering factor on the road to its dissemination since the MS community operated the instruments at room temperature for over a century, and therefore switching to very low temperature operation does not attract a lot of enthusiasm. One option would be to operate these detectors at somewhat higher temperatures, 5–10 K. At such temperatures, the detector may preserve most of its sensitivity but will certainly lose most of its capability for energy resolution, which is one of the attractive features of this family of detectors.

The question of cryogenic temperature is still the focus of concerted research efforts. The realization that the use of liquid helium to achieve extremely low temperature may hinder the wide dissemination of cryogenic detectors has encouraged the search for more practical alternatives. For example, attempts have been made to use mechanical coolers such as closed-cycle refrigerators, some of which have already been developed by private companies, which can provide temperatures as low as 3 K. Combined with adiabatic demagnetization refrigerator units, such refrigerators will be able to achieve subkelvin temperatures without the need for liquid cryogens. Liquid cryogen–free coolers are currently developed for high-resolution cryogenic detectors for use in X-ray fluorescence analysis. As the semiconductor industry begins to apply this type of detector, closed-cycle systems to cool such detectors are expected to become readily available at affordable costs. Such systems will eventually find their way to TOF–MS systems. It is worth pointing out that implementing a cryogenic group into MS systems requires fairly moderate engineering. Typical cryostats do not take up significant additional space when attached to TOF–MS systems. For example, the cryostat that was used at Lawrence Berkeley National Laboratory to cool a Niobium STJ detector to $\sim$1.2 K was approximately 20 cm in diameter and 40 cm tall (Frank et al., 1999).

Another aspect of cryogenic detectors which could make them more attractive to use in TOF–MS is to increase the effective area through the use of an array form of detection. Such development would have the direct effect of sustained detector performance over a longer lifetime and would limit pulse pileup. An added advantage of using parallel arrays of detectors is that such arrangement may allow multiplexed TOF measurements in applications where high throughput is required. Fast DNA sequencing using TOF–MS is an example of the utility of such multiplexing (Twerenbold, 1996). In such an approach, a laser beam with a high repetition rate could be rapidly stepped across an array of oligomer samples such that each shot can desorb ions from the individual samples on the sample array. Using programmable deflection plates, the ions generated from each sample can be guided to a specific element within the array detector. If the laser firing rate can be made faster than the ion flight time, then we can envisage that the analysis time of a whole array of samples can be comparable to the time required for the analysis of a single sample. Despite the attractiveness of such a scenario, its practical implementation still needs more advanced electronics, ion optics, and software capable of handling the speed of data generation.

1.7. TYPES OF ANALYZERS

For almost a century magnetic and electrostatic sectors have been the favorite means for MS analyses. The substantial decline in the use of EI and FAB ionization in the late 1980s and the birth of commercial ESI and MALDI ionization have resulted in the replacement of sector analyzers with quadrupoles, TOF, and ion traps. In fact, these types of analyzers together with FTICR currently represent the backbone of modern proteomics. It can be stated that ion analysis is the one component of a mass spectrometer which over the last 15 years has experienced the major part of development and innovation compared to the other components of a mass spectrometer. The highly attractive characteristics of TOF and ICR in terms of mass accuracy, sensitivity, high transmission and resolution, and the emergence of a wide range of hybrid configurations have given ion analysis an extra momentum compared to ion detection and ion generation. It is unrealistic to pretend that the physical principles and hardware details of these analyzers can be discussed in a single chapter. On the other hand, a brief description of these individual components is the least we can do here.

1.7.1. Quadrupole Mass Filter

Strictly speaking, a quadrupole can be described as a mass filter which transmits a fairly small range of m/z values. Such transmission does not involve dispersion or focusing, as, for example, in the case of sector analyzers. In other words, a quadrupole can be easily compared with a narrow-bandpass electrical filter that transmits signals within a finite frequency bandwidth. The basic concept of this device is to provide a potential field distribution periodic in time and symmetric with respect to the axis of an ion beam. Such a combination will transmit a selected group of ions while those outside such a group will be deflected away from the axis. An ideal quadrupole assembly would consist of four hyperbolic rods; however, for ease of manufacturing, round cylindrical rods are commonly used. The electric field within such an assembly is created by coupling an opposite pair of rods together and applying the combined direct-current (DC) and radio frequency (RF) potentials, positive for one opposite pair and negative for the other pair of opposite rods. Through proper programming of both potentials and frequency of the RF, the injected ions describe complex trajectories. Some of these trajectories are unstable in that they tend toward infinite displacement from the center of the assembly, losing their charge through collision with one of the rods. On the other hand, ions which are successfully transmitted through the assembly are said to possess stable trajectories to reach the detector. For a given interelectrode spacing, the path stability of an ion with a given m/z value depends on the magnitude of the RF drive frequency and on the ratio of the amplitudes of the RF (V) and the DC (U)

potentials. For example, when $U=0$, the resolution is almost nonexistent and a wide band of m/z values is transmitted. Such a situation is encountered in modern collision cells, where only RF is applied to favor ion confinement for an increased collision path but does not sacrifice the transmission of the product ions. As the ratio U/V is increased, the resolution goes up so that at the stability limit only a single value of m/z corresponds to a stable trajectory, resulting in the transmission of ions of a single m/z ratio, such operation mode has earned the quadrupole the description of mass filter. Scanning the values U and V with a fixed U/V ratio and constant drive frequency or by scanning the frequency and holding U and V at constant values may generate a mass spectrum. Commercial quadrupoles can have a scan range of 1–4000 Da; however, when they are coupled to an ES ion source, which produces multiply charged ions, such a range can easily exceed 100 kDa (depending, of course, on the number of charges). An introductory description of the operation of the quadrupole is given by Lawson and Todd (1971), while for a full account of theory the reader is referred to Dawson and Whetten (1969).

1.7.2. Three-Dimensional Quadrupole Ion Trap

The 3D quadrupole ion trap is closely related to the quadrupole mass filter in that it can be visualized as being a solid of revolution generated by rotating the hyperbolic rod electrodes about an axis perpendicular to the z-axis and passing through the centers of two opposing rods (Dawson and Whetten, 1968). This rotation results in one pair of rods joining up to form a doughnut-shaped ring electrode and the other two forming end-cap electrodes, which are moved closer to each other. The ring electrode and the entrance end-cap and exit end-cap electrodes form a small-volume cavity in which ions can be trapped. Both end-cap electrodes have a small hole in their centers through which the ions can travel.

Early ion traps were mainly used to analyze volatile samples by EI or CI. In this case, ions were formed inside the trapping volume. The increasing demands for the analysis of biological molecules led to the need to interface ESI and MALDI methods to the ion trap. These externally created ions need to be injected into the ion trap and need to be efficiently trapped. In this injection phase ions are focused by an Einzel lens system and allowed into the ion trap during the ionization interval. A gating pulse from (+)/(−) is also used to attract or repel ions toward the entrance end-cap aperture. This phase is highly critical because it has to maximize the number of ions admitted into the trapping volume without causing space charge effects. Failure to find the right compromise between maximized ion intensity and the onset of space charge can be one of the limiting factors in the performance of the trap. The ion trap is typically filled with He at ~1 mTorr. Collisions with this gas reduce the kinetic energy of the injected ions and serve to quickly contract ion trajectories toward the center of the trap, enabling more efficient trapping of the injected ions.

The isolation of a particular m/z value can be accomplished in different ways. One approach includes the combined manipulation of the DC and RF potentials to bring the ion to an apex of the stability diagram where all other m/z values will be unstable (Dawson et al., 1969; Mather et al., 1978). The other method is to scan the amplitude of the RF voltage in forward-then-reverse manner while applying a resonance signal (Louris et al., 1987; Kaiser et al., 1989). This allows the ejection of ions with m/z values greater than the ion to be isolated followed by ejection of ions having m/z values smaller than the ion of interest.

To perform CID, the amplitude of the resonance signal can be adjusted to induce collisions with the He damping gas rather than ejection from the ion trap (Louris et al., 1987). An estimated 10,000 low-energy collisions are sufficient to transfer enough energy into a peptide to cause its fragmentation in a manner similar to that obtained in CID in a triple quadrupole. While performing CID, the amplitude of the RF signal, which sets the stability of the isolated ion, has to be carefully set to eject all ions below a specific m/z, a working condition which also limits the acquired fragmentation information associated with low m/z values. This means that to obtain a complete set of b- and y-type ions it is necessary to use multiple stages of analysis (MS^n).

1.7.3. Linear Ion Trap

The 3D quadrupole ion trap has a number of characteristics which made it one of the main instruments in current proteomic analysis. The same trap, however, has a number of limitations associated with its relatively low trapping efficiency for externally generated ions ($<10\%$) and space charge effects due to a relatively small volume to which the ions are injected. Attempts have been made to address these limitations through the use of linear ion trap MS.

It has been recognized for several years that ions can be trapped within a linear ion trap and mass selectively ejected in a direction orthogonal to the central axis of the trap via radial excitation techniques (Beir and Syka, 1994; Welling et al., 1998). There are a number of advantages associated with ion trapping in a linear ion trap: This type of trap has a high ion beam acceptance since there is no quadrupole field along the axis of the beam. Ions admitted into a pressurized linear quadrupole undergo multiple collisions which reduce their axial energy prior to encountering the end electrodes, thereby enhancing trapping efficiency. The larger volume of the pressurized linear trap relative to the 3D conventional ion trap means that more ions can be stored before the onset of the space charge effects. Radial containment of ions within a linear trap results in a strong focusing effect along the center of the trap, a behavior different from that encountered in the conventional ion trap, where the focusing is achieved at a specific point.

In an interesting study by Campbell et al. (1998) a comparison is made between the ion capacities of a linear (20–cm-long) and a commercial 3D (trap radius 0.707) ion trap. The authors concluded that the ion capacity of the linear trap was an order of magnitude higher than the 3D trap. This comparison did not take into account the difference between the two configurations in the presence of the collision gas. In the case of the linear trap the collisional damping affects the ion cloud radially, while in the 3D trap the effect influences both the radial and the axial dimension of the cloud.

In a recent article Hager (2002) described two experimental configurations based on the ion path of a traditional triple quadrupole. In the first arrangement a pressurized collision cell placed between two conventional quadrupole mass filters was used as the linear trap. In such a configuration product ion scans are accomplished by selecting the precursor ion using the first RF/DC transmission quadrupole. Fragment ions are generated by accelerating the precursor ion into the pressurized collision cell where the fragment ions and the remaining precursor ions are trapped and then axially scanned out of the ion trap through an RF-only quadrupole to the detector. This configuration bears a close resemblance to conventional triple-quadrupole instrument with the additional capability of ion trapping. This means that functions like multireaction monitoring (MRM) and constant neutral loss scans can be performed, functions which cannot be performed on conventional 3D ion traps.

Hager (2002) observed that this configuration was found to be limited by the space charge at approximately the same ion density level reported for the 3D trap. However, due to the considerably larger linear ion trap capacity, it was estimated that nearly an order of magnitude more ions can be extracted from the linear trap compared to its 3D counterpart when both operated at the same ion density.

The second arrangement described by the same author (Hager, 2002; Hager and Yves Le Blanc, 2003) uses the final quadrupole (Q3) as the linear trap (see Figure 1.6). In such a configuration the product ion spectra are obtained by using the first quadrupole to select the precursor ion, which is then accelerated to the pressurized transmission collision cell. Ions emerging from this cell are trapped in the Q3 linear ion trap and subsequently mass selectively scanned out of the trap to the detector. Similar to the first configuration, all of the standard triple-quadrupole scan modes can be effected by this configuration. It is worth noting that this configuration is now commercially available under the name Q-Trap (Applied Biosystems/MDS Sciex, Concord, Ontorio, Canada). In a recent study Sandra et al. (2004) have used this type of mass spectrometer to study protein glycosylation, where the MS^3 capability proved to be very useful. Needless to say, for a rational assessment of the capabilities of this newly introduced instrument we need to wait for further experimental data.

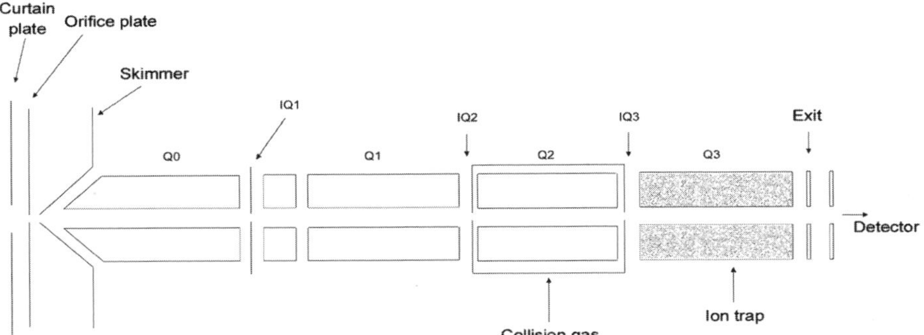

FIGURE 1.6. Schematic representation based on ion path of triple quadrupole, where Q3 can be operated as either RF/DC quadrupole or linear ion trap with mass selective axial ejection (from Hager and Yves Le Blanc, 2003, with permission).

Such data may allow us to evaluate the additional advantages of this trap compared to the well-engrained triple quadrupole. Having said that, linear ion traps are already making an impact on the performance of other instruments, for example, their use for external ion accumulation in conjunction with FT–ICR systems.

1.7.4. Time of Flight

Stephens (1946) first described the principle of using TOF for mass separation, while Wiley and McLaren (1955) subsequently developed a prototype of the modern TOF mass spectrometer. The extensive use of the TOF in modern MS can certainly be attributed to the commercialization of MALDI. A further momentum was given to the same analyzer by the introduction of orthogonal acceleration TOF to be used with both MALDI and above all with ESI.

The basic principle which governs ion analysis by TOF can be summarized by the following simple relations: Ions accelerated through a potential V acquire a velocity v of magnitude

$$v = \left(\frac{2neV}{m}\right)^{1/2} \tag{1.1}$$

where ne is the charge and m is the ion mass. In a drift region with length L, the ion flight time t will be

$$t = \frac{L}{v} = L\left(\frac{m}{2neV}\right)^{1/2} \tag{1.2}$$

The difference in transit time, Δt, for two ions of mass m_1 and m_2 will be

$$\Delta t = \frac{L(\sqrt{m_1} - \sqrt{m_2})}{\sqrt{2neV}} \tag{1.3}$$

Also the mass resolution can be approximated by $m/\Delta m = t/2\ \Delta\text{t}$.

In TOF–MS analysis fast electronics are used to translate the time difference (Δt) into m/z values of the detected ions. Equation (1.3) is obviously valid for both linear and reflector TOF, since the only difference between the two modes is the longer drift length L which is achieved by applying an adequate electric field gradient in the opposite direction of the incident ion beam.

The performance of the TOF analyzers used in today's MS has been enhanced by a number of recent developments: First, It has been known for a while that ion injection into the drift region has a direct effect on the time resolution of the TOF analyzer. Ideally such injection should be in the form of an ion punch that is neither spatially dispersed nor inhomogeneous in kinetic energy. An important step to achieve such a goal has been realized through the introduction of delayed extraction, which was originally described by Wiley and McLaren (1955) and became a common feature on most commercial MALDI–MS instruments. Second, another major development

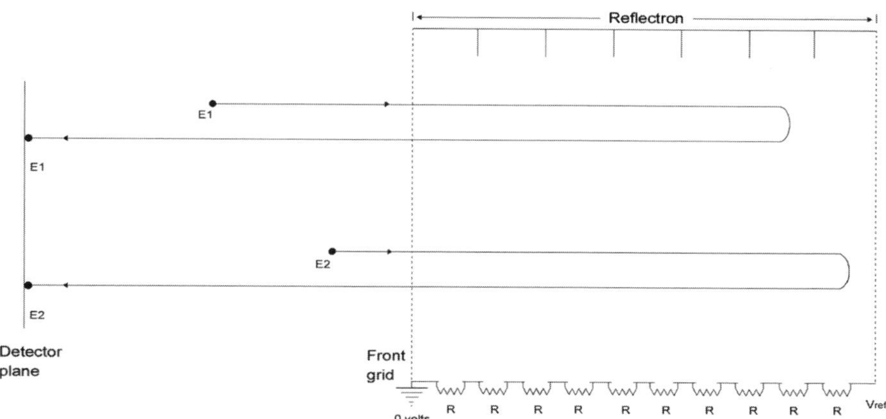

FIGURE 1.7. Schematic representation illustrating principle of energy focusing with reflectron.

which improved the performance of TOF is the introduction of the reflectron, which had a direct effect on the resolution and rendered the TOF instrument a key player in current proteomic analyses. Equation (1.3) clearly shows that the time resolution of two masses is directly proportional to their individual path length prior to their detection. Of course, the length of the flight time is not enough to allow efficient energy focusing of ions having the same mass. This difference in energy could have its origin in the ion source or in the phase of ion injection to the drift tube.

One way of increasing the path length and achieving efficient energy focus is to create a gradient electric field which can reshape the ion trajectories according to the initial kinetic energy of the same masses, which happen to possess different kinetic energy within the linearly drifting ion beam. A simple illustration of how the reflectron works is given in Figure 1.7. This figure shows two ions having the same mass but one of them with kinetic energy E_1 while the other has a higher energy E_2. The reflectron is simply a voltage applied between two grids which are separated by a series of resistors (R), thus creating a gradient electric field between V_0 and V_{ref}. In this simplified scenario it can be appreciated that the mass with energy E_2 will penetrate deeper into the reflectron and thus will cover a longer path compared to that with energy E_1 and thus will be caught up by the slower mass, which penetrated a shorter distance into the reflectron.

1.7.5. Fourier Transform Ion Cyclotron Resonance

Although FT–ICR–MS is considered a recent addition to the tools of modern proteomics, the basic principle of this approach was first introduced by Lawrence and Livingston (1931). It was demonstrated that charged particles would follow a spiral path if an appropriate RF voltage was applied to semicircular electrodes and if the ions were also constrained by a magnetic field. This description was given in the celebrated equation

$$\omega_c = \frac{qB}{m} \tag{1.4}$$

where m and q are the mass and charge of the ion, B is the magnetic field, and ω_c is the unperturbed ion cyclotron frequency. A remarkable feature of Equation (1.4) is that all ions of a given mass-to-charge ratio, m/q, have the same ICR frequency independent of their velocity. This property makes ICR particularly useful for MS, because translational energy focusing is not essential for the precise determination of m/q. Another attractive feature of FT–ICR–MS is that excitation produced by applying a spatially uniform electric field oscillating (or rotating) at or near the cyclotron frquency of ions of a particular m/z value can be used to perform three different functions: to accelerate ions coherently to a larger (and thus detectable) orbital radius; to increase ion kinetic energy above the threshold for ion dissociation and/or ion molecule reaction; and to accelerate ions to a cyclotron radius larger than the radius of the ion-trapping space whereby ions can be removed (ejected) from the cell.

A major advance in the practical use of ICR for MS applications came with the introduction of a pulsed technique together with a specially designed analyzer cell (McIver, 1970, 1971), while a further development by Comisarow and Marshall (1974) was a FT method, so that an entire mass spectrum could be obtained in the same time, which was previously required to generate only a single peak. Mathematical principles and mass spectrometry applications of FT–ICR have been clearly described in a recent review by Marshall et al. (1998).

Early FT–ICR instruments were mainly used with EI ionization, where ions were formed within the cell itself; however, recent needs in proteomic analyses resulted in the development of experimental arrangements in which ions generated by MALDI or ESI are formed outside the cell and injected for analyses. Regardless of where the ions are formed, the sequence of events of ion detection and the conversion of a time domain signal to a full mass spectrum are the same. In current FT–ICR–MS instruments, ions which can be generated by MALDI or ESI are subjected to a simultaneous RF electric field and a uniform magnetic field, causing them to follow spiral paths in an analyzer chamber. When a spatially uniform RF electric field (excitation) resonant with the ion cyclotron frequency of a selected packet of ions with a given m/z ratio is excited, ions within such a packet will absorb energy, thus increasing their velocity and their orbital radius to coherent values, which allows their detection.

A schematic diagram of a cubic trapped-ion cell, which is commonly used for FT–ICR–MS analyses, together with the sequence of events leading to the mass spectrum is given in Figure 1.8. The formed ions undergo an orbital motion (cyclotron motion), such motion being due to the thermal energy of the ions and the applied magnetic field B. Ion motion parallel to the magnetic flux lines is restricted by electrostatic potentials applied to the trapping plates. The polarity of the trapped ions is determined by the polarity of the DC potential applied to the trapping plates. To detect these trapped

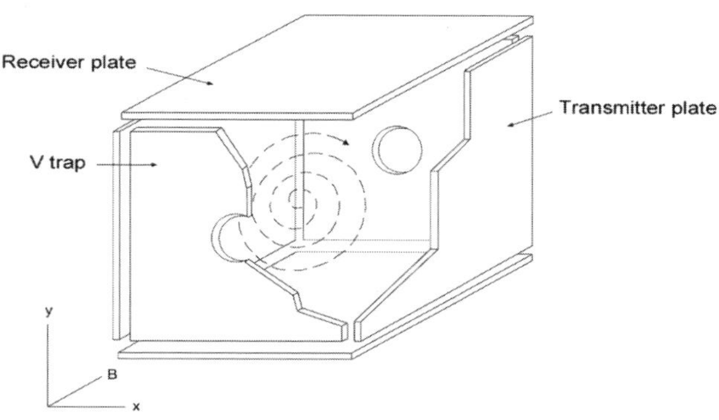

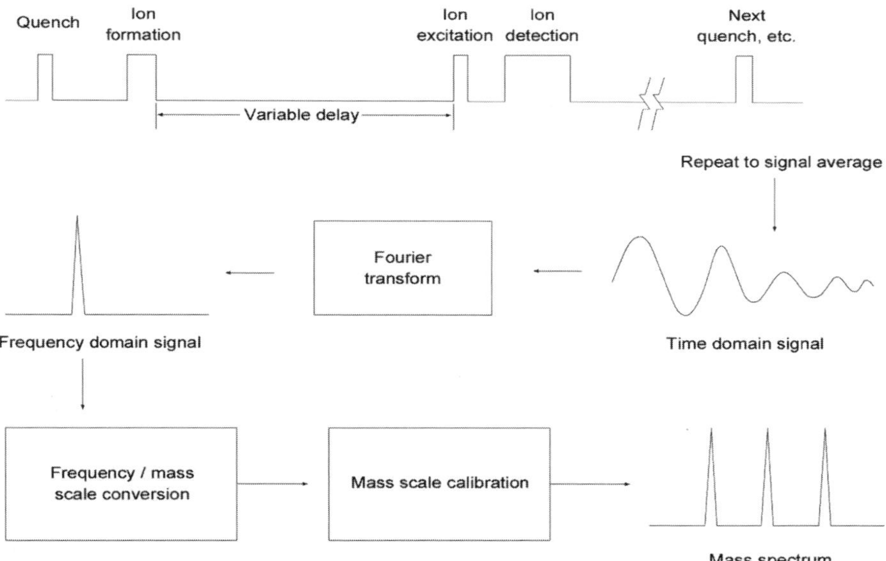

FIGURE 1.8. Schematic representation of cubic ion cyclotron resonance cell and FTMS sequence of ion formation, detection, and conversion of time domain signal to mass spectrum.

ions, the cyclotron motion of the desired ions in the cell are coherently excited to larger circular orbits (Comisarow, 1980). For an ion having a charge-to-mass ratio q/m, the resonance frequency f (in hertz) is related to the magnetic field strength B expressed in tesla (1 T $= 10^4$ G), by the expression

$$f = (1.5357 \times 10^7)\frac{qB}{m} \tag{1.5}$$

The excitation of ions is achieved by an electric field with a sweeping frequency applied to the transmitter plates of the cell, forcing all the ions of a given mass into larger coherent orbits or in phase as the swept frequency goes through the resonant value of the ion cyclotron frequency. The excitation frequency is swept very fast (typically 1 ms). The cyclotron motion of the ions persists after the excitation, and their coherent motion induces an analog signal at the resonance frequency in the receiver plates. The analog signal is amplified, converted to a digital signal, and added into the memory of the computer. A quench pulse at the end of each scan clears the cell from the previously detected ions. This sequence of events can be repeated as fast as 100 times per second. The pulsing and recording of induced signals can be repeated as often as may be required to obtain an adequate signal-to-noise ratio. The resultant summed digital data are then subjected to a Fourier transformation to produce a frequency domain spectrum. The amplitude of each frequency component will correspond to the number of ions of the related frequency. A frequency-to-mass conversion then yields the corresponding spectrum.

1.8. HYBRID ANALYZERS

By consulting current literature on MS, it is not difficult to realize that the dissemination of hybrid analyzers is gaining momentum. One of these hybrid configurations, which has been rapidly embraced by the analytical community in general and practitioners of proteomics in particular, is the quadrupole time of flight (Q-TOF). This hybrid analyzer combines the high performance of TOF for both MS and MS–MS analysis with the well-accepted and widely diffused ESI and MALDI techniques. A brief description of this and other hybrid analyzers is given below.

1.8.1. Quadrupole Time of Flight

The abbreviation QqTOF is commonly used to describe this hybrid analyzer, where Q refers to a mass-resolving quadrupole, q refers to an RF-only quadrupole or hexapole collision cell, and TOF refers to the time-of-flight section. The development of this configuration followed closely the development of the so-called ESI–TOF arrangement, where the principle of orthogonal ion injection to a TOF analyzer was applied. The QqTOF can be regarded either as the addition of an RF/DC resolving quadrupole to an ESI–TOF or a triple quadrupole in which the third mass filter is replaced by a TOF analyzer. The principle and other detailed theoretical treatment of orthogonal TOF have been given in a number of reviews (Guilhous et al., 2000; Yefchak et al., 1990). The following section with the help of Figure 1.9 gives a limited summary of certain MS functions which can be performed on this type of instrument.

To operate the system in the TOF–MS mode, the mass filter Q1 is operated in the RF-only mode so that it serves merely as a transmission element, while the TOF analyzer is used to record the mass spectra. The resulting spectra benefit from the high resolution and mass accuracy of the TOF instrument and also from their ability to record all ions in parallel without scanning. In principle, it is still possible to

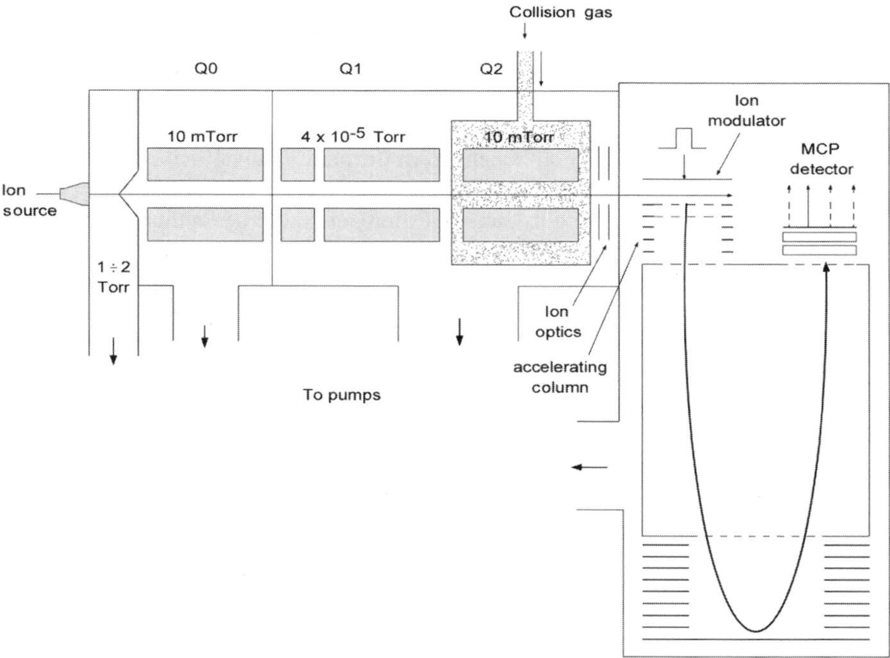

FIGURE 1.9. Schematic diagram of tandem QqTOF mass spectrometer. (From Chernushevich et al., 2001, with permission.)

perform Q1 scans for the MS analysis by using the TOF section as a total ion current detector only. However, owing to the advantages of the TOF spectra, this operational mode is used for Q1 calibration and tuning purposes. In general, ions are sampled from the ion source through an additional RF ion guide (Q0) placed between the resolving quadrupole (Q1) and the ion source. This additional component is used for collisional cooling and focusing of the ions entering the instrument. Both Q0 and Q2 are operated in the RF-only mode: The RF field creates a potential well that provides radial confinement of the precursor and/or fragment ions. Since the RF-only quadrupoles are normally operated at a pressure of several mTorr, they provide both radial and axial collisional damping of ion motion. The ions are thermalized in collisions with neutral gas molecules, reducing both the energy spread and beam diameter and resulting in better transmission into and through both the quadrupole and TOF components. After leaving the RF quadrupoles, ions are reaccelerated in the axial direction to the necessary energies with near-thermal energy spreads.

In the MS/MS mode, Q1 is operated in the mass filter mode to transmit only the parent ion of interest, typically selecting a mass window from 1 to 3 Da wide depending on whether the transmission of the full isotopic cluster is required. The ion is then accelerated to an energy of between 20 and 200 eV before it enters the collision cell Q2 where it undergoes CID. After the first few collisions with neutral gas molecules (usually argon or nitrogen), the resulting fragment ions (in addition to the remaining parent ions) are collisionally cooled and focused. This step is even more important in QqTOF instruments than it is in triple quadrupoles because the TOF analyzer is much more sensitive to the "quality" of the incoming ion beam than is Q3 in a triple-quadrupole instrument.

Both sensitivity and resolution benefit from the additional collisional focusing in the pressurized collision cell. After leaving the collision cell, ions are reaccelerated to the required energy and focused by ion optics into a parallel beam that continuously enters the ion modulator of the TOF analyzer (see Fig 1.9). Initially the modulator region is field free, so ions continue to move in their original direction in the gap. A pulsed electric field is applied at a frequency of several kilohertz across the modulator gap, pushing ions in a direction orthogonal to their original trajectory into the accelerating column, where they acquire their final energy of several kiloelectron volts per charge. From the accelerating column, ions arrive in the field-free drift space, where TOF mass separation occurs. The ratio of velocities (or energies) in the two orthogonal directions is selected such that ions reach the ion mirror and then the TOF detector without requiring an additional deflection in the drift region, which could affect the mass resolution. A single-stage ion reflector provides compensation for the initial energy and spatial spread of the ions. In this arrangement, ions originating at different vertical positions in the extraction gap of the modulator are focused onto a horizontal plane at the detector entrance. The detector is made of two microchannel plates in a chevron configuration. The electrostatic mirror and the ion detector are similar to those used with pulsed ionization sources (e.g., MALDI), apart from differences associated with a larger beam size (usually a few centimeters in the y direction, determined by the modulator aperture). Another major difference results from the requirement that the quadrupole section should be kept at low, near-ground voltages. Therefore, positive ions are accelerated from ground to high negative voltages in the TOF section of the QqTOF instrument, rather than from high positive voltages to ground, as in MALDI–TOF mass spectrometers. This "inverted" configuration requires an extra shield within the TOF chamber to create a field-free drift region floated at high potential, but this shield must be designed carefully, since even a small penetration of an external field can substantially influence the performance of the TOF.

1.8.2. Ion Mobility–TOF

Ion separation based on ion mobility within drift tubes has been around for over 40 years (Kaneko et al., 1966), yet only recently has it been introduced to the world of large-molecules MS. Ion mobility MS can be thought of as gas-phase electrophoresis, which in combination with well-established analyzers could enhance the separation and analysis of complex mixtures. This approach is based on the principle that individual components in a mixture of ions can be separated by mobility differences in a drift tube with inert gas (typically helium) at a pressure ranging from 1 to 200 Torr under the influence of a weak electric field (5–200 V/cm). Recently there have been a number of examples on the use of ion mobility separation coupled to TOF (Srebalus et al., 1999, 2002; Hoaglund et al., 1998; Srebalus Barnes et al., 2002) analyzers. Because the flight time in the TOF analyzer (~100 μs) is much shorter than the residence time in the drift tube (~10 ms), it is possible, in principle, to obtain simultaneous determination of the ion mobility and *m/z* of a drifting ion.

The application of ion mobility combined with MS analyzers has been demonstrated by a number of groups. Analysis of combinatorial libraries (Srebalus et al., 1999, 2002) of peptides and quantification of peptides in isotopically labeled protease digests (Kindy et al., 2002) are two representative examples of such applications. A combination of LC/ion mobility/TOF–MS has been used (Srebalus et al., 2002) to characterize a combinatorial peptide library designed to contain 4000 peptides. Despite a limited number of reports on the combined use of ion mobility with traditional MS analyzers, such combination can be looked at as an additional dimension to LC separation and at the same time as an additional analysis component which can be easily added to existing MS instrumentation. Both characteristics are bound to facilitate the analysis of complex mixtures, particularly in proteome analyses.

1.8.3. Linear Ion Trap–FT–ICR

Setting aside the price of the instrumentation there is no doubt that FT–ICR continues to provide unrivaled mass resolution and sensitivity. For further enhancement of the dynamic range without sacrificing the resolution, there have been various attempts to trap ions prior to their injection to the ICR cell. One of the early attempts (Senko et al., 1997) used an external octapole trap followed by gated trapping to accumulate ions of interest prior to ICR analysis. This approach was later refined by employing a linear quadrupole ion trap, which expanded the dynamic range to allow the detection of 10 zmol (Belov et al., 2001 a,b). These authors have argued that, to gain most benefits from such an arrangement, externally trapped ions should be detected across the desired *m/z* range without significant bias. One way of achieving this is selective ion accumulation in a quadrupole interface/ion guide of an FT–ICR mass spectrometer. To prevent overfilling the external trap, and thus avoid early onset of space charge effects, higher abundance ions can, after initial measurements, be preselected and ejected, thus trapping only lower abundance species and potentially increasing the overall dynamic range. However, the same authors noted that significant increase in the ion accumulation time characterized by higher populations of low-abundance ions in the external trap could be accompanied by a pronounced discrimination against certain *m/z* values or their fragments. This situation was exacerbated by higher ion currents provided by the use of an electrodynamic ion funnel in the ESI–ICR interface (Belov et al., 2000). Like any hybrid arrangement the overall performance of the system is influenced by the working parameters of the individual components, external ion accumulation–ICR being no exception. It has been demonstrated (Belov et al., 2001a,b) that the effectiveness of ion accumulation in an external linear RF quadrupole trap is affected by several parameters: accumulation and storage times; the set of DC potentials applied to the rods, entry and exit plates of the quadrupole used to confine ions axially, and the frequency and amplitude of the RF field, constraining ion motion in the radial direction. In such an arrangement the ion accumulation time corresponds to the time interval during which the RF potentials are applied to the collision quadrupole. After trapping ions in the accumulation quadrupole, the kinetic energy of the ions is damped through collisions with gas prior to their transfer to the ICR cell. The reduction in the kinetic energy spread is required if an efficient gated trapping is to be achieved.

Apart from increasing the dynamic range of ICR–MS, external ion trapping, if properly implemented, can significantly increase the duty cycle. This is because the accumulation and ion detection are spatially separated, the next packet of ions can be accumulated, while the previous packet is being analyzed, providing a duty cycle that can approach 100%. Other benefits of external ion accumulation can also include improved signal-to-noise ratio and an enhanced resolving power.

1.8.4. Ion Trap–TOF

The combinations of TOF with a 3D ion trap and TOF with a linear ion trap have been demonstrated by Chien et al. (1994) and Campbell et al., (1998), respectively. The scope of both arrangements is to combine the storage capability of the trap with the high-resolution, speed, and high-mass capabilities of the TOF. The coupling of a linear rather than a 3D ion trap was found to offer a number of operating advantages: Since there is no quadrupolar electric field in the z-direction, the ion injection and extraction are more efficient. In the linear trap, losses associated with the filling and emptying of the trap are less than those encountered in the 3D trap. Because of its large trapping volume, the linear trap has a greater capacity and less susceptibility to space charge effects. Another advantage is the fact that the linear trap can be operated in a number of MS mode analyses, including MS^n offered by the 3D trap.

1.9. TANDEM MASS SPECTROMETRY

The term MS–MS is commonly perceived as a multistage analysis involving precursor ion selection, fragmentation, and detection. Before the diffusion of various trapping techniques to obtain such types of data, MS–MS analysis could only be conducted on multianalyzer instruments. The recent advances in electronics, versatile software packages, and unprecedented dissemination of trapping techniques have allowed the acquisition of MS–MS data on single-component instruments. The MS–MS data generated by multianalyzer mass spectrometers can now be obtained by single-analyzer mass spectrometers. Postsource decay (PSD) analysis generated by MALDI-TOF and analysis by ion trap and FT–ICR–MS are two examples of such changing situation. The PSD measurements are generally conducted on a MADLDI–TOF mass spectrometer equipped with delayed extraction and reflectron facilities. For a number of years this was the only way to get sequencing information from analysis conducted on MALDI instruments. However, over the last two years, commercial instruments which can perform MALDI–MS–MS analysis have started to appear on the market. Further considerations regarding PSD and conventional MS–MS analysis are given below.

1.9.1. Postsource Decay

Postsource decay analysis can be considered an extension to MALDI–MS that provides structural information derived from fragmentation pathways caused by unimolecular decay and/or collision-induced dissociation with background neutrals in the field-free region just beyond the ion source. The term PSD is commonly used to describe ion fragmentation in MALDI; however, in the MS world, the phenomenon of ion fragmentation of internally excited short-lived ions within field-free regions has been known for a long time under the name *metastable ions* (Cooks et al., 1973). The origin of such effect in MALDI can be partially linked to one of the proposed ionization mechanisms. It has been argued that following irradiation of the crystals (matrix plus analyte) particles are ejected in a jetlike manner (Quist et al., 1994; Spengler and Bökelmann, 1993), forming a forward-peaking cone of neutral and charged particles. The main steps in acquiring a PSD spectrum can be summarized as follows: Ions contained in such a plume are accelerated in a two-stage acceleration system, which usually employs DE. After leaving the ion source, all ions would have the same nominal kinetic energy, and most of them would be intact precursor molecular ions which have acquired internal excitation energy by various mechanisms, including gas-phase collisions, laser irradiation, and thermal processes. During their flight through the field-free drift region they have sufficient time for PSD into their fragment ions. These product ions will continue to move with the same velocity of their precursor ions but will have lower kinetic energy because of their lower masses. The ion reflector TOF instruments used as a device for flight time compensation of the initial energy distributions can be used as an energy analyzer and thus as a mass analyzer for PSD ions. Owing to their mass-dependent kinetic energies, PSD ions are reflected at different positions within the reflector and thus have mass-dependent total flight times through the instrument.

In a typical PSD analysis on a MALDI–TOF, a complete product ion spectrum has to be acquired in several segments. This is because an ion reflector can only analyze energies with sufficient resolution within a limited range (in more recent MALDI–TOF–TOF instruments such segmentation is no longer necessary). In general, part of the product ion spectrum appears as well-resolved signals accompanied by a broad interval of lower mass signals. To analyze these lower mass ions, the potential of the reflector has to be lowered in steps until all product ions have been imaged with sufficient resolution. The acquired segments are then stitched and calibrated by the software to produce the complete fragment ion spectrum of a given precursor ion.

The detection of PSD ions in reflector instruments can be coaxial or off-axis depending on the geometry of the instrument. The latter mode of detection (off-axis) has the advantage of blanking most unwanted secondary ions or electrons formed at the grids of the reflector. However, this mode of detection requires larger detectors, since PSD ions deflected at different positions within the reflector can laterally spread over an area, which can easily exceed standard-size detectors. On the other hand, coaxial detection is possible with simpler instrumental setup within the same flight tube and with a smaller detector. However, this mode of detection has to live with interference from secondary ions and electrons.

Despite the recent increase in commercial instruments which can provide low- and high-energy MALDI–MS–MS measurements, PSD remains a powerful method for peptide sequencing. This observation is supported by the fact that the newly marketed MALDI–TOF–TOF instruments list PSD capability among the main functions which such instruments can perform.

Paradoxically, optimal performance of modern MALDI mass spectrometers requires the contemporary application of both DE and reflector TOF. Such a requirement means that the application of DE will inevitably reduce the PSD yield. Kaufmann et al. (1996) has conducted a careful study involving over 30 peptides under various experimental conditions. The authors reported that in most measurements DE application had a substantial effect on the global PSD ion yield. However, in another study (Stahl-Zeng et al., 1996) it was argued that for a variety of peptides the use of DE had no significant qualitative or quantitative effect on the acquired PSD spectra. The same authors added that the decrease in the ion fragmentation under DE conditions was compensated for by an improved signal-to-noise ratio. It has to be pointed out that the latter group has used a gridless ion mirror, which was not the case for the analysis reported by Kaufmann et al. (1996).

Isolation of the precursor ion is another critical step in PSD analysis. This step is commonly achieved either by the use of beam blanking or through ion gating. Although both modes may differ in the way they are built and in their physical position, the basic principle is the same. In some commercial instruments the ion gate can be thought of as a camera shutter that is normally closed and only opens when the precursor and therefore its fragments arrive to allow them to pass through to the reflectron. Following the passage of the desired mass window, the shutter closes to prevent the transmission of other masses. This seems a simple and straightforward process, but in reality it involves a fairly sophisticated and fast electronics to coordinate and execute the right voltage at precisely the right moment. In simple terms the ion gate consists of an array of alternately charged wires (e.g., ±1000 V). When the voltage is applied, the ion beam is deflected by a large enough angle to prevent the ions reaching the detector. The voltage is switched off long enough to allow the precursor ion and associated fragments undeflected passage through the gate to the reflector and its detector. The timing for the ion gate switching can be derived from the firing of the laser with the mass of interest being entered through the software controlling the experimental settings of the PSD acquisition function. Increase in the PSD yield can be enhanced through the combination of matrix choice (e.g., matrices which favor high-velocity ion formation) and the introduction of postsource collisional activation. We have to bear in mind, however, that gas introduction may enhance fragmentation, but it may also result in losses due to ion scattering within the incident ion beam.

1.9.2. MS–MS Measurements

Performing tandem mass spectrometry measurements at low and relatively high kinetic energy has experienced little change since its early introduction over 25 years ago. What have actually changed are the configurations of the instruments on which such measurements can be performed. The triple quadrupole, which consists of two mass filters (both DC and RF voltages are applied) separated by RF-only quadrupole, holds a special place in the historical development of MS–MS. Its introduction for analytical work by Yost and Enke (1978) and its subsequent commercialization did much to popularize the advantages of MS–MS in the analysis of mixtures. The popularity of the triple quadrupole for MS–MS analyses is related to the capability of this configuration to perform various forms of such analyses. These scan functions include daughter ion scans, where a specific m/z value is selected by the first quadrupole and injected into the collision cell. The postcollision quadrupole is then scanned to analyze the remaining incident ion beam and collision-induced fragment ions. In the parent ion scan a reversed approach is adapted, where the first quadrupole is scanned passing all ions within a given m/z range to the collision cell and tuning the third quadrupole to a specific m/z daugter ion, which may derive from the m/z values transmitted by the first quadrupole. In a third type of scan function, the first and the third quadrupoles are scanned in tandem but with m/z shift corresponding to a particular neutral loss (e.g., –COO, OH, SO_2). Despite an impressive progress in MS instrumentation, triple-quadrupole mass spectrometers remain the most suitable configuration for performing all three functions by using relatively simple and straightforward software backing.

Limiting ourselves to commercially available instruments, there are at least 10 configurations ranging from triple quadrupole to TOF–TOF and FT–ICR on which MS–MS analyses can be performed. Regardless of the type of instrument, CID is almost a common step to enhance the information acquired in MS–MS measurements. Collision-induced dissociation can be obtained through collisions with gas or solid surfaces. Collision with inert gas is the most common method to generate CID, and therefore we find it useful to give this method further consideration.

1.9.3. Collisional Activation

Collision-induced dissociation is a common step in most MS–MS analyses. Setting aside the influence of the resolution and the intensity of the primary (parent) ion beam, the efficiency of the CID can be influenced by a number of factors associated with the collision process itself. Energy of collision (mainly kinetic energy) of the incident ion is one of these parameters. Most quadrupole and ion traps are optimized for use with ions of relatively low kinetic energy (<100 eV), while sector instruments and some emerging TOF–TOF instruments can be operated at collision energies around or above 1 keV (Medzihradszky et al., 2000). In either case the maximum energy which can be transferred from kinetic to internal excitation energy (responsible for the dissociation) can be described by the approximated relationship.

$$E_{\mathrm{cm}} = E_{\mathrm{lab}} \cdot \frac{M_{\mathrm{t}}}{M_{\mathrm{p}} + M_{\mathrm{t}}} \tag{1.6}$$

Where E_{lab} is the collision energy in the laboratory frame of reference, which is a good approximation of the kinetic energy of the precursor ion, and M_{t} and M_{p} are the masses of the target and precursor ions, respectively. Equation (1.6) indicates that, when the precursor ion possesses kinetic energy in the kiloelectron-volt range, as in the case of sector instruments and some TOF–TOF configuration, the energy transfer is far more than that obtained in the electron-volt range. The direct consequence of this difference is the accession of a major number of fragmentation pathways in the first case. For example, fragments occurring remote from the site of charge are generally considered to require high activation energy and are not observed under collision conditions most commonly employed in quadrupole collision cells.

Another parameter which can influence the outcome of CID analysis is the identity and the number density of the collision gas [M_t in Equation (1.6)]. Manufacturers of mass spectrometers tend to recommend a specific collision gas; Ar, for example, is a very popular choice. If Equation (1.6) is taken seriously, then other atomic targets such as Xe or Kr should perform better but at a higher cost. The influence of the target gas on the energy transfer in the sector and in quadrupole instruments has been described in various works (Sheil and Derrick, 1986; Matsuo et al., 1983; Curcuruto and Hamdan, 1993; Hamdan and Brenton, 1989).

Sequential mass spectrometry (MS^n) is an important component in current proteomic analyses, since it can provide more comprehensive information on various fragmentation pathways. This approach was demonstrated almost 15 years ago when hybrid sector instruments were used to generate such data (Schwartz et al., 1990). Cooks and co-workers (Schey et al., 1989) have used a pentaquadrupole to demonstrate the potential of sequential MS in shedding further information on various fragmentation pathways of peptides. In three different regions of a hybrid instrument (electric sector/magnetic sector/RF collision cell/quadrupole mass filter) CID was also used to generate MS^4 analysis (Tomer et al., 1988; Alexander and Boyd, 1989; O'Lear et al., 1987; Ballard and Gaskell, 1992). These early measurements, which were conducted on *tandem-in-space* instruments, have provided enough data to validate the MS^n approach and provided enough evidence on the additional advantages of this approach compared to the conventional MS–MS approach. The same studies, however, have underlined serious limitations of conducting MS^n measurements on tandem-in-space instruments. The main limitation was found to be the low sensitivity due to excessive ion losses during the various stages of transmission from one region to another and the poor fragmentation efficiency. The limitations associated with low sensitivity have encouraged the search for an alternative approach to perform MS^n analysis. Quadrupole ion traps, which can be described as *tandem-in-time* instruments, were found to be the ideal alternative for performing this type of sequential analysis (McLuckey et al., 1991). In current proteome research the 3D ion trap is one of the instruments which is commonly indicated to perform MS^n analyses involving complex peptide mixtures. Multiple stages of ion dissociation (MS^n) in quadrupole ion traps can be performed by a variety of methods, including forward and reverse scans which sandwich a specific m/z value of interest between the limiting values set by two scans or by using broadband notch techniques (Louris et al., 1990; McLuckey et al., 1991; Julian and Cooks, 1993). The kinetic energy of the selected ion population is then increased by applying a voltage resonant with the frequency of the precursor ion, causing more energetic collisions with the He bath gas. By subjecting the ions to hundreds of low-energy collisions, the internal energy of the ion is increased until fragmentation occurs. If a single or a very narrow resonance frequency is used for excitation, the fragment ions produced are no longer excited, as their velocities will have changed. Broadband excitation methods will subject the fragment ions to additional energetic collisions. A dissociation product can also be trapped within the ion trap volume for dissociation. An MS^3 experiment can be used to create structurally important fragment ions of a particular dissociation product from the ions produced in an MS^2 experiment. In contrast to single-frequency excitation methods, fragments produced in the collision cell of a triple-quadrupole mass spectrometer continue to be subjected to collisions until they exit the cell. Consequently, fragmentation patterns for peptides caused by single-frequency excitation in the ion trap instrument can be slightly different from those observed on a triple-quadrupole mass spectrometer.

1.10. CURRENT MS INSTRUMENTATION IN PROTEOME ANALYSES

Over the last 15 years triple quadrupoles, ion-trapping instruments, and MALDI–TOF have been the key players in MS-based proteome analyses. The same instruments with various improvements are still central to diverse proteomic activities. However, the last three years have witnessed the birth of various hybrid instruments which are giving additional capabilities to already existing instruments in terms of higher resolution, higher sensitivity, and an enhanced flexibility in implementing certain scan functions. The driving force behind what can be described as a new generation of hybrid instruments is the impressive variations which ion analyzers have experienced over the last few years. The various MS configurations in current use for various proteomic analyses are schematically represented in Figure 1.10, while a much longer list can be viewed at http://www.hyphenms.nl. In the following sections some comments regarding both well-established and newly introduced instruments are given.

1.10.1. MALDI–TOF

This type of instrument can be used to perform three types of MS analyses. First, analyses in the linear mode are commonly used to measure the molecular masses of intact proteins. Second, the reflector mode of operation can be used to analyze peptide mixtures, data commonly termed PMF. In the third mode of operation, PSD is used to gain peptide sequencing. Although the instrument can be used to analyze samples of different origins, MALDI–TOF has been frequently used for the analysis of gel-separated proteins. The same type of arrangement has been used to perform SELDI, which is attracting increasing interest as a tool for screening biological samples to identify biomarkers associated with various pathologies (see Section 2.1.2).

As a result of its simplicity, excellent mass accuracy, high resolution and sensitivity, MALDI–TOF is still much used to identify proteins by what is known as peptide mapping, also referred to as PMF. In this approach, proteins are identified by matching a list of experimental peptide masses with the calculated list of the peptide masses of each entry in a database. This type of analysis requires fairly pure samples, and therefore it is commonly used in conjunction with protein fractionation with one- or two-dimensional gel electrophoresis.

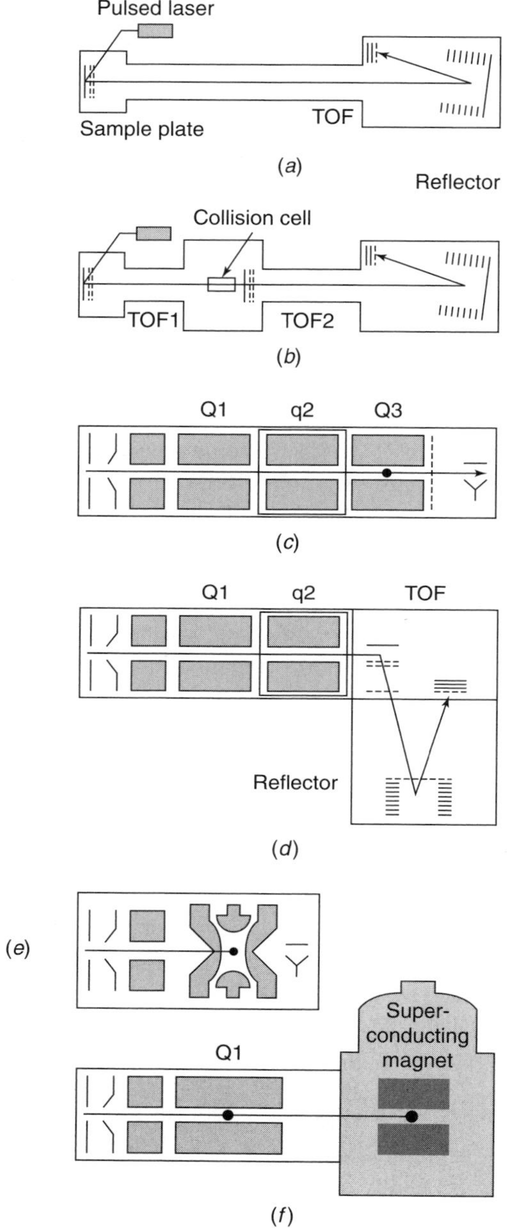

FIGURE 1.10. Various MS configurations in current use in proteome research: (*a*) reflector TOF; (*b*) MALDI–TOF–TOF; (*c*) triple quadrupole or linear trap; (*d*) Qq-TOF; (*e*) 3D ion trap; (*f*) FT–ICR–MS.

1.10.2. MALDI–TOF–TOF

It is commonly acknowledged that protein identification based on peptide CID spectra are more informative than those obtained by PMF. This is because CID spectra can be scanned against comprehensive protein sequence databases using one of a number of different algorithms. For instance, the *peptide sequence tag* approach extracts a short, unambiguous peptide sequence from the peak pattern that, when combined with the mass information, is a specific probe to determine the origin of the peptide (Mann and Wilm, 1994). In the cross-correlation method, peptide sequences in the database are used to construct theoretical mass spectra, and the overlap or cross-correlation of these predicted spectra with the measured mass spectra determines the best match (Eng et al., 1994). In the third approach, probability-based matching, the calculated fragments from peptide sequences in the database are compared with the observed peaks. Based on this comparison, a score is calculated which reflects the statistical significance of the match between the spectrum and the sequences contained in a database (Perkins et al., 1999).

An instrument which has become commercially available uses MALDI–TOF–TOF (Medzihradszky et al., 2000; Bienvenut et al., 2002; Suckau et al., 2003). The two main manufactures of this type of instrument are Applied Biosystems (Framingham, Massachusetts) and (Bruker Daltonik Gmbh, Bremen, Germany). One of the main advantages of this configuration is its capability to provide MS–MS data which can be generated at a relatively high (1–1.2 keV) collision energy. The main components of this instrument are the ion source, a short low-resolution TOF, the collision cell, and a high-resolution second TOF. The first TOF is generally operated at voltage (<5 kV) which is too low for MALDI; therefore, an Einzel lens can be added immediately after the first TOF to achieve higher transmission of low-energy ions from the ion source to the collision cell.

A commercial version of the configuration is called MALDI–LIFT–TOF–TOF (Suckau et al., 2003). According to the manufacturer (Bruker Daltonik Gmbh), the instrument can perform various functions to investigate the fragmentation of peptides. These include PMF, laser-induced dissociation, PSD, and high-energy CID. The MS–MS data generated by this type of instrument as well as those generated by the MALDI–Q-TOF configuration will certainly help to close the gap between MALDI and ESI for protein identification. In other words, the identification of proteins based on the peptide sequence *(de novo sequencing)* is no longer limited to data generated by ESI–MS–MS. Futhermore, the use of TOF–TOF brings further advantages such as faster scan rates and high-energy CID data, which can provide specific structural information (e.g., rich in immonium ions) and side-chain loss fragment ions for isomeric identification (e.g., leucine vs. isoleucine). To give a more rational comparison between the capability of MALDI–MS–MS and ESI–MS–MS in providing data suitable for de novo sequencing, more experimental data by the first approach are required. However, based on the limited data available, the following observations can be made: First, there is evidence that the data generated by MALDI–MS–MS are superior to those generated by PMF. Second, MALDI–MS–MS data, particularly those generated by CID, can be influenced by a number of parameters, including the identity of the investigated peptide. For example, arginine-containing peptides tend to yield side-chain fragment ions that are characteristic of high-energy CID. On the other hand, lysine-containing peptides tend to resemble those obtained by PMF (Medzihradszky et al., 2000). The type of collision gas and matrix can also influence the extent and pathways of fragmentation. For example, the use of a "cool" matrix such as DHB was found to suppress decomposition compared to a "hotter" matrix such as α-cyano-4-hydroxy cinnamic acid (CHCA). The influence of the matrix on the initial velocity and energy of ions formed by MALDI has been described in Section (1.5). Unlike the PSD performed on conventional MALDI–TOF instruments and that typically combines a number of separate segments of different mass ranges, MALDI–TOF–TOF allows the acquisition of a whole fragment ion spectrum in a single acquisition at a fixed reflectron voltage.

1.10.3. FT–ICR–MS

The strengths of FT–ICR are high resolution, high sensitivity, mass accuracy, and relatively wide dynamic range. In spite of their enormous potential, the expense, operational complexity, and low peptide fragmentation efficiency have limited its routine use in proteomic research. There is no doubt that capabilities in proteome analyses of this high-resolution and high-sensitivity technique have been enormously extended through its combination with external ion sources such as ESI and MALDI. Electrospray ionization FT–ICR provides an enhanced performance mass analysis that includes high mass resolution, mass range, and mass accuracy. For example, unit resolution has been achieved for proteins larger than 110 kDa (Kelleher et al., 1997), and high-resolution full mass spectra are generated routinely for molecules ranging from 10 to 30 kDa (Valaskovic et al., 1996). The increasing need for the analysis of complex protein mixtures has underlined the need to couple LC to FT–ICR. However, this coupling has created a number of problems, owing primarily to the mismatch between the desired working pressure within the source and the analyzer. Specifically, the lengthy pump downtimes necessary for an optimized FT–ICR performance come at the expense of poor chromatographic resolution and reduced duty cycle. Attempts to overcome such limitations included gated trapping (Alford et al., 1986) to capture ions before excitation and detection. Gated trapping consists of reducing the electric potential on the entrance and cap electrodes to ground to admit ions to the cell, then rapidly returning the electrode to normal trapping potential before data acquisition. Senko et al. (1997) described the accumulation of electrosprayed ions in an octapole ion guide for the subsequent injection by a second octapole to the center of a superconducting magnet, where ions are captured in a penning trap by gated trapping. Nanoscale LC/FT–ICR–MS of small peptides at the femtomole level was performed.

In another variation on ion accumulation (Ostrander et al., 2000) a magnetic field-focusing central trapping electrode ion accumulation cell for capillary LC–ESI/FT–ICR was described. In such an arrangement the ESI source together with the accumulation cell were located within the magnetic field to confine the radial motion of the ions, thus eliminating the need for elaborate focusing optics to transport the ions to the low-pressure analyzer cell for analysis.

Interfacing a MALDI source to the FT–ICR to perform surface-induced dissociation of peptides was recently reported by Laskin et al. (2004). This type of analysis may help to address some of the limitations associated with the combination of MALDI with FT–ICR–MS. Although such combination offers very high mass resolution, mass accuracy, and multiple-stage tandem mass spectrometry (MS^n), the same combination tends to give disappointing fragmentation of ions generated by MALDI. These poor fragmentation patterns are commonly obtained using conventional ion activation such as sustained off-resonance irradiation or infrared multiphoton dissociation.

Laskin et al. (2004) have described the use of an intermediate-pressure MALDI source combined with FT–ICR to perform surface-induced dissociation. This experimental setup was tested for the sequencing of peptides, which are known to produce poor fragmentation patterns using conventional FT–ICR–MS ion activation modes. The authors attributed the rich CID spectra to a combination of two effects: At collision energies above 40 eV, the shattering (fast decay) of peptide ions on solid surfaces opens up a variety of dissociation channels, and second, collision with the stiff diamond surface provides an efficient mixing between primary reaction channels that are dominant at low internal energies and extensive fragmentation at high internal excitation that results from shattering on such surface.

Another recent development associated with FT–ICR–MS is its coupling to the ESI source through a quadrupole 2D ion trap to facilitate unbiased external ion accumulation (Belov et al., 2001a). In a number of earlier works (Senko et al., 1997; Wang et al., 2000) it has been demonstrated that the sensitivity, dynamic range, and duty cycle provided by FT–ICR can be improved by ion trapping and accumulation in a linear multipole trap positioned externally to the cell. In a recent elaboration (Belov et al., 2001b) the authors utilized a *dynamic range enhancement* applied to a mass spectrometry (DREAMS) algorithm, based on the use of ion ejection from a linear quadrupole device external to the FT–ICR. In this arrangement, ion accumulation was accomplished by resonant RF-only dipolar ejection based on the ion intensities of the previous MS spectrum. The authors claimed that the use of this algorithm resulted in 40% increase of detected peptides from a yeast extract relative to that obtained in the absence of external selective ion accumulation. These and other works pointed out in this chapter strongly suggest that the coupling of this type of ion trap to FT–ICR is likely to enhance the sensitivity and overall performance of this trapping technique, particularly in proteome analysis.

1.10.4. ION Mobility–MS

Despite an impressive improvement in 2D chromatography, high-throughput analysis of complex peptide mixtures containing thousands of components remains a challenging task. Over the last few years a number of attempts have been made to include another dimension of separation based on postsource ion mobility separation. This approach, called gas-phase electrophoresis, can provide an efficient means for m/z separation of ions leaving the ion source. The capability of this method has been demonstrated as an additional dimension to LC separation in the analysis of complex peptide mixtures generated by combinatorial chemistry (Srebalus Barnes et al.,

2002). Because of the principle on which this method is based, ion mobility analyses are commonly effected between the ion source and the analyzer. A variation on this approach has been reported by Gillig et al., 2000), who used a high-pressure MALDI source to generate the ions within the drift tube itself, which was used to separate *m*/*z* values prior to their TOF analysis. So far the capabilities of ion mobility separation have been demonstrated on a limited scale, yet the potential of this separation method is still to be exploited, particularly in the analysis of extremely complex peptide mixtures. One obvious area of application is the coupling of the method to FT–ICR–MS, where *m*/*z* separation combined with high mass accuracy could provide the answer for high-throughput analysis without the need for prior LC separation.

1.11. CURRENT MS-BASED PROTEOMICS

In current MS-based proteomics, the MS component has to interact closely with other components to deliver meaningful data. These components include different separation techniques, bioinformatics, and various chemical protocols to label or enrich certain components within the investigated mixture. A commonly used sequence of steps to effect such analyses is given in Figure 1.11. Before discussing in more detail

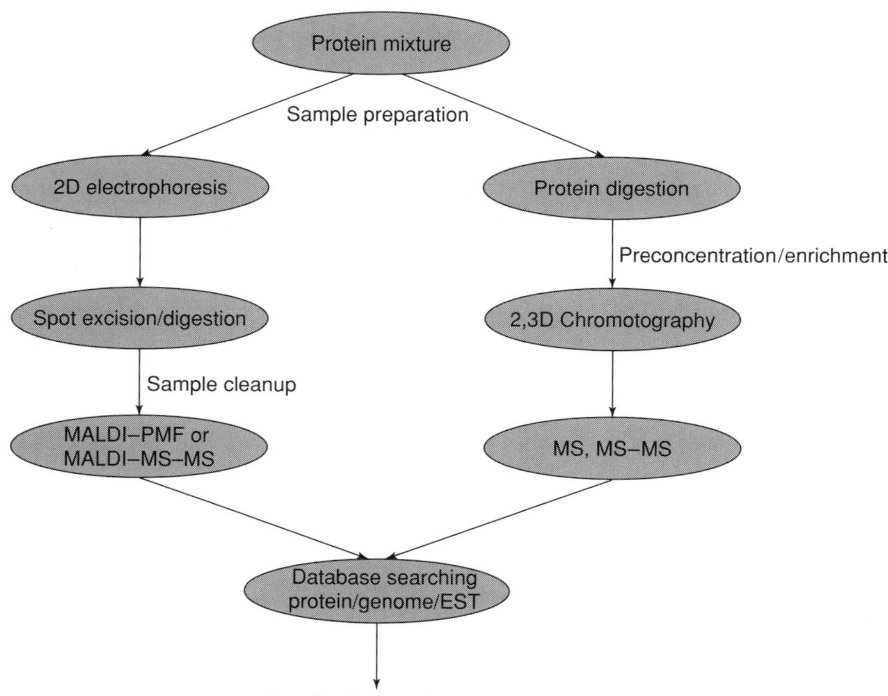

FIGURE 1.11. Schematic representation of various steps in MS-based strategy for protein characterization within a mixture.

the various components in the figure, the following general observations might be of interest: (i) No method or instrument on its own is capable of identifying and quantifying the components of a complex protein sample. Rather, individual components for separating, identifying, and quantifying the derived polypeptides as well as tools for integrating and and analyzing all the data generated must be used in concert. (ii) Protein mixtures of considerable complexity can now be routinely characterized in some depth. One measure of technical progress is the number of proteins identified in each study. Such numbers can now reach into the thousands for suitable complex samples. (iii) Closer interaction between the world of biology and mass spectrometry has highlighted a number of areas where proteomics is called upon to provide solutions to specific biological questions. These questions include protein quantification, including posttranslational modifications, protein interaction maps, and protein function and location. A brief description of MS-based approaches to address some of these emerging challenges has been briefly presented in Chapter 2. Some of these approaches will also be discussed here, but first we ought to have a closer look at the various components which make a workflow for protein analyses.

1.11.1. Delivering Peptides to Ion Source

The method used to introduce the sample to the ion source is an important aspect for the use of MS in proteomic analyses. Moving and manipulating small quantities of proteins from the laboratory bench to the mass spectrometer should be conducted in a way which minimizes sample losses. Most current proteomic analyses are conducted on either MALDI or ESI ion sources. The MALDI source has been mainly used for the ionization of protein digests derived from protein separated by 2D and in some cases 1D gel electrophoresis. The process of spot picking, cutting, digesting, and depositing on various sample plate formats, including 96-well plates, is fully automated, including the deposition of 1–2 μL in each well and the introduction of the plate to the MALDI ion source. Both off-line and on-line LC separations are also used to prepare samples destined for MALDI analysis. Although on-line sample delivery to a MALDI source has been demonstrated over 10 years ago, it has not yet attracted the attention of manufacturers of MALDI instruments. There have been various examples of the direct introduction of liquid to a MALDI ion source (Li et al., 1993; Wall et al., 2002). As mentioned above, this approach has not found its way to a wide use; therefore, we will not give further details on this method. Instead, we will only discuss sample introduction to MALDI through the use of sample plates. There is good evidence to suggest that the front isolation procedures can have the most significant impact on the outcome of MALDI-based analyses. For example, sensitivity of the overall procedure is usually determined more by the purification strategy than by the sensitivity of the ionization process per se. In various proteomic analyses, protein, purification starts with a whole-cell lysate and ends with a gel-separated protein band or spot. The MALDI–MS analysis of such band/spot is carried out on peptides obtained after enzymatic digestion. In special cases, the intact proteins are analyzed to gain more accurate determination of the molecular weight. In principle, any of the classical separation methods, such as centrifugation, column chromatography, and affinity-based procedures, can precede the final step of gel electrophoresis. As long as the proteins of the lysate can be adequately resolved, it is best to minimize the number of separation steps. Generally, silver-stained amounts are required for successful MS identification of proteins (5–50 ng or 0.1–1 pmol for 50-kDa protein). This does not mean that higher sensitivities cannot be achieved. In this kind of analysis it is important to minimize contamination caused by the presence of keratins which are introduced by dust, chemicals, handling without gloves, and so on, as the keratin peptides can easily dominate the spectrum. In MALDI-based analyses, most detergents and salts are eleminated in the gel-washing procedure. Neverthless, the protein should be as concentrated as possible in the gel to avoid excessive background signals in the MS analysis. Pooling of spots is not necessarily advantageous, as both protein and background will be increased. Crosslinkers and harsh oxidizing agents should be avoided as they interfere with the extraction of peptides from the gel. It is also highly recommended to reduce and if possible fully alkylate the investigated proteins prior to any electrophoretic step. Failure to do so may result in a number of artifacts, including the formation of large aggregates which hamper an efficient transfer of proteins from the first to the second sodium dodecyl sulfate (SDS) dimension. To enhance the quality of analysis, peptides are often desalted and concentrated through the use of short reverse-phase columns (50 nL–1 μL packing volume).

By definition, an ESI ion source generates gas-phase ions from liquid drops. Such liquids and the analytes are commonly delivered by LC systems interfaced to the ion source. Although LC has been around for a long time, its use in conjunction with mass spectrometry has been given an unprecedented impetus by the increasing needs of modern proteomics. Although numerous methods for coupling LC to MS have been explored, it is electrospray ionization that has transformed LC/MS–MS–MS into a routine laboratory practice sensitive enough to analyze peptides and proteins at levels which have attracted the interest of biologists. As mentioned above, ESI requires a continuous flow of liquid, and signal strength is almost independent of flow rate. To enhance sensitivity, in the analysis of complex peptide mixtures, research efforts have been focused on coupling nanoscale LC at submicroliter flow rates to the highly sensitive version of ES (nano-electrospray). Currently, detection limits of a few femtomoles of peptide material loaded on the column make this ionization technique suitable for the detection of low-abundance proteins in cell lysate.

In a typical LC/MS–MS-MS experiment, the peptide mixture is eluted from a reverse-phase column to separate the peptides by hydrophobicity and is fed to the ion source for ionization followed by high-efficiency ion transfer to the rest of the mass spectrometer for MS or MS–MS analysis and detection.

The need to generate reliable LC/MS–MS data has encouraged the use of different columns and different combinations of such columns. Multidimensional chromatography is commonly used despite the fact that most published works are based on the use of two different chromatographic approaches. A number of two-dimensional chromatography systems which are in current use in combination with ESI–MS are as follows:

- Size exclusion coupled with reverse-phase (RP) LC
- Ion exchange coupled to RP–LC

- RP–LC coupled to (25–60°C) RP-LC
- RP–LC coupled to capillary electrophoresis (CE)
- Metal affinity chromatography coupled to CE
- Separation based on ion mobility with or without LC

One- and two-dimensional chromatography coupled to MS is schematically presented in Figures 1.12*a*,*b*. Direct infusion (no prior separation) of a sample into an ES ion source is one of the easiest methods to introduce the sample directly into the ion source. In this mode, the sample is simply introduced into a continuous liquid stream (typically a mixture of organic and aqueous liquid such as 50 : 50 methanol–water) via an injection valve. Flow rates are usually between 0.5 μL/min and several microliters per minute. in this mode samples have to be substantially free of salt and detergent. Sample quality can be improved by including a reverse-phase packing loop within the injector valve. In the case of nanoESI no pumps are needed for this mode of operation. Low flow rates in the range of few nanoliters and pulled glass capillaries can be used. However, this method has a limited application in the analysis of mixtures unless extremely high-resolution instruments are used and at the same time a specific class of molecules within the analyzed mixture is sought. A good example of this type

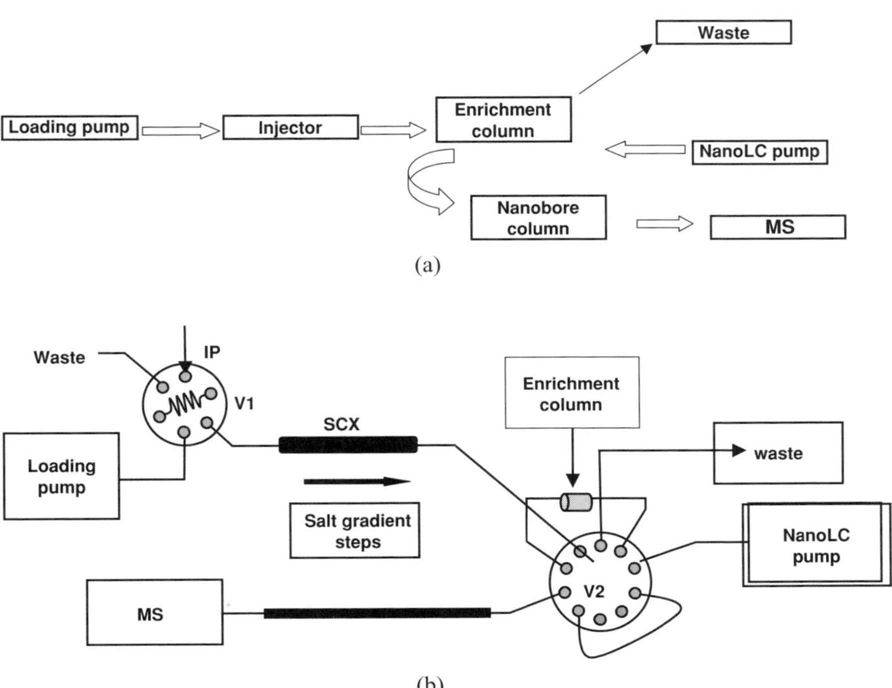

FIGURE 1.12. (*a*) One- and (*b*) two-dimensional chromatography coupled to mass spectrometry commonly used in proteome analyses.

of analysis has recently been reported by Chalmers et al. (2004). The authors have used high-field (9.4-T) FT–ICR–MS to determine the location of phosphorylation of tryptic peptides derived from protein kinase A. Structural information was maximized by the use of electron capture dissociation and/or IR multiphoton dissociation.

1.11.2. Peptide Sequencing and Database Searching

One of the major advances in biological MS is the development of algorithms for the identification of proteins by MS data matched to a database. A number of tools have been developed that specifically consider experimentally measured protein attributes, such as amino acid composition, sequence tags, and other MS-derived data. Table 1.3 gives a list of protein identification tools available on the Web. With the availability of the complete sequence of an increasing number of model species, the peptide-sequencing problem, formerly a holy grail in biological mass MS, has been shifted to database correlation, enabling automation and the scaling up of proteomic experiments. At the present time there are different types of MS data which can be used for protein identification by database searching.

1.11.3. Peptide Mass Fingerprinting

Peptide mass fingerprinting data are generally generated by reflector MALDI–TOF analysis of enzymatically degraded proteins. The acquired peptide masses are then compared to the theoretically expected tryptic peptide masses for each entry in the database, which can rank various proteins according to the number of peptide matches. More sophisticated scoring algorithms take the mass accuracy and the percentage of the protein sequence covered into account and attempt to calculate a level of confidence for the match (Berndt et al., 1999; Perkins et al., 1999; Eriksson et al., 2000). Other considerations can also be included in the search, such as the fact that larger peptides are less frequent in the database and therefore should be given more weight when matched. In this kind of analysis the mass accuracy of the detected peptides has a strong influence on the specificity of the search—for example, when a mass accuracy in the range 10–20 ppm requires at least five peptide masses to be matched to the protein and 15% of the protein sequence needs to be covered to allow unambiguous identification. After a match has been found, a second search is performed to correlate remaining peptides with the database sequence of the match, taking into account possible modifications. Generally, PMF is used for the rapid identification of single proteins but can also identify the components of simple protein mixtures consisting of several proteins within a roughly comparable amount. For example, databases can be searched iteratively by excluding the peptides associated with an unambiguous match (Jensen et al., 1997).

Protein characterization based on PMF data can have a number of limitations: When the measured masses are searched against a large sequence database, there is a higher chance of randomly matching unwanted protein entries. In the case of very small or very large proteins the acquired PMF may not be suitable for a successful database search. In the case of small proteins, the required number of peptides for

TABLE 1.3. Some Protein Identification Tools Available on the Web

Type of Spectrum	Program Name	Search Database	Internet URL Address	References
PMF	ALDentE	SWISS-PROT, TrEMBL	http://www.expasy.org/tools/aldente/	Tuloup et al., 2002
PMF, MS/MS	Mascot	OWL, NCBInr, SWISS-PROT, MSDB, EST	http://www.matrixscience.com	Perkins et al., 1999
PMF, MS/MS	Protein Prospector	NCBInr, dbEST, SWISS-PROT, Genepept, OWL	http://prospector.ucsf.edu	Clauser et al., 1995
PMF, MS/MS	Prowl	NCBInr, dbEST, nr	http://129.85.19.192/prowl/pepfragch.html	Zhang and Chait, 2000; Fenyo et al., 1998
PMF	PeptIdent	SWISS-PROT TrEMBL	http://www.expasy.org/tools/peptident.html	Binz et al., 1999
MS/MS	Sonar MS/MS	NCBInr, SWISS-PROT, dbEST	http://65.219.84.5/service/prowl/sonar.html	Field et al., 2002

matching might not be sufficient. On the other hand, in the case of large proteins the number of theoretical peptides is so huge that a portion of them may randomly match any spectrum. If the excised 2D spot happens to have more than one protein, the acquired spectrum will be fairly complicated and the database search may give the wrong identification. If the protein sequence under investigation is not listed in the database, then a number of possibilities may manifest: Random matches are not correlating with high scores and no significant hit is obtained or there is a protein match that corresponds to a homologues, which means that additional information such as origin of species or tissue may contribute to the identification.

1.11.4. Searching with MS–MS Data

Databases can also be searched by MS–MS data generated by CID. As well as the mass of the individual peptide, these data contain structural information related to its sequence, which render this type of data more specific and more discriminating. Searching databases using MS–MS data encompasses various approaches, some of which are considered here. The peptide sequence tag (Mann and Wilm, 1994) relies on the concept that nearly every MS–MS spectrum contains at least a short series of fragment ions that unambiguously specifies a short amino acid sequence. In most cases as few as two amino acids can be combined with the start mass and the end mass of the series, which specify the exact location of the sequence in the peptide and the known cleavage specificity of the enzyme. Such a peptide sequence tag will then retrieve from the database one or a few sequences whose theoretical fragmentation pattern is matched against its experimental counterpart. This approach can be fully automated and is highly specific, especially when performed using instruments with a high mass accuracy such as ICR–FT and TOF-based configurations.

In another and more ambitious approach Eng et al. (1994) have devised a method to evaluate amino acid sequences pulled from the database by reconstruction of a model tandem mass spectrum and its comparison with the observed tandem mass spectrum using a cross-correlation function (SEQUEST). In essence, this method determines the extent of agreement between the experimental and reconstructed spectra. Another score indicates how differently the next most similar sequence in the database fits the spectrum. Although this method can be highly automated, the sequences need to be verified manually unless the score is very high.

1.11.5. Databases for MS Data Search

Three types of sequence databases can be searched by MS data: (i) Some protein databases contain the known set of full-length protein sequences extracted from the major sequence repositories. For example, nrdb, maintained at the European Bioinformatics Institute, contains an impressive number of sequences. Peptide mass finger printing and MS–MS data can search both these databases. Matches can quickly be followed up via links to annotated protein databases such as SwissProt or Yeast Protein Database (Costanzo et al., 2000). (ii) Expressed sequence tag (EST) databases such as dbEST at the National Center for Biotechnology contain millions of short one-pass sequences from random sequencing of cDNA libraries. These can be searched with appropriate software. Even though EST databases are error prone and cover only a part of the gene, many groups have demonstrated that virtually all proteins encountered in proteomics projects can be correlated to their respective EST entries based on minimal MS information (Mann, 1996; Neubauer et al., 1998). (iii) Genome databases can also be searched with MS data. Surprisingly, it has been found that peptides can be matched in the raw genome sequence, without any assumptions about the reading frame or likely coding regions and without translation into amino acid sequence (Mann et al., 2001). The advantages of searching databases of completely sequenced genomes directly are that each peptide must be present by definition, that genome sequences are of very high quality, and that often the MS data can help to define the structure of the gene, such as start and stop.

1.12. RECENT ACHIEVEMENTS AND FUTURE CHALLENGES

Published works over the last three years reveal unmistakable new and exciting developments in the area of MS-based proteomics. These new developments are associated with the instrumentation and the applications. Some of these new developments have been cited in various parts of this chapter. However, we find it relevant to underline specific developments and their correlation to future progress.

1.12.1. Current Applications

Most readers who followed MS-based proteomic applications over the last 15 years have no difficulty in acknowledging the significant contribution which MS has given to the various facets of proteome research. These recent successes have at the same time highlighted new technical challenges and underlined the need for more breakthroughs, which demand not only new generation of instrumentation but also new approches based on close interaction between MS, chemistry, and biology. Current literature shows that such necessary interaction is already underway in a number of challenging proteomic tasks:

- Large-scale protein–protein interactions
- Probing protein activity on a proteome scale
- Profiling the global tyrosine phosphorylation to allow a better understanding of signal transduction pathways
- Protein interaction maps on multicellular organisms
- High-throughput protein expression, quantification, and localization to allow reliable assignment of various protein functions
- Refinement of proteomic approaches for the detection/quantification of post-translational modifications
- Improvement and refinement of technologies which normally interact with MS, including chromatography, sample preparation, and the chemistry of labeling and enrichment

- Moving from prototype models to models capable of resolving real-word challenges
- Increasing the sensitivity of MS detection and its use with other analytical approaches
- Increased and more focused clinical proteomics to respond to an increasing need for reliable biomarkers and safer drugs

The above list of current challenges in proteome research is by no means exhaustive, yet it may give the reader a reasonable indication of the diversity of the tasks to which MS can contribute. It also shows that MS cannot be considered in isolation from technologies with which this technique has to interact. In the rest of this chapter we will attempt to discuss briefly the various items in this list with the goal of shedding some light on future developments of these elements.

It is becoming more evident that at the biochemical level proteins rarely act alone; rather they interact with other proteins to perform various cellular tasks. Therefore, unraveling the ways in which proteins interact with each other on a genomewide scale is becoming one of the main goals of current and more likely future proteomics research. Over the last three years, large-scale protein interaction maps have been published only for yeast (Uetz et al., 2000; Ito et al., 2001; Schwikowski et al., 2000) and bacteria (Rain et al., 2001). Within two years of these early works, a recent paper by Giot et al. (2003) has reported the largest protein interaction map to date and the first genomewide study for a multicellular organism, the fruit fly *Drosophila melanogaster*. A key to this impressive success is the development of computational and statistical methodologies to extract relevant protein–protein interactions. Currently there are a number of strategies for high-throughput protein interaction, including the MS-based method. In this approach individual proteins are tagged and used as "hooks" to biochemically purify whole protein complexes. This method uses tandem affinity purification (Rigaut et al., 1999) followed by high-throughput MS protein complex identification (Gavin et al., 2002; Ho et al., 2002). This approach has a number of limitations which certainly will benefit from further refinement in the tagging procedure and the washing steps, which tend to eliminate loosely associated components. This method also has to be backed by more powerful computational and statistical methodologies to identify complexes under various conditions. Comparison of the data generated by this method with those generated by genetic interactions, correlated messenger RNA (mRNA) expression, and in silico predictions through genome analyses is also necessary if the full potential of this MS-based protocol is to be realized.

1.12.2. Signal Transduction Pathways

Protein tyrosine kinases and protein tyrosine phosphatases play a key role in cell signaling. Functional profiling of the tyrosine phosphoproteome is likely to lead to the identification of novel targets for drug discovery and provide a basis for novel molecular diagnostic approaches. The ultimate aim of current MS-based phosphoproteomic approaches is the comprehensive characterization of the phosphoproteome.

However, current methods are not yet sensitive enough to allow routine detection of a large percentage of tyrosine-phosphorylated proteins, which are generally of low abundance. Having said that, a number of MS-based strategies have exploited the fact that receptor-mediated signaling pathways generally utilize Ser/Thr or tyrosine phosphorylation of cellular proteins to relay the signal from the membrane to the nucleus. Receptors possessing intrinsic tyrosine kinase activity such as epidermal growth factor (EGF) or platelet-derived growth factor (PDGF) receptors are therefore highly suitable for proteomic approaches. In this case, antiphosphotyrosine antibodies are used to enrich all of the tyrosine-phosphorylated substrates followed by 1D electrophoresis to resolve various proteins. Bands that are found in growth factor treated but not untreated cells are simply excised and analyzed by MS. Several components of the signal transduction pathway of the EGF receptor have been identified by such an approach (Pandy et al., 2000a,b). The same strategy can also be applied to the global study of signaling pathways of T-cell, B-cell, and cytokine receptors. An alternative method is to use 2DE to resolve the proteins of a cell lysate followed by Western blot and MS analysis. The drawback of this approach is that, unless the proteins are enriched, it is not easy to identify key signaling molecules because of their inherently low abundance. Strategies based on derivatization and affinity purification for the detection and quantification of phosphorylated proteins have shown a highly promising potential for the detection of these low-abundance events (see Section 3.4.1.).

1.13. CONCLUDING REMARKS

Mass spectrometry is the core component of modern proteomics. Recent progress in instrumentation will undoubtedly continue to grow and automation will make it possible to generate data that will exceed those of genomics. Progress on the instrumentation side has the scope of addressing a number of demanding proteomics goals, some of which are briefly discussed below.

1. As the interactions in which a given protein participates are likely to correlate with the protein's functional properties, protein interaction maps are frequently utilized to probe in a systematic fashion the potential biological role of proteins of unknown functional classification (Tong et al., 2002; Schwikowski et al., 2000). Uncovering the topology of the identified protein interaction networks is equally relevant, since they may reflect the cell's higher level functional organization. Unraveling the ways in which proteins interact with each other and the topology of such interaction on a genomewide level is certainly one of the major goals of proteomics. Large-scale protein interactions for both single and multicellular organisms have already been published. These same results, however, have highlighted the difficulty of such a task, particularly if we want to extend such an approach to more complex organisms. To appreciate the complexity of the task related to protein interaction maps, it is sufficient to consider the recent work by Giot et al. (2003) regarding protein interaction map for the fruit fly *D. melanogaster*. The enormity of such a task was discussed in a recent article (Uetz and Pankratz, 2004). Some of the following observations are taken

from this article. Building such a map required a *tour de force* of polymerase chain reaction (PCR) to amplify all 14,000 predicted *D. melanogaster* open reading frames (ORFs), of which more than 12,000 worked successfully. These PCR products were then cloned into two-hybrid bait (protein of interest) and prey (interactors) vectors, yielding roughly 11,000 clones each. The authors also commented that a key factor for the success of the work was the development of computational and statistical methodologies to extract relevant protein–protein interactions.

2. Another main goal of proteomics is to characterize and develop "circuit maps" of cellular signaling pathways in normal and diseased cells. For example, it is now recognized that defective, hyperactive or dominating signal pathways may drive cancer growth, survival, invasion, and metastasis (Liotta and Kohn, 2001). Mapping of the information flow through signaling pathways in normal and cancer cells may serve as a means to identify key alterations that occur during tumor progression and thus provide targets for rational molecular targeted drug design. Until recently, efforts to elucidate the activation of signaling pathway events have relied mainly on either gene microarrays/gene-based analysis and bioinformatic tools or 2DE coupled with immunoblotting to detect phosphorylation. The realization that gene transcription profiling does not accurately reflect posttranslational modifications, including protein phosphorylation, which is one of the main drivers in cellular signaling processes, has underlined the need for proteomic-based approaches. Some of these promising approaches to tackle such a task include reverse-phase protein microarrays (Paweletz et al., 2001) and MS-based methods in conjunction with derivatization strategies and affinity purification (Oda et al., 2001; Zhou et al., 2001; Goshe et al., 2002).

3. Another major challenge of modern proteomics is to ascribe the cellular functions of proteins expressed by a given cell. Such knowledge will inevitably accelerate the validation of disease-relevant protein targets, which can help in the development of safer and innovative therapies. Identification of new disease-specific targets has been so far limited to proteins present on the cell surface. A better understanding of proteins and biological networks that lie below the cell's surface will no doubt provide a stronger rationale for the early assessment of target validity. Approaches are being developed or implemented to allow direct protein manipulations which go beyond a simple visualization of up and down regulation of target proteins. One of these approaches is chromophore-assisted laser inactivation (CALI) (see Section 2.8). Using such an approach, a specific site in a protein can be inactivated in a time-dependent and localization-restricted manner. The use of single-chain variable fragment antibodies (scFvs) as specific probes for CALI or FALI (fluorophore-assisted light inactivation) could be used to immunoprecipitate the validated protein that could then be identified by MS. Such a strategy would take advantage of the ever-increasing databases provided by proteomics; furthermore, such a strategy has the potential for multiplex high-throughput analysis.

4. As has been pointed out by Phizicky et al. (2003), the promise of proteomics is the precise definition of the function of every protein in the cell and how that function changes in different environmental conditions, with different modification states of the protein, in different cellular locations, and with different interacting

partners. Recent literature leaves no doubt that the employment of a battery of newly developed approaches has resulted in tremendous progress toward the realization of such a promise. Having said that, there is still a pressing need for both additional high-throughput technologies and further refinement to existing technologies, including MS-based approaches together with more powerful computational methods, to handle and analyze ever-increasing large data sets.

5. Another challenge facing the proteomics community is the need to coordinate and organize their efforts with particular attention to a closer interaction with those focused on biological and biomedical problems. A significant step in this direction has been achived through the birth of the Human Proteome Organization (HUPO, http://www.hupo.org). This organization was founded to regroup scientists in the public and private sectors engaged throughout the world in various aspects of proteomics (Hanash, 2003). HUPO is in a position to provide an important coordinating role which has proclaimed a number of initial goals for wordwide proteomics research: definition of the plasma proteome, proposals for an in-depth proteomics investigation of specific cell types, formation of a consortium to generate antibodies to all known human proteins, development of new technologies, and formation of an informatics infrastructure. These initial goals have attracted further suggestions by various research groups actively engaged in proteomics. For example, Tyers and Mann (2003) suggested that other goals should be added to the initial list, including cataloging the primary structure of all proteins, mapping all organelles that can be purified, and generating protein interaction maps of model organisms for both comparative proteomics and integration with on-going functional genomics projects.

REFERENCES

Alexander, A. J., Boyd, R. K. (1989) *Int. J. Mass Spectrom. Ion Processes* **90,** 211.

Alexandrov, M. L., Gall, L. N., Krasnov, N. V., Nnikolaev, V. I., Panvlenko, V. A., Shkurov, V. A., Baram, G. I., Grachev, M. A., Knorre, V. D., Kusner, Y. S. (1984) *Bioorg. Khim.* **10,** 710, in Russian.

Alford, J. M., Williams, P. E., Trevor, D. J., Smalley, R. E. (1986) *Int. J. Mass Spectrom. Ion Process.* **72,** 33.

Ballard, K. D., Gaskell, S. J. (1992) *J. Chem. Soc. Mass Spectrom.* **3,** 644.

Barber, M., Bordoli, R. S., Sedgwick, R. D., Tetler, L. W. (1981) *Organic Mass Spectrom.* **16,** 256.

Beavis, R. C., Chait, B. T. (1991) *Anal. Chem.* **62,** 1836.

Beir, M. E., Syka, J. E. P. (1994) U.S. Patent 5, 420,425.

Belov, M. E., Gorshkov, M. V., Udseth, H. R., Anderson, G. A., Tolmachev, A. V., Prior, D. C., Harkewicz, R., Smith, R. D. (2000) *J. Am. Soc. Mass Spectrom.* **11,** 19.

Belov, M. E., Nikolaev, E. N., Alving, K., Smith, R. D. (2001b) *Rapid Commun. Mass Spectrom.* **15,** 1172.

Belov, M. E., Nikolaev, E. N., Anderson, G. A., Udseth, H. R., Conrads, T. P., Veenstra, T. D., Masselon, C. D., Gorshkov, M. V., Smith, R. D. (2001a) *Anal. Chem.* **73,** 253.

Berndt, P., Hobohm, U., Langen, H. (1999) *Electrophoresis* **20,** 3521.

Biemann, K. (1988) *Biomed. Environ. Mass Spectrom.* **16,** 99.

Bienvenut, W. V., De'on, C., Pasquarello, C., Campbell, J. M., Sanchez, J-C., Vestal, M. L., Hochstrasse, D. F. (2002) *Proteomics* **2,** 868.

Binz, P. A., Wilkins, M., Gasteiger, E., Bairoch, A., Appel, R., Hochstrasser, D. F. (1999), in Kellner, R., Lottspeich, F., Meyer, H. (Eds.), *Internet Resources for Protein Identification and Characterization*, Wiley-VHC, pp. 277–299.

Buriak, J. M., Allen, M. J. (1998) *J. Am. Chem. Soc.* **120,** 1339.

Campbell, J. M., Collings, B. A., Douglas, D. J. (1998) *Rapid Commun. Mass Spectrom.* **12,** 1463.

Canham, L. T. (1997) in Canham, L. T. (Ed.), *Properties of Porous Silicon*, Institute of Electrical Engineers, London, pp. 83–88.

Caprioli, R. M., Fan, T., Cottrell, J. S. (1986) *Anal. Chem.* **58,** 2949.

Caprioli, R. M., Moore, W. T., Martin, M., DaGue, B. B., Wilson, K., Moring, S. (1989) *J. Chromatogr.* **480,** 247.

Chait, B. T., Shpungin, J., Field, F. H. (1984) *Int. J. Mass Spectrom. Ion Process.* **58,** 121.

Chalmers, M. J., Hakansson, K., Johnson, R., Smith, R., Shen, J., Emmett, M. R., Marshall, A. G. (2004) *Proteomics.* **4,** 970.

Chernushevich, I. V., Loboda, A. V., Thomson, B. A. (2001) *J. Mass Spectrom.* **36,** 849.

Chien, B. M., Michael, S. M., Lubman, D. M. (1994) *Int. J. Mass Spectrom. Ion Process.* **131,** 149.

Clauser, K. R., Hall, S. C., Smith, D. M., Webb, J. W. (1995) *Proc. Natl. Acad. Sci. USA.* **92,** 5072.

Comisarow, M. B. (1980), in Quale, A. (Ed.), *Advances in Mass Spectrometry*, Vol. 8B, Heyden and Son, London, p. 1698.

Comisarow, M. B., Marshall, A. C. (1974) *Chem. Phys. Lett.* **25,** 282.

Conrads, T. P., Issaq, H. J., Veenstra, T. D. (2002) *Biochem. Biophys. Res. Commun.* **290,** 885.

Cooks, R. G., Beynon, J. H., Caprioli, R. M., Lester, G. R. (1973) *Metastable Ions*, Elsevier, Amsterdam.

Costanzo, M. C., Hogan, J. D., Cusick, M. E., Davis, B. P., Fancher, A. M. (2000) *Nucleic Acid Res.* **28,** 73.

Cullis, A. G., Canham, L. T., Calcott, P. D. J. (1997) *J. Appl. Phys.* **82,** 909.

Curcuruto, O., Hamdan, M. (1993) *Rapid Commun. Mass Spectrom.* **7,** 989.

Dass, C., Desiderio, D. M. (1987) *Anal. Biochem.* **163,** 52.

Dawson, P. H., Hedman, J., Whetten, N. R. (1969) *Rev. Sci. Instrum.* **40,** 1444.

Dawson, P. H., Whetten, N. R. (1969) *Adv. Electron. Electron Phys.* **27,** 58.

Dempster, A. J. (1918) *Phys. Rev.* **11,** 316.

Devienne, F. M., Grandclement, G. (1966) C. R. Acad. Sci. **262,** 696.

Dole, M., Mack, L. L., Hines, R. L., Mobley, R. C., Ferguson, L. D., Alice, M. B. (1968) *J. Chem. Phys.* **49,** 2240.

Ehring, H., Karas, M., Hellenkamp, F. (1992) *Org. Mass Spectrom.* **27,** 427.

Eng, J. K., McCormack, A. L., Yates III, J. R. (1994) *J. Am. Soc. Mass Spectrom.* **5,** 976.

Eriksson, J., Chait, B. T., Fano, D. (2000) *Anal. Chem.* **72,** 999.

Fenyo, D., Qin, J., Chait, B. T. (1998) *Electrophoresis*, **19,** 998.

Field, H. I., Fenyö, D., Beavis, R. C. (2002) *Proteomics* **2,** 36.

Frank, M., Labov, S. E., Benner, W. H. (1999) *Mass Spectrom. Rev.* **18,** 155.

Gavin, A. C., Bösche, M., Krause, R., Grandi, P., Marzioch, M., Bauer, A., Schultz, J., Rick, J. M., Michon, A. M., Cruciat, C. M., Memor, M., Höfter, C., Schelder M., et al., (2002) *Nature* **415,** 141.

Gillig, K. J., Ruotolo, B., Stone, E. G., Russell, D. H. (2000) *Anal. Chem.* **72,** 3965.

Giot, L., Brouwer, C., Chaudhuri, A., Kuang, B., Li, Y., Hao, Y. L., et al., (2003) *Science* **302,** 1727.

Goshe, M. B., Veenstra, T. D., Panisko, E. A., Conrads, T. P., Angeli, N. H., Smith, R. D. (2002) *Anal. Chem.* **74,** 607.

Guilhous, M., Selby, D., Mlynski, V. (2000) *Mass Spectrom. Rev.* **19,** 65.

Hager, J. W. (2002) *Rapid Commun. Mass Spectrom.* **16,** 512.

Hager, J. W., Yves Le Blanc, J. C., (2003) *Rapid Commun. Mass Spectrom.* **17,** 1056.

Hamdan, M., Brenton, A. G. (1989) *Int. J. Mass Spectrom. Ion Process.* **87,** 343.

Hanash, S. (2003) *Nature* **422,** 226.

Ho, Y., Gruhler, A., Heilbut, A., Bader, G. D., Moore, L., Adams, S-L., Miller, A., Taylor, P., et al., (2002) *Nature* **415,** 180.

Hoaglund, C. S., Valentine, S. J., Sporleder, C. R., Reilly, J. P., Clemmer, D. E. (1998) *Anal. Chem.* **70,** 2236.

Honig, R. E. (1958) *J. Appl. Phys.* **29,** 549.

Huisman, T. H. J. (1989) *Hemoglobin* **13,** 223.

Ingram, V. M. (1959) *Biochem. Biophys. Acta* **36,** 402.

Ito, T., Chiba, T., Ozawa, R., Yoshida, M., Hattori, M., Sakaki, Y. (2001) *Proc. Natl. Acad. Sci. USA* **98,** 4569.

Ito, Y., Takeuchi, T., Ishi, D., Goto, M. (1985) *J. Chromatogr.* **346,** 161.

Jennings, K. R., Mason, R. S. (1983) in McLafferty, F. W. (Ed.), *Tandem Mass Spectrometry,* Wiley-Interscience, New York, p. 197.

Jensen, O. N., Podtelejnikov, A. V., Mann, M. (1997) *Anal. Chem.* **69,** 4741.

Juhasz, P., Vestal, M. L., Martin, S. A. (1997) *J. Am. Soc. Mass Spectrom.* **8,** 209.

Julian, Jr., R. K., Cooks, R. G. (1993) *Anal. Chem.* **65,** 1827.

Kaiser, Jr., R. E., Louris, J. N., Amy, J. W., Cooks, R. G. (1989) *Rapid Commun. Mass Spectrom.* **3,** 225.

Kaneko, Y., Megill, L. R., Hasted, J. B. (1966) *J. Chem. Phys.* **45,** 3741.

Karas, M., Bahr, U., Strupat, K., Hellenkamp, F., Tsarbopoulos, A., Pramanik, B. N. (1995) A *nal. Chem.* **67,** 675.

Karas, M., Hillenkamp, F. (1988) *Anal. Chem.* **60,** 2299.

Kaufmann, R., Chaurand, P., Kirsch, D., Spengler, B. (1996) *Rapid Commun. Mass Spectrom.* **10,** 1199.

Kelleher, N. L., Senko, M. W., Siegel, M. M., McLafferty, F. W. (1997) *J. Am. Chem. Soc. Mass Spectrom.* **8,** 380.

Kindy, J. M., Taraszka, J. A., Regnier, F. E., Clemmer, D. E. (2002) *Anal. Chem.* **74,** 950.

Kraus, H. (2002) *Int. J. Mass Spectrom.* **215,** 45.

Klauser, K. R., Hall, S. C., Smith, D. M., Webb, J. W., Andrews, L. E., Tran, H. M., Epstein, L. B., Burlingame, A. L. (1995) *Proc. Natl. Acad. Sci. USA* **92,** 5072.

Kruse, R. A., Li, X., Bohn, P. W., Sweedler, J. V. (2001) *Anal. Chem.* **73,** 3639.

Laiko, V. V., Baldwin, M. A., Burlingame, A. L. (2000) *Anal. Chem.* **72,** 652.

Laskin, J., Beck, K. M., Hache, J. J., Futrell, J. H. (2004) *Anal. Chem.* **76,** 351.

Lawrence, E. O., Livingston, M. S. (1931) *Phys. Rev.* **37,** 1707.

Lawson, G., Todd, J. F. J. (1971) *Cem. Brit.* **8,** 373.

Lewis, W. G., Shen, Z., Finn, M. G., Siuzdak, G. (2003) *Int. J. Mass Spectrom.* **226,** 107.

Li, X., Bohn, P. W. (2000) *Appl. Phys. Lett.* **77,** 2572.

Li, L., Wang, A. P. L., Coulson, L. D. (1993) *Anal. Chem.* **65,** 493.

Liao, P-C., Allison, J. (1995) *J. Mass Spectrom.* **30,** 408.

Liotta, L. A., Kohn, E. C. (2001) *Nature* **411,** 375.

Louris, J. N., Brodbelt, J. S., Cooks, R. G., Glish, G. L., van Berkel, G. J. (1990) *Int. J. Mass Spectrom. Ion Process.* **96,** 117.

Louris, J. N., Cooks, R. G., Syka, J. E. P., Kelley, P. E., Stafford, Jr., G. C., Todd, J. F. J. (1987) *Anal. Chem.* **59,** 1677.

Macfarlane, R. D. (1983) *Anal. Chem.* **55,** 1247A

Macfarlane, R. D., Bunk, D., Mudgett, P., Wolf, B. (1991), in Desiderio, D. M. (Ed.), *Mass Spectrometry of Peptides*, CRC Press, New York, pp. 3–63.

Macfarlane, R. D., Torgerson, D. F. (1976a) *Int. J. Mass Spectrom. Ion Phys.* **21,** 81.

Macfarlane, R. D., Torgerson, D. F. (1976b) *Science* **191,** 920.

Mann, M. (1996) *Trends Biochem. Sci.* **21,** 494.

Mann, M., Hendrickson, R. C., Pandey, A. (2001) *Annu. Rev. Biochem.* **70,** 437.

Mann, M., Wilm, M. S. (1994) *Anal. Chem.* **66,** 4390.

Marshall, A. G., Hendrickson, C. L., Jackson, G. S. (1998) *Mass Spectrom. Rev.* **17,** 1.

Mather, R. E., Waldren, R. M., Todd, J. F. J. (1978) *Dyn. Mass Spectrom.* **5,** 71.

Matsuo, T., Matsuda, H., Aimota, S., Shimonishi, Y., Higuchi, T., Mruyama, Y. (1983) *Int. J. Mass Spectrom. Ion Process.* **46,** 423.

McIver, Jr., R. T. (1970) *Rev. Sci. Instrum.* **41,** 555.

McIver, Jr., R. T. (1971) Ph. D. Thesis, Stanford University, Stanford, CA.

McLuckey, S. A., Glish, G. L., Van Berkel, G. J. (1991) *Int. J. Mass Spectrom. Ion Process.* **106,** 213.

Medzihradszky, K. F., Campbell, J. M., Baldwin, M. A., Falick, A. M., Juhasz, P., Vestal, M. L., Burlingame, A. L. (2000) *Anal. Chem.* **72,** 552.

Meng, C. K., Mann, M., Fenn, J. B. (1988) *Z. Phys. D* **10,** 361.

Milner, P. F., Clegg, J. B., Weatherall, D. J. (1971) *Lancet* **10,** 729.

Minard, R. D., Chin-Fatt, P. D., Jr., Curry, A. G. Ewing, (1988) in *Proceedings of the Thiry-Sixth ASMS Conference on Mass Spectrometry and Allied Topics*, San Francisco, June 5–10.

Morris, H. R., et al. (1978) *J. Biol. Chem.* **253,** 5155.

Munson, M. S. B., Field, F. H. (1966) *J. Am. Chem. Soc.* **88,** 2621.

Neubauer, G., King, A., Rappsilber, J., Calvio, C., Watson, M., et al. (1998) *Nat. Genet.* **20,** 46.

Oda, Y., Nagasu, T., Chait, B. T. (2001) *Nat. Biotechnol.* **19,** 379.

Ogata, Y. H., Kato, F., Tsuboi, T. Sakka, T. J. (1998) *Electrochem. Soc.* **145,** 2439.

O'Lear, J. R., Wright, L. G., Louris, J. N., Cooks, R. G. (1987) *Org. Mass Spectrom.* **22,** 348.

Ostrander, C. M., Arkin, C. R., Laude, D. (2000) *J. Am. Soc Mass Spectrom.* **11,** 592.

Pandy, A., Fernandez, M. M., Steen, H., Blagoev, B., Nielson M. M., et al. (2000a) *J. Biol. Chem.* **275,** 38633.

Pandy, A., Podtelejnikov, A. V., Blagoev, B., Bustelo, X. R., Mann, M., Lodish, H. F. (2000b) *Proc. Natl. Acad. Sci. USA* **97,** 197.

Paweletz, C. P., Charboneau, L., Bichsel, V. E., Simone, N. L., et al. (2001) *Oncogene* **20,** 1981.

Perkins, D. N., Pappin, D. J., Creasy, D. M., Cotrell, J. S. (1999) *Electrophoresis* **20,** 3551.

Phizicky, E., Bastiaens, P. I. H., Zhu, H., Snyder, M., Fields, S. (2003) *Nature* **422,** 208.

Prenni, J. E., Shen, Z., Trauger, S., Chen, W., Siuzdak, G. (2003) *Spectroscopy* **17,** 693.

Prome, D., Prome, J. C., Blouquit, Y., Lacombe, C., Rosa, J., Robinson, J. D. (1987) *Spectrosc. Int. J.* **5,** 157.

Quist, A. P., Huth-Fehre, T., Sundquist, B. U. R. (1994) *Rapid Commun. Mass Spectrom.* **8,** 149.

Rain, J. C., et al. (2001) *Nature* **409,** 211.

Reinhoud, N. J., Niessen, W. M. A., Tjaden, U. R., Gramberg, L. G., Verheji, E. R., van der Greef, J. (1989) *Rapid Commun. Mass Spectrom.* **3,** 348.

Rigaut, G., Shevchenko, A., Rutz, B., Wilm. H., Mann, M., Séraphin, B. (1999) *Nat. Biotechnol.* **17,** 1030.

Roepstorff, P., Fohlman, J. (1984) *Biomed. Mass Spectrom.* **11,** 601.

Sandra, K., Devreese, B., Van Beeumen, J. (2004) *Am. Soc. Mass Spectrom.* **15,** 413.

Schey, K. L., Schwartz, J. C., Cooks, R. G. (1989) *Rapid Commun. Mass Spectrom.* **3,** 305.

Schwartz, J. C., Wade, A. P., Enke, C. G., Cooks, R. G. (1990) *Anal. Chem.* **62,** 1809.

Schwikowski, B., Uetz, P., Fields, S. (2000) *Nat. Biotechnol.* **18,** 1257.

Seki, S., Kambara, H., Naoki, H. (1985) *Org. Mass Spectrom.* 20, 18.

Senko, M. W., Hendrickson, C. L., Emmett, M. R., Shi, S. D-H., Marshall, A. G. (1997) *Am. Soc. Mass Spectrom.* **8,** 970.

Sheil, M., Derrick P. J. (1986), in Gaskell, S. J. (Ed.), *Mass Spectrometry in Biomedical Research,* Wiley, Chichester, pp. 251–264.

Spengler, B., Bökelmann, V. (1993) *Nucl. Instr. Meth. Phys. Res.* **B.82,** 379.

Spengler, B., Kirsch, D., Kaufmann, R. (1992) *J. Phys. Chem.* **96,** 9678.

Srebalus, C. A., Li, J., Marshall, W. S., Clemmer, D. E. (1999) *Anal. Chem.* **71**, 3918.

Srebalus Barnes, C. A., Hilderbrand, A. E., Valentine, S. J., Clemmer, D. E. (2002) *Anal. Chem.* **74,** 26.

Srebalus, C. A., Hilderbrand, A. E., Valentine, S. J., Clemmer, D. E. (2002) *Anal. Chem.* **74,** 26.

Stephens, W. E. (1946) *Phys. Rev.* **69,** 691.

Stahl-Zeng, J., Hellenkamp, F., Karas, M. (1996) *Eur. J. Mass Spectrom.* **2,** 23.

Stewart, M. P., Buriak, J. M. (1998) *AngewChem. Int. Ed.* **37,** 3257.

Suckau, D., Resemann, A., Schuerenberg, M., Hufnagel, P., Franzen, J., Holle, A. (2003) *Anal. Bioanal. Chem.* **376,** 952.

Sunner, J. (1993) *Org. Mass Spectrom.* **28,** 805.

Taflin, D. C., Ward, T. L., Davis, E. J. (1989) *Langmuir* **5,** 376.

Takagi, S. T., Iwai, Y., Kaneko, M., Kimura, N., Kobayashi, A., Matsumoto, S., Ohtani, K., et al. (1983) *Nucl. Inst. Methods* **215,** 207.

Tanaka, K., Ido, Y., Akita, S., Yoshida, T. (1987), paper presented at the Second Japan-China Symposium on Mass Spectrom., Osaka, Japan.

Tanaka, K., Ido, Y., Akita, S., Yoshida, Y., Yoshida, T. (1988) *Rapid Commun. Mass Spectrom.* **2,** 151.

Thomson, B. A., Iribarne, J. V. (1979) *J. Chem. Phys.* **71,** 4451.

Tomer, K. B., Guenat, C. R., Deterding, L. J. (1988) *Anal. Chem.* **60,** 2232.

Tong, A. H. Y., Drees, B., Nardelli, G., Bader, G. D., et al. (2002) *Science* **295,** 321.

Twerenbold, D. (1996) *Nucl. Instrum. Methods* **A 370**, 253.

Twerenbold, D., Gerber, D., Gritti, D., Gonin, Y., Netuschill, A., Rossel, F., Schenker, D., Vuilleumier, J-L. (2001) *Proteomics* **1,** 66.

Tuloup, M., Hoogland, C., Binz, P-A., Appel, R. D. (2002) Abstracts, Swiss Proteomics Society, Lausanne.

Tyers, M., Mann, M. (2003) *Nature* **422,** 193.

Uetz, P., Giot, L., Mansfield, T., Judson, R. S., Knight, J. R., Lockshon, D., et al. (2000) *Nature* **403,** 623.

Uetz, P., Pankratz, M.J. (2004) *Nat Biotechnol.* **22,** 43.

Valaskovic, G. A., Kelleher, N. K., McLafferty, F. W. (1996) *Sciencs* **273,** 1199.

Wall, D. B., Finch, J. W., Cohen, S. A. (2002) *Genet. Eng. News.* **22,** 30.

Wang, S. L., Ladingham, K. W. D., Singhal, R. P. (1996) *Appl. Surface Sci.* **93,** 205.

Uetz, P., Pankratz, M. J. (2004) *Nat Biotechnol.* **22,** 43.

Wei, J., Buriak, J. M., Siuzdak, G. (1999) *Nature* **399,** 243.

Welling, M., Schuessler, H. A., Thompson, R. I., Walther, H. (1998) *Int. J. Mass Spectrom. Ion Process.* **172,** 95.

Wiley, W. C., McLaren, I. H. (1955) *Rev. Sci. Instrum.* **26,** 1150.

Williams, D. H., Bradely, C., Bojesen, G., Santikaran, S., Taylor, L.C. E. (1981) *J. Am. Chem. Soc.* **103,** 5700.

Wilm, M., Mann, M. (1996) *Anal. Chem.* **68,** 1.

Wilm, M. S., Mann, M. (1994) *Int. J. Mass Spectrom. Ion Process.* **136,** 167.

Wiza, J. L. (1979) *Nucl. Instrum. Methods* **162,** 587.

Yamashita, M., Fenn, J. B. (1984) *J. Phys. Chem.* 88, 4451.

Yefchak, G. E., Schultz, G. A., Allison, J., Enke, C. G., Holland, J. F. (1990) *J. Am. Soc. Mass Spectrom.* **1,** 440.

Yost, R. A., Enke, C. G. (1978) *J. Am. Chem. Soc.* **100,** 2274.

Zenobi, R. (1997) *Chimia* **51,** 801.

Zenobi, R., Knochenmuss, R. (1998) *Mass Spectrom. Rev.* **17,** 337.

Zhan, Q., Wright, S. J., Zenobi, R. (1997) *J. Am. Soc. Mass Spectrom.* **8,** 525.

Zhang, W., Chait, B. T. (2000) *Anal. Chem.* **72,** 2482.

Zhou, H., Watts, J. D., Aebersold, R. (2001) *Nat. Biotechnol.* **19,** 375.

Zhu, L., Parr, G. R., Fitzgerald, M. C., Nelson, C. M., Smith, L. M. (1995) *J. Am. Chem. Soc.* **117,** 6048.

2

PROTEOMICS IN CANCER RESEARCH

2.1. INTRODUCTION

The sequencing of the human genome and that of numerous pathogens has provided the field of proteomics with a sequence-based framework for mining various proteomes. This scientific leap has opened the door for intense activities in applying proteomics to gain a better understanding of disease, discover new biomarkers for early diagnosis, and eventually find more efficient and safer therapeutics for a wide range of devastating pathologies. Both proteomic and genomic efforts throughout the 1990s have underlined two aspects strictly related to current aspirations to achieve such objectives. First, new technologies amenable to automation and having higher resolution and higher sensitivity are needed. Second, to gain a more comprehensive understanding of the molecular bases which govern the mechanism(s) of various diseases, complementary knowledge generated by genomic and proteomic approaches should be sought. The second aspect has been consolidated particularly in cancer research by the realization that genetic defects may ultimately lead to tumor cell survival by altering the functional proteins that confer survival advantages on the tumor cells (Hanahan and Weinberg, 2000; Hunter, 1998). Furthermore, cancer growth, invasion, and metastasis might be the result of domination or perturbation in the normal network of cell-signaling proteins (Liotta and Kohn, 2001). Therefore, we can envisage that studies based on genomic and DNA microarrays can highlight potential genetic defects that may cause disruptions in cell-signaling pathways. Protein microarrays and other proteomic approaches, on the other hand, are expected to furnish crucial information about the functional state of these disrupted pathways. Such integrated

Proteomics Today. By Mahmoud Hamdan and Pier Giorgio Righetti
ISBN 0-471-64817-5

efforts are likely to result in the development of the cellular proteomic network for an individual patient's tumor and will hopefully reveal potential drug targets and diagnostic molecules for disease prognosis or treatment (Blume-Jensen and Hunter, 2001; Hunter, 2000; Celis and Gromov, 2003). Furthermore, such molecular integration may provide a comprehensive view of the disrupted cellular machinery governing diseases.

In the proteomic field the search for high-throughput and sensitive technologies resulted in some sort of realignment of concepts, and working principles matured in genetic and DNA studies to generate protein microarrays. Today protein microarray assays allow the identification and quantification of a large number of target proteins from a minute amount of a sample within a single experiment. Some posttranslational modifications of protein networks can also be profiled by employing protein microarrays by comparing the proportion of total and activated (e.g., phosphorylated) protein. This type of data in general reflects the state of information flow through a protein network. In other words, monitoring the total and phosphorylated proteins over time, before and after treatment, or between control and pathological states may allow us to infer the activity levels of the proteins in a particular pathway in real time (Liotta et al., 2001; Petricoin et al., 2002c).

The high price in terms of deaths and quality of life which various forms of cancer are exacting from our society has attracted immense research activities to identify biomarkers for early diagnosis of these devastating diseases and to facilitate the identification of more efficient therapeutic targets. Currently, there are two main approaches to tackle such a task: the first is based on large-scale analysis of gene expression at the RNA level, and the second approach is based on the use of various proteomic methods to assess protein expression and when possible to assign function(s) to such proteins. Currently, 2D polyacrylamide gel electrophoresis (PAGE) (O'Farrel, 1975), often referred to as gel-based proteomics, 2D liquid chromatography, and protein biochips in particular protein microarrays and SELDI in combination with mass spectrometry and artificial neural network algorithms are among the main proteomic tools used to search for protein biomarkers and drug target discovery, particularly in the field of cancer research. Although most of the technologies cited in this chapter have been described elsewhere in this book, some general comments and brief technical description of these methods relevant to the material of this chapter are in order.

2.1.1. Two-Dimensional Gel Electrophoresis

The capability of 2DE to separate thousands of proteins in a single analysis made it a major player in the profiling of protein expression in various diseases. Over the last 20 years it has emerged that the investigation of cancerous tissues is a highly suitable area for the application of such an approach. This is because in many cases it is possible to make a direct comparison of proteins expressed in normal and diseased tissue from the same patient. The comparison of 2DE separation of such tissues can highlight proteins that are present in greater or lesser quantities and new proteins which may only be expressed in cancerous cells.

Existing literature cites various forms of cancer in which the application of this approach gave specific protein differences in normal versus diseased tissues. For example, in many countries, lung cancer is the number one oncological disease. Often its diagnosis is so late that no curative measure exists. In such a dramatic scenario, any progress in the diagnostic or prognostic evaluation and in the modes of therapy would be highly beneficial. The potential of 2DE in this kind of situation has been demonstrated over 10 years ago by Okuzawa et al. (1994), who examined 14 lung tumors of various histopathological types and found a variation in the expression of several proteins which correlated with different histological types. They also found a protein which was significantly overexpressed in primary lung adenocarcinomas compared to small-cell lung carcinomas, squamous cell lung carcinomas, metastatic lung adenocarcinomas from colon and rectum, and normal tissue. The authors concluded that their approach could provide reliable information on tumor-specific protein markers. A year later Hirano et al. (1995) investigated the relationship between the histopathological findings in primary lung malignancies and expression of a number of unidentified polypeptides. Samples from patients with primary lung cancer were examined by 2DE, which revealed that at least 16 polypeptides were associated with histopathological features. The authors also suggested that the classification of lung carcinomas based on this approach was likely to be more precise than that based on morphology alone. Other forms of cancer which have been investigated during the 1980s and 1990s include brain (Hanash et al., 1985), thyroid (Lin et al., 1995), breast (Wirth et al., 1987; Giometti et al.,1997; Rasmussen et al., 1997), colon (Ward et al., 1990), kidney (Sarto et al., 1997), and bladder (Celis et al., 1996).

Although 2DE is commonly described as labor intensive, low throughput, and a technique which requires a relatively high amount of sample, the most recent literature shows that such a technique is still preferred for large-scale separation of complex protein mixtures. In fact, this technique currently offers the best possible resolution because it separates proteins according to two independent physicochemical parameters: isoelectric point and size. Furthermore, the gel acts as an efficient fraction collector that stores the protein molecules until picked. In the area of cancer research, 2DE is currently applied in two forms. The first and more conventional approach compares protein expression in two different samples (e.g., one normal and the other pathological) by generating a minimum number (three to five) of 2D map replicates for each sample. Following staining and optical density scanning, the remaining work is handled by one of the commercially available statistical packages, such as Melanie, PDQuest, Z3, Z4000, Phoretix, and Progenesis. These analyses generate two master 2D maps, one for the normal and another for the pathological samples. Subsequent analysis by the same software would compare these two maps to identify proteins which have experienced measurable up or down regulation in their intensity. A schematic representation of the main steps in such a protocol is given in Figure 2.1.

The second form of 2DE, termed differential in-gel electrophoresis (DIGE), was introduced by Ünlü et al. (1997). This version promises to significantly improve the speed and reproducibility of conventional 2DE. Basically, two pools of proteins derived from two different states are labeled with fluorescent dyes, one with Cy3 and the other with Cy5. These two dyes are mass and charge matched and possess distinct

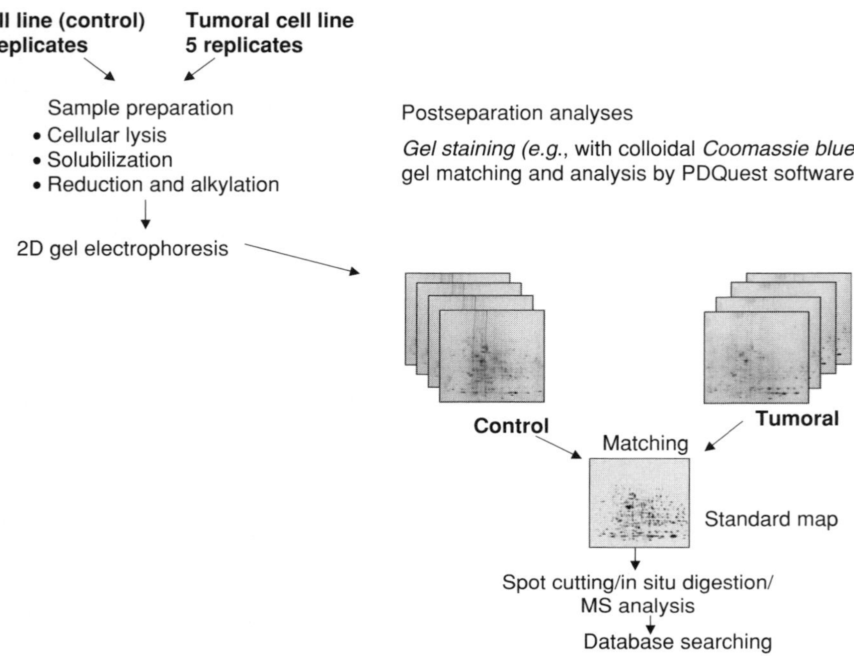

FIGURE 2.1. Main steps in analysis of two samples, control (sample 1) and tumoral (sample 2), using sequence sample preparation, 2DE separation, comparison, matching, and generation of standard maps for two samples using PDquest software and MS characterizations of spots, which experienced up or down regulation.

excitation and emission spectra. Equal concentrations of these labeled pools are mixed and coseparated on the same gel. The 2DE gel pattern is then visualized by imaging the gel with a fluorescence scanner by sequential excitation of the fluorescence dyes.

Despite impressive improvements over the last 15 years, 2DE is still considered a low-throughput technique which requires a relatively large amount of sample. Such shortcomings have encouraged the search for other approaches which may address the speed and the dynamic range of this approach.

2.1.2. Surface-Enhanced Laser Desorption Ionization

Surface-enhanced laser desorption ionization (Hutchens and Yip, 1993) utilizes solid supports or chips (1–2 mm in diameter) made of aluminum or stainless steel and coated with specific matrices ("bait"). The baits are comprised of standard chromatographic supports which can be hydrophobic, cationic, anionic, or normal phase. Biochemical bait molecules may include purified proteins, ligands, receptors, antibodies, enzymes, or DNA oligonucleotides (see Fig. 2.2). Each surface is designed to retain proteins according to some of their general or specific physiochemical properties. Generally, chemically active surfaces can retain whole classes of proteins, whereas surfaces to

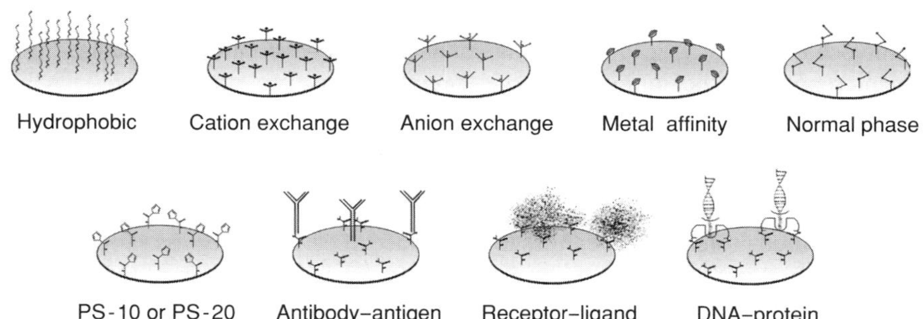

FIGURE 2.2. Commonly used bait surfaces for SELDI chips: chemically modified surfaces (upper raw) and biochemichally modified surfaces (bottom raw). (Adapted from Issaq et al., *Anal. Chem.*, 2003, with permission.)

which a biochemical agent, such as an antibody or other type of affinity reagent, is coupled are designed to interact specifically with a single, target protein. The advantage of biochemically active surfaces is that they can be used to investigate specific molecular recognition mechanisms such as antibody–antigen, enzyme–substrate, receptor–ligand, and protein–DNA. In the SELDI approach, samples derived from serum, plasma, intestinal fluid, urine, cell lysate, and cellular secretion products are directly applied to the bait surface (~0.5 μL). After washing to remove unbound proteins, the proteins that have interacted with the bait surface are then analyzed by MALDI–TOF–MS. Recently there have been initial attempts to integrate SELDI MS with artificial neural network algorithms to validate biomarker correlation with the progression of various forms of cancer (Ball et al., 2002; Milan et al., 2003). Artificial neural networks (ANNs) are powerful statistical tools that have been utilized for the prediction of clinical outcomes in prostate (Babaian and Zhang, 2001) and ovarian cancer (Snow et al., 2001). They are analogous to human neurons in which they learn patterns associated with particular phenotypes through the process of iterative learning, that is, trial and error. Other bioinformatic approaches that have been adopted in order to analyze proteomic data include "genetic algorithms" with self-organizing cluster analysis (Petricoin et al., 2002a,c) and multivariate logistic regression analysis (Li et al., 2002). It is worth noting that the application of pattern recognition software could be useful not only to identify biomarkers associated with a particular disease states and thus enable attention to be focused upon a relatively selected group of molecules with the highest potential of therapeutic value but also to enable the development of novel assays capable of identifying expression patterns from unknown samples to predict the likelihood of response to chemotherapeutic agents.

2.1.3. Protein Microarrays

Unlike DNA microarrays, which provide a measure of gene expression (namely RNA levels), protein microarrays are expected to address various characteristics of proteins that can be altered by disease. These include, on the one hand, determination of their levels and, on the other hand, determination of their selective interactions with other biomolecules (Hanash, 2003). Profiling studies of disease tissue that have used protein microarrays are beginning to emerge. A reverse-phase protein array approach that immobilizes almost the whole repertoire of a tissue's proteins has been described by Paweletz et al. (2001). A potential clinical application of protein microarrays would be the identification of proteins that induce an antibody response in autoimmune disorders (Robinson et al., 2002). This type of array was demonstrated by attaching several hundred proteins and peptides to the surface of derivatized glass slides. Arrays were incubated with patient's serum, and fluorescent labels were used to detect autoantibody binding to specific proteins in autoimmune diseases, including systematic lupus erythematosus and rheumatoid arthritis.

It is becoming more evident that one of the main challenges in making biochips for the global analysis of protein expression is the current lack of comprehensive sets of genome-scale capture agents such as antibodies. One obstacle to be overcome in this particular area is the capability to capture numerous posttranslational modifications that may be crucial to their functions, yet these modifications are generally not captured by using either recombinant proteins or antibodies that do not distinctly recognize specific forms of a protein. One approach for comprehensive analysis of proteins in their modified forms is to array proteins isolated directly from cells and tissues following protein fractionation schemes (Madoz-Gurpide et al., 2001). It is worth noting that fractions which react with specific probes can be easily handled by existing chromatographic and gel-based separation techniques for resolving their individual protein constituents, which can then be analyzed by various MS techniques.

Currently, antibody microarrays are attracting considerable attention by both the public and private sectors. Several aspects of microarray technology make it well suited to cancer research because of the low-volume requirements and a multiplexed detection capability of microarrays that make optimal use of precious clinical samples. These assays are rapid and highly amenable to automation, which makes them ideal for biomarker studies. An important choice in the use of antibody microarrays is the detection format. Figure 2.3 gives three detection formats described by Haab (2003).

2.1.4. Getting More Than Just Simple Change in Protein Expression

Although data obtained by various expression proteomics strategies have functional relevance by providing altered levels or posttranslational modifications of proteins in disease, additional technologies are needed for more direct functional analysis. For example, we need to be able to analyze protein complexes and their disruption in disease, to assay in a high-throughput fashion the activity of various classes of proteins, and to gain reliable information on the levels and activities of individual proteins in a cellular context so as to determine their role in different biological processes and disease states.

Various strategies are currently in use to study protein complexes and protein–protein interactions; however, so far, such strategies have been applied to disease investigations on a fairly limited scale. It is becoming clear that a comprehensive understanding of these interactions is required if we have to unravel the mechanisms

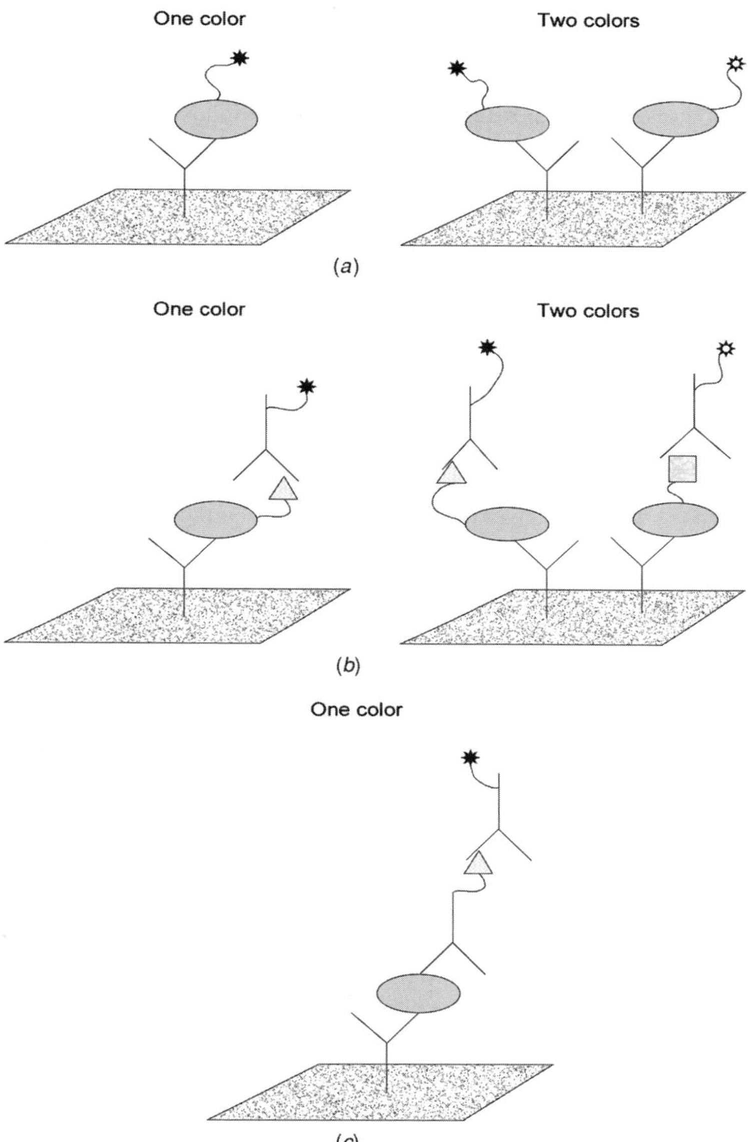

FIGURE 2.3. Schematic representation of commonly used antibody microarray formates. (*a*) Direct labeling with a fluorophore. Immobilized antibodies on a planar substrate capture proteins (represented by the oval) that are labeled with a fluorophore, such as Cy3 or Cy5 (represented by the star). Two different protein pools can be labeled with different fluorophores and co-incubated on the arrays (shown in right portion of (a)). (*b*) Direct labeling with a hapten tag. Proteins are labeled with a tag such as biotin or digoxigenin (represented by the triangle and square), and captured proteins are detected by incubation of labeled antibodies targeting the tag. Multiple samples can be co-incubated on the arrays by labeling with different tags and detecting with multiple labeled antibodies. (*c*) Paired-antibody sandwich assays. Unlabeled antigens are captured by an immobilized antibody and detected by a labeled secondary antibody. Usually the secondary antibodies are biotinylated and detected by a labeled antibiotin antibody. (Adapted from Haab, *Proteomics*, 2003, with permission.)

which govern individual cellular pathways and their interlinks. Recently, Gavin et al. (2002) and Ho et al. (2002) have given a representative example on how such a difficult task can be handled. Both groups have used a similar approach in which individual proteins are tagged and subsequently used to pull down associated proteins which are then characterized by various MS analyses. Each group was able to characterize a substantial number of multiprotein complexes in the budding yeast *Saccharomyces cerevisiae*. In two earlier studies, Uetz et al. (2000) and Ito et al. (2001) have applied the yeast "two-hybrid" assay in a high-throughput fashion for mapping pairwise protein interactions in *S. cerevisiae*. In another approach, Zhu et al. (2001) have developed a microarray method in which purified, active proteins from almost the entire yeast proteome were printed onto a microscope slide at high density such that thousands of protein interactions could be assayed simultaneously.

It is commonly acknowledged that large-scale efforts to characterize protein complexes are generally rate limited by the need for almost pure preparation for each complex. In the works of Gavin et al. (2002) and Ho et al. (2002) protein complexes were purified by a strategy depicted in Figure 2.4. In such a strategy, the first step is to attach tags to hundreds of different proteins, creating what is known as the bait proteins. Second, DNA encoding these bait proteins is introduced into the yeast cells, allowing the tagged proteins to be expressed in cells and to form complexes with other proteins. Using the tag, each bait protein and associated complex were fished out and subjected to MS analysis. In their study, Gavin et al. (2002) have identified over 1400 proteins within 32 complexes; almost 91% of these complexes contained at least one protein of previously unknown function. Furthermore, most of these complexes were found to contain a component in common with at least one other multiprotein assembly, an observation which may suggest coordinating cellular functions into a higher order network of interacting protein complexes.

Using a similar approach, Ho et al. (2002) constructed an initial set of 725 yeast bait proteins from which they identified over 3617 interactions involving over 1578 different proteins. The authors described interaction networks assembled around the protein kinase Kss1, a known component of pathways involved in mating and filamentous growth. Figure 2.5 depicts two kinase-based signalling networks in *S. cerevisiae*.

2.1.5. Laser Capture Microdissection

Our ability to localize specific alterations in tumor DNA, RNA, and proteins can be hampered by our inability to adequately isolate specific cell types from pathological specimens. Over the years a host of techniques have been employed for such a scope, including cell scraping and affinity column purification (Franzen et al., 1995) and manual microdissection of tissues (Berger, 1980; Radford, 1983). While these techniques could provide pure cell lines for evaluation of intracellular processes, the same techniques can result in the loss of relevant information regarding elements, which may confer a unique phenotype upon an individual tumor. For example, the tumor microenvironment of a carcinoma consists not only of the malignant epithelial component but also of the surrounding stroma and normal tissue. These distinct microcompartments use receptors, cell junctions, and inter- and intracellular signaling

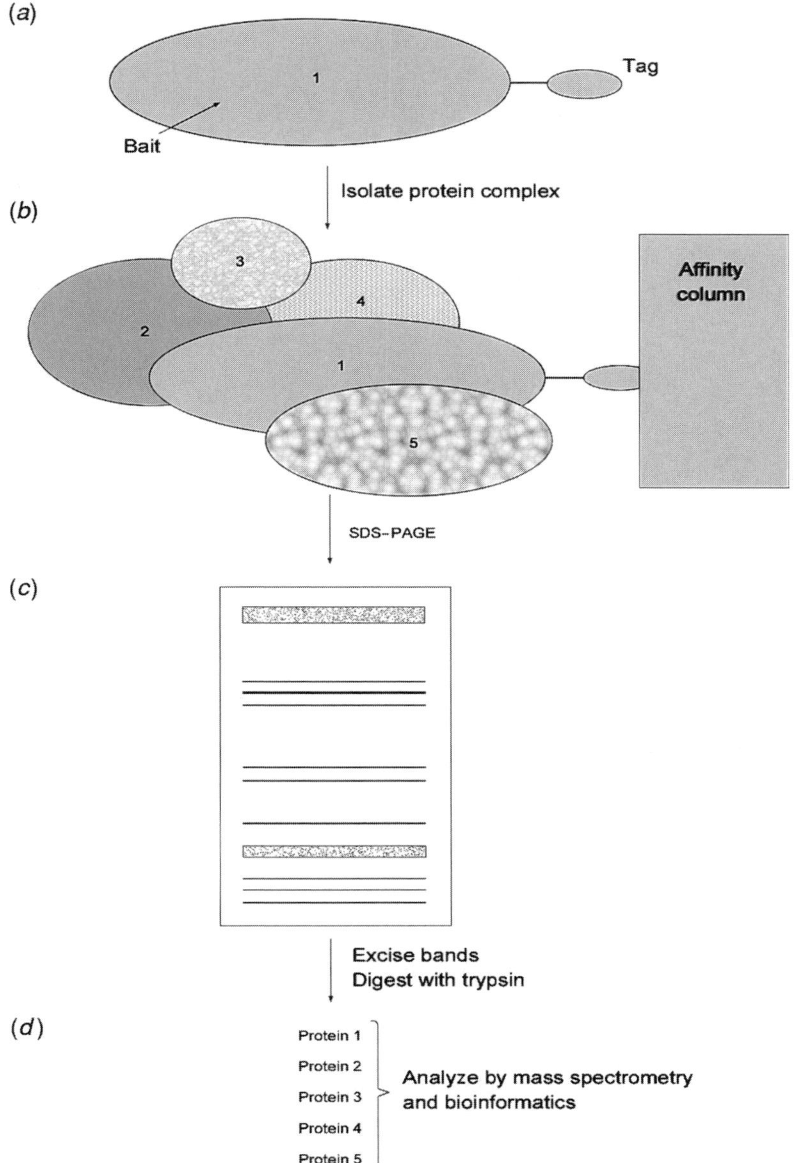

FIGURE 2.4. Coprecipitation/MS strategy for large-scale analysis of protein interactions used by Gavin et al. (2002) and Ho et al. (2002). (*a*) An "affinity tag" is first expressed, attached to a target protein. (*b*) Bait proteins are systematically precipitated, along with any associated proteins, on an "affinity column." (*c*) Purified protein complexes are resolved by one-dimensional SDS-PAGE. (*d*) Proteins are excised from the gel, digested with the enzyme trypsin, and analyzed by mass spectrometry. (Adapted from Kumar and Snyder, *Nature*, 2002, with permission.)

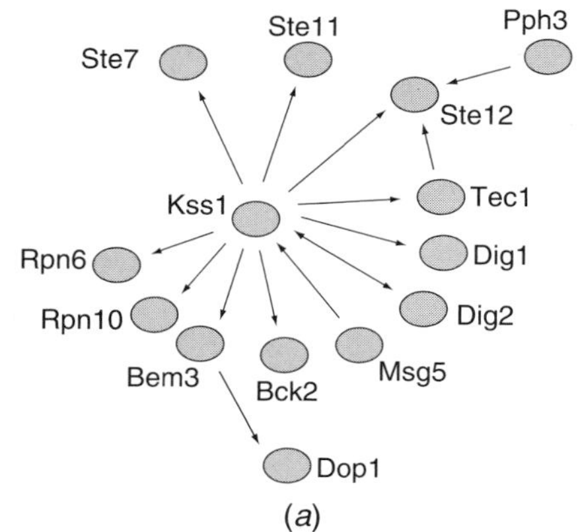

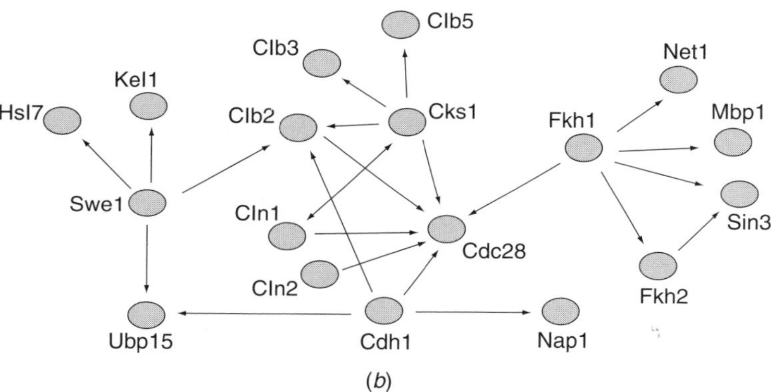

FIGURE 2.5. Two kinase-based signaling networks: (*a*) interaction diagrams for Kss1 complexes and (*b*) interaction diagrams for Cdc28 and Fkh1/2 complexes. (Adapted from Ho et al., *Nature*, 2002, with permission.)

molecules to allow tumor cells to communicate with their surroundings and play an active role in their own control or progression (Liotta et al., 2001). Removing a subpopulation of these cells for growth in an in vitro system interrupts potentially important cell–cell and cell–matrix interactions that may affect tumor behavior, thus giving the researcher a misleading impression of the actual in vivo tumor composition and physiology.

The advent of laser capture microdissection (LCM), first introduced by Emmert-Buck et al. (1996), has enhanced our ability to remove specific subpopulations of cells from frozen or ethanol-fixed tissues under direct microscopic visualization. These tissues can be microdissected either stained or unstained. In fact, coupling rapid immunohistochemical staining techniques with LCM may allow for more accurate microdissection of cell subsets (Fend et al., 1999). Under direct microscopic visualization LCM permits rapid one-step procurement of selected human cell populations from a section of complex, heterogeneous tissue.

In this technique, a transparent thermoplastic film (ethylene vinyl acetate polymer) is applied to the surface of the tissue section on a standard glass histopathology slide; a CO_2 laser pulse then specifically activates the film above the cells of interest. Strong focal adhesion allows selective procurement of the targeted cells. Multiple examples of LCM transfer and tissue analysis, including PCR amplification of DNA and RNA, and enzyme recovery from transferred tissue were described by the authors who first developed this technique (Emmert-Buck et al.,1996). A schematic diagram summarizing the main steps in the LCM approach is given in Figure 2.6.

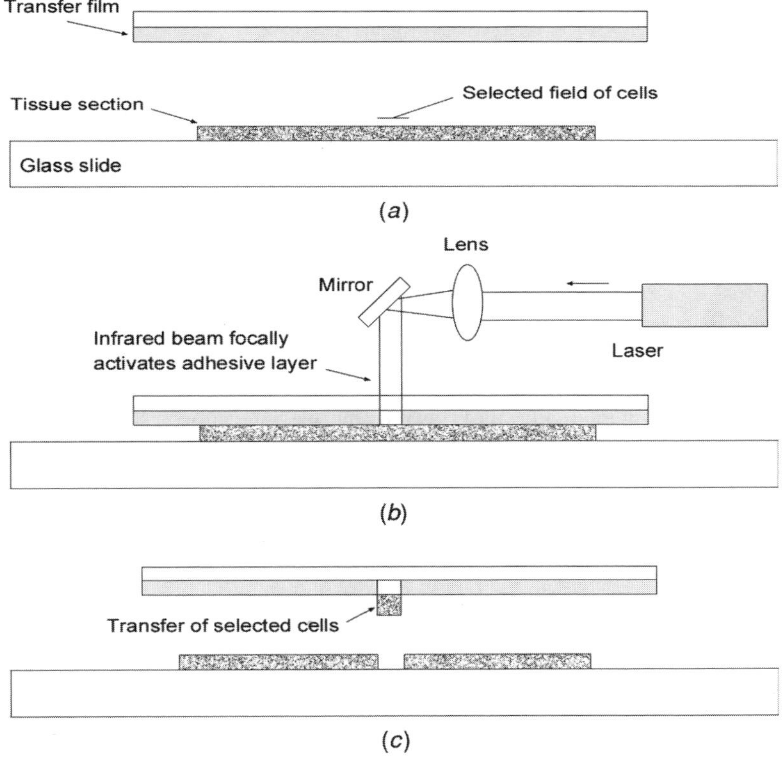

FIGURE 2.6. Schematic representation of main components of LCM arrangement. (*a*) Transparent EVA thermoplastic film is applied to the surface of a routine tissue section mounted on a glass slide. The tissue-EVA film sandwich is viewed under a microscope, and the cells of interest are positioned in the center of the field. (*b*) A focused laser beam coaxial with microscope optics is pulsed to activate the film causing it to become focally adhesive and fuse to the selected underlying cells in the tissue section. (*c*) When the EVA film is removed from the tissue section, the selected cells remain adherent to the film surface. The film is then placed directly into the DNA, RNA, or enzyme buffer. The cellular material detaches from the film and is ready for standard processing. (Adapted from Emmert-Buck et al., *Science*, 1996, with permission.)

Ornstein et al. (2000) sought to objectively evaluate the impact of a tumor microenvironment on its proteome. They used LCM to isolate pure prostate tumor epithelium and stroma from prostatectomy specimens. Subsequently, protein profiles obtained by 2DE of microdissected cell lysates, undissected whole cryostat section lysates, and immortalized cell lines from the same patients were compared. Additional comparison was also carried out on protein profiles from two commercially available prostate cancer cell lines. The findings of this study provided objective clarifications of several important issues. First, stromal and epithelial components of tumors may share less than half of the same proteins. Second, proteins differentially expressed in normal and malignant prostate tissue were solely associated with the epithelial compartment. Third, microdissection does not alter a protein's integrity. The same study provided evidence that protein profiles of in vivo and in vitro cells vary significantly with as little as 25% overlapping. One further point which should be underlined is that analysis of tissue biopsies is by far more complicated than the analysis of fluids due to the heterogeneous nature of the samples. In many of the current proteomic approaches the acquisition and preparation of sections of tissues associated with various malignancies can substantialy influence the outcome of various analysis. Two-dimensional electrophoresis and MS imaging are two approaches which can greatly benefit by applying LCM. In the field of oncology, the genetic and proteomic analysis of microscopic premalignant lesions has important clinical implications. These microscopic lesions, which represent an intermediate step in tumor progression from normal cells to manifest, can offer strong indications as to the nature of the fundamental alterations that underlie cancer development.

Published works dealing with the investigation of various forms of cancer are understandably generated by the use of more than one technique. To introduce a reasonable sequence in the material of this chapter, we have chosen some representative types of cancer in which various types of technology have been applied, including DNA microarrays.

2.2. PANCREATIC DUCTAL ADENOCARCINOMA

Pancreatic ductal adenocarcinoma (PDAC) is one of the leading causes of cancer deaths in the western world, accounting for 40,000 deaths per year in Europe and 28,000 deaths per year in the United States (Perkin et al., 2001; Greenlee et al., 2001). Pancreatic cancer occurs with a frequency of around nine patients per 100,000 individuals. Currently, the only curative treatment is surgery, but only 10% to 20% of patients are suitable for surgery at the time of presentation (Yeo et al., 1997), and of this group, only about 20% who undergo such surgery are alive after 5 years (Hawes et al., 2000). Such frightening statistics can be attributed to a number of causes. First, there are currently no effective biomarkers useful for early detection of pancreatic cancer. Second, this type of cancer is highly resistant to both chemotherapy and radiation therapy (Greenlee et al., 2001). Third, sufficient information on the molecular bases responsible for the main characteristics of this devastating disease is absent.

To gain a better understanding of the molecular basis and cellular mechanisms in pancreatic cancer, a number of approaches have been used, including large-scale serial analysis of gene expression (SAGE), complementary DNA (cDNA) microarrays, and proteomic profiling using 2DE in combination with MS or SELDI.

2.2.1. Analyses Based on Chip Technology

The quest for the identification of biomarkers for pancreatic cancer has followed two main approaches, gel based and chip based, the latter encompassing cDNA microarrays, large-scale serial analysis of gene expression, and SELDI MS.

A known characteristic of pancreatic ductal carcinoma is the paucity of neoplastic cells which are commonly embedded in a densely desmoplastic stroma. One way of bypassing such intrinsic characteristic is to use LCM to obtain fairly pure and homogenous neoplastic and normal ductal cells. Laser capture microdissection in combination with cDNA microarrays was used by Crnogorac-Jurcevic et al. (2002) to profile differentially expressed genes associated with pancreatic cancer. The authors identified a number of differentially expressed genes, comprising cell cycle and growth regulators, invasion regulators, signaling, and development molecules. As well as overexpressed genes already known to be associated with pancreatic cancer, the same study provided a new list of genes which could be implicated in the same disease. These include the overexpressed genes *ABL2*, *Notch4*, and *SOD1* and a down-regulated DNA repair gene *XRCC1*. Various functions of these genes are fairly recognized, yet their potential as possible biomarkers is yet to be demonstrated. Copper–zinc–superoxide dismutase (*SOD1*) is known to be involved in the elimination of certain radicals; it is also considered as a key enzyme for the protection of certain cells from damage induced by reactive oxygen species (Bankson et al., 1993). Although the activity of *SOD1* seems to be dependent on the type of tumor, increased levels of this enzyme were found in both human and mouse leukemia cells compared to mature normal blood cells. It has also been hypothesized that the reduced levels of this enzyme may result in the accumolation of O^{2-}, which could induce eventual mitochondrial membrane damage and increased release of cytochrome *c*, thus contributing to apoptosis. It is interesting to note that targeting of this molecule has been proposed as a potential therapeutic approach for the selective killing of malignant cells (Huang et al., 2000). The DNA repair gene *XRCC1* was found to be down regulated, this gene being known to be involved in sister chromatid exchange and repair of single-stranded DNA breaks. It recognizes DNA breaks, binds them, and recruits other components of the repair machinery, such as polynucleotide kinase and DNA ligase III (Whitehouse et al., 2001). It is worth noting that decrease in *XRCC1* transcript expression in pancreatic adecarcinoma tends to correlate with decreased expression of its protein product(s), a correlation that has been detected by cDNA measurements as well as in immunohistochemical data reported by Crnogorac-Jurcevic et al. (2002). Although the mechanism of down regulation of this gene is still to be elucidated, it is likely that the loss of its function could enhance the genotoxic effects of exposure to carcigenes under various environmental conditions, which may accelerate the accumulation of various genetic abnormalities. Crnogorac-Jurcevic et al. (2002) have pointed out that the use of pure cell populations of both normal and malignant cells acquired by LCM and their examination by cDNA microarrays provided enough data to implicate 15 differentially expressed genes in pancreatic adenocarcinoma. The same study demonstrated that the variety of these genes could imply multiple pathways in the development of this disease.

In a recent study by Ryu et al. (2002) serial analysis of gene expression was used to analyze gene expression profiles in short-term cultures of normal pancreatic ductal epithelium and primary pancreatic cancer tissue. A total of 294,920 tags representing 77,746 genes in ten SAGE libraries were performed. The authors reported that several genes were overexpressed in pancreatic cancer cells compared to normal epithelium. Some overexpressed genes, such as *S100A4*, prostate stem cell antigen, carcinoembryonic antigen-related cell adhesion molecules, and mesothelin could represent potential diagnostic markers. The relevance of *S100A4* overexpression is further discussed in the following section.

It is commonly argued that one of the challenging problems in pancreatic cancer is how to distinguish gene expression associated with adenocarcinoma from that associated with chronic pancreatitis. Such differentiation would be a key factor in current attempts to understand the molecular basis of both forms in particular and pancreatic cancer in general. An attempt in this direction has been recently reported by Logsdon et al. (2003). These authors used oligonucleotide-directed microarrays to profile gene expression in pancreatic adenocarcinoma, pancreatic cancer cell lines, chronic pancreatitis, and normal pancreas. Molecular profiling of these samples revealed a clear difference in a large number of differentially expressed genes between pancreatic cancer and normal pancreas but far fewer differences in gene expression between pancreatic cancer and chronic pancreatitis. The absence of a substantial difference in the latter case was attributed to shared stroma in the two diseases. To specifically identify genes expressed in neoplastic epithelium, the authors deliberately selected genes more highly expressed (more than twofold, $p < 0.01$) in adenocarcinoma compared with both normal pancreas and chronic pancreatitis and which at the same time were highly expressed in pancreatic cancer cell lines. Some of the microarray data, as in the case of S100P and 14-3-3$_{\xi}$. were further validated by quantitative–reverse transcription–PCR. Furthermore, immunocytochemical measurements were used to support the localization of 14-3-3$_{\xi}$. S100P, S100A6, and β4 integrin in neoplastic cells in pancreatic tumor. Given the comprehensive and specific characteristic of this study, it is worth considering some of its conclusions and their possible impact on future attempts to understand the molecular basis of this devastating disease: Three of the molecules which were highly and specifically expressed in pancreatic adenocarcinoma are members of the S100 protein family, namely, S100A6, S100A11, and S100P. It is interesting to note that an earlier profiling study using global gene expression technology (Iacobuzio-Donahue et al., 2002) has associated the differential expression of S100P with pancreatic cancer, similar association for S100A11 being also reported by Han et al. (2002). In a recent report Rosty et al. (2002b) used the National Center for Biotechnology Information Serial Analysis of Gene Expression database together with measurements using reverse transcriptase–PCR to demonstrate the overexpression of S100A4 in pancreatic ductal adenocarcinoma. These findings are in apparent contrast with the findings reported by Logsdon et al. (2003).

Such contrast is regarding the specificity of the expression, the latter work reporting significantly higher levels of S100A4 in adenocarcinoma compared with normal pancreas, yet no statistically significant difference between its levels in adenocanciroma and in chronic pancreatitis. On the contrary, the same authors reported higher expression of S100sA6, A11, and P in pancreatic cancer compared to chronic pancreatitis. This deduction was further supported by immunocytochemistry, which confirmed the specific localization of S100A6 and S100P in neoplastic epithelium.

The apparent disagreement between the findings of Logsdon et al. (2003) and Rosty et al. (2002b) regarding the specificity of expression of S100A4 in pancreatic adenocarcinoma merits further considerations. Calcium-binding proteins form a large family involved in numerous functions ranging from the control of cell cycle progression and cell differentiation to enzyme activation and regulation of muscle contraction (Kligman and Hilt, 1988; Shafer and Heizmann,1996). Throughout the past few years, the S100 family has emerged as an important group of proteins with the capacity to promote invasiveness and metastasis of many human neoplasmas. Recent studies have shed some light on the mechanisms of action of S100A4 protein and implied a possible prognostic role in human neoplasmas (Lukandin and Georgiev, 1996; Barraclough, 1998; Sherbet and Lakshmi, 1998). S100A4 protein has 101 amino acids ($M_r \sim 11.5$ kDa). Initial cloning experiments performed by screening cDNAs obtained from cultured cell lines before and after growth simulation (Goto et al., 1988; Linzer and Nathans, 1983) as well as the gene isolation from metastatic tumor cell lines (Ebralidze et al., 1989) already suggested a possible link between the expression of this protein and cancer progression. These initial findings have been subsequently supported by the demonstration of a marked up regulation of S100A4 at the mRNA and protein level in murine NIH3T3 fiberblasts. Complementary DNA array measurements by El-Rifai et al. (2001) showed an overexpression of the same protein in gastric adenocarcinoma. Setting aside the difference between the two studies by Rosty et al. (2002b) and Logsdon et al. (2003) regarding the specifity of expression of S1004 in pancreatic adenocarcinoma, it is likely that further evidence regarding the involvement of this protein in human cancer may help in transferring research results into clinical applications.

Two other molecules which experienced up regulation in the study by Logsdon et al. (2003) are 14-3-3_{ξ} and β4 integrin. The latter was highly and specifically expressed in neoplastic cells of pancreatic adenocarcinoma. Integrins are dimeric proteins composed of noncovalently bound α- and β-subunits that mediate cellular adhesion and have been implicated in the progression and spread of various forms of cancer. Molecule 14-3-3_{ς} was found to be highly expressed in pancreatic cancer adenocanciroma. The material presented in this chapter will show that the down or up regulation of one or more members of the 14-3-3 family have been reported by different techniques and by different research groups; therefore it would be useful to give a brief mention of the known biological functions of this group of proteins. The crystal structures of 14-3-3 proteins (Xaio et al., 1995; Liu et al., 1995) show that they exist as dimers, with residues 5–21 in one monomer forming contacts with residues Ser 58–Glu 89 in the opposing monomer. The role of these proteins was unclear up to the mid-1990s when they started to gain acceptance as a novel type of chaperone protein that modulate interactions between components of signal transduction pathways.

Further studies have also shown that phosphorylation of the binding partner or even the 14-3-3 proteins themselves is important in regulating these interactions. Clearly, there are still many unresolved issues regarding the physiological functions of this family of proteins. Many of the proposed and reported interactions with other proteins had appeared to be unrelated. In other words, there seems to be too many proteins of distinct functions associated with 14-3-3, which implies that many interactions are probably nonphysiological. Having said that, over the last few years some common ground has started to emerge. For example, it is generally agreed that 14-3-3 acts as an adapter to mediate interaction between signal transduction complexes, including some that are involved in regulating cell cycle events (for more details see Aitken, 1996). Recent results obtained by emerging technologies which use activity-based protein profiling and strategies for protein–protein interactions are bound to shed more light on the role of this family of proteins in cancer research.

2.2.2. SELDI Analysis of Pancreatic Ductal Adenocarcinoma

The complexity of the molecular basis which governs the development of various forms of cancer and the multiplicity of the genes which may participate in the various mechanisms render the use of different analytical approaches compulsory rather than optional. Pancreatic ductal adenocarcinoma is a representative case where DNA microarrays, 2DE, and SELDI proteomic analyses provided what can be considered complementary data regarding the involvement of specific proteins in this pathology. Identification of some members of S100 calcium-binding proteins and various members of the 14-3-3 family are two such examples. Given that SELDI is based on a different principle, its use is likely to provide additional information which can be easily missed by DNA microarrays.

In a recent study Rosty et al. (2002a) used SELDI technology (Ciphergen Biosystems, Fremont, California) to investigate differentially expressed proteins in the pancreatic juice of patients suffering from pancreatic adenocarcinoma. Samples were taken from 22 patients, 15 of them suffering from pancreatic adenocarcinoma, while the other 7 patients were suffering from other forms of pancreatic diseases. The arrays were run twice with two different laser intensities to achieve better resolution for low- and high-molecular-mass proteins. The α-cyano-4-hydroxycinaminic acid matrix was used for the detection of $m/z < 5000$, whereas sinapinic acid was used for $m/z > 15{,}000$. The authors reported that up to 140 proteins per spot were detected in the mass range 2000–200,000 Da. One of the main findings of this study was the identification of a 16,573-Da protein, which was mainly present in the cancer pancreatic juice samples. Following SWISS-PROT and TrEMBL database search, this molecular weight was assigned to a secreted form of the human HIP/PAP-I protein. For further confirmation of this differentially expressed protein, a SELDI immunoassay with specific anti-CRD/HIP polyclonal antibody was performed on 12 pancreatic juice samples, which confirmed the identity of this protein. Based on this study, the authors listed a number of deductions which are worth considering: The levels of HIP/PAP-I protein were found to be 24-fold higher in pancreatic juice with pancreatic adenocarcinoma as compared with patients having other pancreatic conditions, including chronic pancreatitis. These levels were significantly higher than those measured in serum (Motoo et al., 1999; Cerwenka et al., 2001), suggesting that serum is a relatively less sensitive medium for detecting markers of pancreatic adenocarcinoma. Comparing the findings of this study with existing data, HIP/PAP-I is found to be a secreted C-type lectin protein (Lasserr et al., 1994), originally identified as a PAP-I released by acini during acute pancreatitis (Keim et al., 1991). Furthermore, functional studies have indicated that HIP/PAP-I may be involved in adhesion of tumor cells to extracellular matrix proteins (Christa et al., 1996) and in the protection of pancreatic cells from apoptosis during oxidative stress (Ortiz et al., 1998). The strong acinar expression of HIP/PAP-I protein in areas adjacent to infiltrating carcinoma compared to the low-rate expression of the same protein in the neoplastic cells themselves measured by immunohistochemical labeling indicates that the main source of this protein in pancreatic juice are the acini. Because *pancreatic* adenocarcinomas are generally characterized by a prominent stromal reaction, biomarkers derived from the stromal reaction or from pancreatic acini may prove to be a useful diagnostic tool.

Three recent studies (B.-L. Adam et al., 2002a; Qu et al., 2002; Petricoin et al., 2002a) have also reported the use of SELDI to search for new markers associated with prostate cancer. Observations and comments regarding the conclusions of these studies are given in the final remarks of this chapter.

2.2.3. Protein Profiling Following Treatment with DNA Methylation/Histone Deacetylation Inhibitors

In the last decade, aberrations in DNA methylation patterns have been accepted to be a common feature of human cancer (Esteller, 2002; Jones and Baylin, 2002; Sato et al., 2003a,c). Currently, there is a large body of evidence to suggest that DNA methylation in eukaryotes generally inhibits the expression of certain genes. Such an effect is generally attributed to the methylation of cytosine at specific sites, which can interfere with the binding of protein factors, including some known to be required for RNA synthesis or chromatin structure.

It is generally assumed that transcriptional efficiency is correlated with the relative activities of histone acetyltransferases and histone deacetylases. Acetylation is believed to cause the separation of the basic N-termini of histones from DNA that then becomes more accessible to transcription factors (Tsukiyama and Wu,1997; Gregory and Hörz, 1998). Thus, histone acetylation seems to lead to gene activation, while histone deacetylation leads to a tighter histone–DNA interaction and, accordingly, to gene repression. The role of histone acetylation and deacetylation in the genesis of cancer has been discussed by Jacobson and Pillus (1999).

The importance of DNA methylation in the transcriptional silencing of cancer-associated genes is increasingly recognized (Jones and Baylin, 2002). For example, a variety of tumor supressors, growth regulators, and mismatch repair genes, including those encoding *p16*, *preproenkephalin*, and *hMLH1*, are inactivated by promoter region hypermethylation in benign and malignant pancreatic neoplasmas (Ueki et al., 2000, 2001; Sato et al., 2002, Fukushima et al., 2002). In the case of *p16*, for example, silencing of its expression by methylation is an effective alternative to its mutation or deletion, an event found in up to 20% of both xenografted (Sorio et al., 2001; Schutte et al., 1997) and primary pancreatic cancer (Kouzarides, 1999). Although the expression of several thousand genes may be controlled by methylation, only a handful have been studied and implicated in pancreatic cancer (Ueki et al., 2000). There is also an accumulating evidence to suggest that the DNA methylation inhibitor 5′-aza-2′-deoxycytidine, either alone or in combination with the histone deacetylation inhibitor trichostatin A, can reactivate the transcription of a number of genes associated with various types of cancer (Sato et al., 2003a; Grassi et al., 2003; Belinsky et al., 2003; Bender et al., 1998).

Cellular gene modulation by DNA methylation and histone acetylation can influence malignant cell transformation and is also responsible for the silencing of DNA constructs introduced into mammalian cells for therapeutic or research purposes. These observations have been tested in a recent work by Grassi et al. (2003). The neuroblastoma cell line U87 was stably transfected with a cytomegalovirus promoter–driven reporter gene construct whose expression was then analyzed following treatment with DNA methylation inhibitor 5′-aza-2′-deoxycytidine or histone deacetylation inhibitor trichostatin A (TSA). The authors concluded that both substances reactivated the silenced cytomegalovirus promoter but with different reaction kinetics. The simultaneous use of the two inhibitors resulted in cooperative rather than synergistic reactivation, which suggests a cell type–specific role for DNA methylation and histone deacetylation in the regulation of gene expression.

Another aspect on which aberrations in the DNA methylation patterns can impact is the analysis of human cancer cell lines, which are commonly used in basic cancer research to understand the behavior of primary tumors. Human cancer cell lines are a widely adopted experimental tool, and their use in cancer research has several advantages; for example, a wide spectrum of tumor types is commercially available and they are easy to culture in vitro, yielding significant amounts of high-quality RNA and DNA. Given that most current knowledge of the genetic alterations present in human cancer is derived from in vitro analysis of cancer cell lines, it is essential to provide careful assessment of DNA methylation patterns in human cancer cell lines used around the world. An attempt in this direction has been made by Paz et al. (2003). These authors analyzed 70 widely used human cancer cell lines of 12 different tumor types for CpG island promoter hypermethylation of 15 tumor suppressor genes, global 5-methylcytosine genomic content, chemical response to the demethylating agent 5′-aza-2′-deoxycytidine, and their genetic haplotype for methyl group metabolism genes. The main conclusions of this study were that a specific profile of CpG island hypermethylation exists for each tumor type, allowing its classification within hierarchical clusters according to the originating tissue. Cancer cell lines have higher levels of CpG island hypermethylation than primary tumors. Given current efforts to establish translational cancer research, these data assume a particular value.

2.2.4. Proteomic Profiling of PDAC Following Treatment with Trichostatin A

The effect of TSA, a potent inhibitor of histone deacetylases, has been recently explored by Cecconi et al. (2003b). The authors used 2DE and the PDQuest software package to generate 2D master maps of pancreatic adenocarcinoma cell lines before and after treatment with TSA. This comparison revealed that out of 700 detected spots there were 51 polypeptide chains which have experienced up or down regulation, 22 of which were identified by MALDI–TOF–MS fingerprinting. The authors reported alterations in protein expression ranging from two- to eight-fold. Among the 22 proteins identified by MS, of particular interest are the two down-regulated proteins nucleophosmin and transitionally controlled tumor protein (TCTP) as well as the up-regulated proteins stathmin and programmed cell death protein 5 (PDCD5). In the absence of other proteomic data obtained under similar experimental conditions, it is too early to assess the importance of these proteins in the understanding of this pathology; however, it is useful to consider some of their biological functions, which have been pointed out in various published works.

Nucleophosmin (NPM) is a ubiquitously expressed nuclear phosphoprotein that continuously shuttles between the nucleus and the cytoplasm. One of its suggested roles is in ribosomal protein assembly and transport and also as a molecular chaperone that prevents proteins from aggregating. Evidence is accumulating that the *NPM* gene is involved in several tumor-associated chromosome translocations and in the oncogenic conversion of various associated proteins (Colombo et al., 2002; Okuda, 2002; Yang et al., 2002). This protein appears to be present in most human tissues, with, especially, substantial expression in pancreas and testis yet it a poor expression in lung (Shackleford et al., 2001). Interestingly, a fusion protein containing the amino-terminal 117 amino acid portion of NPM joined to the entire cytoplasmic portion of the receptor tyrosine kinase ALK (anaplastic lymphoma kinase) has been found to be involved in oncogenesis in the case of non-Hodgkin's lymphoma (Bischof et al., 1997).

The same proteomic study by Cecconi et al. (2003b) reported that TCTP has experienced threefold down regulation. This polypeptide has been implicated in tumor reversion, that is, in the process by which some cancer cells lose their malignant phenotype. In a recent study, Tuynder et al. (2002) have shown that TCTP is strongly down regulated in the reversion process of human leukemia and breast cancer cell lines. Based on this observation, they hypothesized that tumor reversion is a biological process in its own right, involving a cellular reprogramming mechanism, overriding genetic changes in cancer, which triggers an alternative pathway leading to suppression of tumorigenicity. The same protein was also recently found, for the first time, in rat and human testes (Guillame et al., 2001). Interestingly, the mRNA of TCTP was also recently reported to be overexpressed in human colon cancer Chung et al. (2000).

The PDCD5 (also designated as TFAR19) is another protein which was found to be up regulated by a factor of 4. This is a recently discovered protein involved in the regulation of cell apoptosis (Liu et al.,1999; Chen et al., 2001; Rui et al., 2002). The level of this protein in cells undergoing apoptosis is significantly increased compared with normal cells. Thus, its up regulation in TSA-treated cell lines, as reported by Cecconi et al. (2003b), is rather consistent with their own finding of apoptotic cell death of Paca44 following TSA treatment.

Stathmin (oncoprotein 18, OC18) is another protein which experienced up regulation (eightfold) following TSA. Stathmin is a *p53*-regulated member of a novel class of microtubule-destabilizing proteins known to promote microtubule depolymerization during interphase and late mitosis (Alli et al., 2002). Thus, high levels of stathmin could induce growth arrest at the G2-to-mitotic boundary (Cassimeris, 2002; Mistry and Atweh, 2002). This again is highly consistent with the observation that Paca44 showed a cell cycle arrest at the G2 phase. It is of interest to note that overexpression of this protein, via its effect of inhibiting polymerization of microtubules, permits increased binding to these structures of vinblastin, a well-known chemotherapeutic agent, during treatment of human breast cancer (Alli et al., 2002). Due to its effect of inhibiting cell proliferation via a mitotic block, the up regulation of stathmin appears to be consistent with the antitumoral activity of TSA.

2.2.5. Proteomic Profiling of PDAC Following Treatment with 5′-aza-2′-deoxycytidine

In an attempt to investigate a possible correlation between DNA methylation, transcriptional regulation of several matrix metalloproteinase (MMP) genes, and the invasive phenotype of pancreatic adenocarcinoma, Sato et al. (2003b) have assessed the invasiveness of this type of cancer following the treatment of five different cell lines with the DNA methylation inhibitor 5′-aza-2′-deoxycytidine (DAC). Using a modified Boydon chamber in vitro invasion assay (Albini et al., 1987), the authors reported that four of the five examined cell lines have exhibited an increased invasiveness. The authors also deduced that the DNA methylation could influence the expression of MMP genes, and therefore the use of methylation inhibitors such as DAC might stimulate the invasion of pancreatic cancer by reactivating invasion-promoting genes.

In another study Sato et al. (2003a) have used high-throughput oligonucleotide microarrays to analyze global changes in gene expression profiles in pancreatic cancer cell lines following exposure to DAC and/or trichostatin A. The main conclusions of this study were as follows: After DAC treatment over 475 genes were markedly (more than fivefold) induced in pancreatic cancer cell lines but not in nonneoplastic epithelial cell lines. The methylation status of 11 of these genes were examined in a panel of 42 pancreatic cancers, and all 11 genes were aberrantly methylated in pancreatic cancer but rarely in 10 normal pancreatic ductal epithelia. The authors concluded that a substantial number of genes are reactivated by DAC treatment and many of them may represent novel targets for aberrant methylation in pancreatic carcinoma.

In a recent article (Cecconi et al., 2003a) the effect of DAC on the proteomic profiling of pancreatic ductal carcinoma cell lines has been investigated by 2DE combined with PDQuest statistical analysis and MALDI–TOF MS. Comparing standard 2D maps of the total protein extract before and after DAC treatment showed that, of 700 spots visualized in the pH range 3–10, a total of 45 polypeptide chains were differentially expressed following DAC treatment. Thirty-two were down regulated and 13 up regulated; most of these proteins were characterized by MALDI–TOF–MS. Based on the observed alteration in the expression of these proteins following DAC treatment, the authors listed a number of deductions which are worth considering: Two proteins, cofilin and profilin 1, were found to be silenced. Cofilin is an actin-depolymerizing protein found widely distributed in animals and plants. Actin filaments are a major cytoskeletal component of eukaryotic cells. Rapid changes in the levels of polymeric F-actin and monomeric G-actin may lead to morphological changes that occur in spatially and temporally controlled fashion in response to environmental signals in living cells. A family of related actin-binding proteins [actin-depolymerizing factor (ADF)/cofilin] has been implicated in the control of actin-based motility in response to signaling. These conserved proteins, initially characterized by their ability to depolymerize F-actin, possess the unique property among other G-actin-binding proteins to be regulated by reversible phosphorylation. It is interesting to note that in many situations cited in the literature the activity of ADF/cofilin was inhibited by phosphorylation of a serine in the amino terminal region, while rapid dephosphorylation seems to occur in response to stimuli (Carlier and Pantaloni, 1997; Carlier et al., 1997). Interestingly, prevention of spontaneous cofilin dephosphorylation in the jurkat cell line by the serine phosphatase inhibitor okadaic acid was accompanied by apoptosis (Samstag et al., 1994; Nebl et al., 1996).

Kanai et al. (2001) also reported strongly reduced mRNA expression for cofilin following DAC treatment of human stomach and colorectal cancer cells. On the other hand, in pancreatic adenocarcinoma cell lines, Sinha et al. (1999b) found overexpression of cofilin, a result which might appear to be in contrast with the findings of Cecconi et al. (2003a). However, it is of interest to note that this up regulation occurs only in cell lines which have been selected for their chemoresistance to multidrug treatment. Thus, it is reasonable to speculate that suppression of cofilin might lead to cancer regression, whereas its up regulation could confer strong resistance to drug treatment.

Another silenced protein reported in the study by Cecconi et al. (2003a) is profilin. This is a small actin-binding protein that is involved in diverse functions such as maintaining cell structure integrity, cell mobility, growth factor signal transduction, and tumor cell metastasis (Hu et al., 2001; Feldner and Brandt, 2002).

In the same study the down-regulated proteins accounted for 32 of a total of 45 differentially expressed proteins; the relevance of some of these polypeptides is briefly discussed here. The authors reported that coactosin-like protein (CLP) has experienced a 22-fold decrease. This is a human filamentous actin (F-actin) binding protein (Provost et al., 2001b) which binds to actin filaments with a stoichiometry of 1 : 2 (CLP–actin subunits). Additionally, it binds to 5-lipoxygenase in a 1 : 1 molar ratio (Provost et al., 2001a). A recent report by Nakatsura et al. (2002) indicated CLP as a tumor-associated antigen, suggesting this protein is a candidate for a vaccine for immunotheraphy of cancer patients. Thus, its strong supression upon DAC treatment suggests a possible role of this drug in tumor regression.

Peptidyl-prolyl cis–trans isomerase A (PPIA) is thought to control mitosis by binding to cell cycle regulatory proteins and altering their activity. It appears that the gene coding for this protein is almost exclusively overexpressed in aggressive metastatic or chemoresistant tumors (Meza-Zepeda et al., 2002). In addition, PPIA has been recently proposed as a novel marker candidate for hepatocellular carcinoma (Lim et al., 2002). It is interesting to point out that, following DAC treatment, all main isoforms of this protein have exhibited various degrees of down regulation ranging from 2- to 22-fold (Cecconi et al., 2003a). Cystatin B (also called stefin B) belongs to the cystatin superfamily of cysteine protease inhibitors and target cysteine proteases, such as cathepsin B. Cystatin B has been implicated in malignant progression (Calkins et al., 1998), and alterations in its expression, processing, and localization have been observed at various levels in malignant human tumors (Kos and Lah, 1998). Progressively higher levels of cystatin B have been associated, for example, with short survival in patients with colorectal cancer (Kos et al., 2000). In human squamous cell carcinoma of the lung and in a number of other cancer types, cystatin B has been found to be significantly increased as compared to normal tissues and proved to be a prognostic factor (Ebert et al., 1997). This latter observation together with marked down regulation (15-fold) following DAC treatment as reported by Cecconi et al. (2003a) suggests that this particular protein may have a role to play in maintaining tumor cell growth. Down regulation of more than one isoform was also observed for Rho GDP the Ras (Rhomologons guanosine triphospotase) dissociation inhibitor (Rho GDI-2). This is a cytosolic protein that participates in the regulation of both the GDP/GTP cycle and the membrane association/dissociation cycle of Rho/Rac. This family of proteins is in general overexpressed in breast tumors and human bladder cancer and their increased synthesis correlates with malignancy (Seraj et al., 2000; Fritz et al., 2002). In another study, Kovarova et al. (2000) have reported, by 2D map analysis followed by MALDI–MS, that Rho GDI proteins were significantly down regulated in CEM T-lymphoblastic leukemia cell lines after treatment with bohemine.

Annexin 1 is a member of a family of calcium- and phospholipid-binding proteins related by amino acid sequence homology. Annexins 1 and 2 are substrates for protein tyrosine kinases. Both these proteins appear to be involved in mitogenic signal transduction and cell proliferation. Existing evidence indicates that overexpression of annexins occurs in a variety of malignancies and well correlates with their progression (De Coupade et al., 2000; Wu et al., 2002a,b; Rhee et al., 2000; Pencil and Toth, 1998).

Among the up-regulated proteins reported by Cecconi et al. (2003a) are the isoforms of superoxide dismutase (6- to 11-fold), which is one component of the antioxidant system, and is located in the mitochondrion. It has recently been shown that superoxide dismutase (*SOD*) could be a tumor suppressor in human pancreatic cancer to the point of suggesting delivery of the *SOD* gene for suppression of pancreatic cancer growth (Weydert et al., 2003; Cullen and Lockyer, 2002).

2.3. PROTEOMIC ANALYSIS OF HUMAN BREAST CARCINOMA

With a lifetime currently estimated at 1 in 8, breast cancer is the most commonly diagnosed neoplasm in women, second only to lung cancer. Currently it is estimated that over 40,000 women are likely to die of breast cancer each year (Landis et al., 1999). Similar to other cancer types, breast tumorigenesis is a multistep process starting with benign, then atypical hyperproliferation, progressing into in situ invasive carcinomas, and culminating in metastatic disease. Although rapid, high-throughput proteomic techniques and gene expression profiling assays are currently being developed and applied to breast cancer; the current methods for the detection of both benign and malignant breast tumors are based on mammography. These methods, however, have a number of limitations: First, X rays may induce its own carcinogenesis and, second, what may be more serious is the fact that a breast tumor has to reach certain dimensions (few millimeters) before it can be detected by mammography, which means that the tumor is detected at a fairly late stage in its development. Such delayed detection also impacts on the following steps in clinical practice, where the surgical removal of the tumor is followed by its classification as a malignant or benign using histology. Such classification is then used to decide the type of treatment and prognosis. Of course, we know now that breast cancer has different forms depending on the cellular and histological origin of the cancer cells and the evolution of the disease. This means that the postsurgical classification of the type of tumor, which is based on tumor size and inflammation, histoprognostic grading, and node involvement, might not be accurate enough for deciding the treatment and prognosis. Such limitations, which are associated with the well-established mammography approach, together with the devastating effects of breast cancer have encouraged the search for new approaches which may provide a better understanding of the molecular basis of cancer cell growth and the eventual development of new therapeutic strategies. Treatment based on the inhibition of tyrosine kinase receptors is an example of such emerging strategies. Such an approach is well illustrated with ErbB2, a tyrosine kinase receptor overexpressed in more than 20% of breast tumors (Hondermarck, 2003). Specific inhibition of ErbB2 using herceptin, a truncated blocking antibody directed against it, has been successfully developed and has now entered into clinical practice (Colomer et al., 2001). Currently, gene expression profiling and various proteomic approaches represent the main source of information on molecular alterations involved in the initiation and progression of various types of breast cancer. Furthermore, the well-engrained 2DE and the emerging chip technology, in particular SELDI in conjunction with ANN algorithms, are providing relevant information on potential early biomarkers for breast cancer. Some examples of the contribution of these approaches are considered in this section.

2.3.1. Two-DE Analysis in Breast Cancer

The use of 2DE to resolve proteins in the serum of cancer patients has been demonstrated as early as 1974 (Wright, 1974). A few years later (Westly and Rochefort, 1980) the same technique was used to detect a secreted glycoprotein in human breast cancer cell lines, which was identified, with specific antibodies, to be the protease cathepsin D. In 1989, Maloney et al. (1989) used 2DE in conjunction with computer-based statistical analysis, thus identifying a number of polypeptides which were different in cancerous and normal breast epithelial cells. A better characterization of these proteins was latter reported by Trask et al. (1990), who demonstrated that a number of keratins (K5, K6, K7, and K17) were associated with normal cells, whereas tumor cells produced other forms of keratins (K8, K18, and K19). A period extending between the mid-1990s and lasting to the present day has witnessed the contribution of large-scale 2DE analysis which resulted in the establishment of 2DE databases of human breast epithelial cell proteins: http://www.anl.gov/BIO/PMG/projects/index_hbreast.html,

Giometti et al. (1997); http://www.ludwig.edu.au/ipsl/data-bases/MDA-MB231.asp, Rasmussen et al. (1997); and http://www.bio-mol.unisi.it/2d/2d.html., Bini et al. (1997). Over the last 10 years the 2DE approach has generated a considerable amount of valuable data for breast cancer research; however, the translation of these data to clinical practice is still to be demonstrated. This rather disappointing deduction has a number of good reasons; one in particular is the fact that only a limited number of individual protein modifications commonly divide normal cells from their cancerous counterparts. This subtle difference is by no means limited to proteomic analysis, a similar situation is encountered at the genomic level, where a limited number of molecular modifications affecting oncogenes and suppressor genes are required to transform normal cells into cancerous ones (Haber, 2000). This limited number of molecular modifications has two direct consequences. First, so far, no technique has been able to provide an unmistakable marker for the early detection of this pathology. Second, finding drugs which specifically target breast cancer cells has not been achieved yet, and therefore patients continue to receive treatments which are known to have heavy side effects.

Attempts to improve the capabilities of 2DE in cancer analysis included more use of LCM, more efficient labeling, and of course more frequent use of MS for the characterization of the separated proteins. Some recent applications of such an approach are considered here.

Molecular alterations accompanying ductal carcinoma in situ (DCIS), the earliest detectable form of cancer, represent an attractive target for the exploration of breast cancer origin and alterations likely to influence its latter development. In a recent study, 2DE, in conjunction with LCM and MS, has been used by Wulfkuhle et al. (2002) to investigate protein profiling of matched normal ductal/lobular units and DCIS in human breast. In their study the authors used 10 sets of 2D gels containing either matched normal ductal/lobular units or DCIS from either whole-tissue sections or up to 100,000 laser-captured microdissected epithelial cells. The resulting gels were silver stained and up to 315 spots were analyzed by MS. The authors reported that 57 proteins showed differential expression between matched normal ductal/lobular cells and DCIS. Densitometric measurements showed variation in the expression from a modest 2-fold up to 400-fold. The expression pattern of a number of these proteins was further validated through the use of an immunohistochemical (IHC) protocol. Based on these measurements, a number of deductions were made, some of which are considered here: DCIS lesions exhibited increased expression of barbed-end capping proteins, the Arp3 component of the actin nucleation complex, the depolymerization protein cofilin, the actin-bundling protein L-plastin, and profilin. It is interesting to note that both cofilin and profilin, which are overexpressed here, were found to be fully silenced in a recent 2DE study (Cecconi et al., 2003a) which examined pancreatic ductal carcinoma cell lines following exposure to DAC treatment. The authors compared their findings with those based on LCM combined with high-density oligonucleotide arrays (Luzzi et al., 2001) and concluded that the only overlapping identification between the two studies was limited to lactoferrin. Furthermore, a SAGE study which included two DCIS samples yielded a list of differentially expressed genes none of which coincided with those identified in the proteomic study by Wulfkuhle et al. (2002). The authors attributed the lack of overlapping to two possible reasons. The first reason is the relatively limited dynamic range of 2DE, although the use of LCM-enriched cell populations and the capability of the same study to detect low-level proteins such as RhoGDI seem to rule out such a deduction. Second, the absence of the overlapping may be related to a previously documented observation which showed that mRNA levels do not necessarily correlate to functional proteins.

The same study noted the overexpression of a number of 14-3-3 chaperone proteins, an overexpression worth further consideration (see also Section 2.2.1). This family of highly conserved protein forms (α, β, δ, σ, and ξ), having $M_r \sim 25$–30 kDa, is expressed in all eukaryotic cells that play a role in the regulation of signal transduction pathways implicated in the control of cell proliferation, differentiation, and survival (Fu et al., 2000). This family of proteins is known to associate directly or indirectly with proliferative signal-transducting proteins such as Protein Kinase C (PKC), extracellurar regulated kinase Kinases (MEKs), P13 kinase, and RAF. For example, RAF-1 activation by 14-3-3 can lead to either cell cycle arrest or cell proliferation (Tzivion et al., 1998; Roy et al., 1998). It has recently been suggested that gene expression of 14-3-3σ is silenced in breast cancer cells (Ferguson et al., 2000) and that transfixion of breast cancer cells with the σ-form could inhibit kinase activity and thus cell cycle progression (Laronga et al., 2000). Recently, Vercoutter-Edouart et al. (2001) have used a narrow-range pH gradient in a 2DE study to investigate the distribution of the sigma form of 14-3-3 in tumor biopsies and deduced that the level of the sigma form was systematically down regulated in the investigated biopsies. It is also interesting to note that at the mRNA level it was shown that gene expression of 14-3-3σ was 7–10 times lower in breast cancer cells compared to their normal counterparts, an observation which was attributed to massive hypermethylation of the sigma form (Ferguson et al., 2000). In another study Vercoutter-Edouart et al. (2001) have used 2DE in combination with MALDI–TOF–MS to profile the expression of 14-3-3σ breast cancer cell lines, primary breast carcinoma, and normal breast epithelial cells. These authors concluded that this particular protein was down regulated in breast cancer cells compared to normal breast epithelial cells. Sinha et al. (1999a) reported overexpression of 14-3-3σ in chemoresistant human adenocarcinoma of the pancreas.

The differential in-gel electrophoresis (DIGE) originally developed by Ünlü et al. (1997) promises to significantly improve the speed and reproducibility of conventional 2DE. In this approach, two pools of protein extracts are labeled covalently with fluorescent cyanine dyes, Cy3 and Cy5, respectively. These labeled proteins are mixed and separated on the same 2D gel. The resulting 2D gel patterns can be rapidly imaged by the fluorescence excitation of either Cy3 or Cy5 dyes (see Chapter 3 for more details).

Recently, Somiari et al. (2003) have used DIGE in combination with MALDI–TOF–MS to assess protein profiles associated with four different phases of human infiltrating ductal carcinoma (IDCA) and to compare them with the proteins expressed in a nonneoplastic tissue obtained from a female donor with no personal or family history of breast cancer. These authors reported that the samples studied were from different tumor stages, grades, and ages of the donors. To maximize the total amount of protein loaded on each gel, whole sections were used instead of cells obtained by laser microdissection. In summary, after loading a total of 175 μg of protein from each sample, four gels were run in parallel and each gel contained Cy dye–labeled proteins from one test sample (IDCA) and one reference sample (histologically normal). Gel images were generated and subjected to statistical analysis. Each pair of protein spots generated from the Cy3 (test) and Cy5 (reference) channels were converted to 3D representations and displayed to show the relative peak volumes, height, and area of each spot. A few hundred spots of the four IDCA gels were picked up and analyzed by MALDI–MS or LC/MS–MS. Based on these measurements, the authors listed a number of proteins which have experienced up or down regulation in the IDCA samples compared to the histologically normal sample. A number of these proteins are known to be involved in breast and other forms of cancer. For example, gelsolin, a protein that interacts with actin and regulates actin polymerization, was observed at significantly lower levels in two of the IDCA samples. The down regulation of this protein was also reported in earlier studies (Winston et al., 2001; Asch, 1999) that correlated such down regulation to progression in breast carcinoma. β-Actin was found 75% more abundant in IDCA samples. This finding is rather interesting for a number of reasons. First, β-actin is considered a housekeeping protein and its expression would be expected to be comparable in the normal and diseased samples. Second, the same measurements showed that gelsolin, which binds to β-actin, was down regulated, an observation which contradicts the up regulation of β-actin reported in the same study. Another protein involved in cell migration and proliferation that was identified as differentially abundant in IDCA is lucamin. This is a member of a small leucine-rich proteoglycan family which plays an important role during embryonic development, tissue repair, and tumor growth (Ping et al., 2002). Lumican genes and proteins have been found to be overexpressed in pancreatic cancer tissues, breast carcinoma, uterine cervical cancer cells, and benign prostatic hyperplasia (Ping et al., 2002; Leygue et al., 2000; Naito et al., 2002; Luo et al., 2002).

The combination of DIGE, LCM, and MS–MS was also applied by Zhou et al. (2002) to compare protein expression in esophageal carcinoma cells and normal cells. The authors reported the identification of over 1000 proteins in each sample, among which 58 spots were up regulated by at least threefold, while over 100 spots were down regulated by more than threefold. In addition to previously identified down-regulated protein annexin I, the authors reported the up regulation of tumor rejection antigen (gp96), found in esophageal squamous cell cancer.

2.3.2. Proteomic Profiling of Breast Cancer Cell Membranes

Proteomic analysis of plasma membrane proteins can be considered an initial step in the search for novel tumor marker proteins expressed during the various stages of cancer progression. This is because cancer cell plasma membranes are rich in known drug and antibody targets as well as other proteins known to play key roles in the abnormal signal transduction processes required for carcinogenesis. Membrane proteins still give disappointing results in 2DE separation. This is normally attributed to poor solubility, precipitation during first-dimension Isoelectric focusing (IEF) employing

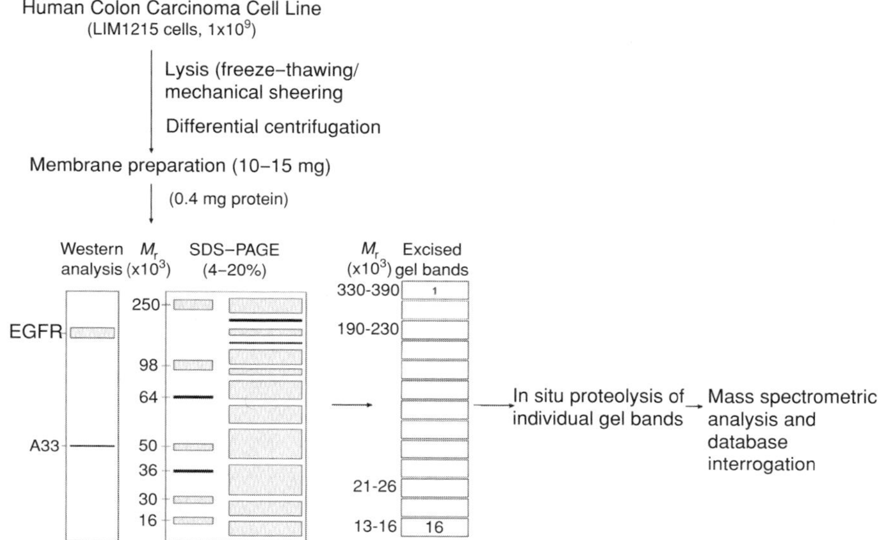

FIGURE 2.7. Schematic representation of strategy used for proteomic analysis of enriched membrane preparation from human colorectal LIM1215 cells. (Adapted from Simpson et al., *Electrophoresis*, 2000, with permission.)

immobilized pH gradients, and inefficient transfer of hydrophobic proteins from the 1D to the 2D gel. Another reason for the poor separation of this class of proteins is their charge heterogeneity due to numerous glycosylated extracellular domains, which results in their focusing over a wide pH range during the IEF phase. To bypass the charge separation step in 2DE and rely solely on size separation, Simpson et al. (2000) used the experimental approach depicted in Figure 2.7 to investigate proteins from an enriched membrane preparation of the human colorectal carcinoma cell line LIM1215. Following fractionation by sodium dodecyl sulfate–polyacrylamide gel electrophoresis (SDS–PAGE), the unstained gel slices were subjected to in-gel tryptic digestion and the resulting peptide mixtures were examined by reverse-phase LC coupled to MS. The acquired CID spectra were used to interrogate various databases for protein identification.

In a fairly recent and well-planned proteomic study P.J. Adam et al. (2003) investigated the protein content of breast tumor cell membranes. To enhance the possibility of obtaining the highest number of protein identifications, the authors took a number of steps. First of all, to simulate clinical heterogeneity of breast cancer, the authors used multiple cell lines with different molecular pathologies. Additionally, tumor-derived cell lines were used to ensure enrichment for cancer cell–specific plasma membrane proteins. Membrane preparations were resolved on 1D rather than 2D gels to avoid discrimination against insoluble hydrophobic membrane proteins. The separated proteins were digested and the resulting peptide mixtures were examined by MALDI–TOF and MS/MS analysis. This resulted in the identification of a number of trans–plasma membrane or plasma membrane–associated proteins. According to the authors, a number of these proteins were of unknown functions. Three of these proteins were assigned as breast cancer membrane proteins, BCMP11, BCMP84, and BCMP101.

2.3.3. Proteomic Analysis on Selected Tissue Samples

A combination of LCM together with on-line LC/MS–MS has been used for protein profiling of breast cancer cell line SKBR-3 (Wu et al., 2003). In Wu et al.'s study, captured cells were isolated in a dehydrated and reduced state and solubilized with a denaturing buffer. Following dilution the protein mixture was enzymatically digested and the resulting peptide mixture was fractionated by reverse-phase LC and analyzed on an ion trap mass spectrometer. The authors used the direct-analysis approach of Haynes and Yates (2000), in which the protein extract from the captured cells was reduced, alkylated, digested with trypsin, and separated by a single dimension of reverse LC. Such an approach has the advantage of minimizing sample losses and reducing contamination for very small sample amounts. The authors estimated the total protein extract to be 5 μg. Protein identification in this study is based on MS–MS spectra of the tryptic peptides and the subsequent search of genomic and proteomic databases. Among the proteins which have been identified by this study are human receptor protein kinase (HER-2 or ERBB-2) and related kinases HER-3 and HER4, fibroblast growth factor receptor variants (FGFR-2 and -4), and T-lymphoma invasion and metastasis-inducing protein 1 (TIAM1). This latter is a guanine nucleotide exchange factor which has been designated as a possible target of heregulin-beta 1 signaling and related to inhibition or promotion of cell migration in a cell type–dependent manner (L. Adam et al., 2001). Other proteins listed in the same study include the oncoproteins n-myc and a-myb from the c-*myc* gene and the *ras*-related proteins RAB-9L and RAB-6B.

Another recent study which targeted specific cell lines has been described by 0'Neil et al. (2003). This study targeted MCF10 breast epithelial cell lines and used two separation techniques followed by MS analysis. Intact membrane proteins were first separated by hydrophobicity using nonporous reverse-phase LC to generate distinct chromatographic profiles of the various protein components. Fractions of eluent were then separated using SDS–PAGE which yielded a distinct banding pattern. The combination of liquid phase with gel phase overcomes certain difficulties associated with membrane protein precipitation and provides a valid strategy for the isolation and characterization of traditionally underrepresented classes of proteins. The characterization of the separated proteins was achieved through the combination of MALDI–MS fingerprinting and MALDI/MS–MS.

The MCF-10 cell line model includes a series of cell lines that depict the neoplastic progression of breast epithelial cells (Heppner et al., 2000). It is assumed that the progression of these cells from a normal to a metastatic phenotype is reflected in changes in cell surface protein profiles. In their study the authors examined membrane extracts from five cell lines corresponding to normal, premalignant, DCIS and metastatic phenotypes.

A number of cell surface proteins have been identified including α-6 and α-3 integrin. Integrins are heterodimeric cell surface adhesion and signaling proteins proposed to work in cooperation with growth factors regulating gene expression, cell proliferation and differentiation, and cell mobility (Hemler et al., 1996; Watts, 2002; Buckly et al., 1998). It is interesting to note that alterations in integrin expression have been linked to the pathogenic nature of both benign and neoplastic cell growth; in particular, expression of α6β1 integrin was found to be involved in increased metastatic potential in prostate carcinoma cells (Robinovitz et al., 1995) and was linked to a decreased survival rate for breast cancer patients (Friedrichs et al., 1995). Similarly, the α3β1 integrin has been correlated with invasiveness and neoplastic progression in melanoma cells (Natali et al., 1993; Melchiori et al., 1995). Logsdon et al. (2003) reported specific highly expressed β4 integrin in neoplastic cells of pancreatic adenocarcinoma compared to normal and chronic pancreatitis.

2.4. PROTEOMIC PROFILING OF CHEMORESISTANT CANCER CELLS

Cancer cell resistance to chemotherapy is multifactorial and can be affected by the cell cycle stage and proliferation status, biochemical mechanisms such as detoxification, cellular drug transport, or DNA replication, and repair mechanisms. It is now well accepted that resistance to chemotherapy is an important factor in the failure of much cancer treatment. Such failure can have different forms: for example, some tumors are intrinsically resistant and never respond to cytostatic drug treatment, whereas others initially respond well but eventually regrow to become resistant. This acquired resistance is still difficult to decipher, it may result from genetic mutations induced by the administered antitumor drug, or it may represent the selection of preexisting resistant cell populations in the malignant tumor. The integration of 2DE with sensitive MS analysis and modern bioinformatic tools has greatly fostered its application to monitor alterations in protein expression after exposure of cells to chemotherapeutic agents. Combined with artificial network algorithms SELDI MS is also emerging as a useful tool to assess cellular response toward certain chemotherapeutic agents. To appreciate the contribution of 2DE in combination with MS to this particular area, we have chosen a number of recent studies in which protein alterations in various forms of cancer have been monitored following the use of different therapeutic agents.

2.4.1. Protein Alterations in Pancreas Carcinoma Cells Exposed to Anticancer Drug

Tumors of the pancreas are characterized by a high intrinsic capability to develop chemoresistance toward cytotoxic drugs, which is among the causes of ineffective treatment. Several groups have used 2DE with and without MS to monitor alterations in protein expression following exposure to well-known chemotherapy drugs (Möller et al., 2001, 2002; Sinha et al., 1999a,b, 2003). The response of pancreas tumor cells after exposure to the anthracyclin daunorubin (DRC), a well-known antitumor drug for chemotherapy, was investigated by 2DE in conjunction with MALDI–TOF–MS (Möller et al., 2002). Alteration in various protein expression was monitored at different dose concentrations of the drug, thus simulating a situation close to clinical chemotherapy. A number of proteins have shown significantly increased levels following DRC treatment; the significance of some of these overexpressed proteins is worth considering: Three of these proteins, TCP-1, HSP60, and Grp-78, belong to the same family, the chaperones, known to be involved in folding and rearrangement of misfolded proteins (Hendrick and Harti, 1995). In addition to these known functions, recent reports showed that HSP60 and HSP70 play a direct role in modulating the activity of proteins engaged in apoptotic pathways (Jaattela et al., 1998; Xanthoudakis et al., 1999). TCP-1 exhibits certain substrate specificity for tubulins and actins (Roobol et al., 1999) and has been considered to be involved in the assembly of microtubules prior to mitosis (Ursic and Culberston, 1991). Another overexpressed protein, Drg-1, is related to cell differentiation and morphological alterations and has been identified in various tumor cells (van Belzen et al., 1997). The authors noted that the expression of this protein was inversely proportional to the dose concentration of DRC.

The use of 2DE in conjunction with MS to examine drug-resistant human adenocarcinoma of the pancreas has also been investigated by Sinha et al. (1999a). To accomplish this, a cell line derived from an epithelial carcinoma of the pancreas was cultured in the presence of two cytostatic drugs, daunorubicin and mitoxantrone, to yield two chemoresistant cell lines. To assess differences in protein expression between the parental cell line and their chemoresistant counterparts, 2DE and image analysis of the fractionated proteins using PDQuest software were used. Proteins of interest were excised and subjected to microsequencing. The authors reported that three proteins, epidermal fatty acid–binding protein, cofillin, and 14-3-3-σ (stratifin), were overexpressed in chemoresistant cell lines, with cofillin being present in both multidrug-resistant and chemoresistant cell lines.

In a more recent study Sinha et al. (2003) have used 2DE in conjunction with MALDI MS to investigate chemoresistance development in melanoma cell lines. To probe molecular factors potentially associated with the drug-resistant phenotype of malignant melanoma, the authors used a panel of human melanoma cell variants exhibiting low and high levels of resistance to four commonly used anticancer drugs in melanoma treatment. This study revealed that in the neutral and weakly acidic range (pH 4–8) a total of 14 proteins showed alterations in their expression, whereas 20 proteins were differentially expressed in the basic region (pH 8–11). The authors grouped these differentially expressed proteins in three main groups, which are briefly considered here. Various members of the group of molecular chaperones have shown overexpression in certain chemoresistant cells. The authors reported that in various chemoresistant variants there was an increased expression of the small stress protein HSP27, whereas in most cell lines up regulation of HSX70 and HSP60 isoforms were registered. It is interesting to note that existing literature (Sarto et al., 2000) shows that upon cell stimulation HSP27 becomes a frequent target of phosphorylation. On the other hand, HSP70 has long been recognized as one of the primary heat shock proteins in mammalian cells. Both this protein and HSP27 have been frequently associated with the inhibition of apoptosis induced by different chemotherapeutics, particularly those that target topoisomerase II enzymes, such as anthracyclins and etoposide. Indeed, a recent study (Creagh et al., 2000) has shown that the production of reactive oxygen species by these drugs plays an important role in their induction of apoptosis, while treatment with antioxidants increases cellular resistance to these agents. The study by Sinha et al. (2003) has also listed other proteins with possible roles in drug detoxification, metabolism, and regulation of apoptotic pathways.

2.4.2. Proteomic Profiling of Cervix Squamous Cell Carcinoma Treated with Cisplatin

The chemotherapeutic agent cisplatin is commonly used for the treatment of ovarian and testicular carcinomas as well as for squamous cell carcinomas (Loehrer and Einhorn, 1984; Green et al., 2001; Giaccone, 2000). This agent is known to target DNA, with which it forms both interstrand and intrastrand crosslinks (Easterman, 1991; Crul et al., 2002). As such, cisplatin is a cytotoxic agent which causes cell death via genotoxic damage and interferences with replicate processes in proliferating cells (Andrews and Howell, 1990). Despite its effectiveness, a major limitation in its clinical use is the development of resistance. It has been reported that low levels of cisplatin resistance readily emerge in treated patients, which may be sufficient for treatment failure (Mandic et al., 2002). A rapid emergence of low levels of cisplatin resistance has also been documented in human tumor xenografts (Andrews et al., 1990). A number of mechanisms for such resistance have been suggested, such as reduced drug accumulation (Gately and Howell, 1993), increased detoxification through cellular thiolos (Kondo et al., 1995), and reduced DNA platination (Easterman et al., 1988).

In a recent study Castagna et al., 2004 have used 2DE in conjunction with MALDI–MS to examine four distinct samples associated with the cervix squamous cell carcinoma cell line A431. These samples included a control cell line and a cisplatin-resistant cell line, and each of these two cell lines was exposed to 1 h equitoxic cisplatin exposure. As well as 2DE analysis, some of the separated proteins were also examined by Western blot using specific polyclonal antibodies. Expression analysis via RT-PCR was also used to evaluate the expression levels of some members of the 14-3-3 family. Although the mechanisms that could confer drug resistance have been studied in a variety of ways (Helmbach et al., 2001), there is still a paucity of proteomic analysis regarding chemoresistance. Sinha et al. (1999a,b, 2001) have conducted systematic proteomic analyses of various diseases, including analysis related to cisplatin chemoresistance. The results described by Castagna et al. (2004) can be considered an extension to the work of Sinha's group. This is because the two groups used different cell lines and because of a different approach of comparison followed by Castagna et al. (2004), where not only were the parental cell line and its resistant subline compared but also the comparison was extended to drug-treated cells via a four-way confrontation.

Castagna et al. (2004) reported significant modulation in constitutive chaperones, so-called heat shock cognate (HSC) proteins as well as the chaperones induced by environmental stresses, nicknamed heat shock proteins (HSPs), in essentially all the four comparisons. Three members of this superfamily were modulated: HSC (M_r = 71 kDa) protein (present in no less than three isoforms) was up regulated in resistant cells and was induced by cisplatin in A431 cells; the HSP (M_r = 90 kDa) was expressed after cisplatin exposure of parental cells; the mitochondrial form HSP (M_r = 60 kDa) was up regulated in A431 cells exposed to cisplatin. Although in general these proteins have the function of ensuring proper folding of newly synthesized proteins, they also appear to have an antiapoptotic function. In particular, HSP71 and HSP90 have been reported by Ciocca et al. (1992) and Sinha et al. (2003) to be involved in the resistance of diverse tumor cell lines to different antitumor drugs, including cisplatin. Takayama et al. (1997) have demonstrated that both the inducible form (HSP70) and the constitutive form (HSC71) of the 70-kDa chaperone can interact, via the Bag1 protein, with several proteins which participate in signal transduction and apoptosis, including the Bcl-2 protein. According to these authors, the transient complex chaperone–Bag1–Bcl-2 could facilitate the insertion of the antiapoptotic protein into the mitochondrial membrane, thus preventing apoptosis. Molecular chaperones could, additionally, prevent apoptosis via other mechanisms, such as the "unfolded protein response," a protection mechanism activated by the endoplasmic reticulum under stress conditions, in which the synthesis and proper folding and/or postsynthetic modification of proteins inside the endoplasmic reticulum could be compromised (Mendic, 2003). A relationship between cisplatin resistance/response and expression of HSPs has also been previously documented with conventional approaches as well as with differential expression techniques (Righetti et al., 2002).

Calcium-binding proteins comprise another group of proteins that manifest significant modulation as a response to cisplatin. For example, calmodulin was found to be down regulated in resistant cells and after cisplatin exposure of parental cells. Calmodulin is able to bind four Ca^{2+} ions with high affinity and contains ubiquitination and phosphorylation sites able to modulate its Ca^{2+} affinity. Increasing evidence suggests that calmodulin is a key regulator of a large number of phosphatase and kinase enzymes that could be implicated in modulation of survival/cell death pathways. The regulated enzymes include extracellular-signal-regulated kinase (ERK)/mitogen-activated kinase (MAPK), effectors of the receptor "small GTYP-binding Ras," the latter involved in the fine regulation of cellular proliferation and differentiation. According to Cullen and Lockyer (2002), calmodulin participates in the regulation of transduction, as mediated by the Ras protein, via complexing to the GRF1 and GRF2 proteins (which induce the GDP–GTP exchange on the Ras surface) and to the CAPRI protein (which stimulates Ras to hydrolyze bound GTP). In turn, the Ras protein, in addition to transducing survival signals via its action on the system ERK/MAPK, can also activate the MEKK1 protein, whose truncated form stimulates, in turn, the proapoptotic kinase cascade JNK/SAPK. The latter induces the expression of Bak (a proapoptotic protein) while inactivating Bcl-2 and Bcl-x_L, twoantiapoptotic proteins residing onto the mitochondrial membrane (Mendic, 2003). The precise role of calmodulin in modulating cisplatin sensitivity remains to be defined, since calmodulin antagonists have been shown to revert cisplatin resistance and to increase cisplatin sensitivity in selected model systems (Kikuchi et al., 1990). Calumenin, another protein that is involved in regulation of cell Ca^{2+} levels, was modulated. This protein is inserted into the endoplasmic reticulum and contains seven, low-affinity, calcium-binding sites. Although its precise biological role is still uncertain, its presence exclusively in the resistant cells could suggest an important role on the onset of chemoresistance. The release of Ca^{2+} from the endoplasmic reticulum under stress conditions is a well-known mechanism for inducing apoptosis, based on the enzymatic activation of caspase 12, as mediated by calpain, a Ca^{2+}-dependent cysteine-protease (Mendic, 2003). Thus, it could be reasonable to assume that the much-increased levels in cisplatin-resistant cells could be responsible for calcium sequestering and therefore reduced apoptosis. A similar modulation of calumenin has been reported in head–neck carcinomas (Wu et al., 2002b).

Two important antiapoptotic proteins (Bcl-2 and 14-3-3) were also found to be strongly modulated in cisplatin-treated cells. The protein Bcl-2 forms homo- and heterodimers with other components of the same family (Bax, Bad, Bak, and Bcl-x_L) and with other species (e.g., Apaf-1, a constituent of the aptosome and Raf-1). Increased levels of Bcl-2 have also been associated with a marked decrement of Ca^{2+} concentration in the endoplasmic reticulum (Foyouzi-Youssefi et al., 2000). The reduction of Ca^{2+} levels could prevent cytochrome *c* release and caspase activation by the aptosome (Szalai et al., 1999).

2.5. SIGNAL PATHWAY PROFILING OF PROSTATE CANCER

Prostate cancer is the most commonly diagnosed noncutaneous malignancy in the United States and the second leading cause of cancer death (Howe et al., 2001). The prostate-specific antigen (PSA) test is currently the best overall serum marker for prostate cancer. Nevertheless, such a test lacks specificity (Djavan et al., 1999; Pannek and Partin, 1998), limiting its use as an early detection biomarker. Gene expression alone cannot determine the activation (phosphorylation) state of in vivo signal pathway checkpoints (Paweletz et al., 2001). Aberrations in the regulation of these pathways may be a key factor in carcinogenesis. Reverse-phase protein arrays (Paweletz et al., 2001) allow the study of the dynamic proteome of human cancer. This approach is further enhanced when used in conjunction with LCM. Such a combination facilitates the examination of the relative states of several key phosphorylation checkpoints in pathways involved in prosurvival, mitogenic, apoptotic, and growth regulation pathways involved in the progression from normal prostate epithelium to invasive prostate cancer. Although immunohistochemistry can provide indications on the abundance of cellular protein antigens in tissue, it fails to detect subtle quantitative changes in multiple classes of proteins taking place simultaneously within an individual cell type. Such changes may have a role to play in the slow progression of precancerous lesions over a number of years. In particular, changes in the activation status of signal pathway circuits that regulate downstream cell cycle progression and prosurvival can generate an imbalance, which ultimately results in the loss of cell growth control and the net accumulation of neoplastic cells. Small reductions in the apoptotic rate, which normally balances cell death against the cell population birth rate, will cause cell population growth. Growth is further stimulated if progression through the cell cycle is perpetuated rather than being checked as it might be during differentiation. In contrast to previous antibody, ligand, or heterogeneous tissue fragment arrays, which mostly incorporate the use of single probes, the reverse-phase protein array immobilizes whole-protein lysates from histopathologically relevant cell populations procured by LCM. The array is designed to capture various stages of microscopic progressing cancer lesions within individual patients. Furthermore, each patient set is arrayed in miniature dilution curves to facilitate accurate quantification and enlarge the dynamic range. Paweletz et al. (2001) applied this microarray to analyze, in human tissue, the state of checkpoints for prosurvival and growth regulation at the transition from histologically normal epithelium to prostate intraepithelial neoplasia (PIN) and into prostate carcinoma at the invasion front.

Basically, histophathologically relevant cell populations are microdissected and lysed in a suitable lysing buffer, and approximately 3 nL of that lysate is arrayed with a pin-based microarrayer onto glass-backed nitrocellulose slides at defined positions. These applications result in 250–350-μm-wide spots each containing the whole cellular repertoire corresponding to a given pathological state that has been captured. Subsequently, each slide can be probed with an antibody that can be detected by fluorescent, calorimetric, or chemiluminescent assays. Over 1000 individual cellular lysates can be accommodated on a 20 × 30-mm slide with 1 μm of lysate.

The main deductions drawn by Paweletz et al. (2001) were as follows. First, an increase in the ratio of phosphorylated Akt to total Akt will suppress downstream apoptosis pathways through intermediate substrates such as GSK3. Additionally, reduction in apoptosis will shift the balance of cell birth and death rates favoring accumulation of cells within the prostate gland. Piling up of cells within the gland is a major pathological hallmark of PIN. Prosurvival is required for migrating cells to resist the proapoptotic signals which take place during the disruption of integrin-mediated adhesion to extracellular matrix molecules. In parallel, transient ERK activation and augmentation of prosurvival pathways may be associated with cellular migration. Activation of Akt, a substrate of phosphatidylinositol 3-kinase (PI3K) can therefore promote cell motility and survival as the invading cancer cells leave the gland and invade the stroma. In a more recent study the same group (Grubb et al., 2003) have used multiplexed reverse-phase protein microarrays coupled with LCM to study the states of signaling changes during disease progression from normal prostate epithelium to invasive prostate cancer. Focused analysis of phosphospecific endpoints revealed changes in cellular signaling events through disease progression and between patients. Furthermore, molecular pathways believed to be relevant for cell survival and progression from normal to epithelium to invasive carcinoma were assessed directly from human tissue specimens. The results obtained from this study have substantiated and expanded deductions based on an earlier study (Paweletz et al., 2001) in which a similar technology was used.

Recently, reverse-phase protein microarrays have been used to examine signal pathways of ovarian cancer from human tissue specimens (Wulfkuhle et al., 2003).

The majority of cases related to ovarian cancer are diagnosed at an advanced stage, and current treatment modalities do not provide significant hope for curing the disease. Altered expression or function of a number of kinases and phosphatases has been reported earlier and is thought to participate in the development and progression of ovarian cancer. For example, the PI3K pathway, associated with cell survival, is activated in a significant measure in a number of ovarian cancers and is implicated in the growth and invasion of tumor (Hu et al., 2000; Shayesteh et al., 1999; Philp et al., 2001). Genes encoding subunits of PI3K and its downstream target, AKt2, are amplified in primary ovarian tumors and overexpression of the AKt2 kinase is associated with aggressive, advanced-stage tumors (Bellacosa et al., 1995). Wulfkuhle et al. (2003) have used reverse-phase protein array technology coupled with LCM and phosphospecific antibodies to examine the activation status of several key molecular "gates" involved in cell survival and proliferation signaling in human ovarian tumor tissue. This study resulted in a number of relevant deductions which are worth summarizing here: The levels of activated ERK1/2 varied considerably in tumors of the same hystotype, but no significant differences between hystotypes were observed. The same study showed that advanced-stage tumors had slightly higher levels of phosphorylated ERK1/2 compared to early-stage tumors. The activation status of AKt and glycogen synthase kinase 3β, key proteins and indicators of the state of the PI3K/AKt prosurvival pathway, also showed some variation within each histotype than between the hystotypes studied.

2.6. EMERGING ROLE OF FUNCTIONAL AND ACTIVITY-BASED PROTEOMICS IN DISEASE UNDERSTANDING

At the biochemical level, proteins rarely act alone; rather they interact with other proteins to perform a given cellular task. Although data obtained by various expression proteomics strategies have functional relevance by detecting changes in protein abundance, such measurements offer only an indirect readout of dynamics in protein activity. This means that numerous postranslational forms of protein regulation, including those governed by protein–protein interactions, remain undetected. Technologies which may address such limitations will no doubt provide a new dimension to the way the data related to proteins expression are interpreted today. The recent impressive growth of genomic data highlighted the urgent need for a systematic and high-throughput proteomic approach to decipher protein signaling and protein–protein interactions.

Currently, the most popular method for detecting protein–protein interactions is the yeast two-hybrid approach (Ito et al., 2001; Uetz et al., 2000). Interaction mapping using this approach has now been applied to several organisms, including two large-scale efforts for the budding yeast *S. cerevisiae*. In a yeast two-hybrid assay, pairs of proteins to be tested for interaction are expressed as fusion proteins ("hybrids") in the yeast: One protein is fused to a DNA-binding domain, the other to a transcriptional activator domain. Any interaction between them is detected by the formation of a functional transcription factor. This pioneering approach has a number of advantages and drawbacks. Regarding the first aspect, it is an in vivo technique; it is independent of endogenous protein expression, it can detect transient and unstable interactions, and it has a good resolution to allow interaction mapping. The drawbacks are that only two proteins are tested at a time, which means no cooperative binding is included, and it takes place in the nucleus, which means that many proteins are not in their native compartment (von Mering et al., 2002).

Another approach to probe protein–protein interactions uses affinity purification of proteins followed by MS characterization. This method was used by Gavin et al. (2002) and Ho et al. (2002) for the identification of a number of protein complexes from *S. cerevisiae.* These investigations were able to provide relevant information regarding multiprotein assemblies within the cell that has not been identified by earlier two-hybrid analyses. Like the yeast two-hybrid data sets, a comparison of the data obtained from the two MS-based studies also showed relatively poor agreement between them; furthermore, little overlap between the yeast two-hybrid interaction data sets and the MS data sets was observed (Bader, 2001).

Zhu et al. (2001) have provided a convincing demonstration that proteome microarrays can be used to screen for protein–protein interactions. The authors reported that 39 calmodulin-interacting proteins were identified by adding biotinylated calmodulin to a chip printed with the majority of the yeast proteome. Within this set, several expected calmodulin-binding proteins as well as 33 novel calmodulin-interacting proteins were revealed. In a single experiment, a new consensus binding site for calmodulin was defined, an achievement that would not have been possible without the comprehensive screening enabled by the proteome microarray. Several of the interactions with calmodulin were missed in the large-scale yeast two-hybrid and/or affinity purification–MS studies, including the well-known interaction of calmodulin with calmodulin kinase.

Another potentially general method to monitor protein–protein interactions is based on the use of fluorescence resonance energy transfer (FRET). Interaction between two proteins can be imaged by detecting FRET between donor and acceptor fluorescent tags attached to the interacting proteins, which takes place nonradiatively (Wouters et al., 2001). The FRET approach has two attractive characteristics. First, it can be used to perform measurements in living cells, which allows the detection of protein interactions at the location in the cell where they normally occur, in the presence of the normal cellular milieu. This has been demonstrated for the binding of Grb2 to activated epidermal growth factor receptors (Sorkin et al., 2000) and the hormone-induced binding of coactivator proteins to nuclear receptors (Llopis et al., 2000). Second, transient interactions can be followed at high temporal resolution in single cells. One potential application of this technology is the investigation of a given class of posttranslational modifications.

One of the early methods to be used for protein complex purification and subsequent identification was described by Rigaut et al. (1999) and designated as a tandem affinity purification (TAP) tag. Basically, the TAP method involves the fusion of the tag to the target protein and the introduction of the construct into the host cell or organism, maintaining the expression of the fusion protein at or close to its natural level. The fusion protein and associated components are then recovered from cell extracts by affinity selection on an immunoglobalin G (IgG) matrix. Following washing, TEV protease is added to release the bound material. The elute is incubated with calmodulin-coated beads in the presence of calcium. This second affinity step is required to remove the TEV protease as well as any traces of contaminants remaining after the first affinity selection. Other strategies for protein complex purification and MS characterization have been described in other parts of this chapter.

So far such strategies have been applied to cancer investigations in a limited fashion. However, more recently, there have been intense activities that underline the potential of probing functional interaction and activity-based protein profiling in understanding various diseases. Chemical approaches for functional probing of proteome is a flourishing area which demonstrated its validity through various applications.

2.7. ACTIVITY-BASED PROTEIN PROFILING

The ability to profile classes of proteins on the basis of changes in their activity would greatly contribute to both a reliable assignment of protein function and a possible identification of potential pharmaceutical targets. Liu et al. (1999) described the chemical synthesis and application of an active-site directed chemical probe for visualizing dynamics in the expression and function of an entire enzyme family, the serine hydrolases. Biotinylated fluorophosphonate (FP-biotin) probes were reacted with crude tissue extracts and the pattern of expression of these enzymes was monitored. In a latter work (Jassani et al., 2002) the same approach was applied to quantitatively compare enzyme activities across a panel of human breast and melanoma cancer cell lines. The authors reported that a comprehensive analysis of the activity, subcellular distribution, and glycosylation state for the serine hydrolase superfamily resulted in the identification of a cluster of proteases, lipases, and esterases that distinguished cancer lines based on tissue of origin. This study revealed a number of relevant features regarding the expression and interactions of this family of enzymes. For example, nearly all of these enzyme activities were down regulated in the most invasive cancer lines examined, which instead up regulated a distinct set of secreted and membrane-associated enzyme activities. These invasive-associated enzymes included urokinase, a secreted serine protease with a recognized role in tumor progression, and a membrane-associated hydrolase KIAA 1363, for which no previous link to cancer had been made. Based on these and other experimental observations within the same study, the authors suggested that the invasive cancer cells examined share discrete proteomic signatures that are more representative of their biological phenotype than cellular heritage, highlighting the possibility that a common set of enzymes may sustain the progression of tumors from a variety of origins and thus represent an attractive route for the diagnosis and treatment of cancer. There is no doubt that these data represent a valuable source of information which, when considered with data obtained by other approaches, will indeed increase our capability in the search for reliable methods for cancer diagnosis and treatment. Having said that, further data involving a higher number of samples from different origins will be needed before a realistic assessment of such data can be proposed.

In another variation of the chemical probe approach, G.C. Adam et al. (2002) described the use of trifunctional chemical probes for the detection and identification of enzyme activities in complex proteomes. The trifunctional probe contains three components: The first component is for target detection (e.g., rhodamine), the second for target isolation (e.g., biotin), and the third is a sulfonate ester reactive group, permitting visualization and affinity purification of labeled proteins by a combination of in-gel fluorescence and avidin chromatography. This strategy was used to identify several protein targets, including the integral membrane enzyme 3β-hydroxysteroid dehydrogenase/Δ5 isomerase and the cofactor-dependent enzymes platelet type phosphofructokinase and type II tissue transglutaminase. The authors reported that the latter two enzymes showed significantly up-regulated activities in the invasive estrogen receptor negative [ER(−)] human breast cancer cell line mda-mb-231 compared to its noninvasive ER(+) breast cancer lines MCF7 and T-47D.

2.8. PROBING PROTEIN FUNCTIONS USING CHROMOPHORE-ASSISTED LASER INACTIVATION

Chromophore-assisted laser inactivation (CALI) can be described as an approach for targeted functional proteomics. This technique uses a combination of dye-coupled ligands and laser irradiation to specifically inactivate the function(s) of targeted proteins. It was developed as a method to determine the in situ function of proteins in cellular processes (Jay, 1988). Basically, a dye molecule such as malachite green is coupled to a chosen ligand (e.g., antibody) which in turn binds to the protein of interest. The complex is irradiated with a laser light. This light is not absorbed readily by cellular components and, hence, they are not damaged. However, this laser light is efficiently absorbed by the dye molecule, which in turn generates hydroxyl radicals that selectively inactivate the protein bound by the dye-labeled ligand (Liao et al., 1994). Unbound dyes do not cause significant damage because the effects of CALI are spatially restricted to within 15 Å (Linden et al., 1992). Figure 2.8 depicts the principle of CALI. There are two settings of the CALI approach: macro-CALI, which addresses the total contents of a microtiter plate, and micro-CALI, in which a microscope is used to focus the laser beam onto a single cell or to subcellular compartments (Ilag et al., 2000).

Chromophore-assisted laser inactivation is highly suitable for addressing in situ function of cellular proteins. This approach provides an excellent temporal and spatial resolution in functional inactivation since the loss of function commonly occurs when and where the laser is applied. For micro-CALI the irradiated area may be within part of a cell, such that CALI-treated regions may be compared with the nonirradiated regions of the same cell. The CALI approach has the potential to be applied to address a protein's role in a specific cultured human cell type that is affected by a given disease. Therefore, it is unnecessary to extrapolate the protein's role from genetic deletion experiments performed in other species.

Like any other technique, CALI has drawbacks, some of which are considered here: Although CALI can demonstrate a correlation between the acute loss of function and a change in cellular processes, it does not provide an evidence for a direct link. This approach may not inactivate the protein(s) in situations where the distance between

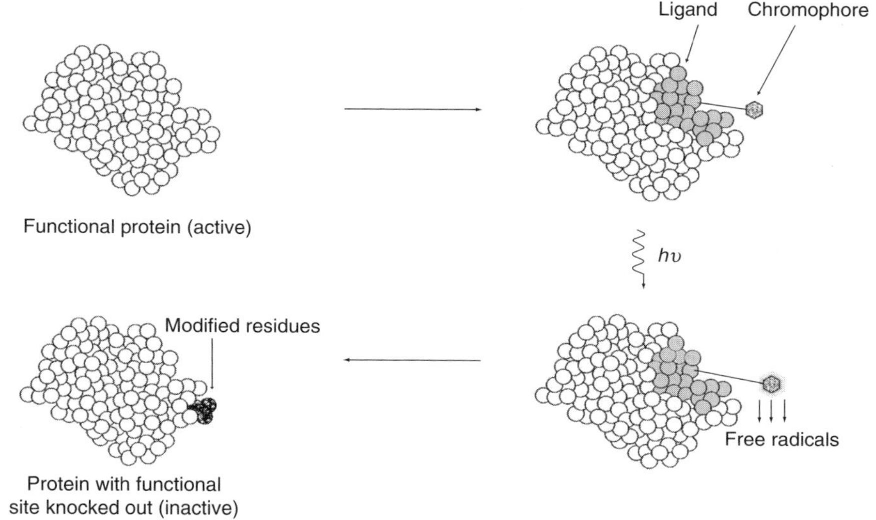

FIGURE 2.8. Basic principle of CALI. (Adapted from Ilag et al., *Drug Devel. Res.*, 2000, with permission.)

the antibody binding site and the domains required for the function of the protein in question is too large or where the reaction of such domains to hydroxyl radical damage is too weak. One further drawback of this technique is associated with the lifetime of the inactivation; in other words, the loss of function is temporary and cannot be compared with the permanent loss of function generated, for example, by genetic knockout technology. More detailed descriptions of these and other drawbacks have been given by Ilag et al. (2000).

Another variation of the CALI, termed fluorophore-assisted light inactivation (FALI), has been described by Beck et al. (2002). This method uses both coherent (laser) and diffuse (ordinary lamp) light sources to target fluorescein-labeled probes to inactivate specific proteins. Unlike CALI, which relies on hydroxyl-mediated damage, FALI relies on singlet oxygen as one of the major components for achieving protein inactivation. Other differences include the half maximal radius of damage, 15 Å for CALI and about 40 Å for FALI, and that CALI uses malachite green isothiocyanate as a coupling dye, while FALI uses fluorescein.

2.9. ROLE OF PROTEIN–TYROSINE KINASES

Protein–tyrosine kinases play a central role in cellular signal transduction, regulating many activities of direct relevance to a host of diseases, including various forms of cancer. Inappropriate activation of tyrosine kinase signaling by mutation, overexpression, or chromosomal rearrangement is often reported in human cancers.

Accordingly, there is a considerable interest in proteomic efforts to profile and identify tyrosine-phosphorylated proteins under various physiological conditions. The approval of ST1571 (Novartis, Basil, Switzerland) by the U.S. Food and Drug Administration for the treatment of chronic myeloid leukemia has been heralded as a landmark achievement in drug development (Fletcher, 2001). Such approval has dispelled a long-held myth that it was not feasible to develop selective inhibitors of key cell-signaling molecules as safe and effective medicines. Of course, it is too early to tell whether such a single success is the start of a new era in drug development or simply an isolated case. There is no doubt that the abundance and ubiquity of kinase targets and the incomplete understanding of their role in various cellular events still represent a number of formidable challenges which have to be tackled before a complete understanding of such events can be achieved.

There is the expectation that functional profiling of the tyrosine phosphoproteome is likely to lead to the identification of novel targets for drug discovery and provide the basis for novel molecular diagnostic approaches. This is by no means an easy task given that protein phosphorylation is one of the most prevalent mechanisms for covalent modifications, reflected in as many as one-third of eukaryotic gene products. Conventional techniques of site identification by MS or ^{32}P-labeling coupled with Edman sequencing have proved to be inefficient when dealing with highly complex mixtures of proteins or peptides and still require relatively pure starting samples. Currently there are various efforts to address the difficult task of the functional profiling of phosphoproteome; a brief description of these methods is given below.

Mass spectrometry–based methods aimed at comprehensive characterization of the phosphoproteome have been described by Zhou et al. (2001) and Oda et al. (2001). The first strategy starts with a proteolic digest that has been reduced and alkylated to neutralize possible reactivity by the –SH groups of the Cys residues. Following N– and –C-termini protection, phosphoramidate adducts at phosphorylated residues are formed by carbodiimide condensation with cystamine. The free sulfhydryl groups produced from this step are covalently captured onto glass beads coupled to iodoacetic acid. This step is then followed by elution with trifluoroacetic acid, which regenerates phosphopeptides for MS analysis. When tested on yeast extracts, this strategy showed that 80% of the recovered peptides was phosphorylated.

The strategy described by Oda and Chait (2001) starts with a protein mixture in which Cys reactivity is removed by oxidation with performic acid. Base hydrolysis is then used to induce β-elimination of phosphate from phosphoserine and phosphothreonine followed by the addition of ethanedithiol to the alkene. The resulting free sulfhydryls are coupled to biotin, allowing purification of phosphoproteins by avidin affinity chromatography. Following elution of the captured phosphoproteins and proteolysis, enrichment of phosphopeptides is conducted by performing a second round of avidin purification. Both approaches have been tested against yeast extracts, where the detected proteins were among the most highly expressed. This means that both approaches have to be improved and refined before they reach the point to be considered valid methods for the investigation of global protein phosphorylation in cells. One major challenge for both approaches is the capability of routine detection of tyrosine-phosphorylated proteins, which are generally present at low abundance.

Understanding protein–protein interactions associated with protein–tyrosine kinases is one way of gaining more specific information on their role in cellular signal transduction and their contribution to various regulatory activities relevant to certain diseases. It is well recognized that nonreceptor tyrosine kinases such as Src bind via their Src homology (SH)[1]3 and SH2 protein binding domains to a large number of substrates, including cytoskeletal proteins, enzymes, and adaptor molecules (Shokat., 1995; Pellicena and Miller, 2001, 2002). This stable interaction with substrates is important for the function of these kinases because of the relatively weak enzymatic activity and modest substrate specificity of their catalytic domains (Pellicena and Miller, 2002; Songyang et al., 1995). It is also worth noting that many substrates of normal and transforming tyrosine kinases have been identified; however, in most cases the relative contribution of the individual substrates to a given biological event remains difficult to identify. Such difficulty derives from the fact that kinases generally phosphorylate many different substrates upon activation, and no experimental approach is currently available to induce phosphorylation of a single substrate without a simultaneous phosphorylation of others. Two recent articles (Fujirawa et al., 2002; Sharma et al., 2003) have described a strategy, termed *functional interaction trap* (FIT), for forcing pairwise interaction of two proteins in vivo as a means to elucidate their functional interaction. The authors argued that such a strategy could be adapted to induce specific substrate phosphorylation. The strategy is based on the idea of replacing SH3 and SH2 domain-mediated interactions with an engineered, highly specific protein-binding interface, thereby forcing a substrate of choice to bind the kinase and thus promoting its efficient phosphorylation. It was hoped that the specificity of the reaction would be conferred by the engineered binding interface, thus allowing tyrosine phosphorylation of the protein of choice within the cell. The authors demonstrated the validity of this approach on a model system, and therefore more real-life examples are needed before such an approach can be rationally evaluated.

There are of course other approaches for the profiling of phosphorylation, including immobilized metal affinity chromatography (Anderson and Porath, 1986) and far Western blotting (Nollau and Mayer, 2001; Machida et al., 2003).

2.10. CONCLUDING REMARKS AND FUTURE PROSPECTS

The completion of the draft sequence of the human genome by the public consortium (Lander et al., 2001) and a commercial entity (Venter et al., 2001) revealed a lower than expected number of genes. Such a lower number has generated a certain degree of optimism regarding future efforts to determine the function(s) of these genes. This optimism was quickly dampened by the emerging evidence that future challenges will be the identification of functions associated with gene products, in other words proteins, which represent the main executors of biological functions and have been estimated to be magnitudes greater in number and complexity than genes.

The authors realize that even a chapter dedicated to the various attempts to decipher the complexity of cancer cannot give a fair and unbiased overview of the immense proteomic activities which are currently employed in this particular area. Such admission of guilt has a number of good reasons. First, the current rate at which data are generated and the variation in methods employed to obtain them are such as to frustrate attempts at giving an updated and complete account of such activities. Second, to give more space to comments and observations which may interest a wider audience, we have limited the examples chosen to a recent and fairly narrow time window, thus illustrating the various technical approaches and underlining the type of information gleaned from such studies. The alternative would have been a longer list of examples, which would have given the presented material a repetitive character. Third, given the close interaction between proteomics and genomics, it was inevitable to make some sort of comparison between the information obtained by the two approaches. Despite the limited number of examples considered, it is evident that, regardless of the technological approach, the main objective of these studies, whether proteomic or genomic, remains the identification of reliable biomarkers which can be used for the early diagnosis of the disease and for the eventual design of efficient therapeutic targets. Careful consideration of the presented examples and the technology behind them reveals that at the present time there is not a single technological platform which can tackle the complexity of such a task. In the rest of this chapter an attempt is made to highlight the current contribution of various proteomic approaches to disease understanding and the future role of what is considered today as emerging technologies.

There is no doubt that, over the last few years, genomic research has revolutionized our conception of disease and drug discovery. A convincing example of such contribution is the use of DNA microarrays to probe the expression of thousands of genes and even the entire gene repertoire of a single organism. There are many examples in the literature demonstrating the value of these whole-genome arrays for identifying genes that can experience up and down regulation during development and in response to environmental stimuli (Horak and Snyder, 2002) and in several disease states (Nutt et al., 2003; Ramaswamy et al., 2001). Having said that, approaches based on DNA microarrays suffer from a number of drawbacks which necessitate the complement of proteomic approaches. Existing studies have highlighted two of these drawbacks. First, without performing a great deal of further experiments, it is quite difficult to assess the biological significance of hundreds of differentially expressed genes observed in a single measurement. The main reason behind such difficulty is that the measured relative variation in the RNA level between two or more states may only be indirectly related to one or more proteins and in some cases such relative change is not at all related (Gygi et al., 1999). Second, the data obtained by DNA microarrays do not provide direct information about the function(s) of the gene products or how they might interact with one another.

The emerging needs of drug discovery started to place a greater emphasis on the development of molecular-targeted therapeutics and therefore on the need for a precise characterization of the activity status of pathways and targets of interest in individual tumors beyond levels of gene amplification and total protein expression. Gene expression alone cannot determine the activation (phosphorylation) state of in vivo signal pathways (Williams et al., 1997). Aberrations in the regulation of these pathways may play a key role in carcinogenesis. Shortcomings associated with DNA microarray technology have put in focus the need for proteomic approaches to provide further information to tackle what can be described as a complex scenario.

On the proteomic side we have highlighted a number of techniques currently employed in the search for protein biomarkers associated with various forms of cancer. At this stage, it is natural to draw some conclusions and to underline the current achievements and future potential of some of the approaches described in this chapter.

A recent article by Celis and Gromov (2003) is a good starting point for such an attempt. These authors argued that, while some progress has been made in disease proteomics, the field is still in its infancy. Knowledge of the sequence of the human genome has provided a framework for genomic approaches to unravel disease processes. Similar knowledge of the human proteome is currently lacking. Developing a comprehensive knowledge framework of the proteome is considerably more complex than sequencing the human genome. Ideally, such a proteome framework would encompass knowledge of all human proteins, from their sequence to their post translational modifications, to their interactions among each other, to their cellular and subcellular distribution, and to their temporal pattern of expression. Although such an exhaustive framework will not materialize in the foreseeable future, more modest goals may well be within reach.

1. A technology described in this chapter that merits further consideration is SELDI. This approach was introduced over 10 years ago, yet its contribution to the field of cancer research, and in particular to the search for new markers, has only started a couple of years ago. The sudden explosion of this approach on the proteomic scene has attracted both enthusiasm and skepticism; before attempting to rationalize such contradicting views, it is reasonable to ask where does the impetus behind this approach derive from? A partial answer to such question is twofold. First, the approach is robust and highly amenable to automation and handles crude biological samples. Second, and probably a more influential factor which pushed SELDI into the forefront of methods in the search for new biomarkers, is the realization that the pipeline of approval of new markers is drying up (Anderson and Anderson, 2002; Ward and Henderson, 1996). At this point it is useful to remind ourselves that the SELDI approach goes against many of the current trends in proteomics, where the name of the game is not only the identification of thousands of proteins in a single measurement but also the identification of their posttranslational modifications and interaction between them. In effect, what SELDI provides is a simple low-resolution pattern based on measured m/z values of various proteins within a mixture. If that is the case, then how can we justify the current vigorous investments in the technique? This question is by no means a veiled criticism of the approach; on the contrary, it is meant to stimulate a further and deeper look at what the technique is delivering now and what an improved SELDI can deliver in the future. Among the recent debates regarding this particular argument, we would like to cite a recent exchange of opinion among scientists who use SELDI as well as other approaches in the search for new disease markers, which hopefully can be translated into clinical practice (Diamandis, 2003; Petricoin and Liotta, 2003) In his paper Diamandis (2003) compared three

recent SELDI reports on prostate cancer. According to the author, the first report by Adam et al. (2002) claimed 83% sensitivity at 97% specificity for prostate cancer detection whereas Petricoin et al. (2002b) claimed 95% sensitivity at 78% to 83% specificity and Qu et al. (2002) reported 97% to 100% sensitivity and 97% to 100% specificity. Considering that at a similar sensitivity the PSA has a specificity of ~25%, Diamandis described all three SELDI results as "impressive"; at the same time the author raised a number of legitimate queries regarding these results. Some of these queries are also shared by the authors of this book, and therefore, we find it relevant to give more space to what Diamandis had to say regarding these measurements: First, Diamandis expressed surprise at the similar outcome between the three studies in terms of specificity and sensitivity, although they have been conducted using different chips and monitoring different m/z values. Second, B.-L. Adam et al. (2002a) and Qu et al. (2002) used the same type of chip for serum extraction and the same MS instrument, yet only two out of nine monitored m/z values were the same in the two works. One of these peaks (m/z 7820) was described by both B.-L. Adam et al. (2002a) and Qu et al. (2002) as distinguishing between healthy and patients with benign prostatic hyperplasia but not between cancer and noncancer patients. Third, Diamandis (2003) raised a question regarding the sensitivity of SELDI when dealing with low-level markers which happen to be present in a medium such as human serum, where a wide range of high-abundance noninformative molecules will surely be present. The author argued that current knowledge still indicates that the best distinguishing serum protein for patients with prostate cancer is PSA. The free antigen has $M_r \sim 27775$ Da (Bedzyk et al., 1998), which puts it within the m/z range of the current SELDI analysis. In the three studies cited by Diamandis, PSA was not identified as a distinguishing molecule. An earlier SELDI MS study by Wright et al. (1999) reported the detection of both free and complexed PSA in various biological fluids and tissue extracts, including seminal plasma, prostatic extracts, and serum. However, the authors qualified these findings by admitting that there was not the certainty that the assigned masses are actually associated with PSA or with other molecules which happen to have the same molecular mass of the free/complexed antigen. Based on these observations, Diamandis (2003) asked why both B.-L. Adam et al. (2002a) and Qu et al. (2002) could not identify PSA as a distinguishing molecule? Without going to further details, the bottom line of Diamandis's argument is this: The SELDI analysis cited by the author may have yielded molecules present at concentrations manyfold higher (milligrams per liter) than the classic cancer biomarkers known to be present at concentrations of micrograms per liter. A statement of this kind is of course difficult to verify one way or another for two simple reasons. First, currently there are no published works on a systematic and careful assessment of SELDI sensitivity and reproducibility when dealing with human serum samples. Second, the types of "bait" used to cover chip surfaces are still in evolutionary phase, which makes it difficult to point out the ideal bait for SELDI analyses dedicated to the search for so-far-unknown biomarkes. Last, but certainly not the least, is the question of detecting low-abundance components in a medium containing a number of high-abundance species. Existing literature makes it clear that whether we are dealing with ionization efficiency or separation efficiency (gel, column, or capillary) the influence of dominant components in a mixture can never be easy to overcome. In fact, the whole idea behind coupling various separation methods to MS is to reduce the influence of abundant components, which in turn will favor the detection of less abundant components. The same thing can be said for 2DE, where various approaches of prefractionation and more frequent use of narrow pH gradients are normally employed to favor the detection of low-abundance proteins. Given the current speed at which SELDI is developing, we do not exclude that, in the near future, SELDI analysis will incorporate some sort of prefractionation together with the use of chip surfaces which target specific molecular entities.

The above comments by Diamandis (2003) have attracted a response from Patricoin and Liotta (2003), who were in partial agreement with these statements and made a number of suggestions which could improve the reliability and reproducibility of the current SELDI approach. In response to Diamandis's concerns regarding poor reproducibility, sensitivity, and the questionable specificity of the biomarker patterns detected by SELDI compared to traditional clinical tests, Petricoin and Liotta (2003) advanced the following arguments. First, because cancer cells themselves are deranged host cells, it is unlikely that we will find a true cancer-specific molecule. On the other hand, the complex proteomic signature of the tumor's host microenvironment may be unique and may constitute a biomarker amplification cascade. The specificity of this microenvironment may be mirrored by a catalog of low-molecular-mass proteins and peptides, including specifically cleaved or otherwise modified proteins produced in sufficient abundance to be detected by current MS platforms. Second, sources of variability and potential bias could arise from differences at one clinic and between different clinics. To reduce this bias and variations, the authors proposed a standard operating procedure which encompasses quality control, quality assurance for sample collection, handling, and shipping, a procedure which the authors claim is already functioning in their laboratories. Third, the authors agree that current low-resolution SELDI–MS platforms are not capable of delivering the kind of accuracy and reproducibilty required for clinical testing. The authors reported that high-resolution hybrid Q-TOF instruments are already in use in their respective laboratories.

By considering the above comments and others in the current literature, it is not difficult to deduce that SELDI is an emerging technology which has the potential to provide a substantial contibution to attempts to discover specific and reliable biomarkers for various forms of cancer. However, before such potential can be translated into routine clinical practice, a number of steps have to be implemented. First, on the instrumentation side, we need to employ high-resolution, high-sensitivity MS instruments and to improve and extend the capabilities of current baits used for protein extraction from clinical samples. Second, and possibily more urgent, is the need to establish a rigorous and standarized procedure for sample collection, handling, and analysis. In the absence of translational SELDI research, such a standarized procedure will be difficult to achieve at the present. Given the potential of this technique to become a worldwide clinical tool for marker discovery and validation, a closer involvement of international organizations in shaping up standarized procedures is a necessity rather than an option. We are seeing an early attempt along these lines, where the World Health Organization in partnership with the American Red Cross, the Food and Drug Administration, and NIST is developing serum/plasma reference standards (Petricoin and Liotta, 2003). It is evident that the availability and worldwide distribution of these standards is an important step toward more rigorous quality assurance and a more reliable calibration of the individual assays.

2. Protein interactions comprise another argument which we have touched upon in this chapter. We tried to highlight the emerging role of protein–protein interactions in current attempts to decipher the complexity of molecular basis which governs directly or indirectly various forms of diseases, particularly various forms of cancer. Before giving further thought to this central argument, we ought to remind ourselves that the examples and references presented in the chapter have underlined two relevant aspects. First, at the biochemical level, proteins rarely act alone; rather they interact with other proteins to perform various cellular tasks (Alberts,1998). Second, protein–protein interaction maps have the potential to reveal many aspects so far unknown about the complex regulation network, which in turn will provide a better understanding of various diseases.

We know, now, that the total number of human genes does not differ substantially from the number of genes of the nematode worm *Caenorhabditis elegans*, which implies that the complexity of the proteomic world is partly associated with the contextual combination of the gene products (mainly proteins) and their interactions (Lander et al., 2001). The importance of the in vitro probing of protein–protein interactions using immunoprecipitation in conjunction with MS was demonstrated over 10 years ago (Zhao and Chait, 1994). The utility of this approach was illustrated in the discovery of new downstream regulated patterns for the epidermal growth factor receptor (Pandey et al., 2000). Traditionally, protein interactions have been studied individually through the use of genetic, biochemical, and biophysical methods. The speed and output of these early methods could not be sustained in the face of an increasing number of newly discovered proteins, thus creating the need for high-throughput interaction–detection methods. In a recent excellent review von Mering et al. (2002) gave a comparative assessment of various strategies employed in the investigation of protein–protein interactions. The following observations are based on the contents of the same review. Careful examination of about 80,000 interactions between yeast proteins obtained by five high-throughput methods for detecting protein interactions revealed that less than one-third of this number was supported by more than one method. The scarce overlapping was attributed to a number of reasons, including false positives and/or the failure of one or more of these methods to detect certain types of interactions. Other groups have also cited both causes. For example, Gavin et al. (2002) and Ho et al. (2002), who used the affinity tag approach and MS to investigate protein complexes, reported a significant number of false-positive interactions as well as the failure to identify many known associations. Another aspect has been highlighted by von Mering et al. (2002) and may have an impact on both current and future strategies regarding protein interactions. These authors assessed the possible existence of bias toward a particular class of the detected proteins. Starting with statistics showing that none of the existing methods covers more than 60% of the proteins in the yeast model, the authors raised the question of whether there are common biases as to which proteins are covered. The authors identified three areas where the high-throughput interaction data are indeed biased. In the absence of genomewide measurements of protein abundance in yeast, the authors used mRNA levels as a crude substitute. A plot of interaction coverage versus mRNA abundance showed that most protein interaction data sets are heavily biased toward proteins of high abundance. However, the data sets associated with the two genetic approaches, two-hybrid and synthetic lethality, appear relatively unbiased. This is rather surprising if we consider that both approaches are largely independent of endogenous protein levels. Furthermore, both methods are capable of detecting transient or indirect reactions. The same authors deduced that the investigated data sets were biased toward particular cellular locations of interacting proteins, for example, toward mitochondial proteins in the case of the in silico predictions. Considering these and other analyses, von Mering et al. (2002) suggested that to increase interaction coverage and improve confidence in the detected/predicted protein interactions as many complementary methods as possible should be used.

At the present time it is reasonable to state that strategies for the investigation of protein interactions (Rigaut et al., 1999; Gavin et al., 2002; Ho et al., 2002; Uetz et al., 2000; Ito et al., 2001, Zhu et al., 2001) have so far generated a substantial volume of exciting data; however, setting aside some of the shortcomings of these strategies, such as false positives and negatives, there is still much to be done before we have a semicomprehensive knowledge of functional pathways within even a model organism such as yeast. In fact, recent publications indicate that efforts to achieve such objectives are already underway. In a recent paper Uetz and Pankratz (2004) commented that the protein interaction map for the fruit fly *Drosophila melanogaster* published by Giot et al. (2003) promises to facilitate functional analyses of many eukaryotic proteins. The latter paper contained two novel elements. First, prior to its publication, large-scale protein interaction maps were limited to two simple systems, yeasts (Uetz et al., 2000; Ho et al., 2001; Schwikowski et al., 2000) and bacteria (Rain et al., 2001). The work by Giot et al. (2003) is the first to describe a vast protein interaction map based on genomewide study of a multicellular organism. Second, another key element in the same study is the development of combined computational and statistical methodologies capable of gleaning the relevant protein–protein interactions partaining to the investigated multicellular organism. Building the protein interaction map of the fruit fly is the result of *a* tour de force of PCR to amplify all 14,000 predicted *D. melanogaster* open reading frames, of which more than 12,000 worked successfully. The present authors have no intention to go into all the details of this complex and pioneering effort by Giot et al. (2003); at the same time we recommend the consultation of the original work together with recent publications related to the argument of high-throughput protein interaction networks (Bader et al., 2004; Loppe and Holm, 2004).

3. *No smoking gun yet*. The data presented in this chapter were generated by a host of technologies ranging from the well-engrained 2DE to more recent reverse-phase protein microarrays. We can state at this point that, regardless of the age or the degree of sophistication of the technology employed, the main contribution which such data

provide is the aquisition of information on the modulation of proteins/genes which are eventually linked to one type or another of diseases. As far as the area of cancer research is concerned, in particular the discovery of new markers, we still do not have what is commonly termed "smoking gun" evidence for a reliable marker to be used clinically. Here we have to point out that the term "reliable" tends to have a meaning within a research enviroment different from that accepted in its clinical counterpart. For example, on more than one occasion in this brief chapter we have encountered the term "potential biomarker," which refers to certain protein/gene which has experienced substantial differentiation under certain disease conditions. Understandably, for this potential marker to become an accepted marker in clinical practice, a certain period of time and further experiments and tests are needed. This phase shift between markers at the research level and their translation to clinical practice cannot be considered the only parameter to assess the quality and the contribution of a given technology to the overall scenario of the search for biomarkers, particularly in cancer research. Although the question of reliable markers will always remain complex and less straightforward than other issues, the future contribution of emerging technologies which can provide comprehensive information on protein–protein interactions and signaling pathways should reduce the gap between research findings and the clinical adoption of such findings.

There is no doubt that biomarker discovery in cancer is one of the main objectives of current research activities. However, it is reasonable to hypothesize that another major task is to predict response patterns to existing therapies, which will in turn help to design more efficient and less damaging therapeutics. To achieve such tasks, we need predictions based on individual genetic factors that might influence parameters such as drug metabolism and the genetic makeup of a specific tumor. Microarray techniques that screen gene expression tumor patterns have given initial demonstration on their capability to distinguish subgroups with different clinical outcomes from a fairly large number of apparently clinically homogeneous groups of patients (Azadeh et al., 2000; Shipp et al., 2000). In this complex scenario, input from emerging proteomic approaches cannot be and should not be excluded, particularly if we want to have a more comprehensive view of such response patterns. In this particular area, the characterization and development of circuit maps of cellular signaling pathways in normal and cancerous cells are examples of such needed input. Relatively established proteomic approaches and emerging high-throughput strategies capable of providing precise information on protein identity, cell location, interaction with other proteins, possible modifications, and function(s) will no doubt also enhance present capabilities to decipher a wide range of complex aspects of cancers.

To end this chapter on an optimistic note, the authors remind the readers that many of the problems facing emerging proteomic technologies have been faced and almost resolved in the genomic field. A good example is DNA microarrays, where only a few years back such arrays were lacking comprehensive content, required expensive equipment, and had limited applications. As the utility of the technology for understanding biology became more evident, the demand for comprehensive, yet less expensive DNA arrays grew. Today, DNA microarray technology is an integral part in most laboratories engaged in functional genomics. An interesting parallel can be drawn between the development of DNA microarrays and their protein counterparts. Recent developments which have been briefly cited in this chapter have shown that a number of technical barriers holding back the technology are being surmounted. Comprehensive protein content has either been created or is in the process of being created, the production of whole-proteome microarrays has been put into practice, and a number of useful applications have been demonstrated. The authors are fairly confident that the next generation of protein microarrays and related technology is not far off, which will no doubt contribute to a better understanding of disease and hopefully provide us with the means to predict and cure various forms of cancer.

REFERENCES

Adam, B-L., Qu, Y., Davis, J. W., Ward, M. D., Clements, M. A., Cazares, L. H., Semmes, O. J., Schellhammer, P. F., Yasui, Y., Feng, Z., Wright, Jr., G. L. (2002a) *Cancer Res.* **62,** 3609.

Adam, G. C., Sorensent, E. J., Cravatt, B. F. (2002b) *Mol. Cell. Proteomics* **1.10**, 828.

Adam, L., Vadlamudi, R.K., McCrea, P., Kufyyamar, R. (2001) *J. Biol. Chem.* **276,** 28443.

Adam, P. J., Boyd, R., Tyson, K. L., Fletcher, G. C., Stamps, A. R., Hudson, L., Poyser, H. R., Redpath, N., Griffiths, M., Steers, G., Harris, A. L., Patel, S., Berry, J., Loader, J. A., Townsend, R. R., Daviet, L., Legrain, P., Parekh, R., Terrett, J. A. (2003) *J. Biol. Chem.* **278,** 6482.

Aitken, A. (1996) *Trends Cell Biol.* **6,** 341.

Alberts, B. (1998) *Cell* **92,** 291.

Albini, A., Iwamoto, Y., Kleinman, H. K., Martin, G. R., Aaronson, S. A., Kozlowski, J. M., McEwan, R. N. (1987) *Cancer Res.* **47,** 3239.

Alli, E., Bash-Babula, J., Yang, J-M., Hait, W. N. (2002) *Cancer Res.* **62**, 6864.

Anderson, N. L., Anderson, N. G. (2002) *Mol. Cell Proteomics* **1,** 845.

Andersson, L., Porath, J. (1986) *Anal. Biochem.* **154,** 250.

Andrews, P. A., Howell, S. B. (1990) *Cancer Cells* **2,** 35.

Andrews, P. A., Jones, J. A., Varki, N. M., Howell, S. B. (1990) *Cancer Commun* **2,** 93.

Asch, H. L. (1999) *Breast Cancer Res. Treat.* **55,** 177.

Azadeh, A. A., Eisen, M. B., Davis, R. E., Ma, C., Lossos, I. S., Rosenwald, A., Boldrick, J. C., Sabet, H., Tran, T., Yu, X., et al. (2000) *Nature* **403,** 503.

Babaian, R. J., Zhang, Z. (2001) *Mol. Urol.* **5,** 175.

Bader, G. D., (2001) *Nucleic Acids Res.* **29,** 242.

Bader, J. S., Chaudhuri, A., Rothberg, J. M., Chant, J. (2004) *Nature Biotechnol.* **22,** 178.

Ball, G., Milan, S., Holding, F., Allibone, R. O., Lowe, J., Ali, S., Li, G., McCardle, S., Ellis, I. O., Creaser, C., Rees, R. C. (2002) *Bioinformatics* **18,** 395.

Bankson, D. D., Kestin, M., Rifai, N. (1993) *Clin. Lab. Med.* **13,** 463.

Barraclough, R. (1998) *Biochim. Biophys. Acta* **1448,** 190.

Beck, S., Sakurai, T. A., Eustace, B. K., Beste, G., Schier, R., Rudert, F., Jay, D. G., (2002) *Proteomics.* **2,** 247.

Bedzyk, W. D., Larsen, B., Gutteridge, S., Ballas, R. A. (1998) *Biotecnol. Appl. Biochem.* **27,** 249.

Belinsky, S. A., Klinge, D. M., Stidley, C. A., Issa, J. P., Herman, J. G., March, T. H., Baylin, S. B. (2003) *Cancer Res.* **63,** 7089.

Bellacosa, A., De Feo, D., Godwin, A. K., Bell, D. W., et al. (1995) *Int. J. Cancer* **64,** 280.

Bender, C. M., Pao, M. M., Jones, P. A. (1998) *Cancer Res.* **58,** 95.

Berger, S. J. (1980) *J. Biol. Chem.* **255,** 3128.

Bini, L., Magi, B., Marzocchi, B., Arcuri, F., et al. (1997) *Electrophoresis* **18,** 2832.

Bischof, D., Pulford, K., Mason, D. Y., Morris, S. W. (1997) *Mol. Cell Biol.* **17,** 2312.

Blume-Jensen, P., Hunter, T. (2001) *Nature* **6835,** 355.

Buckly, C. D., Rainger, G. A., Bradfield, P. F., Nash, G. B., et al. (1998) *Mol. Membr. Biol.* **15,** 167.

Calkins, C. C., Sameni, M., Koblinski, J., Sloane, B. F., Moin, K. (1998) *J. Histochem. Cytochem.* **46,** 745.

Carlier, M. F., Pantaloni, D. (1997) *J. Cell Biol.* **269,** 459.

Carlier, M., Laurent, V., Santolini, J., Melki, R., Didry, D., Xia, G. X., Hong, Y., Chua, N. H., Pantaloni, D. (1997) *J. Cell Biol.* **136,** 1307.

Cassimeris, L. (2002) *Curr. Opin. Cell. Biol.* **14,** 18.

Castagna, A., Antonioli, P., Astner, H., Hamdan, M., Righetti, S. C., Perego, P., Zunino, F., Righetti, P. G. (2004) *Proteomics* **4,** 3246.

Cecconi, D., Astner, H., Donadelli, M., Palmieri, M., Missiaglia, E., Scarpa, A., Hamdan, M., Righetti, P. G. (2003a) *Electrophoresis* **24,** 4291.

Cecconi, D., Scarpa, A., Donadelli, M., Palmieri, M., Hamdan, M., Astner, H., Righetti, P. G. (2003b) *Electrophoresis* **24,** 1871.

Celis, J. E., Gromov, P. (2003) *Cancer Cell* **3,** 9.

Celis, J. E., Ostergaard, M., Basse, B., Celis, A., Lauridsen, J. B., Ratz, G. P., Anderson, I., Hein, B., Wolf, H., Orntoft, T. F., Rasmussen, H. H., (1996) *Cancer Res.* **56**. 4782.

Cerwenka, H., Aigner, R., Bacher, H., Werkgartner, G., el-Shabrawi, A., Quehenberger, F., Mischinger, H. J. (2001) *Anticancer Res.* **21,** 1471.

Chen, Y., Sun, R., Han, W., Zhang, Y., Song, Q., Di, C., Ma, D. (2001) *FEBS Lett.* **509,** 191.

Christa, L., Carnot, F., Simon, M. T., Levavasseur, F., Stinnakre, M. G., Lasserre, C., Thepot, D., Clement, B. (1996) *Am. J. Physiol.* **271,** G993.

Chung, S., Kim, M., Choi, W., Chung, J., Lee, K. (2000) *Cancer Lett.* **156,** 185.

Ciocca, D. R., Fuqua, S. A., Lock-Lim, S., Toft, D. O., Werlch, W. J., McGuire, W. L. (1992) *Cancer Res.* **52,** 3648.

Colombo, E., Marine, J. C., Danovi, D., Falini, B., Pelicci, P. G. (2002) *Nat. Cell Biol.* **4,** 529.

Colomer, R., Shamon, L. A., Tsai, M. S., Lupu, R. (2001) *Cancer Invest.* **19,** 49.

Creagh, E. M., Sheehan, D., Cotter, T. G., (2000) *Leukemia* **14,** 1161.

Crnogorac-Jurcevic, T., Efthimiou, E., Nielsen, T., Loader, J., Terris, B., Stamp, G., Baron, A., Scarpa, A., Lemoine, N. R. (2002) *Oncogene* **21,** 4587.

Crul, M., van Waardenburg, R. C. A. M., Beijnen, J. H., Schellens, J. H. M. (2002) *Cancer Treat. Rev.* **28,** 291.

Cullen, P. J., Lockyer, P. J. (2002) *Nature Rev.* **3,** 339.

de Coupade, C., Gillet, R., Bennoun, M., Briand, P., Russo-Marie, F., Solito, E. (2000) *Hepatology* **31,** 371.

Diamandis, E. P., *Clin. Chem.* **49,** 1272 (2003).

Djavan, B., Zlotta, A., Kratzik, C., Remzi, M., Seitz, C., Schulman, C. C., Marberger, M. (1999) *Urology* **54,** 517.

Easterman, A. (1991) *Cancer Treat. Res.* **57,** 233.

Easterman A., Schutte, N. (1988) *Biochemistry* **27,** 4730.

Ebert, E., Werte, B., Julke, B., Kopitaer-Jerala, N., Kos, J., Lah, T., Abrahamson, M., Spiess, E., Ebert, W. (1997) *Adv. Exptl. Med. Biol.* **421,** 259.

Ebralidze, A., Tulchinsky, E., Grigorian, M., Afanasyeva, A., Senin, V., Revazova, E., Lkanidin, E. (1989) *Genes Dev.* **3,** 1086.

El-Rifai, W., Frierson, Jr., H. F., Harper, J. C., Powell, S. M., Knuutila, S. (2001) *Int. J. Cancer.* **92,** 832.

Emmert-Buck, M. R., Bonner, R. F., Smith, P. D., Chuaqui, R. F., Zhuang, Z., Goldstein, S. R., Weiss, R. A., Liotta, L. A. (1996) *Science* **274,** 998.

Esteller, M. (2002) *Oncogene* **21,** 5427.

Feldner, J. C., Brandt, B. H. (2002) *Expt. Cell Res.* **272,** 93.

Fend, F., Emmert-Buck, M. R., Chuaqui, R., Cole, K., Lee, J., Liotta, L. A., Raffeld, M. (1999) *Am. J. Pathol.* **154,** 61.

Ferguson, A. J., Evron, E., Umbricht, C. B., Pandita, T. K., Chan, T. A., Hermeking, H., et al. (2000) *Proc. Natl. Acad. Sci. USA.* **97,** 6049.

Fletcher, L. (2001) *Nature* **19,** 599.

Foyouzi-Youssefi, R., Arnaudeau, S., Borner, C., Kelley, W. L., Tschopp, J., Lew, D. P., Demaurex, N., Krause, K. H., (2000) *Natl. Acad. Sci. USA.* **97,** 5723.

Franzen, B., et al. (1995) *Electrophoresis* **16,** 1087.

Friedrichs, K., Ruiz, P., Franke, F., Gille, I., et al. (1995) *Cancer Res.* **55,** 901.

Fritz, G., Brachetti, C., Bahlmann, F., Schmidt, M., Kaina, B. (2002) *Br. J. Cancer* **87,** 635.

Fu, H., Subramanian, R. R., Masters, S. C. (2000) *Ann. Rev. Pharmacol.Toxicol.* **40,** 617.

Fujiwara, K., Poikonen, K., Aleman, L., Valtavaara, M., Saksela, K., Mayer, B. J., (2002) *Biochemistry* **42,** 12729.

Fukushima, N., Sato, N., Ueki, T., Rosty, C., Walter, K. M., Wilentz, R. E., Yeo, C. J., Hruban, R. H., Goggins, M. (2002) *Am. J. Pathol.* **160,** 1573.

Gately, D. P., Howell, S. B. (1993) *Br. J. Cancer* **67,** 1171.

Gavin, A-C., Bösche, M., Krause, R., Grandi, P., Marzioch, M., Bauer, A., Schultz, J., Rick, J. M., Michon, A. M., Cruciat, C. M., Memor, M., Höfter, C., Schelder M., et al. (2002) *Nature* **415,** 141.

Giaccone, G. *Drugs* (2000) **59,** 37.

Giometti, C. S., Williams, K., Tollaksen, S. L. (1997) *Electrophoresis* **18**. 573.

Giot, L., Brouwer, C., Chaudhuri, A., Kuang, B., Li, Y., Hao, Y. L., et al. (2003) *Science* **302,** 1727.

Goto, K., Endo, H., Fujiyushi, T. (1988) *J. Biochem. (Tokyo)* **103,** 48.

Grassi, G., Maccaroni, P., Meyer, R., Keiser, H., Ambrosio, A. D., Pascale, E., Grassi, M., Kuhn, A., Di Nardo, P., Kandolf, R., Küpper, J. H. (2003) *Carcinogenesis* **24,** 1625.

Green, J. A., Kirwan, J. M., Tierney, J. F., Symonds, P., Fresco, L., Collingwood, M., Williams, C. J. (2001) *Lancet* **358,** 781.

Greenlee, R. T., Hill-Harmon, M. B., Murray, T., Thun, M. (2001) *CA Cancer J. Clin.* **51,** 15.

Gregory, P. D., Hörz, W. (1998) *Eur. J. Biochem.* **251,** 9.

Grubb, R. L., Calvert, V. S., Wulkuhle, J. D., Paweletz, C. P., Linehan, W. M., Phillips, J. L., Chuaqui, R., Valasco, A., Gillespie, J., Emmert-Buck, M., Liotta, L. A., Petricoin, E. F. (2003) *Proteomics* **3,** 2142.

Guillame, E., Pineau, C., Evrad, B., Dupaix, A., Moertz, E., Sanchez, J. C., Hochstrasser, D. F., Jegou, B. (2001) *Proteomics* **1,** 880.

Gygi, S. P., Rochon, Y., Franza, B. R., Aebersold, R., (1999) *Mol. Cell Biol.* **19,** 1720.

Haab, B. B. (2003) *Proteomics* **3,** 2116.

Haber, D. (2000) *N. Engl. J. Med.* **343,** 1566.

Han, H., Bearss, D. J., Browne, L. W., Calaluce, R., Nagle, R. B., Von Hoff, D. D. (2002) *Cancer Res.* **62** 2890.

Hanahan, D., Weinberg, R. A. (2000) *Cell* **1,** 57.

Hanash, S. (2003) *Nature* **422,** 226.

Hanash, S. M., Gagnon, M., Seeger, R. C., Baier, L. (1985) *Prog. Clin. Biol. Res.* **175,** 261.

Haynes, P. A., Yates III, J. R. (2000) *Yeast* **17,** 81.

Hawes, R. H., Xiong, Q., Waxman, I., Chang, K. J., Evans, D. B., Abbruzzese, J. L. (2000) *Am J. Gastroenterol.* **95,** 17.

Helmbach, H., Rossmann, E., Kern, M. A., Shandendorf, D. (2001) *Int. J. Cancer Res.* **93,** 617.

Hemler, M. E., Mannion, B. A., Berditchevski, F. (1996) *Biochem. Biophys. Acta* **1287,** 67.

Hendrick, J. P., Harti, F. U. (1995) *FASEB J.* **9,** 1559.

Heppner, G. H., Miller, F. R., Shekhar, P. V. M. (2000) *Br. Cancer Res.* **2,** 331.

Hirano, T., Frinzen, B., Uryu, K., Okuzawa, K., Alaiya, A. A., Vanky, F., Rodrigues, L., Ebihara, Y., Kato, H., Auer, G. (1995) *Br. J. Cancer* **72,** 840.

Ho, Y., Gruhler, A., Heilbut, A., Bader, G. D., Moore, L., Adams, S-L., Miller, A., Taylor, P., et al. (2002) *Nature* **415,** 180.

Hondermark, H. (2003) *Mol. Cell Proteomics* **2.5,** 281.

Horak, C. E., Snyder, M. (2002) *Funct. Integr. Genomics* **2,** 171.

Howe, H. L., Wingo, P. A., Thun, M. J., Ries, L. A., Rosenberg, H. M., Feigal, E. G., Edwards, B. K. (2001) *Natl. Cancer Inst.* **93,** 824.

Hu, E., Chen, Z., Fredrickson, T., Zhu, Y. (2001) *Expt. Nephrol.* **9,** 265.

Hu, L., Zaloudek, C., Mills, G. B., Gray, J. G., Jaffe, R. B. (2000) *J. Clin. Cancer Res.* **6,** 880.

Huang, P., Feng, L., Oldham, E. A., Keating, M. J., Plunkett, W. (2000) *Nature* **407,** 390.

Hunter, T. (2000) *Cell.* **1,** 113.

Hunter, T. (1998) *Philos. Trans. R. Soc. London. Biol. Sci.* **1368,** 583.

Hutchens, T. W., Yip, T. T. (1993) *Rapid Commun. Mass Spectrom.* **7,** 576.

Iacobuzio-Donahue, C. A., Maitra, A., Shen-Ong, G. L., Van-Heek, T., Ashfaq, R., Meyer, R., Walter, K., Berg, K., Hollingsworth, M. A., Cameron, J. L., Yeo, C. J., Kern, S. E., Goggins, M., Hruban, R. H. (2002) *Am. J. Pathol.* **160,** 1239.

Ilag, L. L., Ng, J., Jay, D. G. (2000) *Drug Dev. Res.* **49,** 65.

Issaq, H.J., Conrads, T.P., Prieto, D.A., Tirumalai, R., Veenstra, T.D. (2003) *Anal. Chem.* **149A.**

Ito, T., Chiba, T., Ozawa, R., Yoshida, M., Hattori, M., Sakaki, Y. (2001) *Proc. Natl. Acad. Sci. USA* **98,** 4569.

Jaattela, M., Wissing, D., Kokholm, K., Kallunki, T., et al. (1998) *EMBO J.* **17,** 6124.

Jacobson, S., Pillus, L. (1999) *Curr. Opin. Genet. Dev.* **9,** 175.

Jassani, N., Liu, Y., Humphrey, M., Cravatt, B. F. (2002) *Proc. Natl. Acad. Sci. USA* **99,** 10335.

Jay, D. G. (1988) *Proc. Natl. Acad. Sci. USA* **85,** 5454.

Jones, P. A., Baylin, S. B. (2002) *Natl. Rev. Genet.* **3,** 415.

Kanai, Y., Ushijima, S., Sato, Y., Nakanishi, Y., Sakamoto, M., Hirohashi, S. (2001) *J. Cancer Res. Clin. Oncol.* **127,** 697.

Keim, V., Iovanna, J. L., Rohr, G., Usadel, K., Dagorn, J. C. (1991) *Gastroenterology* **100,** 775.

Kikuchi, Y., Iwano, I., Myauchi, M., Sasa, H., Nagata, I., Kaki, E. (1990) *Gynecol. Oncol.* **2,** 199.

Kligman, D., Hilt, D. C. (1988) *Trends Biochem. Sci.* **13,** 437.

Kondo, Y., Kuo, S. M., Watkins, S. C., Lazo, J. S. (1995) *Cancer Res.* **55,** 474.

Kos, J., Krasovec, M., Cimerman, N., Nielson, H. J., Christensen, I. J. E., Brünner, N. (2000) *Clin. Cancer Res.* **6,** 505.

Kos, J., Lah, T. T. (1998) *Oncol. Reports* **5,** 1349.

Kouzarides, T. (1999) *Curr. Opin. Genet. Dev.* **9,** 40.

Kovarova, H., Hajduch, M., Korinkova, G., Halada, P., Krupickova, S., Gouldsworth, A., Zhelev, N., Strand, M. (2000) *Electrophoresis* **21,** 3757.

Kumar, A., Snyder, M. (2002) *Nature* **415,** 123.

Lander, E. S., Linton, L. M., Birren, B., Nusbaum, C., Zody, M. C., Bawldwin, J., Devon, K., Dewar, K., Doyle, M., Fitzhugh W., et al. (2001) *Nature* **409,** 860.

Landis, S., Murray, T., Bolden, S., Wingo, P. A. (1999) *CA Cancer J. Clin.* **49,** 8.

Lappe, M., Holm, L. (2004) *Nature Biotechnol.* **22,** 98.

Laronga, C., Yang, H. Y., Neal, C., Lee, M. H. (2000) *J. Biol. Chem.* **275,** 23106.

Lasserr, C., Simon, M. T., Ishikawa, H., Diriong, S., Nguyen, V. C., Christa, L., Vernier, P., Brechot, C. (1994) *Eur. J. Biochem.* **224,** 29.

Leygue, E., Snell, L., Dotzlaw, H., Troup, S., Hitchcock, T-H., Murphy, L. C., Roughley, P. J., Watson, P. H. (2000) *J. Pathol.* **192,** 313.

Li, J., Zhang, Z., Rosenzweig, J., Wang, Y. Y., Chan, D. W. (2002) *Clin. Chem.* **48,** 1296.

Liao, J. C., Roider, J., Jay, D. G. (1994) *Proc. Natl. Acad. Sci. USA* **91,** 2659.

Lim, S. O., Park, S. J., Kim, W., Park, S. G., Kim, H. J., Kim, Y. I., Sohon, T. S., Noh, J. H., Jung, G. (2002) *Biochem. Biophys. Res. Commun.* **291,** 1031.

Lin, J. D., Huang, C. C., Weng, H. F., Chen, S. C., Jeng, L. B. (1995) *J. Chromatogr. B Biomed Appl.* **667,** 153.

Linden, K. G., Liao, J. C., Jay, D. G. (1992) *Biophys. J.* **62,** 956.

Linzer, D. I., Nathan, D. (1983) *Proc. Natl. Acad. Sci. USA* **80,** 4271.

Liotta, L. A., Kohn, C. E. (2001) *Nature* **411,** 375.

Liotta, L. A., Kohn, E. C., Petricoin, E. F. (2001) *JAMA* **18,** 2211.

Liu, D., Bienkowska, J., Petosa, C., Collier, R. J., Fu, H., Liddington, R. (1995) *Nature* **376,** 191.

Liu, H., Wang, Y., Zhang, Y., Song, Q., Di, C., Chen, G., Tang, J., Ma, D. (1999) *Biochem. Biophys. Res. Commun.* **254,** 203.

Liu, Y., Patricelli, M. P., Cravatt, B. F. (1999) *Proc. Natl. Acad. Sci. USA* **96,** 14694.

Llopis, J., Westin, S., Ricote, M., Wang, J., Cho, C. Y., Kurokawa, R., et al. (2000) *Proc. Natl. Acad. Sci. USA* **97,** 4363.

Loehrer, P. J., Einhorn, L. H. (1984) *Ann. Int. Med.* **100,** 704.

Logsdon, C. D., Simeone, D. M., Binkley, C., Arumugam, T., Greenson, J. K., Giordano, T. G., Misek, D. E., Hanash, S. (2003) *Cancer Res.* **63,** 2649.

Lukandin, E. M., Georgiev, G. P. (1996) *Microbiol. Immunol.* **213,** 171.

Luo, J., Dunn, T., Ewing, C., Sauvageot, J., Chen, Y., Trent, J., Isaacs, W. (2002) *Prostate* **51,** 189.

Luzzi, V., Holtschalg, V., Watson, M. A. (2001) *Am. J. Pathol.* **158,** 2005.

Machida, K., Mayer, B. J., Nollau, P. (2003) *Mol. Cell Proteomics* **2.4,** 215.

Madoz-Gurpide, J., Wang, H., Misek, D. E., Brichory, F., Hanash, S. M. (2001) *Proteomics* **1,** 1279.

Maloney, T. M., Paine, P. L., Russo, J. (1989) *J. Breast Cancer Res.* Treat. **14,** 337.

Mandic, A. Viktorsson, K., Strandberg, L., Heiden, T., Hansson, J., Linder, S., Shoshan, C. M. (2002) *Mol. Cell Biol.* **22,** 3003.

Melchiori, A., Mortarini, R., Carlone, S., Marchisio, P. C., et al. (1995) *Expt. Cell Res.* **219,** 233.

Mendic, A. (2003) PhD Thesis, Karolinska Institute, Stockholm.

Meza-Gurpide, J., Wang, H., Misek, D. E., Brichory, D. E., Hanash, S. (2001) *Proteomics* **1,** 1279.

Meza-Zepeda, L. A., Forus, A., Lygren, B., Dahlberg, A. B., Godager, L. H., South, A. P., Marenholz, I., Lioumi, M., Florenes, V. A., Maelansmo, G. M., Serra, M., et al. (2002) *Oncogene* **21,** 2261.

Milan, S., Ball, G., Hombuckle, J., Holding, F., Carmichael, J., Ellis, I., Ali, S., Li, G., McArdle, S., Creaser, C., Rees, R. C. (2003) *Proteomics* **3,** 1725.

Mistry, S. J., Atweh, G. F. (2002) *Mt. Sinai J. Med.* **69,** 299.

Möller, A., Malerczyk, C., Völker, U., Stöppler, H., Maser, E. (2002) *Proteomics* **2,** 697.

Möller, A., Soldan, M., Völker, U., Maser, E. (2001) *Toxicology* **160,** 129.

Motoo, Y., Satomura, Y., Mouri, I., Mouri, H., Ohtsubo, K., Sakai, J., Fujii, T., Taga, H., Yamaguchi, Y., Watanabi, H., Okai, T., Sawabu, N. (1999) *Dig. Dis. Sci.* **44,** 1142.

Naito, K. Ishiwata, T., Kurban, G., Teduka, K., et al. (2002) *Int. J. Oncol.* **20,** 943.

Nakatsura, T., Senju, S., Ito, M., Nishimura, Y., Itoh, K. (2002) *Eur. J. Immunol.* **32,** 826.

Natali, P. G., Nicotra, M. R., Bartolazzi, A., Cavaliere, R., et al. (1993) *Int. J. Cancer* **54,** 68.

Nebl, G., Meuer, S. C., Samstag, Y. (1996) *J. Biol. Chem.* **271,** 26276.

Nollau, P., Mayer, B. J. (2001) *Proc. Natl. Acad. Sci. USA* **98,** 13531.

Nutt, C. L., Mani, D. R., Betensky, R. A., Tamayo, P., et al. (2003) *Cancer Res.* **63,** 1602.

Oda, Y. N., Chait, B. T. (2001) *Nat. Biotechnol.* **19,** 379.

O'Farrell, P. H. (1975) *J. Biol. Chem.* **250,** 4007.

Okuda, M. (2002) *Oncogene* **21,** 6170.

Okuzawa, K., Franzen, B., Lindholm, J., Linder, S., Hirano, T., Bergman, T., Ebihara, Y., Kato, H., Auer, G. (1994) *Electrophoresis* **15,** 382.

O'Neil, K. A., Miller, F. R., Barder, T. J., Lubman, D. M. (2003) *Proteomics* **3,** 1256.

Ornstein, D. K., Gillespie, J. W., Paweletz, C. P., Duray, P. H., Herring, J., Vocke, C. D., Topalian, S. L., Bostwick, D. G., Linehan, W. M., Petricoin III, E. F., Emmert-Buck, M. R. (2000) *Electrophoresis* **21,** 2235.

Ortiz, E. M., Dusetti, N. J., Vasseur, S., Malka, D., Bödeker, H. Dagorn, J-C., Iovanna, J. L., (1998) *Gastroenterology* **114,** 808.

Pandey, A., Podtelejnikov, A. V., Blagoev, B., Bustelo, X. R., Mann, M., Lodish, H. F. (2000) *Proc. Natl. Acad. Sci. USA* **97,** 179.

Pannek , J., Partin, A. W. (1998) *Semin. Urol. Encol.* **16,** 100.

Paweletz, C. P., Charboneau, L., Bichsel, V. E., Simone, N. L., Chen, T., Gillespie, J. W., et al. (2001) *Oncogene* **20,** 1981.

Paz, M. F., Fraga, M. F., Avila, S., Guo, M., Pollan, M., Herman, J. G., Esteller, M. (2003) *Cancer Res.* **63,** 1114.

Pellicena, P., Miller, W. T. (2001) *J. Biol. Chem.* **276,** 28190.

Pellicena, P., Miller, W. T. (2002) *Front. Biosci.* **7,** d256.

Pencil, S. D., Toth, M. (1998) *Clin. Expt. Metastasis.* **16,** 113.

Perkin, D. M., Bray, F. I., Devesa, S. S. (2001) *Eur. J. Cancer* **8,** 4.

Petricoin III, E. F, Ardekani, A. M, Hitt, B. A, Levine, P. J., Fusaro, V. A., Steinberg, S. M., Mills, G. B., Simone, C., Fishman, D. A., et al. (2002a) *Lancet* **359,** 572.

Petricoin III, E., Liotta, L. A. (2003) *Clin. Chem.* **49,** 1276.

Petricoin III, E. F., Ornstein, D. K., Paweletz, C. P., Ardekani, A., Hackett, P. S., Hitt B. A., et al. (2002b) *J. Nat. Cancer Inst.* **94,** 1576.

Petricoin, E. F., Zoon, K. C., Kohn, E. C., Barrett, J. C., Liotta, L. A. (2002c) *Nat. Rev. Drug Discov.* **9,** 683.

Philp, A. J., Campbell, I. G., Leet, C., Vincan, E., Rockman, S. P., Whitehead, R. H., Thomas, R. J. S., Phillips, W. A. (2001) *Cancer Res.* **61,** 7426.

Ping Lu, Y., Ishiwata, T., Asano, G. (2002) *J. Pathol.* **196,** 324.

Provost, P., Doucet, J., Hammerberg, T., Gerisch, G., Samuelsson, B., Radmark, O. (2001a) *J. Biol. Chem.* **276,** 16520.

Provost, P., Doucet, J., Stock, A., Gerisch, G., Samuelsson, B., Radmark, O. (2001b) *Biochem J.* **359,** 255.

Qu, Y., Adam, B-L., Yasui, Y., Ward, M. D., Cazares, L. H., Schellhammer, P. F., et al. (2002) *Clin. Chem.* **48,** 1835.

Radford, D. M. (1983) *Cancer Res.* **53,** 2947.

Rain, T. C., Selig, L., De Reuse, H., Battaglia, V., Reverdy, C., Simon S., et al. (2001) *Nature* **409,** 211.

Ramaswamy, S., Tomayo, P., Rifkin, R., Mukherjee, S., et al. (2001) *Proc. Natl. Acad. Sci. USA* **98,** 15149.

Rasmussen, R. K., Ji, H., Eddes, J. S., Moritz, L. R., Reid, G. E., Simpson, R. J., Dorow, D. S. (1997) *Elecrophoresis* **18,** 588.

Rhee, H. T., Kim, G. Y., Huh, J. W., Kim, S. W., Na, D. S. (2000) *Eur. J. Biochem.* **267,** 3220.

Rigaut, G., Shevchenko, A., Rutz, B., Wilm, M., Mann, M., Séraphin, B. (1999) *Nature Biotechnol.* **17,** 1030.

Righetti, S. C., Gatti, L., Beretta, G. L., Zunino, F., Perego, P. (2002) *Dev. Mol. Pharmacol.* **1,** 167.

Robinovitz, I., Nagle, R. B., Cress. A. E. (1995) *Clin. Exp. Metastasis* **13,** 481.

Robinson, W. H., DiGennaro, C., Hueber, W., Haab, B. B., Kamachi, M., Dean, E. J., Fournel, S., Fong, D., Genovese, M. C., De Vegvar, H. E. N., Skriner, K., Hirschberg, D. L., Morris, R. I., Muller, S., Pruijn, G. J., van Venrooij, W. J., Smolen, J. S., Brown, P. O., Steinman, L., Utz, J. J. (2002) *Nature Med.* **8,** 295.

Roobol, A., Sahyoun, Z. P., Carden, M. J. (1999) *J. Biol. Chem.* **274,** 2408.

Rosty, C., Christa, L., Kuzdzal, S., Baldwin, M., Zahurak, M. L., Carnot, F., et al. (2002a) *Cancer Res.* **62,** 1868.

Rosty, C., Ueki, T., Argani, P., Jansen, M., Yeo, C. J., Cameron, J. L., Hruban, R. H., Goggins, M. (2002b) *Am. J. Pathol.* **160,** 45.

Roy, S., McPherson, R. A., Appoloni, A., Yan, J., Lane, A., Clyde-Smith, J., Hancok, J. F. (1998) *Mol. Cell Biol.* **18,** 3947.

Rui, M., Chen, Y., Zhang, Y., Ma, D. (2002) *Life Sci.* **71,** 1771.

Ryu, B., Jones, J., Blades, N. J., Parmigiani, G., Hollingsworth, M. A., Hurban, R. H., Kern, S. E. (2002) *Cancer Res.* **62,** 819.

Samstag, Y., Eckerskorn, C., Wesselborg, S., Henning, S., Wallich, R., Meuer, S. C. (1994) *Proc. Natl. Acad. Sci. USA* **91,** 4494.

Sarto, C., Binz, P. A., Mocarelli, P. (2000) *Electrophoresis* **21,** 1218.

Sarto, C., Marocchi, A., Sanchez, J. C., Giannone, D., Frutiger, S., Golaz, O., Wilkins, M. R., Doro, G., Cappellano, F., Hughes, G., Hochstrasser, D. F., Mocarelli, P. (1997) *Electrophoresis* **18,** 599.

Sato, N., Fkushima, N., Maitra, A., Matsubayashi, H., Yeo, C. J., Cameron, L., Hruban, R. H., Goggins, M. (2003a) *Cancer Res.* **63,** 3735.

Sato, N., Maehara, N., Su, G. H., Goggins, M. (2003b) J. *Natl. Cancer Inst.* **95,** 327.

Sato, N., Maitra, A., Fkushima, N., van Heek, N. T., Matsubayashi, H., Iacobuzio-Donahue, C. A., Rosty, C., Goggins, M. (2003c) *Cancer Res.* **63,** 4158.

Sato, N., Ueki, T., Fukashima, N., Iacobuzi-Donahue, C. A., Yeo, C. J., Cameron, J. L., Hruban, R. H., Goggins, M. (2002) *Gastroenterology* **123,** 365.

Schutte, M., Hruban, R. H., Geradts, J., Maynard, R., Hilgers, W., Rabindran, S. K., Moskaluk, C. A., Hahn, S. A., et al. (1997) *Cancer Res.* **57,** 3126.

Schwikowski, B., Uetz, P., Fields, S. (2000) *Nature Biotechnol.* **18,** 1257.

Seraj, M. J., Harding, M. A., Zgildea, J. J., Welch, D. R., Theodorescu, D. (2000) *Clin. Expt. Metastasis* **18,** 519.

Shackleford, G. M., Ganguly, A., MacArthur. C. A. (2001) *BMC Genomics* **2,** 8.

Shafer, B. W., Heizmann, C. W. (1996) *Trends Biochem. Sci.* **21,** 134.

Sharma, A., Antoku, S., Fujiwara, K., Mayer, B. (2003) *Mol. Cell. Proteomics* **2.11,** 1217.

Shayesteh, L., Lu, Y., Kuo, W-L., Baldocchi R., et al. (1999) *Nat. Genet.* **21,** 99.

Sherbet, G. V., Lakshmi, M. S. (1998) *Anticancer Res.* **18,** 2415.

Shipp, M. A., Tamayo, P., Angelo, M., Ray, T., Neuberg, D., Last, K., Aster, J., Masirov, J., Lister, A., Golub, T. R. (2000) *Blood* **96,** 222a.

Shokat, K. M. (1995) *Chem. Biol.* **2,** 509.

Simpson, R. J., Connolly, L. M., Eddes, J. S., Pereira, J. J., Moritz, R. L., Reid, G. E. (2000) *Electrophoresis* **21,** 1707.

Sinha, P., Hütter, G., Köttgen, E., Dietel, M., Shadendorf, D. (1999a) *Electrophoresis* **20,** 2952.

Sinha, P., Hütter, G., Köttgen, E., Dietel, M., Shadendorf, D., Lage, H. (1999b) *Electrophoresis* **20,** 2961.

Sinha, P., Poland, J., Kohl, S., Schnölzer, M., Helmbach, H., Hütter, G., Lage, H., Schadendorf, D. (2003) *Electrophoresis* **24,** 2386.

Sinha, P., Poland, J., Schnölzer, M., Celis, J. E., Lage, H. (2001) *Electrophoresis* **22,** 2990.

Snow, P. B., Brandt, J. M., Williams, R. L. (2001) *Mol. Urol.* **5,** 171.

Somiari, R. I., Sullivan, A., Russell, S., Somiari, S., Hu, H., Jordan, R., George, A., Katenhusen, R., Buchowiecka, A., Arciero, C., Brzeski, H., Hooke, J., Shriver, C. (2003) *Proteomics.* **3,** 1863.

Songyang, Z., Carraway, K. L., Eck, M. J., Harrison, S. C., Feldman, R. A., Mohammadi, M., Schlessinger, J., Hubbard, S. R., Smith, D. P., Eng, C., Lorenzo, M. J., Ponder, B. A. J., Mayer, B. J., Cantley, L. C. (1995) *Nature* **373,** 536.

Sorio, C., Bonora, A., Orlandini, S., Moore, P., Cristofori, P., Dal Negro, G., Gaviraghi, G., Falconi, F., Pederzoli, P., Scarpa, A. (2001). *Virchows Archiv.* **438,** 154.

Sorkin, A., McClure, M., Huang, F., Carter, R. (2000) *Curr. Biol.* **10,** 1395.

Szalai, G., Krishnamurthy, R., Hajno'czky, G. (1999) *EMBO J.* **18,** 6349.

Takayama, S., Bimston, D. N., Matsuzawa, S., Freeman, B. C., Aime-Sempe, C., Xie, Z., Morimoto, R. E., Reed, J. C. (1997) *EMBO J.* **16,** 4887.

Trask, D. K., Band, V., Zajchowski, D. A., Yaswen, P., Suh, T., Sager, R. (1990) *Proc. Natl. Acad. Sci. USA.* **87,** 2319.

Tsukiyama, T., Wu, C. (1997) *Curr. Opin. Genet. Dev.* **7,** 182.

Tuynder, M., Susini, L., Prieur, S., Besse, S., Fiucci, G., Amson, R., Telerman, A. (2002) *Proc. Natl. Acad. Sci. USA* **99,** 14976.

Tzivion, G., Luo, Z., Avrueh, J. A. (1998) *Nature* **394,** 88.

Ueki, T., Toyota, M., Skinner, H., Walter, K. M., Yeo, C. J., Issa, J.-P. J., Hruban, R. H., Goggins, M. (2001) *Cancer Res.* **61,** 8540.

Ueki, T., Toyota, M., Sohn, T., Yeo, C. J., Issa, J.-P. J., Hruban, R. H., Goggins, M. (2000) *Cancer Res.* **60,** 1835.

Uetz, P., Giot, L., Mansfield, T., Judson, R. S., Knight, J. R., Lockshon, D., et al. (2000) *Nature* **403,** 623.

Uetz, P., Pankratz, M. J. (2004) *Nature Biotechnol.* **22,** 43.

Ünlü, M., Morgan, M. E., Minden, J. S. (1997) *Electrophoresis* **18,** 2071.

Ursic, D., Culberston, M. R. (1991) *Mol. Cell. Biol.* **11,** 2629.

van Belzen, N., Dinjens, W. N., Diesveld, M. P., Groen, N. A., et al. (1997) *Lab. Invest.* **77,** 85.

Venter, J. C., Adams, M. D., Myers, E. W., Li, P. W., Mural, R. J., Sutton, G. G., Smith, H. O., Yandell, M., C. A. E., Holt, R. A., J. D. Gocayne, Amanatides, P., Ballew, R. M., et al. (2001) *Science.* **291,** 1304.

Vercoutter-Edouart, A. S., Lemoine, J., le Bourhis, X., Louis, H., Boilly, B., Nurcombe, V., Revillion, F., Peyrat, J. P., Handermarck, H. (2001) *Cancer Res.* **61,** 76.

von Mering, C., Krause, R., Snel, B., Cornell, M., Oliver, S. G., Fields, S., Bork, P. (2002) *Nature* **417,** 399.

Ward, Jr., J. B., Henderson, R. E. (1996) *Environ. Health Prospect.* **5,** 895.

Ward, L. D., Hong, J., Whitehead, R. H., Simpson, R. J. (1990) *Electrophoresis* **11** 883.

Watts, F. M. (2002) *EMBO J.* **21,** 3919.

Westly, B., Rochefort, H. (1980) *Cell* **20,** 353.

Weydert, C., Roling, B., Liu, J., Hinkhouse, M. M., Ritchie, J. M., Oberle, L. W., Culleny, J. J. (2003) *Mol. Cancer Therapy* **2,** 361.

Whitehouse, C. J., Taylor, R. M., Thistlethwaite, A., Zhang, H., Karimi-Busheri, F., Lasko, D. D., Weinfield, M., Caldecott, K. W. (2001) *Cell* **104,** 107.

Williams, K. L., Pallini, V. (1997) *Proteome Research: New Frontiers in Functional Genomics,* Springer Verlag, Heidelberg.

Winston, J. S., Asch, H. L., Zhang, P. J., Edge, S. B., Hyland, A., Asch, B. B. (2001) *Breast Cancer Res. Treat.* **65,** 11.

Wirth, P. J., Egilsson, V., Gudnasen, V., Ingvarsson, S., Thorgeirsson, S. S. (1987) *Breast Cancer Res. Treat.* **10**. 177.

Wouters, F. S., Verveer, P. J., Bastiaens, P. I. H. (2001) *Trends Cell Biol.* **11,** 203.

Wright, G. L. (1974) *J. Ann. Clin. Lab. Sci.* **4,** 281.

Wright, Jr., G. L., Cazares, L. H., Leung, S. M., Nasim, S., Adam, B-L., Yip, T. T., et al. (1999) *Prostate Cancer Prostatic Dis.* **2,** 264.

Wu, W., Tang, X., Hu, W., Lotan, R., Hong, W. K., Mao, L. (2002a) *Exp. Metastasis* **4,** 319.

Wu, W., Tang, X., Hu, W., Lotan, R., Ki Hong, W., Mao, L. (2002b) *Clin. Expt. Metastasis* **19,** 319.

Wu, S-L., Hancock, W. S., Goodrich, G. G., Kunitake, S. T. (2003) *Proteomics* **3,** 1037.

Wulfkuhle, J. D., Sgroi, D. C., Krutzsch, H., McLean, K., McGarvey, K., Knowlton, M., Chen, S., Shu, H., Sahin, A., Kurek, R., Wallwiener, D., Merino, M. J., Petricoin, III, E. F., Zhao, Y., Steeg, P. S. (2002) *Cancer Res.* **62,** 6740.

Wulfkuhle, J. D., Aquino, J. A., Calvert, V. S., Fishman, D. A., Coukos, G., Liotta, L. A., Petricoin III, E. F. (2003) Proteomics. **3,** 2085.

Xanthoudakis, S., Roy, S., Rasper, D., Hennessey, T., Aubin, Y., Cassady, R., Tawa, P., Ruel, R., Rosen A., Nicholson, D. W. (1999) *EMBO J.* **18,** 2049.

Xaio, B., Smerdon, S. J., Jones, D. H., Dodson, G. G., Soneji, Y., Aitken, A., Gamblin, S. J. (1995) *Nature* **376,** 188.

Yang, C., Maiguel, D. A., Carrier, F. (2002) *Nucleic Acids Res.* **30,** 2251.

Yeo, C. J., Cameron, J. L., Sohn, T. A., Lillemoe, K. D., Pitt, H. A., Talamini, M. A., Hruban, R. H., Ord, S. E., Sauter, P. K., Coleman, J., Zahurak, M. L., Grochow, L. B., Abrams, R. A. (1997) *Ann. Surg.* **226,** 248.

Zhao, Y., Chait, B. T., *Anal. Chem.* **66,** 3723 (1994).

Zhou, H., Watts J. D., Aebersold, R. (2001). *Nat. Biotechnol.* **19,** 375.

Zhou, G., Li, H., DeCamp, D., Chen S. et al., (2002) *Mol. Cell. Proteomics* **1,** 117.

Zhu, H., Bilgin, M., Bangham, R., Hall, D., Casamayor, A., Bertone, P., Lan, N., Jansen, R., Bidlingmaier, S., Houfek, T., Mitchell, T., Miller, P., Dean, R. A., Gerstein, M., Snyder, M. (2001) *Science* **293,** 2101.

3

CURRENT STRATEGIES FOR PROTEIN QUANTIFICATION

3.1. INTRODUCTION

The complete sequencing of more than 100 genomes, including the human, and the current large-scale genomics have provided proteomics with sequence infrastructure which together with modern MS is giving a new vision to the study of biological systems. Quantitative proteome profiling, defined as the systematic identification of proteins in complex samples and the determination of their abundance or quantitative change, is an important component in the emerging science of systems biology. The role of proteomic approaches for protein quantification has been strengthened by two deductions based on genomic studies: First, mRNA quantities based on gene expression do not always correlate to protein quantities (Anderson and Seilhamer, 1997; Le Naour et al., 2001; Ideker et al., 2001). Second, It is increasingly recognized that co- and posttranslational modifications, including phosphorylation and glycosylation, can determine the localization, activity, and function of proteins and thereby tend to play a key role in controlling various cellular processes. These modifications together with other relevant biological events involving protein complexes can only be determined at the protein level.

Over the last 5 years a host of strategies for protein quantification in complex samples have been described. Most, if not all, are based on the early work of Schena et al. (1995) used to study mRNA expression. Current strategies for protein quantification involve three basic steps: chemical modification or isotope labeling of one or more amino acids, electrophoretic or chromatographic separation of intact proteins or peptide digests, and MS or fluorescence detection. Strategies for the identification and

Proteomics Today. By Mahmoud Hamdan and Pier Giorgio Righetti
ISBN 0-471-64817-5

quantification of modified proteins tend to use additional steps such as enrichment and multistep derivatization (Oda et al., 2001; Goshe et al., 2002; Zhou et al., 2001; Kaji et al., 2003; H. Zhang et al., 2003). A different approach for the identification of various forms of protein modifications which uses multienzyme digestion, multidimensional chromatography, and MS–MS has been also described by MacCoss et al. (2002).

Setting aside the tagging and derivatization steps, the strategies described here use two main approaches for the separation and detection of complex protein samples. The first approach uses 2D gel electrophoresis in conjunction with MS (Patterson and Aebersold, 1995; Oda et al., 1999; Sechi, 2002; Gehanne et al., 2002), while the second approach uses various forms of chromatography in combination with MS–MS (Link et al., 1999; Veenstra et al., 2000; Gygi et al., 1999).

3.1.1. Strategies Based on Labeling a Specific Amino Acid Residue

Most, if not all, of the current methods for protein quantification use an approach which was first pioneered by Schena et al. (1995) in the study of mRNA expression. In that approach, mRNA species in control and experimental samples were labeled with different fluorescent dyes, the two samples were mixed, individual mRNA species were isolated by hybridization, and the relative degree of change in mRNA concentration between the two samples was determined by fluorescent ratio measurements. Current labeling strategies for protein quantification can be divided into two or three main groups, each of which has within it a number of variations. In all MS-based strategies the first and maybe the most critical step is to modify the molecular mass of a specific amino acid so it can be distinguished from its unlabeled counterpart in the detection phase. This can be done in various ways: In one approach stable isotope labeling is used without changing the chemical identity of the amino acid, as in the case of introducing heavy atoms of H, C, N, and O within various functional groups. In a second approach chemical modification with or without stable isotope labeling is used. Alkylating Cys is an example of the first case, while guanidination which transfers C-terminal Lys to homoarginine is an example of the latter.

3.1.2. Isotope-Coded Affinity Tags

This approach was first introduced by Gygi et al. (1999; Gygi and Aebersold, 2000), who used light/heavy chemical reagents to alkylate Cys residues. The Isotope-coded affinity tag (ICAT) reagent consists of three components: The first is a thiol-specific reactive group for alkylating Cys residues; the second component is simply a polyether linker to allow the replacement of eight hydrogen atoms with their corresponding deuterium atoms; and the third component is a biotin group which is used to isolate ICAT-labeled peptides during the chromatography phase. In this strategy proteins from two different cell states are harvested, denatured, reduced, and labeled at cysteines, one of the states with light and the other with heavy ICAT. The samples are then combined and digested with trypsin. The ICAT-labeled peptides are isolated by biotin affinity chromatography and then analyzed by on-line liquid chromatography (LC)–MS and MS–MS. The measured ratio of the ion intensities of an ICAT-labeled pair of peptides

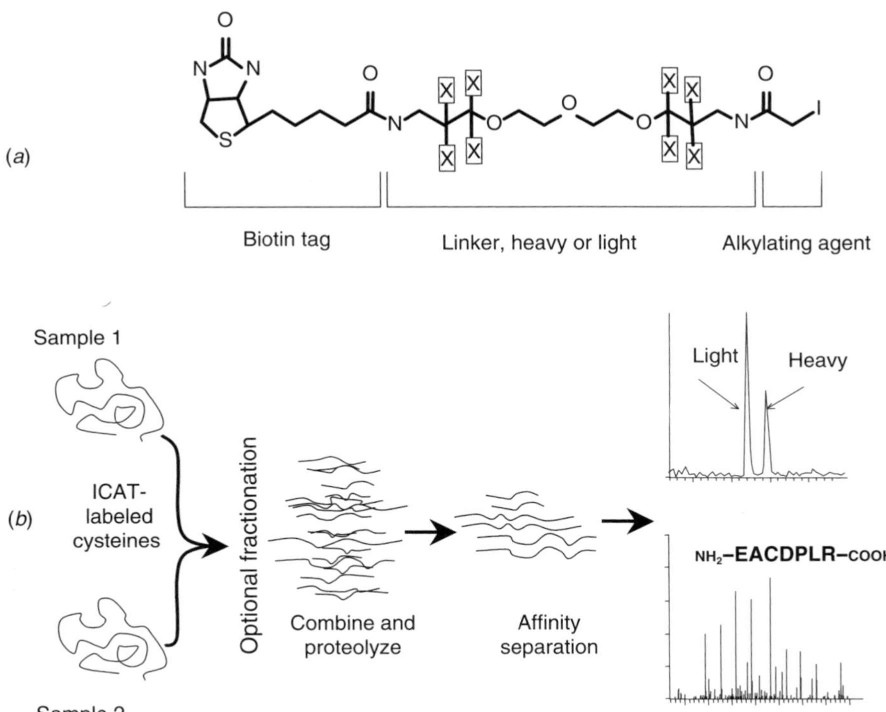

FIGURE 3.1. (*a*) Structure showing three main components of cysteine-selective ICAT reagent; (*b*) ICAT strategy for labeling, separation, and quantification of proteins derived from two different samples. (Adapted from Tureček, 2002, with permission.)

can be used for quantifying the relative abundance of their parent proteins present in the two samples. Furthermore, the MS–MS data can provide amino acid sequences which can be used for database searches and the identification of the detected proteins. The main steps of this protocol are represented in Figure 3.1 (Tureček, 2002). Similar to other existing MS-based methods for protein quantification, the ICAT approach has its strengths and weaknesses. One of the advantages of this approach is that it can be used for proteins harvested from body fluids, cells, and tissues. The alkylation is highly specific and under certain experimental conditions is limited to the thiol groups of Cys residues. Such specificity helps in reducing the complexity of the analyzed mixture, particularly during the LC phase. The chemical structure of this particular reagent and its specific reaction with Cys residues are also responsible for a number of its shortcomings. For example, the size of the ICAT reagent [average molecular weight (MW) = 570.5 Da] can interfere with database-searching algorithms, especially for short peptides (<7 amino acids). The thiol-specific group within the ICAT reagent is of the iodoacetamide type, a class of alkylating reagents which have shown, together with acrylamide-based reagents, rarely to achieve 80% blocking of the SH groups within intact proteins (Galvani et al., 2001*a*,*b*; Hamdan and Righetti, 2002). Moreover, when the reaction times were extended (e.g., overnight), there was no improvement in the alkylation; instead, undesired reactions with Lys residues were detected. If these effects occur with such relatively simple molecules such as iodoacetamide and acrylamide, then one would ask whether a much larger, iodoacetamide-based reagent such as the ICAT would fare better! However, *for* the record we have to point out that Smolka et al. (2001) reported full tagging of some standard proteins, in particular α-lactalbumin. This is rather a surprising result in view of the fact that the solubilization of the tested samples was conducted in the presence of SDS, a reagent known to interfere with the efficiency of the alkylation (Galvani et al., 2001b). It also remains to be seen whether such complete tagging can be achieved in a more complex medium such as a total cell lysate. Another shortcoming of the ICAT protocol has been pointed out (Zhang and Regnier, 2002) which applies to capture techniques that are different from those reported in the initial version of this protocol, which were based on the use of avidin–biotin affinity chromatography. When dealing with a complex tryptic digest of a total cell lysate, one might want to use different chromatographic procedures. For example, isotopically labeled peptides could be separated by ion exchange chromatography followed by a reverse-phase column or by reverse-phase chromatography followed by ion mobility separation. It turns out that, if ICAT-labeled peptides are separated through a reverse-phase column, then a well-known isotope effect takes place by which the deuterated peptide elutes earlier than its nondeuterated counterpart. Such separation causes a variation in the isotope ratio across the two differently eluting profiles of the isoforms, an effect which becomes more pronounced with small peptides. For instance, in the case of a simple, Cys-bearing octapeptide, the resolution was as high as $R_s = 0.74$ (we should recall that baseline resolution corresponds to $R_s = 1.2$). Given that column elute is continuously monitored by MS, obtaining the correct quantitative peak ratios according to the isotope ratios is bound to suffer. In a recent work by Zhang and Regnier (2002) it was reported that the resolution of isoforms in a C18 column exceeded 0.5 for about 20% of the peptides within the investigated digest. A similar chromatographic effect was reported earlier by Chiari et al. (1994). These authors reported baseline separation of benzyl–H_5/benzyl–D_5 alcohol and benzene–H_6/benzene–D_6 brought about by the presence of SDS micelles in the background electrolyte. Clearly, in this micellar electrokinetic process, the residence times of light and heavy species are bound to be different. Conversely, Zhang and Regnier (2002) reported the complete absence of such an isotope effect in the case of peptides marked with ^{13}C- and ^{12}C-succinate, an observation which the authors used to recommend such peptide coding when separation is conducted in C18 columns.

Despite the exclusion of Cys-free peptides, the ICAT approach can yield an impressive coverage of a proteome in a single LC/MS–MS analysis. Such capability was demonstrated for yeast, which has ~30,000 Cys-containing peptides, which provided ~92% coverage of the predicted proteome. Having said that, we must bear in mind that one protein in seven does not contain Cys residues (Vuong et al., 2000), and therefore, it is reasonable to predict that the highly successful yeast example cannot be considered as a common case for everyday analysis. This observation can also be considered a good reason for seeking other approaches to complement the ICAT strategy, particularly in the analysis of complex biological samples. Having said that, the combination of ICAT with microcapillary LC (μLC)–MS/MS and newly developed software tools was used to investigate the role of detergent-resistant lipid raft membrane microdomains in T-cell receptor signaling in the human cell line jukart (von Heller et al., 2003). The acquired MS–MS spectra were submitted to SEQUEST (Eng et al., 1994), for searching protein sequence databases. The compiled data of this search were then submitted to PeptideProphet (Keller et al., 2002), and the combined SEQUEST/PeptideProphet outputs were displayed via the HyperText Markup Language (HTML) interface INTERACT (Han et al., 2001). The authors reported that this experimental approach, including the use of newly developed software tools, substantially reduced the need for manual verification of the results and at the same time gave the user a significant control over the final output of a large-scale proteomic experiment.

In a recent article Griffin et al. (2003) reported a new variation in the way the ICAT approach is used for quantitative proteomics. Basically, the new application uses a highly accurate algorithm, chromatographically fractionated ICAT-tagged peptides, and MALDI–MS–MS. The sequence of steps in this variation was as follows: Affinity-purified ICAT reagent-labeled peptides were separated by μLC, collected onto a MALDI sample plate, and inserted into the mass spectrometer. Each position of the sample plate was surveyed by MALDI–TOF–MS and the data were transferred to be elaborated by the quantification software. Abundance ratios were determined for ICAT-labeled with normal (d_0) and heavy (d_8) atoms detected as peak pairs separated by 8.05 Da for each Cys. Potential single peaks due to unlabeled peptides were minimized using an optimized affinity selection protocol that provided more than 90% enrichment for Cys-containing peptides. The software outputs a list containing the mass and sample plate position of those peptides which showed an abundance ratio that exceeds a user-defined threshold. The precursor ions identified in the above step are then selected for CID and the data can be used for database searching.

Zhou et al. (2002) performed parallel measurements to compare the performance of the ICAT tagging and solid-phase isotope tagging. In the latter approach an *O*-nitrobenzyl-based photocleavable linker was first attached to aminopropyl-coated glass beads by solid-phase peptide synthesis (Holmes and Jones, 1995). Next the isotope tag in the form of leucine containing either seven hydrogen atoms (d_0) or seven deuterium atoms (d_7) was attached to the photocleavable linker, again by solid-phase synthesis. Finally, a sulfhydryl-specific iodoacetyl group was attached. A schematic representation of this approach is given in Figure 3.2. Following a number of parallel measurements involving the ICAT and solid-phase tagging, the authors deduced that the solid-phase tagging was simpler, more reproducible, more efficient, and more sensitive for protein quantification compared to the conventional ICAT approach. Considering the advantages of the solid-phase approach reported by the same authors, it can be appreciated that most of them are related to the identity of the tagging reagent and to the procedure of tagging, in which unlike with the ICAT approach peptide labeling was effected after proteolysis. This step renders the solid-phase approach unsuitable for intact protein labeling, which is destined for gel electrophoresis analysis.

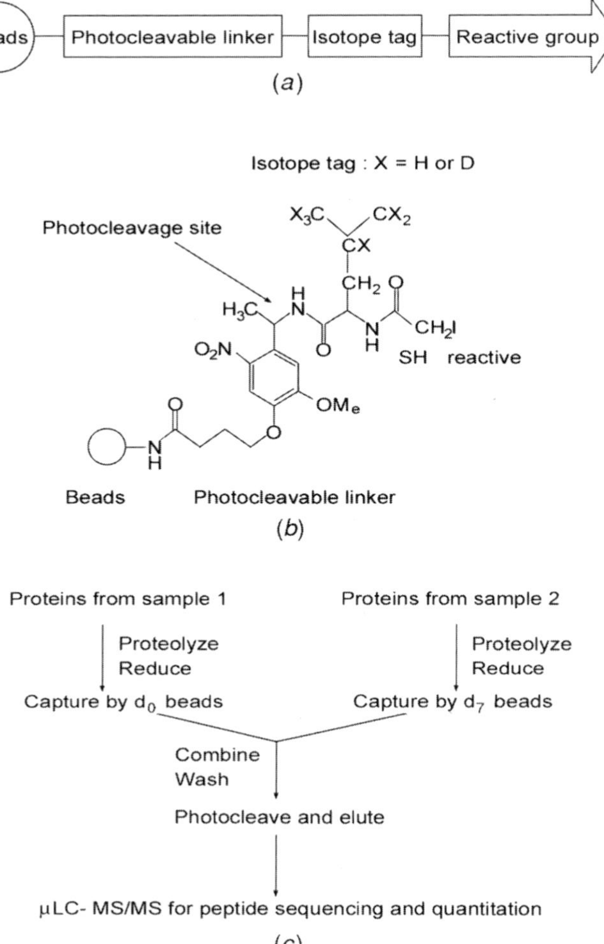

FIGURE 3.2. Schematic representation of solid-phase isotope tagging strategy: (*a*) modular composition of solid-phase isotope tagging reagent; (*b*) chemical composition of sulfhydryl-reactive solid-phase isotope tagging reagent; (*c*) strategy for quantitative analysis of proteins from two different samples. (Adapted from H. Zhou et al., 2002, with permission.)

In a recent article Thompson et al. (2003) described another approach which highly resembles the ICAT method. The authors used the name tandem mass tags (TMTs). In this approach peptides derived from two different pools are tagged with two different synthetic markers which have the same overall mass but different functional components. These tags are designed so that in analysis by collision-induced dissociation the tag is released to give rise to an ion with a specific m/z ratio. The authors reported that the TMT approach is similar in principle to other peptide isotope-labeling techniques, yet according to the same authors, when their method is compared with isotope-labeling techniques, a number of advantages can be pointed out. According to the same authors, TMT-labeled peptides coelute during chromatographic separation, which results in a more reliable quantification. The absence of a particular restriction on the choice of the active functionality to be used with these tags renders their approach suitable for a wide range of peptide separation protocols. The use of MS–MS rather than MS-based detection results in better signal-to-noise ratios and an improved dynamic range.

A stable isotope-labeling method similar to ICAT has been recently described by Hoang et al. (2003). Instead of targeting cysteines, this method uses a membrane impermeant biotinylating reagent which reacts with primary amines, that is, Lysines and the unmodified amino termini. In conjuction with this labeling approach, the authors have introduced a secondary heavy isotope label (D_0- or $^{13}CD_3$-methyl iodide) to allow protein quantification in two different samples. This method was tested on a limited number of commercial proteins and peptides and has the advantage of targeting Lys which occurs at a higher frequency in proteins compared to Cys. Apart from this difference, it is difficult to derive other advantages compared to the well-established ICAT approach.

3.1.3. Two-Dimensional Gel Electrophoresis/MALDI–TOF–MS

This strategy also targets Cys residues for labeling by using commercially available light/heavy acrylamide. Before considering further details of this approach, the following general observations are relevant. Two-dimensional polyacrylamide gel electrophoresis (2D PAGE) is the only method currently available that is capable of separating thousands of proteins in a single analysis. Although O'Farrel (1975) demonstrated such capability about 30 years ago, the extensive use of this powerful technique in conjunction with MS had to wait a considerable time. It is needless to say that the main impetus to such a combination has come from the various needs of genomics and proteomics. Despite the impressive contribution of this combination to protein identification over the last decade, the role of this approach is likely to gain further momentum, particularly in its interaction with functional proteomics (Godovac-Zimmermann and Brown, 2001).

Prior to any electrophoretic step, proteins are normally dissolved in a solubilization cocktail that contains a number of chemicals, including an alkylating agent that specifically reacts with the –SH groups of Cys residues and thus blocks undesired reactions that are known to result in spurious spots which have the direct effect of complicating the final 2D maps and even lead to misleading interpretations and negative database searches. Acrylamide is one of the common alkylating agents and can be purchased as light (d_0-acrylamide) or as heavy (d_3-acrylamide). Sechi and Chait (1998) used this reagent in its light and heavy forms to enhance information gleaned from peptide-mapping experiments. The idea was to use these chemicals to alkylate Cys residues prior to 1D PAGE; separated proteins were then extracted, digested, and examined by MALDI–TOF–MS, which allowed the determination of the Cys content of each separated component. Such an approach demonstrated an added value to the usual database search for protein identification. This experimental deduction was in

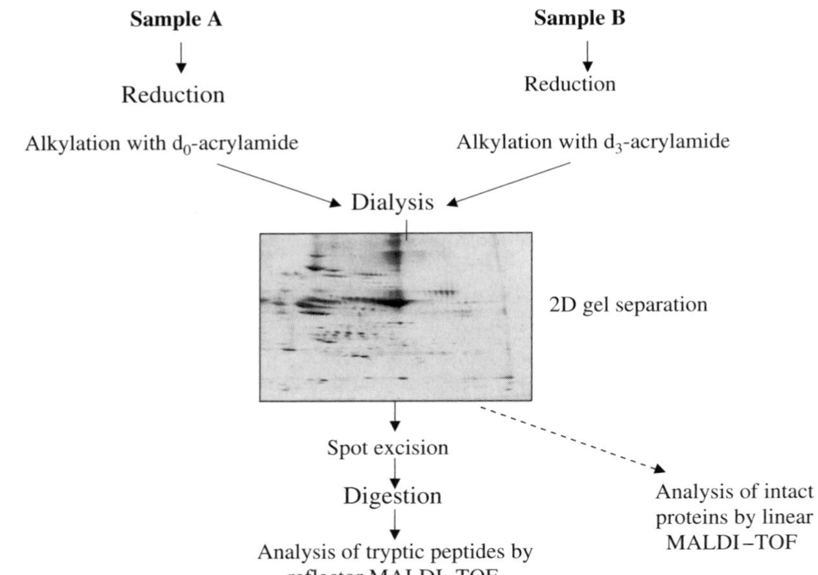

FIGURE 3.3. Schematic representation of main steps in combination of 2DE and MALDI–MS for protein quantification using light and heavy commercial acrylamide to tag proteins in samples A and B. (Adapted from Gehanne et al., 2002, with permission.)

accord with a theoretical study by Fenyo et al. (1998) which showed that a knowledge of the presence or absence of Cys residues in peptides could improve the reliability of protein identification via peptide mapping.

The use of commercial light and heavy acrylamide to alkylate proteins prior to their 2D electrophoretic separation has been reported by two groups, Gehanne et al. (2002) and Sechi (2002); both reports have demonstrated that this procedure combined with MALDI–TOF–MS could be a valid tool for the relative protein quantification in protein mixtures. The basic steps used in this approach are depicted in Figure 3.3. Basically, the relative quantification of the individual proteins in two different samples is achieved by alkylating one sample with d_0-acrylamide, and the second with its d_3 counterpart; the two samples are then mixed with predetermined ratios, dialyzed to remove excess acrylamide, and subjected to 2DE. Following visualization of the separated proteins, each spot can be excised, proteolyzed, and examined by reflector MALDI–TOF. The relative quantification of various proteins was obtained by comparing the relative peak heights within a reflector MALDI spectrum of two adjacent isotopic envelopes that differ by $m/z = 3$.

The application of this approach for quantifying various proteins within the 2D map of rat serum was demonstrated by Gehanne et al. (2002). The map in Figure 3.4. was obtained within the pH 3–10 immobilized pH gradient (IPG) interval of rat serum composed of two fractions; the first (30%) was alkylated with d_0-acrylamide and the second (70%) was reacted with d_3-acrylamide. A representative example of a reflector

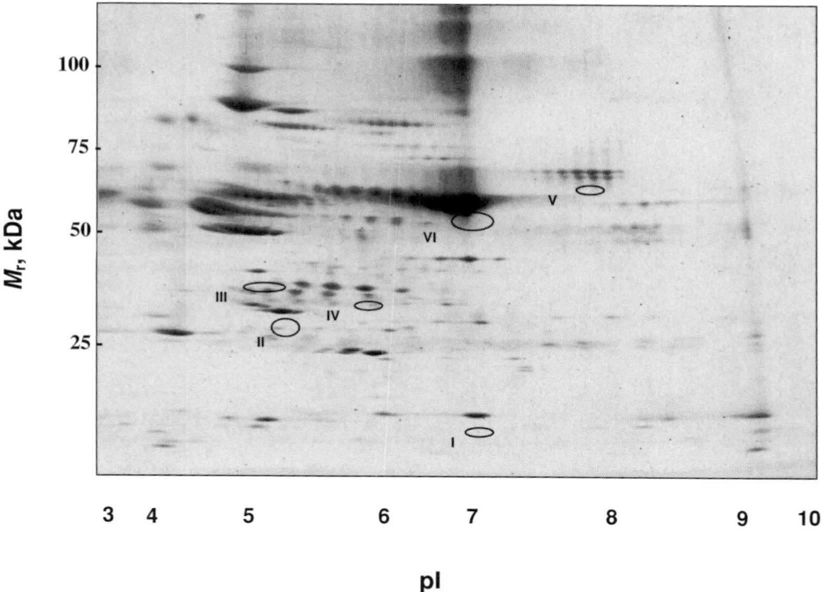

FIGURE 3.4. Two-dimensional map in 3–10 IPG interval of rat plasma composed of two fractions, the first (30%) was alkylated with d_0-acrylamide, and the second (70%) was alkylated with d_3-acrylamide. Circled spots refer to transthyretin (I), apo-lipoprotein (II), apo-lypoprotein A4 (III), 1-macroglobulin (IV), sero-transferrin (V), and albumin (VI). (Adapted from Gehanne et al., 2002, with permission.)

MALDI spectrum that pertains to albumin (spot VI) is given in Figure 3.5*a*–*c*. The spectrum of the entire digest is given in (*a*), whereas (*b*) and (*c*) give two short intervals of the same spectrum and show two isotopic distributions marked *A* and *A** in which a difference of 3 in the m/z values of the corresponding peaks is clearly evident. Database searches yielded the indicated peptides, each of which contains a single cysteine residue. Considering the relative peak heights in both isotopic distributions, a ratio of 34 : 66 was obtained, which is in reasonable agreement with the labeling ratio of 30 : 70 prior to 2D separation.

Quantitative profiling of complex protein mixtures labeled with the acrylamide d_0/d_3 alkylation tag system was also reported by Cahill et al. (2003). The authors deduced that the chemistry of alkylation is quantitative and specific and alkylation by light and heavy acrylamide induces a detectable but insignificant incidence of relative migration differences in 18-cm IPGs in the pH 5.5–6.7 range.

As all existing analytical methods, the use of 2D PAGE in conjunction with MS has its pros and cons; however, before discussing such characteristics, we find it important to underline a certain effect which tends to be underestimated during the phase of sample preparation. The efficiency and completeness of alkylation have a direct effect on both the quality of the final 2D maps and on the eventual quantification of the separated proteins. At this point one is entitled to ask what are the consequences

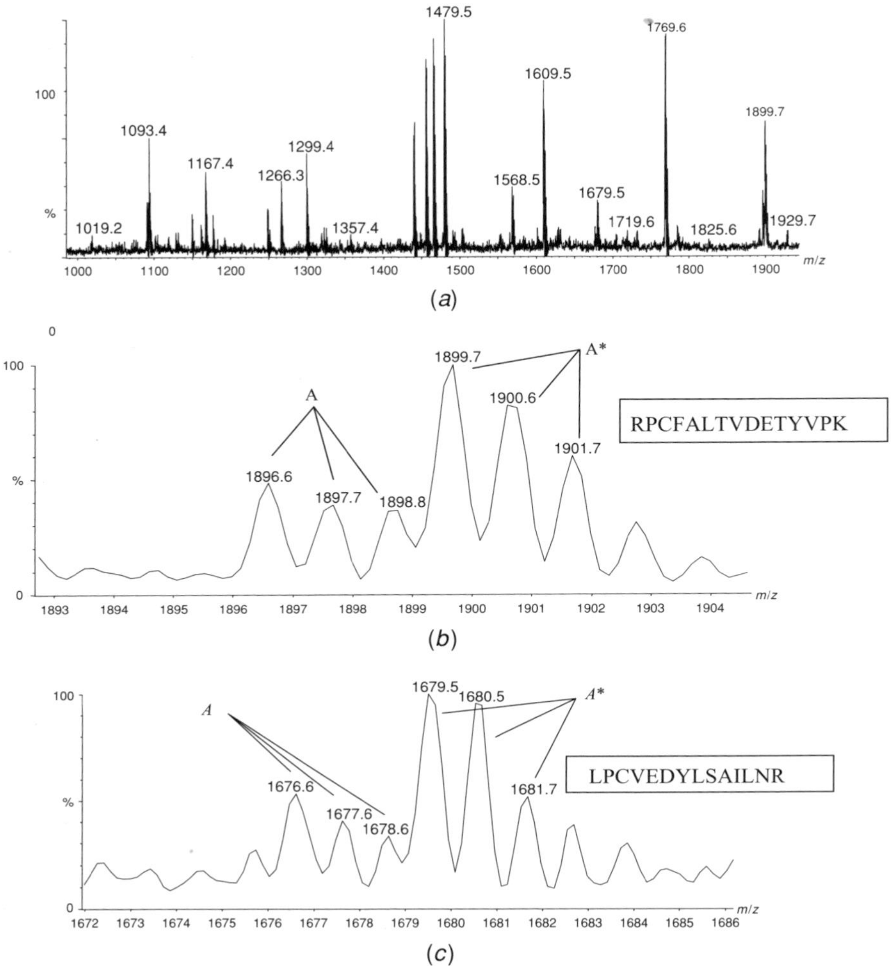

FIGURE 3.5. (*a*) Reflector MALDI mass spectrum of in situ digest of albumin (spot VI) in Figure 3.4. (*b,c*) Two short intervals taken from spectrum of total digest of same spot as in (*a*). (From Gehanne et al., 2002, with permission.)

of poor/incomplete alkylation? A partial answer to this question can be given by the following considerations: First, failure to achieve complete alkylation renders certain –SH groups vulnerable to attack by nonpolymerized chemicals within the gel and protein–protein interaction with the inevitable formation of aggregates. It has been also shown that poor alkylation may result in a considerable loss of spots, particularly in the alkaline gel region, a phenomenon which could manifest when certain proteins, at their isoelectric point (pI), regenerate their disulfide bridges with concomitant formation of macroaggregates which become entangled within the polyacrylamide gel fibers, thus quenching their transfer to the subsequent SDS–PAGE. The formation of aggregates in the alkaline pH region has been demonstrated by Herbert et al. (2001). The authors compared 2D maps of human globin chains with and without alkylation prior to the first dimension and the resulting spots were examined by MALDI–TOF–MS. These results showed that the alkylated sample generated a 2D map containing a single string of spots at ~16 kDa, while in the case of the nonalkylated sample a number of strings in the range 16–70 kDa were observed. The MALDI–MS analyses have revealed that these unexpected high-molecular-weight spots were simply homo- and heteroaggregates of α- and β-globin chains. If we are to consider that such simple polypeptides where one (α-) contains a single Cys residue while the other (β-) contains two Cys residues can generate such large aggregates, then we should take the appropriate steps to prevent their formation, particularly when dealing with complex mixtures such as total cell lysates. The drawbacks of alkylation performed between the first and second dimensions have also been pointed out by Yan et al. (1999). These authors noted that poor alkylation was responsible for the interaction between free acrylamide monomer and certain proteins during the SDS–PAGE. This interaction was responsible for the observation of a major peak, Cys-carboxyamidomethyl, and a minor one, Cys-propionamide, both associated with the same protein. The question of gel-induced modifications provoked by the various components of commonly used matrices has been amply described in a number of recent publications. These components included acrylamide and some of its N-substituted monomers (Bordini and Hamdan, 1999; Bordini et al., 1999a,b; Hamdan et al., 2001); Immobiline chemicals (Bordini et al., 2000), and crosslinkers (Galvani et al., 2000).

Despite these shortcomings, this approach has a number of advantages. The labeling reagents are inexpensive and do not need additional chemical or analytical manipulation following their purchase. Acrylamide is commonly used in the solubilization cocktails used to prepare proteins prior to their separation by 2DE, which means that no additional efforts are required for the labeling protocol. Furthermore, the relatively small size of acrylamide compared with the ICAT renders the first a more efficient alkylating agent, particularly when such reaction is performed on intact proteins. The 2D MALDI–MS approach has the additional advantage of allowing the examination of the intact proteins prior to their digestion, a feature which under certain circumstances may allow the identification of modifications both posttranslational and in gel-induced forms (e.g., phosphorylation, protein–acrylamide complexes).

3.1.4. Labeling Lysine Residues

It has been known for a number of years that MALDI is biased toward species that contain arginine residues over those that contain lysine (Krause et al., 1999). The same study reported that arginine-containing peptides were detected with 4 to 18-fold more intensity than lysine-containing peptides; the increase in intensity is commonly attributed to a higher condensed-phase basicity and/or gas-phase basicity of the arginine-containing peptides. Existing studies in proteome analyses leave no doubt that an enhanced ionization efficiency of tryptic peptides has a substantial influence on the outcome of database searches. This deduction has been exploited by Hale et al. (2000), who proposed a derivatization procedure to facilitate de novo sequencing of lysine-terminated tryptic peptides through the guanidination of the ε-amino group. This reaction converts lysine residues into the more basic homoarginine without modification of the amine terminus of the same residue. Within the same year, three other groups (Beardsley et al., 2000; Brancia et al., 2000; Keough et al., 2000) reported the same approach and applied it for protein analysis. These studies have provided useful information on this reaction, including the experimental conditions that might influence its efficiency. Keough et al. (2000) showed that the guanidination was highly dependent on the duration of the reaction and its pH. An observation which, for example, could explain why the MALDI mass spectra from solutions at pH 13.6 were dominated by hydrolysis products and only solutions at a final pH 12.9 and 12.3 provided significant yields of the desired guanidination.

This derivatization procedure was later adopted by Cagney and Emili (2002), who introduced the term mass-coded abundance tagging (MCAT) and demonstrated its use in combination with LC/MS–MS for the relative quantification of proteins in mixtures. As in the case of ICAT, this approach is based on the principle of modifying a specific amino acid residue and following its course through the use of chromatography coupled to MS and database searching to identify and possibly quantify the precursor protein(s) associated with the two different states. When comparing the initial work on ICAT with MCAT, two qualitative differences can be identified: The first chooses Cys for modification through an alkylation reaction, whereas MCAT uses guanidination to convert C-terminal Lys to homoarginine. In the ICAT approach the reaction is effected on intact proteins, whereas the MCAT applies the reaction on the tryptic peptides. In the ICAT approach the mass difference between two identical peptides (one modified) is 8 Da, whereas the reaction used for the MCAT approach results in a 42-Da difference (in both cases the difference refers to singly charged ions). Apart from these differences, the two approaches adopt a very similar procedure for the relative quantification of proteins present in two different states. In the MCAT approach the two tryptic digests before and after guanidination are combined with a predetermined ratio, separated by reverse-phase capillary LC, and introduced into an ESI source for MS and MS/MS analysis. The resulting full-scan mass spectra are used to obtain the relative abundance of correlated peptides by comparing the intensities of their reconstructed single-ion chromatograms. Because the guanidination reaction is specific to the Lys side chain, it can be exploited to identify y-ions in CID–MS/MS spectra; fragmentation of Lys-containing peptides before and after modification would reveal the y-ions as those that shifted by 42 Da. Furthermore, the problem of partial guanidination can also prove to be advantageous for protein identification (but not for quantification) because database-searching algorithms will detect both modified and unmodified peptides. One must bear in mind, however, that this relative quantification does not take into account the inherent differences in the ionization efficiency between Lys (untreated sample) and homoarginine (treated sample), particularly when MALDI–MS is used. Although the reaction was initially described as being relatively slow to achieve complete modification of the intact proteins, latter studies (Hale et al., 2000) have demonstrated that for unhindered peptides the reaction proceeds to near completion in 1 h at pH 10 and 37°C. Of course a short reaction time is central in any strategy which is designed for high throughput analyses.

Another approach to modify C-terminal Lys residues has been described by Peters et al. (2001). The authors described the use of a multifunctional labeling reagent, 2-methoxy-4,5-dihydro-1 *H*-imidazole, to replace *O*-methylisourea, which is commonly used for the guanidination reaction. Figure 3.6 compares the two reactions used to modify Lys residues. To allow quantitative analysis, four deuterium atoms can be

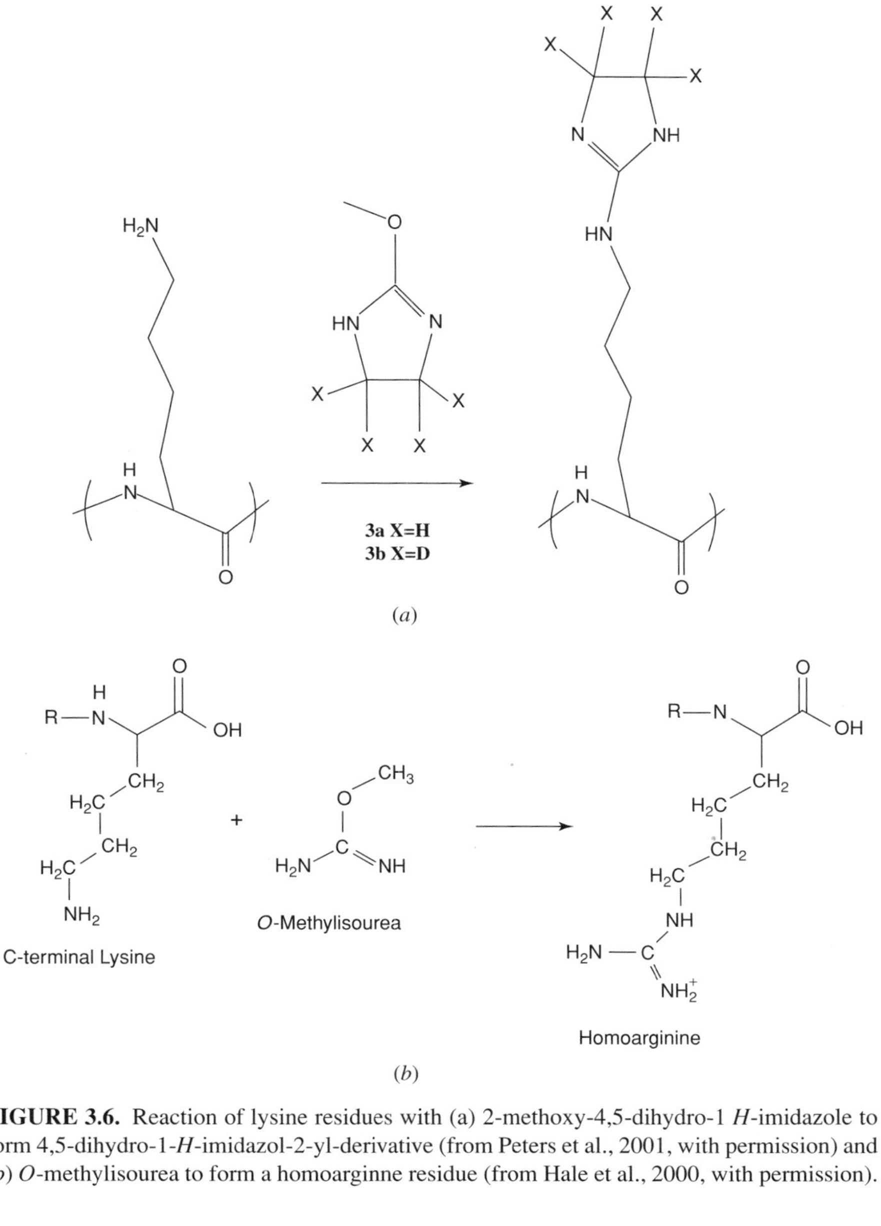

3a X=H
3b X=D

(*a*)

C-terminal Lysine

O-Methylisourea

Homoarginine

(*b*)

FIGURE 3.6. Reaction of lysine residues with (a) 2-methoxy-4,5-dihydro-1 *H*-imidazole to form 4,5-dihydro-1-*H*-imidazol-2-yl-derivative (from Peters et al., 2001, with permission) and (*b*) *O*-methylisourea to form a homoarginne residue (from Hale et al., 2000, with permission).

incorporated during the synthesis of the reagent. As we have indicated earlier, this kind of derivatization favors C-terminal fragment ions, resulting in a simplified tandem mass spectra dominated by the y-ions.

3.1.5. Labeling Tryptophan Residues

In a recent report Kuyama et al. (2003) have described a method for relative protein quantification based on labeling of tryptophan residues through the reaction with light and heavy 2-nitrobenzenesulfenyl chloride (NBSCL–$^{12}C_6$ or NBSCL–$^{13}C_6$). The two versions of this chemical reagent differ by 6 Da, a difference that can be used in MS analysis to distinguish and quantify two identical sequences present in two different protein mixtures. It is not difficult to note that the underlying principle of this approach is identical to that applied in the ICAT and MCAT protocols. The main difference, of course, is in the chemistry of labeling and the choice of the amino acid to be labeled. The authors reported that the NBS reagent exhibited selectivity for the indole ring of tryptophan and could incorporate a sixfold stable isotope labeled benzene nucleus. The main steps in the experimental procedure can be summarized as follows. The tryptophan residues in a denatured protein sample derived from cell state 1 are modified with isotopically light NBS; the equivalent residues derived from cell state 2 are modified with the isotopically heavy NBS. The two samples are combined and passed through a Sephadex LH-20 column to remove excess reagent and other small molecules. Disulfide bridges in the eluting mixture were reduced and alkylated and the mixture was subjected to enzymatic digestion. Peptides containing labeled tryptophan were enriched using the Sephadex LH-20 column and were analyzed by MALDI–TOF–MS or LC–ESI–MS. This approach was tested for fairly simple standard mixtures of peptides to verify the reliability of the quantification and the influence of labeling on the elution times of light- and heavy-labeled peptides. Biologically derived samples were also spiked with known proteins and examined by the same approach. These results demonstrated the validity of the approach (for the examples reported by the authors). Having said that, we stress that this approach, like others in the literature, has still to demonstrate its capabilities when tested with complex Iysates or complex bodily fluids. What this study demonstrates is that the principle is valid but does not provide any evidence that such an approach will fare better compared to existing techniques when it comes to protein quantification in real-life examples. This is by no means a criticism of the approach; it is simply to point out that such an approach is a member of a rapidly growing family of methods each of which has its strengths and limitations. To underline this observation, here are some general considerations. We have pointed out that the ICAT approach fails for Cys-free proteins. An identical situation is encountered in the present approach where it fails to detect tryptophan-free proteins. Furthermore, the scarcity of tryptophan residues in proteins will certainly result in missing a number of proteins higher than those missed in the ICAT protocol. The isotope effect on the accuracy of quantification cannot be ignored in such an approach; as we have mentioned in the case of the ICAT approach, the deuterated peptide elutes earlier than its nondeuterated counterpart. Such separation causes a variation in the isotope ratio across the two differently eluting profiles of the isoforms. In fact, Kuyama et al. (2003) were aware of such an effect and presented a single example to show the coelution of two tryptic peptides labeled with light or heavy NBS reagent.

3.1.6. Methionine-Containing Peptides

Current literature cites two approaches for the analysis (no quantification) of methionine-containing peptides. Both approaches, if combined with some form of isotope labeling, could be used for protein quantification. The first approach uses commercially available methionine-specific beads to covalently attach methionine-containing peptides via bromoacetyl functional groups to a solid support (Weinberger et al., 2002). After removal of the unbound peptides, the methionine-containing peptides are released from the support with β-mercaptoethanol and analyzed by MS. The second approach is based on the isolation of methionine-containing peptides through the use of diagonal chromatography. In this approach, side chains of methionine are oxidized between different chromatographic runs, thus resulting in increased polarity and therefore earlier elution of the oxidized peptides in the reverse-phase LC separation (Gevaert et al., 2002). One obvious drawback of methionine enrichment is that methionine residues are prone to oxidation in the cell or during sample preparation, an effect which interferes with the accuracy of such measurements, particularly if they are to be used for protein quantification.

3.2. GLOBAL INTERNAL STANDARD TECHNOLOGY

The term *global internal standard technology* (GIST) was first introduced by Chakraborty and Regnier (2002). This technology encompasses more than one approach, which have been described by various research groups for protein quantification. The general idea behind such a strategy is to label more than one amino acid and allow the use of different separation methods for the eventual quantification of the labeled proteins.

3.2.1. Acylation

This method exploits the proteolytic cleavage of proteins, which generates a primary amino terminus that, along with the primary amine on lysine residues, can be easily acylated. This method was applied by Chakraborty and Regnier (2002) and tested for the quantification of β-galactosidase in *Escherichia coli*. Prior to proteolysis, the proteins were fully reduced and fully alkylated, a step which enhances the digestion of proteins resistant to proteolysis (Friedman et al., 1970). Following proteolysis, primary amino groups were amino acylated with $[^1H_3]$-*N*-acetoxysuccinimide or $[^2H_3]$-*N*-acetoxysuccinimide succinimide in water, and the sites of acetylation were examined by MS and were established to be at the amino terminus of the investigated peptides and at the ε-amine groups of lysine. This means that tryptic peptides with a carboxy terminal arginine are acetylated only once and should show an increase of

42 Da in their mass, while those with both a free amino terminus and a C-terminal lysine are acylated twice (84-Da increase in mass). This also means that, when the peptides from two different pools labeled with light/heavy reagents are mixed in a 1 : 1 ratio and analyzed by MS, their respective peaks will be shifted by 3 Da (for singly labeled) and 6 Da (for doubly labeled).

In a second variation of the above method, the same group adopted acylation of peptides via light and heavy (deuterated) succinic anhydride (Wang and Regnier, 2001). Here, too, peptides were differentially labeled and those terminating with Arg will exhibit monoisotopic peaks that are spaced apart by 4 Da, whereas those having a Lys at their C-terminus will exhibit a difference of 8 Da. Occasionally some peaks have shown a 12-Da difference which was attributed to serine-containing peptides. Such adducts were eliminated by treating the derivatized peptides with hydroxylamide at pH ~12. The attractive feature of succinic anhydride is that it is inexpensive and can be used directly as purchased. On the other hand, it has the disadvantage of reducing the ionization efficiency in Lys-containing peptides; furthermore, the same reaction renders this type of peptide more acidic, which can impact on the selection of certain chromatographic modes. When such a strategy uses $^{2}H_4$-succinic anhydrade for acylation, it would be desirable in the case of higher molecular weight peptides to have a mass difference between isoforms greater than 3 Da. Such a limitation can be partially addressed by using derivatives of $^{2}H_5$-propionate and $^{2}H_9$-propionate. However, increasing deuterium content of the labeling reagent has its own price. As noted earlier, such an increase in deuterium content aggravates the resolution problem, particularly in reverse-phase LC analysis.

This limited number of examples on the acylation approach indicates that a single labeling procedure can in theory be used to label all peptides in a given digest, regardless of their amino acid composition, excluding the N-terminally blocked proteins. An additional advantage of this approach is the flexibility in applying a wide range of separation techniques for the identification and relative quantification of proteins in mixtures. It is even possible to employ multiple types of selection with the same sample. However, as other existing strategies for protein quantification methods, this one also has some shortcomings. One of such limitation is that acylation reduces the charge on the C-terminal in Lys-containing peptides, particularly when MALDI–MS is used, unless of course such peptides happen to have another basic amino acid in their sequences. Second, as in all labeling procedures, accurate control of the extent and sites of labeling is always a problem to be tackled and the GIST approaches are no exception. For example, when *N*-acetoxysuccinimide is used for acylation, it can react with more than one lysine if present within the same tryptic peptide, an effect which, if not considered, can result in wrong quantitative analyses.

3.2.2. Esterification

The esterification of carboxyl groups of aspartic and glutamic acid residues and at the C-terminus of peptides is another method which belongs to the GIST family. This method was first introduced to address the issue of in vitro labeling of Cys-free peptides, which the already existing ICAT approach could not handle (Goodlett et al.,

2001). In this approach protein mixtures to be compared are digested and the resulting peptides are methylated using either d_0- or d_3-methanol. Methyl esterification converts carboxyl acids such as those in the side chains of aspartic and glutamic acids as well as the carboxyl terminus to their corresponding methyl esters. The two-peptide mixtures marked with d_0 or d_3 are combined with a known ratio and then examined by μLC coupled to MS–MS. There are a number of good reasons why this approach can be considered complementary to the ICAT technique: Not all proteins within a mixture contain Cys residues; some proteomes will have much lower total protein coverage than the 92% reported for yeast. Some Cys residues may contain posttranslational modifications prior to isolation, making them unavailable for alkylation with Cys-specific ICAT reagents; and while one advantage of the in vivo labeling approaches is that an internal standard for all peptides is produced, not all systems of biological interest are amenable to tissue culture growth conditions. It has to be pointed out, however, that the esterification approach has two distinct advantages compared to the ICAT approach: First, the chemistry is simple and does not require difficult-to-produce reagents and, second, the reaction is carried out at the peptide level rather than on the intact proteins. Although the latter is a minor detail, it can have a profound influence on the efficiency of the reaction and hence on the completeness of the information which can be gleaned from such protocol.

3.2.3. Incorporation of Heavy Isotopes

Several groups have used stable isotope-labeled internal standards (e.g., $^{2}H_{1}$ ^{13}C, ^{14}N, ^{18}O) to obtain global quantitative protein profiles. Stable isotopes have been introduced to proteins via metabolic labeling using heavy salts or amino acids (Conrads et al., 2002), enzymatically via transfer of ^{18}O from water to peptides (Oda et al., 1999; Mirgorodskaya et al., 2000; Yao et al., 2001). In principle, isotope tags can be incorporated into proteins during cell growth or after cell lysis. For example, Oda et al. (1999) used this approach in conjunction with LC/MS and SDS–PAGE/MS to identify and quantify a number of proteins in yeast culture. The authors compared two samples: The first was grown on a medium that contained a natural abundance of nitrogen, and another culture was grown on a medium chemically enriched with ^{15}N. At the end of the growing period, the cell pools were processed, combined, and separated by reverse-phase LC and SDS–PAGE and were analyzed by MS. The authors measured protein expression of 42 high-abundance proteins derived from two pools of *Saccharomyces cerevisiae*. In the same study, the authors measured differential phosphorylation states in the yeast proteins.

In another study Smith et al. (2002) used capilary LC combined with accurate mass measurements to obtain quantitative changes in protein abundances. In their study, *Deinococcus radiodurans* was cultured in two media, the first containing the natural abundance of ^{14}N and ^{15}N and the second medium chemically enriched with ^{15}N (>98%). The two cultures were mixed and processed and the combined proteome sample was analyzed by capillary LC coupled to high-resolution FT–ICR MS.

Yao et al. (2001) described stable isotope labeling which could provide quantitative data as well as a comparison between individual proteins from two entire proteome pools or their subfractions. Basically two ^{18}O atoms are incorporated universally into the carboxyl termini of all tryptic peptides during the proteolytic cleavage of all proteins in the first pool. Proteins in the second pool are cleaved analogously with the carboxyl termini of the resulting peptides containing two ^{16}O. The two peptide mixtures were pooled for fractionation and separation and examined by high-resolution MS, which yielded the accurate masses and the isotope ratios of peptide pairs which differ by 4 Da. Relative signal intensities of the paired peptides were used to quantify the expression levels of their respective precursor proteins from proteome pools to be compared.

It is now well established that, in addition to the functional groups on the peptide side chains, two additional potential sites, N- and C-termini, can be exploited as targets for stable isotope tagging. The latter sites have the advantage of being present in nearly all proteolytic peptides, which renders the approach highly attractive for a semiglobal approach in protein quantification. Geng et al. (2000) reported the labeling of tryptic peptides by acetylation of N-termini using acetyl *N*-hydroxysuccinimide or its trideuteriocetylated analogue. Isotope ratios of peptides were determined by MALDI MS. Tagging the C-termini of tryptic peptides has the additional advantage of providing sequence information in mass spectrometers. Schnöelzer et al. (1996) have conducted a systematic study on protease digestion of proteins in highly ^{18}O-enriched water and reported that, while chymotrypsin and Asp-N incorporated only one ^{18}O atom, trypsin, Glu-C, or Lys-C could incorporate two ^{18}O atoms into the C-terminal of the resulting peptides. The same study as well as that reported by Murphy et al. (1979) have shown that ^{18}O labels involving the carboxylate group in peptides and amino acids were resistant to back exchange. This is of course an important characteristic of the reaction because it means that bonds between the incorporated oxygen atoms and the carbonyl carbon are highly stable in analyses involving LC, ESI and MALDI.

3.3. DIFFERENTIAL IN-GEL ELECTROPHORESIS

To assess protein differences between two samples, the conventional 2DE methodology relies on the comparison of at least two different gels. Unfortunately, no two gels can be identical due to inhomogeneity in polyacrylamide gels, electric and pH fields, and thermal fluctuations. Even with the recent advances in spot-matching software together with the maximum attention to experimental details, complete spot matching remains a difficult task which often necessitates the use of multiple 2D gels to analyze a single sample. Most of these shortcomings were eliminated by the introduction of a new methodology by Ünlü et al. (1997), which they termed difference in-gel electrophoresis (DIGE). In this approach, two cyanine dyes (Cy3 and Cy5) are commonly used to label two different samples which are subsequently mixed and run on the same gel. These two dyes are mass and charge matched and possess distinct excitation and emission spectra. In later studies (Knowles et al., 2003; Kondo et al., 2003) a third dye (Cy2) was used as an internal standard to provide a link for intergel comparison and to facilitate a more robust statistical analysis. These dyes undergo nucleophilic substitution with ε-amine groups of lysine residues present within the investigated sequences. Before considering some of the recent applications of the DIGE approach, we ought to consider some relevant experimental observations which have been emphasized by the authors who first introduced this approach: (i) The dyes must have similar M_r and charge. (ii) The dyes must match the charge of the residue that they modify. (iii) The dyes must possess different fluorescence spectra to allow reliable differentiation.

The initial DIGE method (Ünlü et al., 1997) was evaluted by determining the detection limit of labeled bovine serum albumin in *Drosophilia* extract and the identification, isolation, and microsequencing of an inducible, cloned protein from *E. coli* extracts.

The same authors reported that the initial approach was to label all Lysines through the use of excess dye, since incomplete reaction was expected to cause heterogeneity in electrophoretic mobility. However, it was found that carrying out the labeling reaction to completion reduced protein solubility and resulted in protein precipitation before entering the gel, probably due to the replacement of primary amino groups with the hydrophobic cyanine dye. To avoid such an effect, the authors adopted what they termed a minimal labeling scheme in which only 1% to 2% of all the Lysines were labeled. However, this minimal labeling still had a detectable effect on the relative molecular masses of the labeled proteins, particularly in the low-molecular-mass region, such effect being detected when comparing fluorescent with silver-stained images. One further point which has to be considered in this type of analysis is the influence of the labeling dyes on the pI of the investigated proteins. In theory, the replacement of a number of ionizable positively charged primary amino groups with nonionizable quaternary amino groups of the dye should result in shifting the pK_a from about 9 to 14. It can be argued that most commercially available IPG strips are in the pH 3–10 interval, and therefore such displacement in the pK_a of the labeled residues should not have a detectable influence on the analysis; yet if we were to move to a higher pH region, then such effect should be considered.

In a number of recent articles the application of DIGE for the assessment of protein expression and their comparison in different samples has been demonstrated. A recent article (Knowles et al., 2003) described the use of DIGE in a multiplex analysis of two distinct proteomes. As a model system, cerebral cortex tissues were analyzed from neurokinin 1 receptor knockout (NK_1R −/−) and (NK_1R +/+) mice. In this protocol Cy2-labeled sample was run together with all Cy3- and Cy5-labeled sample pairs as an internal standard providing a link for intergel comparisons and for more reliable statistical analysis of the data. Figure 3.7 summarizes the main steps [excluding the use of Cy2 as an internal standard of the protocol used by Knowles et al. (2003)]. These authors also pointed out that protein labeling was limited to about 5% of each protein. MALDI–MS was used to confirm the identity of the proteins, which have experienced up or down regulation. G. Zhou et al. (2002) used DIGE in conjunction with LCM and MS to compare and quantify various proteins in esophageal carcinoma cells and normal epithelial cells. These analyses identified 1038 protein spots in cancer cell lysates and 1088 protein spots in normal cell lysates Among the detected spots, 58 were up regulated by more than threefold and 107 spots were down regulated by

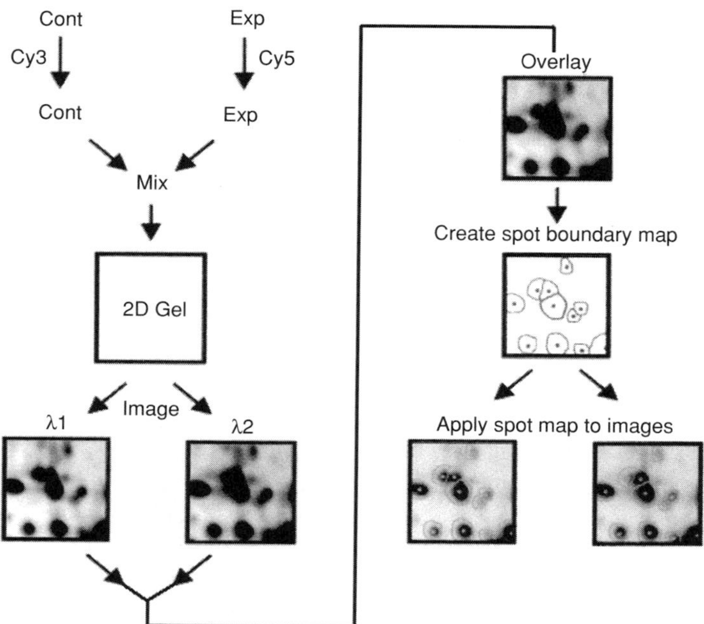

FIGURE 3.7. Schematic representation of various steps in DIGE analysis of two samples, control (Cont) and experimental (Exp) labeled with cyDyes, Cy3, and Cy5 respectively. (Adapted from Knowles et al., 2003, with permission.)

more than threefold in cancer cells. Gharbi et al. (2002) have also used DIGE together with MALDI MS to assess protein expression in a model breast cancer cell system. The authors have investigated ErbB-2-mediated transformation in a model cell line system comprised of an immortalized luminal epithelial cell line and a derivative stably overexpressing ErbB-2 at a similar level to that seen in breast carcinoma.

Recently (Shaw et al., 2003) described a new version of CyDye (Cy3, Cy5) which targets Cys rather than Lys residues to label proteins prior to DIGE analysis. These "new' dyes were recently used by Kondo et al. (2003) in conjunction with LCM for the analysis of specific populations of cells in cancer tissues. These dyes, which are developed by Amersham Biosciences, are supposed to be the answer to a shortcoming of the previous generation which necessitated minimal labeling. As all newly introduced labeling reagents, it is difficult to give a rational assessment of their performance compared to their older and well-tested counterparts. Having said that, some partial assessment of the pros and cons of these new dyes can be given: First, both sets of dyes have comparable M_r values (~0.5–0.6 kDa), and therefore unlabeled proteins will have a higher mobility during 2DE analysis. The extent of the phenomenon will surely influence the reliability of spot excision for subsequent MS analyses. Second, one of the characteristics of the new set of reagents is that they target Cys, which is less prevalent than Lys, and therefore the reaction can be driven to saturation without an excessive use of the reagents (in theory at least). This characteristic, however, has a price. We know that at least one out of seven known proteins do not have Cys in their sequences (Vuong et al., 2000) and therefore they will be excluded from such detection. It is worth pointing out that the same shortcoming was pointed out in the ICAT strategy. It might be argued that in the absence of Cys the next favorite target would be the Lys residues. This might be true. However, considering the pK_a of Cys (8.3) and Lys (10.3) together with the recommended working pH (6.5–7.5), one would ask what sort of efficiency such reaction would have?

Considering the existing literature on the DIGE technique, it can be deduced that such an approach has been developed to facilitate a direct and reproducible comparison between mixtures of proteins derived from cell tissues or bodily fluids. Its main advantage is the elimination of the need for multiple gels and the wider dynamic range of flurescence detection (at least in theory). When running two samples on the same gel, an additional internal control is provided, thus allowing for a faster and reproducible identification of differences in protein composition. Furthermore, once protein differences have been detected, DIGE provides the possibility to scale up to preparative levels for microsequencing. As other existing methods for protein analysis, DIGE requires a number of experimental precautions if its full potential is to be achieved. Such precautions have been pointed out at the start of this section.

Before closing this section, we find it relevant to say a few words regarding the alternative set of dyes cited in the works by Shaw et al. (2003) and Kondo et al. (2003). The idea behind this new set is the opposite to that on which the original DIGE is based. In other words the minimal labeling of Lys residues is replaced by the maximum (saturation) labeling of the Cys residues. A close look at the data presented in both works suggests that the use of the new set of dyes will not add any advantage to the existing performance of this protocol. In an example given by Shaw et al. (2003) the authors compared silver-stained 2D maps with their Cy3/Cy5 Lys and Cy3/Cy5 Cys counterparts. Such a comparison revealed superior sensitivity and a superior spot number in the silver-stained map. The inferior number of spots in the maps labeled with cyanine dyes strongly indicates substantial precipitation of barely soluble proteins. We realize that the absence of sufficient examples on the use of the new set of dyes renders their rational assessment rather difficult. However, the initial data provided in the works of Kondo et al. (2003) and Shaw et al. (2003) suggest that the new variation of dyes is unlikely to have a substantial influence on the performance already achievable by the dyes which react with Lys.

3.4. QUANTIFICATION OF MODIFIED PROTEINS

Although the identification of proteins in complex mixtures is becoming routine, such identification alone provides only a limited insight into protein functions and signaling pathways. Covalently modified proteins provide an important component of protein regulation and protein functions. These modifications (co- or posttranslational) cannot be obtained from protein sequences deduced from nucleotide sequences. Over 200 different modifications have been described (Krishna and Wold, 1993). Many of these modifications, such as phosphorylation and glycosylation, have well-documented roles in signal transduction, regulation of cellular processes, clinical biomarkers, and therapeutic targets. A limited number of recent strategies have demonstrated the potential for large-scale analysis of phosphorylated and glycosylated proteins. Some of these approaches are considered below.

3.4.1. Detection and Quantification of Phosphorylated Proteins

Over the last 20 years there have been various attempts to analyze phosphoproteins. Most of these approaches were applied to the area of phosphopeptides. Iron metal affinity chromatography or phosphopeptide-specific antibodies have been used to enrich phosphopeptides for analysis (Michael et al., 1998). Other methods have used ^{32}P labeling to obtain enrichment before analysis by standard phosphopeptide mapping or by MS (Watts et al., 1994; de Carvalho et al., 1996). Mass spectrometry methods that use characteristic fragment ions to detect phosphorylation have also been used to analyze peptide mixtures (Annan et al., 2001). Although the phosphorylation sites Tyr, Ser, or Thr can be identified using a combination of 2D PAGE with various forms of MS–MS, the detection and quantification of low-abundance regulatory proteins remain extremely difficult task. Radiolabeling with ^{32}P-labeled inorganic phosphate has been the dominant method for detecting changes in protein phosphorylation (Van der Geer and Hunter (1994). Recently, increased efforts have been put into the development of derivatization strategies which use affinity purification and mass spectrometric quantification of phosphorylated proteins. Three of these methods (Oda et al., 2001; H. Zhou et al., 2001; Goshe et al., 2002) use multistep chemical derivatization strategies for the enrichment of phosphopeptides prior to MS analysis.

Oda et al. (2001) have used base-induced elimination of phosphate from Ser and Thr residues to generate a reactive acrylate double bond that is coupled by Michael addition to ethane-1,2-dithiol. This experimental protocol starts with a protein mixture in which cysteine reactivity is removed through the oxidation with performic acid. Base hydrolysis is used to induce β-elimination of phosphate from phosphoserine and phosphothreonine followed by the addition of ethanedithiol to the alkene. The resulting free sulfhydryls are coupled to biotin, allowing purification of phosphoproteins by adivin affinity chromatography. Following elution of phosphoproteins and proteolysis, enrichment of phosphopeptides is carried out by a second round of avidin purification. The tryptic peptide mixture is then analyzed by MALDI–TOF–MS or LC–ESI/MS–MS. In these measurements the presence of the labeled peptides can be recognized by monitoring a characteristic fragment at $m/z = 446$ originating from the biotinylated side chain. This approach can suffer certain interferences which limit its specificity and sensitivity. For example, it requires the blockage of the reactive thiolates of cysteinyl residues via reductive alkylation or performic acid oxidation. Such reactions also result in the oxidation of tryptophan indole rings and of methionine residues to mixtures of sulfoxides and sulfones. Furthermore, the base-induced phosphate elimination is not entirely specific and can also affect serine and threonine glycosylation sites.

The method of Zhou et al. (2001) starts with a proteolytic digest that has been reduced and alkylated to eliminate reactivity of the cysteine residues. Following N-terminal and C-terminal protection, phosphoramidate adducts at phosphorylated residues are formed by carbodiimide condensation with cystamine. The free sulfhydryl groups produced from this step are covalently captured onto glass beads coupled to iodoacetic acid. Elution with trifluoroacetic acid then regenerates phosphopeptides, which are then analyzed by MS. The authors claimed that initial experiments showed that 80% of peptides recovered from yeast extracts were phosphorylated. Figure 3.8 compares the two experimental protocols used by Zhou et al. (2001) and Oda et al. (2001).

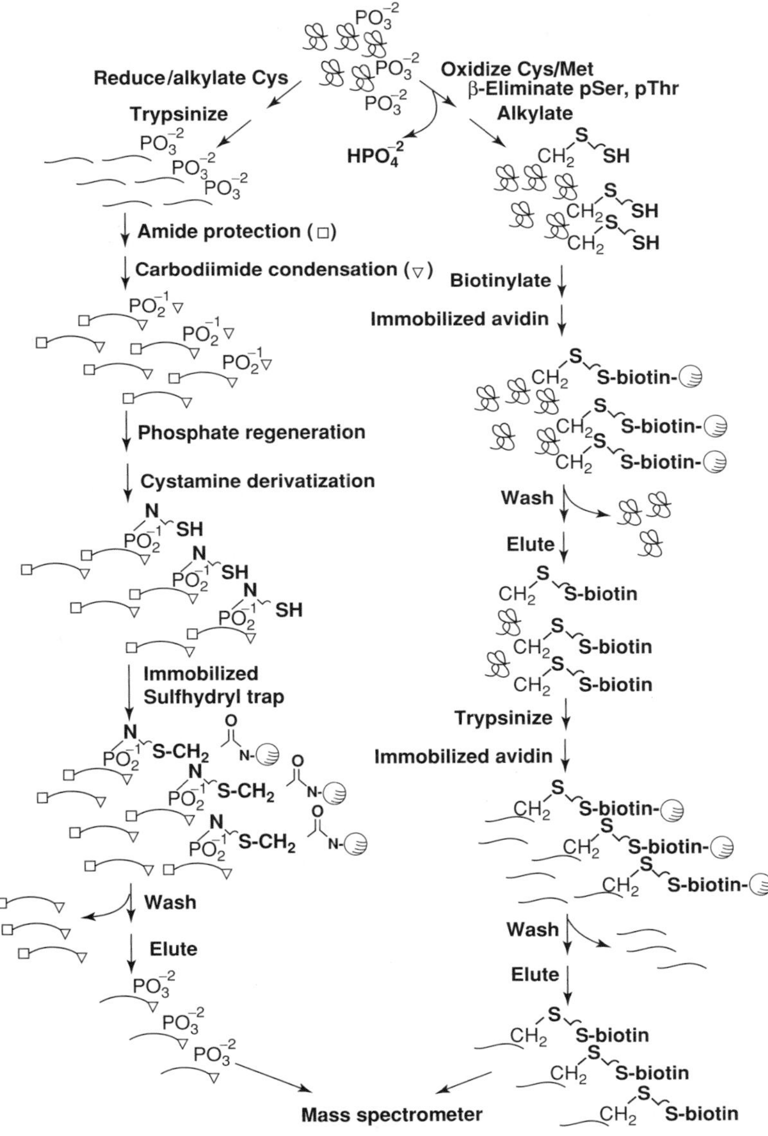

FIGURE 3.8. Strategies for chemical derivatization, purification, and MS characterization of phosphopeptides. Left: protocol of Zhou et al., 2001; Right: Protocol of Oda et al., 2001. (Adapted from Ahn and Resing, 2001, with permission.)

A similar approach by Goshe et al. (2002) used ethane-1,2-dithiol and [1,1,2,2-d_4]ethane-1,2-dithiol for Michael addition to serine and threonine residues following phosphate elimination. The affinity tag was introduced via a biotinyl–PEG–diamine–iodoacetamidyl reagent that reacted selectively with the terminal thiol groups. Alternatively, ethane-1,2-dithiol was first conjugated to biotinyl–PEG–diamine–iodoacetamidyl and the conjugate was then used for Michael addition to dephosphorylated serine. Affinity capture–release was performed with tryptic peptide mixture. Several singly and multiply charged phosphorylated peptides were detected from β-casein as groups of d_0 and d_4 peaks that indicated the potential of the method for quantification.

Another approach for phosphopeptide mapping has been described by Annan et al. (2001) as a multidimensional electrospray MS-based strategy. In this case the multidimensionality refers not to the separation but to the use of more than one MS scanning modes. Basically, during LC/ESI–MS two diagnostic ions, PO^{2-} and PO^{3-}, are monitored to allow the identification and collection of the LC fractions containing phosphopeptide. The authors described this step as the first dimension while in the second dimension the molecular weights of the collected phosphopeptides were measured by performing negative-ion precursor ion scanning for $m/z = 79$. This approach was tested for a number of standard mixtures.

In a recent article MacCoss et al. (2002) described what they termed a "shotgun" approach for the identification of various forms of protein modifications (including phosphorylation) in complexes and in lens tissue. To digest the investigated protein mixtures, the authors used three different enzymes, one that cleaves at a specific site while the other two cleave at nonspecific sites. The mixture of the resulting peptides was separated by multidimensional LC and analyzed by MS–MS. This approach has been applied to a simple protein mixture, Cdc2p protein complexes isolated by affinity tag, and to lens tissue from a patient with congenital cataracts. These results yielded various sites of phosphorylation, acetylation, methylation, and oxidation.

In a recent article Wind et al. (2003) described a combination of 1D gel electrophoresis and laser ablation inductively coupled plasma–mass spectrometry (ICP–MS) with ^{31}P for detection. In this approach a reverse phase at acidic pH is used to separate in-gel digested peptides where covalantly bound phosphorus mainly represents phosphate esters of serine, threonine, and tyrosine. The method has been tested for a standard mixture of myoglobin, α-casein, and reduced fibrinogen. A special washing step was found to be necessary to remove noncovalently bound phosphate. The authors claimed that the ^{31}P signal was found to contain quantitative information regarding both the relative and absolute amounts of phosphorus present in phosphoproteins. Furthermore, normalizing the ^{31}P signal generated by a single laser ablation trace by the total amount of phosphoprotein applied to the gel, a detection limit of 5 pmol was estimated.

3.4.2. Detection and Quantification of Glycosylated Proteins

Protein glycosylation has long been recognized as a common posttranslational modification and has been increasingly recognized as one of the central biochemical alterations associated with various forms of disease. From the analytical point of view, glycosylation is one of the most challenging tasks, simply because carbohydrates are the most abundant and structurally diverse compounds found in nature. Unlike the linear polymers, such as proteins and nucleic acids, oligo and polymeric carbohydrates can form branched structures because the linkage of the constituent monosaccharides can occur at a number of positions. It has been calculated (Laine, 1994) that for a simple hexasaccharide there are in excess of 1.05×10^{12} possible isomeric structures. Luckily nature is far more merciful than most of us tend to think; therefore highly specific biosynthetic pathways which are limited by the available glycosyltransferases render the above number far less prohibiting. Typically, carbohydrates are linked to the side chains of serine or threonine residues (O-linked glycosylation) or of asparagine residues (N-linked glycosylation). Despite the fact that glycoproteins represent an attractive target in the search for biomarkers and new therapeutic targets, an indication of the difficulties associated with the characterization and quantification of glycoproteins is reflected in the limited number (only 172) of experimentally confirmed human glycoproteins listed in the current Protein Information Resources Protein Sequence Database (http://pi.georgetown.edu/pirwww/search/textpsd.shtml) (Kaji et al., 2003).

In two recent articles which appeared in the same issue, two independent groups (H. Zhang et al., 2003; Kaji et al., 2003) described similar strategies for the identification and quantification of N-linked glycoproteins. The first group used a strategy which combines hydrazide chemistry, stable isotope labeling, and MS, and the second group used lactin affinity capture in combination with isotope-coded tagging and MS. The approach by Kaji et al. (2003), termed isotope-coded glycosylation site-specific tagging (IGOT), is based on the lactin column-mediated affinity capture of glycopeptides generated by tryptic digestion of protein mixtures followed by peptide-*N*-glycosidase-mediated incorporation of stable isotope tag ^{18}O specifically into the N-glycosylation site. The tagged peptides are then identified by multidimensional LC coupled to MS. This approach was tested on N-linked high-mannose and and/or hybrid-type glycoproteins derived from an extract of *Caenorhabditis elegans*. The authors reported the identification of 250 glycoproteins, including 83 putative transmembrane proteins, with the simultaneous determination of 400 unique N-glycosylation sites. To demonstrate the potential of the IGOT strategy for protein quantification, the authors processed two peptide aliquots differentially labeled with ^{18}O and ^{16}O and the mixed preparation was examined by LC/MS. The authors noted that, although the isotope distribution of the two-tagged peptides partly overlapped owing to the natural isotopic abundance, both spectra were good enough to permit relative quantification of ^{16}O- and ^{18}O-tagged peptides.

The difference between the above approach and that reported by H. Zhang et al. (2003) lies in the mechanisms by which the glycopeptides are captured to the solid support and in the labeling isotopes, as shown in Figures 3.9*a*,*b* (H. Zhang et al., 2004). The approach by H. Zhang et al. (2003) was tested on two protein mixtures containing three glycoproteins, α-1-antichymotrypsin, α-1-antitrypsin, and α-2-hs-glycoprotein. The identification and quantification was based on μLC/MS–MS spectra in which d_0/d_4 tagged peptide ratios were obtained. To assess the potential of the glycopeptide

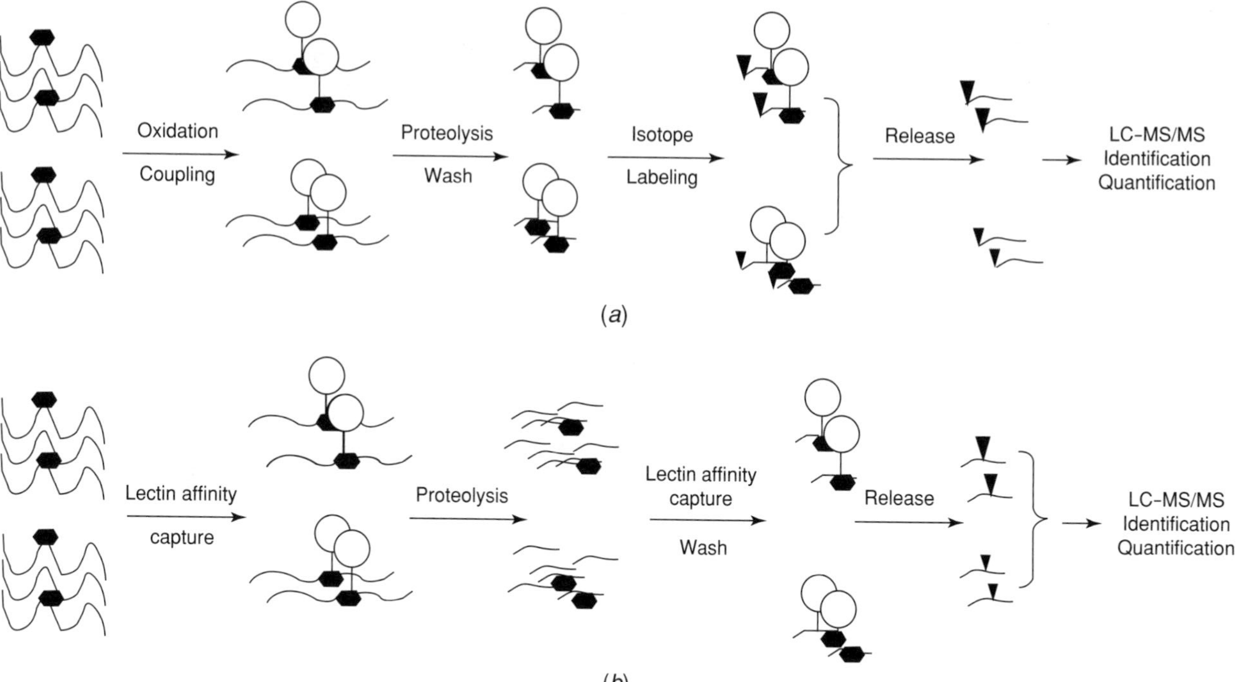

FIGURE 3.9. Schematic diagram of quantitative analysis of N-linked glycopeptides using (*a*) hydrazide chemistry and (*b*) lectin affinity capture. (Adapted from Zhang et al., 2004 with permission.)

capture method for the analysis of cell surface proteins, the same authors used a crude membrane fraction from the LNCaP prostate cancer epithelial cell line to select and identify N-linked glycosylated peptides. This experiment, which used single-dimensional μLC/MS–MS, allowed the identification of 104 unique peptides associated with 64 proteins.

3.4.3. Ubiquitination

The phenomenon of protein ubiquitination has been known for almost 30 years (Goldknopf et al., 1975). The involvement of ubiquitination in processes as diverse as cell cycle regulation, DNA repair, and receptor-mediated endocytosis demonstrates the biological significance of such a phenomenon (Finley, 2001; Ciechanover et al., 2000). Although such biological significance is perceived, its study remains more challenging than other forms of modifications, being inherently difficult because the modification is large (~8 kDa) and because the turnover of ubiquitinated proteins is very rapid, so that steady-state conjugate levels are characteristically low. In two recent reports, Marotti et al. (2002) and Peng et al. (2003) have described the use of MS for the identification of protein ubiquitination sites. The work by Marotti et al. (2002) was the first example on the use of MS for the direct identification of an in vivo ubiquitination site. The same work was a rare example of a G protein–signaling component that undergoes ubiquitination. Peng et al. (2003) have described a proteomics approach to enrich, recover, and identify ubiquitin conjugates from *S. cerevisiae* lysate. This method exploits a characteristic of trypsin proteolysis of ubiquitin-conjugated proteins which produces a signature peptide at the ubiquitination site containing glycine–glycine residues derived from the C-terminus of ubiquitin and that is still covalently attached to the target lysine residue. In the study by Peng et al. (2003), ubiquitin conjugates from a strain expressing 6xHis-tagged ubiquitin were isolated, proteolized with trypsin, and analyzed by multidimensional LC coupled to MS–MS. The authors claimed the identification of 1075 proteins containing 110 specific ubiguitination sites present in 72 ubiquitin–protein conjugates. Furthermore, ubiquitin itself was found to be modified at seven lysine residues. The authors observed that the detected conjugates reported in their study could only represent a subset of the total ubiquitin conjugates. According to the authors, such partial identification can be attributed to two possible causes: First, MS is a concentration-dependent detection method and therefore bias toward more abundant conjugates is likely. Second, fast protein degradation following ubiquitination resulted in the absence of known, short-lived regulators of the cell cycle. One way to facilitate the detection of such fast-degrading proteins is to use some form of chemical or genetic stabilization, neither of which was applied in the study by Peng et al. (2003).

3.5. COMMENTS AND CONSIDERATIONS

Having reviewed a number of strategies for protein quantification in mixtures, it is useful to give further considerations to some of the main elements which have a direct influence on the final outcome of such strategies. Here we will consider two of these elements, method of separation and detection.

3.5.1. Method of Separation

It is commonly acknowledged that MS on its own would not have had the impact it enjoys today in the proteome world without its close interaction with a host of separation technique, in particular 2D gel electrophoresis and 2D LC. Although the power of the first technique as a biochemical separation method has been well recognized since its introduction in the mid-1970s (O'Farrel, 1975), its application nevertheless has become widely recognized in the field of proteomics in the past few years. Such recognition has been accelerated by a number of developments, including improved 2D maps in both resolution and reproducibility, the latter feature directly related to the introduction of immobilized pH gradients instead of carrier ampholyte; MS analysis, chemical microsequencing, and amino acid analysis could be performed on increasingly smaller amounts of gel-separated proteins. Since its introduction, this extremely high-resolution method has witnessed a number of major changes which improved its resolution, reproducibility, and dynamic range. Of course, like any other technique, 2DE is far from being perfect. One of its inherent weaknesses is the difficulty to automate the initial phase of sample preparation and separation. Although the resolution of this technique is superior to other competing methodologies, such resolution is often challenged by the enormous diversity of cellular proteins, and therefore comigration of more than one protein in the same spot is not an unusual event. Another limitation which is not limited to 2DE is the chemical diversity of proteins and their very divergent expression in cells and tissues. This problem can be simply illustrated by the standard human cells where the most abundant protein is often actin, which is present at $\sim 10^8$ molecules per cell. On the other hand, some cellular receptors or transcription factors are probably present at 100–1000 molecules per cell. This means that to detect the latter components within the cell the dynamic range of 2DE has to be 10^5–10^6, which is clearly outside the current range of $\sim 10^4$. The situation can be even worse in certain media such as serum, where albumin is present at 40 mg/mL. There are, of course, other problems related to protein extraction and solubility; such problems are normally associated with certain classes of proteins such as membrane and nuclear proteins. Despite these drawbacks and others, which have been pointed out elsewhere in this book, 2DE still enjoys a central role in various approaches currently used for protein detection and quantification. However, for this technique to capture a wider functional picture of the proteome and to compete with a number of emerging separation approaches, a number of aspects of the technique have to be reassessed. For example, there is an accumulating evidence to suggest that at the present status the technique is incapable of detecting low-abundance (codon bias index less than 0.2) proteins. Such a drawback becomes even more serious when the technique is part of a multistep strategy for protein quantification, since various procedures of sample preparation, protein extraction, and sample introduction to mass spectrometers will inevitably result in further discrimination toward low-abundance proteins. This does not mean that there are no ongoing attempts to improve the detection capability of this approach. Enrichment, prefractionation, and more efficient solubilization cocktails are some examples of such attempts. However, such attempts have not found their way to existing strategies for protein quantification; they remain within the domain of a limited number of groups who use the technique in a stand-alone mode. Having said that, there are a number of recent examples to suggest that some progress is being made. In a recent article by Pederson et al. (2003) the use of highly solubilizing reagents combined with isoelectric fractionation allowed the identification of 780 protein isoforms, representing 323 gene products within the membrane fraction of yeast. Furthermore, the authors reported that 28% were low-abundance proteins and 49% membrane or membrane-associated proteins. Other attempts to observe proteins which are normally missing in 2D analysis have been reported by Molloy et al. (2001) and Zuo et al. (2001). In the first work the authors enriched bacterial outer membrane proteins by removing most nonmembrane proteins with an alkaline pH wash before reconstituting the membrane fraction for 2DE analysis. This fairly simple protocol allowed the identification of over 80% of the expected outer membrane proteins within the separation window of pH 4–7 and molecular weight 10–80 kDa. Zuo et al. (2001) demonstrated a useful strategy to enhance the separation and quantification of difficult proteins. A key feature of the approach is the prefractionation of samples into narrow pI zones using microscale solution isoelectric focusing. The authors reported higher loading capacity on narrow-pH-range gels, which improved the dynamic range, including the detection of low-abundance proteins. More detailed coverage of the role of 2DE as a separation method is given in other parts of this book. At this point, however, we can state that various aspects of this technique have to be improved to maintain its role in current and future strategies for protein quantification. These aspects certainly include more efficient solubilization media and more integration of prefractionation with particular attention to the use of highly narrow pH gradients. Current literature suggests that Improvement in the above elements can improve the detection capability of 2DE by at least one order of magnitude over the current situation of analysis of a total cell extract on a single gel (Rabilloud, 2002).

Having said that, 2DE remains an unsurpassed separation tool for protein mapping in complex mixtures, a tool which also has its shortcomings, including extensive sample handling, nonlinear response factors for most commonly used staining procedures, limited loading capacity, low extraction efficiency of the gel-embedded proteins, a decreasing resolving power for very low and very high molecular masses (below 15 and above 200 kDa), and increasing difficulty in isolating proteins with a pI at the acedic and basic extremes of the pH gradient.

Because of the above shortcomings and the absence of a platform to couple this technique directly to MS, today we have a number of chromatographic techniques competing and trying to replace an approach incriminated on the grounds of its complexity and labor-intensive manipulations. Curiously, chromatographers do not seem to be satisfied with just a 2D approach but seek a multidimensional approach. Link et al. (1999) used different LC columns in conjunction with nanoelectrospray MS–MS to analyze a fairly complex mixture of the peptides derived from multiple-enzyme digestion of the total yeast proteome. Although the number of proteins identified by this approach is rather impressive, it must be pointed out that, in contrast to gel-based analysis, where quantitative data are available, this approach provides a raw list of proteins present in the analyzed sample, without the quantitative aspect. Furthermore, such an approach does not adequately resolve multiple forms of proteins, thus underestimating the complexity of the analyzed proteome. Optiek et al. (1997, 1998) described two different chromatographic approaches for the separation of complex protein mixtures. In one approach, an *E. coli* lysate was injected into a cation exchange column and the eluent was fed stepwise into a reverse-phase column which was connected to an ESI source. In a second system, size exclusion chromatography was coupled to a reverse-phase column and ESI spectrometry. In another variation along these lines, Link et al. (1999) devised an on-line approach by which a fairly complex peptide mixture from digested *S. cerevisiae* ribosome was examined by a biphasic 2D capillary column. In such an arrangement, strong-cation exchange (SCX) resin is placed upstream of the revese-phase portion of the column. Peptides at low pH get bound to the SCX phase and subsequently are stepped off using multiple salt bursts followed by reverse-phase separation of each subset of peptides, which were fed to ESI–MS/MS analysis. These measurements revealed two previously undetected ribosomal components and demonstrated the validity of 2D chromatography as an alternative or at least as a complementary approach to 2DE in the analysis of complex protein mixtures. Other groups have also demonstrated the validity of 2D LC coupled to MS/MS as a possible substitute to gel-based analysis (Gygi et al., 1999; Davis et al., 2001). For the analysis of more complex protein mixtures, such as whole-cell extracts, a number of improvements have been introduced on the LC material buffers which have resulted in a more refined form of multidimensional protein identification technology (MudPIT). Using these improvements and based on the original work by Link et al. (1999), a fairly large scale analysis of *S. cerevisiae* has been reported by Washburn et al. (2001). In these analyses a microcapillary column was packed with two independent chromatography phases, the peptide mixture was loaded into the system, and the eluting components were analyzed by ESI/MS–MS. It can be said that the number of proteins identified by this approach was rather impressive (~1500). Having said that, such a list does not give quantitative evaluation in the real sens; such data simply represent a raw list of proteins present in the analyzed sample.

Because of the inefficiency of the peptide interaction with the SCX caused by competing salts, peptide solutions must first be desalted before being loaded into the two-phase column. This desalting step is usually performed off-line using, for example, a solid-phase extraction column; the peptides in the organic eluent are then vacuum concentrated down to near dryness before being diluted back into the aqueous buffer for loading into the column. To minimize these extra steps, and thus manipulation of the sample, McDonald et al. (2002) added an additional reverse-phase material in the microcapillary upstream of the SCX portion which served as on-line desalting step. This experimental arrangement was described as three-phase MudPIT, which was subsequently used to analyze a fairly complex mixture of proteins associated with bovine brain microtubules. These results were also compared with those acquired using one and two phases, and the authors deduced that the three-phase MudPIT column was superior in both number of proteins and number of peptides that were sampled.

Based on the above examples, the following general observations can be made: First, 1D and 2D LC coupled to MS/MS is a common feature of most commercially available mass spectrometers. Second, There is a general misuse of the term *multidimensional chromatography*, which in our understanding means the coupling of more than two orthogonal approaches which together influence the final output of the analysis. Most of the cited examples would fit nicely under the term 2D *chromatography,* even when three-phase columns are used. This is because only two phases influence the resolution of the separation process; the third phase is simply a desalting step which can be effected off-line or on-line. This comment is not meant as a hair-splitting exercise; on the contrary, it is an attempt to give the correct description of a given approach. If the separation of proteins, which is based on their pI values and their molecular weights, is given the description 2D, then we cannot see how the separation of a mixture of peptides in two different mobile phases can be given the name multidimensional chromatography. Third, it is reasonable to ask in which scenario the use of multiple phases is necessary? If it were only necessary to identify the most abundant proteins within a given mixture, then 1D LC/MS–MS would be preferred because it is faster and more easily automated. We have to bear in mind that the amount of sample which can be loaded onto a three-phase MudPIT column without off-line desalting is limited to about 100 μg without substantial risk of clogging the column.

3.5.2. Detection

The sensitivity, response, and dynamic range are central to any detection method, particularly when such a method is applied to protein quantification in complex mixtures. Most existing strategies for protein quantification rely heavily on MS detection. Being the last phase in such strategies, MS tends to reflect both its own shortcomings as well as those associated with the chemistry of labeling, sample preparation, method of separation, and delivery. To give a rational assessment of the present MS detection capability in protein quantification, we find it relevant to underline a number of general considerations: First, many of us believe that most existing strategies for protein quantification in mixtures are capable of detecting more proteins than is currently reported. One obvious way of demonstrating this is to minimize protein losses during purification, separation, and delivery to the ion source of a mass spectrometer. Second, at the present time neither 2DE nor the combination of various chromatographic/capillary electrophoretic separation methods is capable of delivering to the ion source truly pure individual components of a complex protein mixture without some form of discrimination due to either low concentration and/or the physicochemical characteristics of such components. Therefore assessment of the present MS capability cannot be considered in isolation from such an influence. For instance, the literature associated with protein quantification often cites low femtomoles as the detection limit without specifying that such a limit refers to what certain separation methods can deliver rather than what actually is delivered to the mass spectrometer. To clarify this particular observation, here is a general scenario which MS detection within a given protein quantification strategy may have to deal with. If we are to assume a hypothetical medium which has $\sim 10^8$ cells and protein copies per cell in the range of $10–10^6$, this would translate into 1.6 fmol–160 pmol of proteins. Separating this mixture by 2DE and using silver staining for visualization currently would allow the detection of 160 fmol (1000 copies/cell), which is also sufficient to allow MS analysis both by ESI and MALDI. On the other hand, the lower concentration proteins (10,100 copies/cell) would require radioisotope detection. Most of us agree that, if we can deliver 1.6 fmol ($\sim$0.15 ng of 100-kDa protein), then both ESI and MALDI should be able to deliver reasonable MS and MS/MS spectra. We are not trying to say that current achievable sensitivity of mass spectrometry does not need to be improved. We are simply trying to say that the ultimate sensitivity of current MS cannot be evaluated in isolation from elements such as sample preparation, separation, and efficiency of recovery of the components of a mixture and their introduction to the ion source. This observation can be supported by some considerations based on current literature. For example, in the context of TOF–MS, cryogenic detectors have been used to detect single protein molecules (Twerenbold et al., 1996). Using IR–MALDI and a detector based on secondary ion emission, Berkenkamp et al. (1997) reported that singly charged proteins with masses exceeding 500 kDa and multiply charged proteins with masses approaching 2 MDa could be measured with MALDI–TOF–MS. When this sort of performance cannot be achieved by the same mass spectrometers within a strategy which uses separation and extraction prior to MS analysis, it is highly misleading to single out MS for the failure to detect certain components within a given mixture. It would be more appropriate to attribute the low performance to the strategy as a whole. We are not trying to give unnecessary weight to the terminology; we are simply trying to underline the concept that, in the context of protein quantification within complex mixtures, there are a number of elements to be considered, including, of course, MS as a method of detection. Most readers would agree that at the present time we have two extremely sensitive ionization techniques, ESI and MALDI, yet the absence of efficient extraction, focusing, and detection does not allow us to detect all the ions formed within these sources.

Fluorescence detection is emerging as the main competitor to MS detection in both macro- and microstrategies for protein quantification. This technique intrinsically has a wider dynamic range than existing MS detection; however, the use of this detection method within certain strategies for protein quantification renders it susceptible to certain experimental conditions which reduce its inherent sensitivity. For example, in the DIGE approach (Ünlü et al., 1997) labeling through the use of Cy (2, 3, 5) dyes has to be limited to less than 5% of each protein to allow reliable quantification and reduce excessive changes in the molecular masses and in the pI of the labeled proteins. Preferential labeling of some proteins within a mixture can result in misleading deductions when two groups of proteins are labeled by these dyes. It can be deduced from these observations that to glean the full advantages of this highly sensitive technique we have to optimize certain preparations which in reality have nothing to do with the technique itself but which influence its final performance.

3.6. OTHER APPROACHES

3.6.1. Imaging Mass Spectrometry

Mass spectrometry is not normally conceived as a technique which can provide spatially resolved information. This wrong conception was first challenged by Casting and Slodzian (1962) over 40 years ago. In their work the authors generated ions by bombarding solids with few tens of kiloelectron-volt particles and monitored the ejected ions by a magnetic analyzer. This type of analysis formed the basis for today's SIMS. The same authors argued that it should be possible to build an ion-optical collection system, analogous to a lens used in a light microscope to preserve the spatial relationship of the desorbed ions from sample to detector. The various aspects and phases of development of imaging using MS have been reviewed by Packolski and Winograd (1999). Although current use of MALDI MS to map specific molecules in mammalian tissue sections and other biological samples is not strictly a strategy for protein quantification in such tissues, there are a number of recent examples which demonstrate the potential of this technique for the assessment of protein expression in tissue sections.

Imaging MS is relatively new and utilizes MALDI–TOF–MS and MS–MS to profile and map proteins in thin tissue sections. Over the last few years, various groups have demonstrated that peptide and protein ions can be directly desorbed from cells and tissues using MALDI–MS (Caprioli et al., 1997; Chaurand et al., 1999; L. Li et al., 1999; Stoeckli et al., 2001, 2002; Chaurand and Caprioli, 2002). A review by Li et al. (2000) gave an overview of the technique and its various applications in cell analysis. The use of this technique to image and determine low-molecular-weight pharmaceutical compounds has also been reported. For example, the detection of a low-M_r drug (paclitaxel) in rat liver and human ovarian tumor xenograft tissue has been examined by MALDI–MS–MS using an ion trap mass spectrometer (Troendle et al., 1999). Recently a similar approach has been used by Reyzer et al. (2003) to detect and image two drug candidates in mouse tumor tissue and in rat brain tissue. Both groups have demonstrated that this approach can be successfully applied to obtain spatial information about the localization of low-M_r drug candidates in tissue samples. In this approach the mass spectrometer is used to determine the molecular weights of specific species in the surface layer of the tissues. The experimental procedure involves coating the tissue section, or a blotted imprint of the section, with a thin layer of laser-absorbing matrix followed by sample analysis to produce an ordered array of mass spectra, each representing nominal m/z values and covering a predetermined mass range. Images for each m/z value can then be displayed and used to localize individual compounds within the irradiated tissue. This sequence of events is conceptually represented in Figures 3.10*a*,*b*. In this type of analysis, tissue samples can be prepared using several protocols: direct analysis of fresh frozen sections (Moroz et al., 1999), individual cells, or cluster of cells isolated by LCM (Chaurand et al., 2003) or contact blotting of a tissue on a membrane target steel. Caprioli's group (Capriol et al., 1997; Stoeckli et al., 2001; Chaurand et al., 1999), which is very active in this area, has published a number of works demonstrating the viability of this approach in the investigation

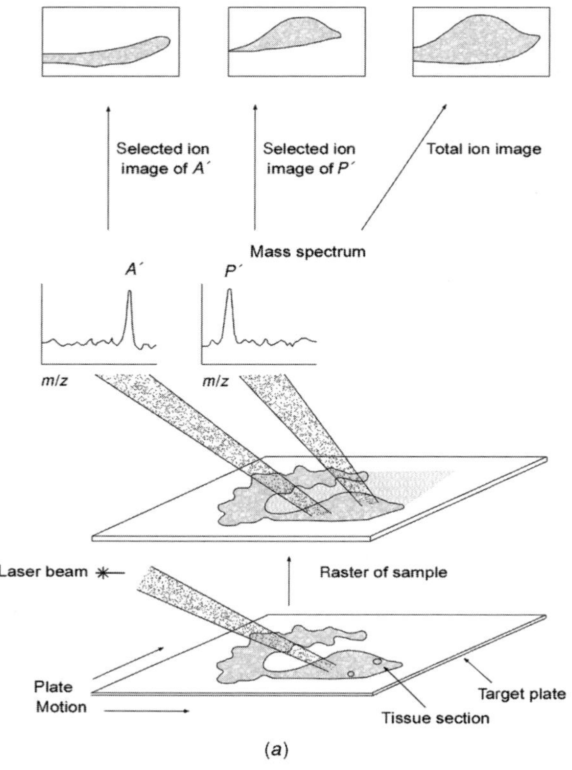

(*a*)

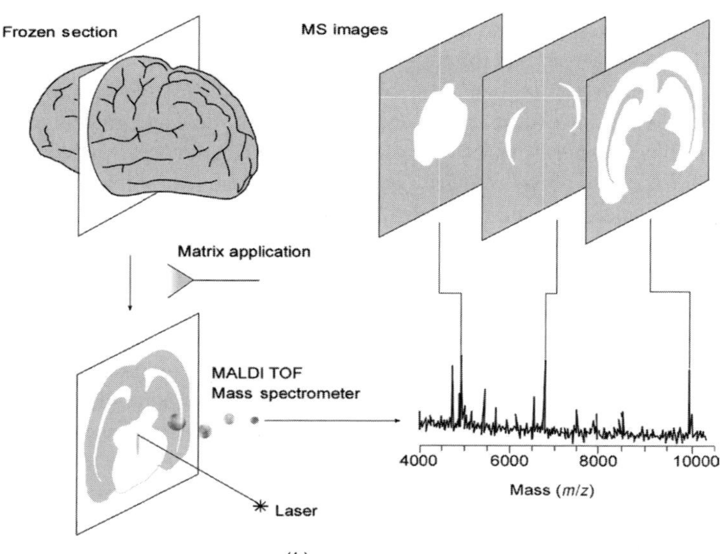

(*b*)

FIGURE 3.10.

of various biological samples. The same group has provided clear indications on the experimental parameters which need to be optimized if meaningful data have to be obtained (Schwarz et al., 2003). For example, the successful analysis of tissue structure as well as single cells less than 10 μm in size require that the laser beam diameter on the target be 1 μm or less, whereas laser beams used in commercial instruments have a beam of few tens of micrometers. Other parameters which have to be addressed are described within this section. In a recent article by Stoeckli et al. (2002) MALDI–MS molecular imaging has been applied to map amyloid β-peptides in brain sections. This study gave highly useful hints on sample preparation prior to MS analysis. It was suggested that to prevent migration of proteins during the matrix-coating process a mild fixation was needed. Such fixation was performed by immersing the MALDI plates into acetone for 30 s followed by air drying at room temperature; storage of the frozen sections was at –80°C. In a second attempt to enhance the signal, the plates containing the sample were cooled to 4°C prior to coating by placing them on a prechilled aluminum block. The sections were coated by directly deposing 50 μL of freshly prepared matrix and allowed crystallization at 4°C.

The use of LCM (see Section 2.1.5) for sample preparation can enhance the capability of imaging MS. Laser capture microdissection permits the isolation of selected cells or groups of cells from a thin tissue section (typically 5–10 μm in thickness). Briefly a narrow laser beam (>3 μm in diameter) is focused onto a heat-sensitive transparent polymer film contacting the section. When the laser beam heats the polymer, it deforms locally and contacts the cell(s) to be microdissected. The cells bind to the polymer and are separated from the section when the polymer is removed. The polymer film may then be mounted on the target plate. Following microspotting of a suitable matrix the cells can be examined by MALDI–TOF–MS.

Recently LCM in conjunction with imaging MS has been used by Chaurand et al. (2003) to profile and image several epididymal proteins in mouse. The authors monitored over 400 proteins and obtained detailed localization of three proteins: retinoic acid binding protein, the epididymal secretory glutathione peroxidase, and the cysteine-rich secretory protein-1. The authors also reported that the overall detection sensitivity of these proteins was similar to that obtained by immunohistochemistry conducted by the same authors.

FIGURE 3.10. (*a*) Conceptual representation of MALDI MS imaging procedure on tissue slice. Tissue in example contains two defined areas, A′ (anterior) and P′ (posterior), containing specific localized proteins A′ and P′, respectively. From data array of raster of laser spots, each containing complete mass spectrum, selected ion images can be created in which molecules can be localized in tissue. (Adapted from Caprioli et al., 1997, with permission.) (*b*) Methodology developed for spatial analysis of tissue by MALDI MS. Frozen sections are mounted on metal plate, coated with UV-absorbing matrix, and placed in mass spectrometer. Pulsed UV laser desorbs and ionizes analytes from tissue and their m/z values are determined using TOF analyzer. From raster over tissue and measurement of peak intensities over thousands of spots, mass spectrometric images are generated at specific molecular weight values. (Adapted from Stoeckli et al., 2001, with permission.)

At this point it is reasonable to state that, despite the limited number of examples on the application of imaging MS, it looks as if such technology will have an important role to play in future developments regarding protein expression and identification of specific markers associated with various pathologies, including various forms of cancer. This technique has already provided some examples on its viability for the direct analysis of drug candidates in tissue. However, before this relatively young approach can realize its full potential, a number of technical difficulties have to be resolved. Such difficulties are strictly related to laser characteristics, tissue preparation, and software and data elaboration. First, laser beam size and its frequency are essential for high spatial resolution and for reasonable analysis times. Reducing the beam size to 1 μm or less would certainly facilitate spatial resolution of the desorbed ions, particulariy from tissue structures and single cells which are less than 10 μm in size. Currently, laser beams on commercial instruments have a diameter of 25–50 μm. Increasing imaging speed through the use of high-repetition-rate lasers (kHz range) in combination with fast data acquisition systems would certainly reduce the current analysis times. Second, improvement on the bioinformatics side is also needed to handle storage, processing, and mining of an enormous amount of data generated from the imaging of tissue sections. Such additional efforts can be justified by the simple consideration that even at medium image resolution thousands of spots are irradiated per single image. Normalization of these spectra and their comparison with those from other sections, mixed-pattern recognition, as well as other data-related tasks represent a formidable challenge which has to be addressed. There are other minor shortcomings associated with the current capability of commercially available MALDI instruments. For example, at the present time and even under delayed extraction conditions the mass resolution required for this type of analysis starts to deteriorate above m/z 50,000 (Bahr et al., 1997). Above this value, peaks become broader, resulting in inevitable low mass accuracy. Most MALDI–TOF instruments use microchannel plates for ion detection, which are known to be more biased toward smaller ions, which have a higher velocity compared to their heavy counterparts present within the same ion beam. Tissue preparation is another critical step which can influence the quality and reproducibility of the analysis. Of course, at present there is not a standard method to perform such preparation, and therefore analysis conducted in different laboratories may provide different data for the same tissue samples. Having said that, there are already some indications on how to prepare such a sample in a manner which should help reproducibility (Schwarz et al., 2003; Stoeckli et al., 2002). Tissue preparation methods should be carefully followed to maintain the integrity of the spatial arrangement of compounds. Mishandling or improperly storing tissue samples in the early stages of sample preparation may result in delocalization and degradations of the molecules of interest. Careful consideration must also be given to how the tissue is treated immediately after surgical removal, how it is sectioned, taking into account temperature, section thickness, fixing, and embedding media, the method used to attach it to the MALDI target, the matrix, and eventual storage conditions.

3.6.2. Microarrays in Protein Quantification

Microarray technology is addressed in this chapter as well as in Chapter 2, which is dedicated to proteomics in cancer research. In the same chapter we gave a number of examples on the use of micraoarrays and chip technology in the search for biomarkers and to investigate protein interactions and signaling pathways. In the present chapter, we are trying to highlight the general potential of this emerging technology in protein quantification.

The protein chip, or protein microarray, is one of the emerging technologies proposed for high-throughput identification and quantification of proteins in complex mixtures. This chip is similar to the oligonucleotide chips used in gene expression profiling (de Wildt et al., 2000). Each chip can have thousands of addressable locations designed to capture proteins by binding to a specific molecular entity immobilized in that location. Current literature indicates that to construct an efficient and practical protein chip (array) a number of conditions have to be met:

1. Generation and isolation of a vast repertoire of molecules/functions which can be specifically recognized by the target proteins
2. Capability to construct homogeneous surfaces and availability of immobilization chemistry to maintain the conformation and binding specificity and to preserve the native function of the immobilized entities
3. High sensitivity and precision of the readout method
4. Data management and interpretation that allow the extraction of the relevant information from data generated by automated or semiautomated methods

Whether, all these elements can be put to work in harmony is one of the main questions which have to be addressed if practical and reliable protein arrays are to become a reality.

The readers are aware that in most existing publications on this particular argument a comparison with DNA arrays is always invoked. As things stand at the present time, such comparison is valid at the concept level but it fails at the level of implementation due to a number of technical difficulties which prevent the mimicking of DNA arrays. Such difficulties can be appreciated by the following considerations. First, in the case of DNA arrays, the recognition molecule and the molecules that are recognized are both nucleic acids and hence their physiochemical properties are similar. In the case of protein arrays, on the other hand, the arrayed molecules can be nonprotein moieties. Immobilization of the arrayed recognition molecules in a site-directed manner and in such a way that they retain their inherent specificity toward targeted proteins may prove more difficult than in the case of DNA. Second, we also have to recall that well-tested and effective methods for immobilizing DNA to glass or nylon without affecting their properties already exist. On the detection side, there are well-tested protocols for the incorporation of radioisotopic and fluorescent tags which do not influence the ability of the arrayed DNAs to bind to their target sequences. On the other hand, the incorporation of the same detection tags into proteins is more difficult to implement since such incorporation must be stoichiometrically reliable and must not influence the ability of such proteins to bind specifically to the arrayed recognition molecules.

Currently, there are two different array formats, planar and suspension. In the planar format, capture reagents are spotted into a slide, filter, well, or other planar surface, where each array element is identified by its location within the array's grid. Target molecules which are captured by the spotted reagents can be detected by laser scanning (Ramdas et al., 2001), MS (Fung and Enderwick, 2002; Rodi et al., 2002), and surface plasmon resonance imaging (Nelson et al., 2001). In a suspension array, on the other hand, capture reagents are immobilized on coded microparticle surfaces, where flow cytometry (Nolan and Sklar, 2002; Taylor et al., 2001) is commonly used for the detection of the targets captured by the various elements of the suspension array. Other methods for reading particle-based arrays include fiber-optic microsensing (Michael et al., 1998; Ferguson et al., 2000), fluorescence (Brenner et al., 2000), and optical imaging (Walton et al., 2002). Careful consideration of current literature reveals an increasing trend toward the application of microarray technology in the various areas of proteomic research, including protein expression and protein quantification. The authors wish to note that the bulk of existing literature on the use of microarrays for protein quantification is mainly dedicated to the hardware and to the chemistry associated with this technology, yet few real-life examples on the application of such technology are just beginning to emarge. This is not surprising in an emerging technology where surface chemistry and the type of capture elements have a substantial influence on its eventual performance. Given the importance of both parameters, we find it relevant to give a brief summary of recent efforts associated with both parameters.

3.6.3. Protein Recognition Molecules

It is well recognized that antibodies are the obvious choice for protein recognition molecules (PRMs), as their primary biological function is molecular recognition; furthermore, there are already well-tested methods for the production of both polyclonal and monoclonal antibodies (Dunbar and Skinner, 1990). Having said that, it is wise to bear in mind that the demands placed on the generation of antibodies to support the construction of protein arrays may be far more demanding than for the conventional applications of antibodies. Such stringent demands have to allow the detection of rare or impure proteins. At present most antibody screening methods require relatively large quantities of protein and do not guarantee the selection by the antibody of targeted proteins present in complex mixtures. Another feature which such arrays should be able to perform is the capability of specific recognition of all possible posttranslational modifications.

Phage-displayed antibodies have the potential of providing an alternative source for PRMs. Phage display technology is based on bacteriophages that have been genetically modified to act as efficient transporters of DNA inserts into their bacterial host(s), where the phages are amplified and packaged to express foreign proteins of interest on their surfaces, including short-chain variable fragments of immunoglobulins

(Clackson et al., 1991; Vaughan et al., 1998; Winter et al., 1994; Hudson et al., 1998; Hoogenboom et al., 1998; Knappik et al., 2000; Marks and Sharp, 2000). Such phage-displayed antibodies may be affinity selected and the clones amplified to produce monospecific antibodies which may then be matured to enhance their properties in particular affinity and specificity to targets. This approach was applied successfully by de Wildt et al. (2000), who developed antibody arrays for high-throughput screening of antibody–antigen interactions. The development was based on the use of robotic picking and high-density grading of bacteria containing antibody genes followed by filter-based ELISA screening for identifying clones that can express binding antibody fragments. The authors reported that, by eliminating the need for liquid handling, they could screen up to 18,342 different antibody clones at a time. Furthermore, the clones were arrayed from master stocks, and the same antibodies could be double spotted and screened simultaneously against 15 different antigens. This setup has a number of advantages. The ability to create double spots and duplicate arrays helped to eliminate false positives and cross-reactive clones. The screening could be applied to recombinant antibodies of different origins. The use of a single round of selection followed by mass array screening enabled the isolation of a wide range of binders to serum albumin and several recombinant human proteins.

3.6.4. Surface Chemistry

It is well recognized that protein activity can be susceptible to derivatization and immobilization chemistry; therefore considerable efforts have been dedicated to the optimization of the solid surfaces in protein arrays. The composition of such surfaces can include glass, porous acrylamide gel, membranes, and other polymeric materials. Of course, high-quality substrates with well-defined and reproducible surface properties and optimized surface chemistry are required to immobilize protein capture homogeneously and to ensure a functional conformation. This means that various strategies have to be considered since the diversity of the targeted proteins can range from known binders to proteins with additional functions such as soluble enzymes or the more delicate-to-immobilize, water-insoluble membrane proteins such as ion channels or G-protein-coupled receptors, which are attracting considerable attention from the pharmaceutical industry. Although it is still difficult to define general immobilization strategies that do not discriminate between proteins, we are witnessing various efforts directed toward the optimization and refinement of protein–solid substrate attachment. A convenient method is to coat the glass surface with a thin nitrocellulose membrane or poly-L-lysine such that proteins can be passively adsorbed to the modified surface through nonspecific interactions (MacBeath and Schreiber, 2000). To achieve more specific and stronger protein attachment, several groups have used reactive surfaces on glass that can covalently crosslink proteins (Zhu et al., 2001). Gold-coated glass surfaces are a variation of the self-assembled monolayer (Houseman et al., 2002). The gold surface is usually covered with a bifunctional thio-alkylene, which has a SH group that reacts with gold and another that reacts with the captured target. The advantage of this appropach is that both MS and surface plasmon resonance can be used as detection methods to monitor the dynamics of the reaction or to identify the captured molecules. It is becoming more evident that the solid support used in protein arrays has profound consequences on the quality of the microarray analysis since it influences not only the efficiency of the target's attachment but also the degree of nonspecific binding. The use of different surfaces has been reported by various groups, which demonstrates that progress is being made in this area. A good example has been reported in Zhu et al. (2001), where the immobilization and functional testing of 5800 yeast proteins immobilized on a chip have been successfully tested for a series of known and new functional interactions with different types of molecules. The authors described the cloning of 5800 open reading frames and overexpressed and purified their corresponding proteins. The protein chips were prepared by printing 6566 protein preparations, representing 5800 different yeast proteins, in duplicate onto glass slides using a commercially available microarrayer. In one of the experiments the authors used aldehyde-treated microscope slides (MacBeath and Schreiber, 2000) in which fusion proteins were attached to the surface through primary amines at their NH_2 termini. In another variation proteins were spotted into nickel-coated slides, in which the fusion proteins were attached through their HisX6 tags. According to the authors, this mode of spotting gave superior signals for protein preparations. In a recent article Kusnezow et al. (2003) gave a fairly detailed assessment on the parameters which may influence the performance of antibody microarrays. The authors investigated the modification of the glass surface, the type and length of crosslinkers, the composition and pH of the spotting buffer, blocking reagents, antibody concentration, and storage procedures. Based on more than 700 individual array experiments, the authors listed a number of generic deductions which we find useful to summarize here. Regarding antibody immobilization on glass surfaces, the 2.5% epoxysilane surface gave the best results in terms of sensitivity, signal-to-background ratio, and spot quality. The highly reactive surface reacted not only with the amino groups but also with other nucleophiles such as alcohol, thiol, and acid groups on the protein surface (Piehler et al., 2000). Other surfaces, including mercaptosilane and aminosilane, used with a relatively long maleinimido–NHS crosslinker gave reasonable results.

Unspecific background signals due to antigen binding in the absence of an antibody is a problem commonly encountered in microarray technology; tackling this problem through chemical modification of the surface coating (MacBeath et al., 1999) is an option which had detrimental effects on signal intensity and in certain cases could enhance background noise. Treating the glass slides with 10 mM mercaptoethanol or cysteamine to block the maleimide group had the same negative effects cited above. Considering this deduction together with those in the current literature, it seems that optimal experimental conditions for an efficient blocking process remain an artwork which has to be optimized and adapted to specific working conditions.

Another interesting example pertaining to the aspect of surface chemistry has been reported by Arenkove et al. (2000). A chip based on acrylamide pads arrayed on silane-treated glass surfaces was used. The device contains $100 \times 100 \times 20$ μm at 200-μm intervals, which formed small reactors. This 3D array has an enhanced capacity, and the presence of acrylamide pads creates a hydrating environment which contributes to protein stability. This type of device has been applied for direct antibody detection and antigen determination such as recombinant hepatitis B virus antigen and alpha-fetoprotein. The drawbacks of this type of chip are associated with the complexity of fabrication and the presence of acrylamide, which may create difficulties with certain reagent delivery systems.

Other applications of protein immobilization chemistry are based on the unique affinity of streptavidin and biotin. Patel et al. (1999) arrayed streptavidin onto solid surfaces to capture biotinylated proteins, while Kambhampati et al. (2001) have used self-assembling streptavidin monolayers to prepare biotinylated gold-coated glass slides for immobilizing biotinylated proteins. Initial results based on this type of surface chemistry showed a good linear response in the detection of a number of proteins. Recently, cocrystallized wild-type and mutated forms of streptavidin to produce a planar array which may be suitable for forming a solid surface for protein chips have been described by Farah et al. (2001).

3.6.5. Some Applications

The application of microarrays in protein quantification is still in its infancy. However, there are a number of applications which indicate the future potential of this technology. For example, beads-based assays were used to determine the concentration of cytokines or antibodies in biological samples taken from patient's serum and cell culture supernatant (Dunbar et al., 2003; De Jager et al., 2003; Chen et al., 1999). A highly specific and sensitive protein microarray assay was used to identify and quantify 24 different cytokines from conditioned media and sera (Huang, 2001). In another fairly complex multiplexed sandwich immunoassay, Scweitzer et al. (2002) have quantified 75 different cytokines. However, it has to be noted that these analyses were not conducted within a single microarray due to cross reactivity of some of the detection antibodies with immobilized capture antibodies and/or cross reactivity with nonspecific analytes.

Other types of microarrays have been designed and used to detect and quantify antibodies in human serum. Hanash and co-workers (cited in Mandoz-Gurpide et al., 2001) created microarrays of protein fractions through chromatographic separations of complex protein mixtures. Twenty protein fractions from the A549 lung cancer cell lines were separated by isoelectric focusing and arrayed. Labeled antibodies were incubated on the arrays to detect specific proteins in the arrayed fractions. In such an approach the authors applied multidimensional chromatography separation, which serially combines different separation phases, including reverse-phase and ion exchange chromatography. These authors managed to obtain as many as 2000 fractions from a single protein pool and used microarrays to detect and quantify various antibodies in human serum.

Haab et al. (2001) have used a set of over 100 defined antigen and antibody pairs to create microarray-based immunoassays. Antibodies or antigen was immobilized on the surface and the corresponding targets (e.g., in serum) were labeled with fluorescent tags. In this procedure one sample was spiked with antigens labeled with Cy5 fluorophore, while a second sample containing a different concentration of the antigens was labeled with Cy3 fluorophore. The two samples were mixed and incubated on the same microarray. This dual-color detection system was able to reveal different concentrations in the range of picomoles of various captured targets. A complementary approach to antibody microarrays is the reverse-phase microarray (Paweletz et al., 2001), in which multiple complex protein mixtures can be spotted into membranes and probed with antibodies targeting specific proteins. This approach has the advantages that many samples can be analyzed in a single experiment, including insoluble proteins. This technique was used by the same authors to analyze proteins collected from resected prostate tumors, identifying changes in apoptosis pathways correlating with the progression of cancer cells.

In adition to measuring serum protein abundances, antibody microarrays have been applied in different studies to determine protein levels from frozen resected tumor tissues (Knezevic et al., 2001). Laser capture microdissection was used to remove selected regions of tumor and stroma from a frozen tissue sample. The isolated proteins were labeled with biotin, applied to antibody microarrays on nitrocellulose, and measured with chemiluminescence. Western blot and immunohistochemistry subsequently validated the antibody microarray measurements. Similar measurements were applied to the cell culture to examine the effect of UV radiation on protein expression in opoptosis signaling pathways. Proteins were isolated from colon carcinoma cells grown in culture before and after exposure to the radiations. The two pools were labeled with either Cy3 or Cy5 fluorophores and incubated together on microarrays containing 146 different antibodies spotted about 10 times each. Measuring the expression of various proteins revealed multiple proteins which experienced up- or down regulation following exposure to the UV radiation. It is interesting to note that most of the altered proteins were apoptosis regulators, consistent with the observation of induced apoptosis in the investigated cells. This type of study may prove useful for studying signaling pathways, particularly if the measurements can be conducted on both phosphorylated and unphosphorylated forms of proteins.

A quantitative method that utilizes a scanning electron microscope for the analysis of protein chips has been recently described by Levit-Binnun et al. (2003). This method is based on the idea of counting target-coated gold particles interacting specifically with ligands or proteins arrayed on a derivatized microscope glass slide and detecting back-scattered electrons. This approach was tested on model systems to quantify interactions of biotinstreptavidin and of an antibody with its cognate hatpin.

3.7. EMERGING ROLE OF MICROFLUIDIC DEVICES

The need for novel analytical approaches which require minimal amount of sample and reduce analysis times has encouraged various groups to develop microfluidic devices which use one or more chromatographic/electrophoretic methods to separate and deliver analytes to mass spectrometers. Over the last 5 years we have witnessed massive efforts on the part of the pharmaceutical industry, academic institutions, and highly specialized biotechnology companies to optimize the construction and performance of such microdevices. The role of such devices has started to manifest in two particular areas, sample preparation and sample delivery to MS. A brief description of the contribution of microfluidic devices in these two areas is provided below.

3.7.1. Sample Preparation and Sample Delivery

Current literature cites a number of examples which demonstrate the potential of microfluidic devices for the automated preparation and sample handling prior to their analysis by MS. Rocklin et al. (2000) have described the combination of micellar electrokinetic chromatography (MECK) with rapid channel electrophoresis (RCE) for 2D separation of peptide mixtures. In such an approach (see Fig. 3.11) the peptides are first separated with MECK and almost at the same time the eluting components are rapidly sampled for separation by RCE and detection by laser-induced fluorescence.

One area in which microfluidic devices can contribute is sample preparation such as desalting and digestion prior to sample delivery to the mass spectrometer. Xu et al. (1998) provided a promising example on how such devices can be used to perform protein microdialysis prior to ESI–MS analysis. Briefly, the approach consists of sandwiching a microdialysis membrane between two microfluidic systems, one containing a 6-μm-wide sample channel and the other a 500-μm-wide buffer channel. The protein solution was pipetted into the system and directed through the channel into the mass spectrometer. During such passage, the background electrolytes are exchanged through the dialysis membrane. During the same process, background electrolytes could be replaced with MS-compatible buffers. The desalting of standard protein mixtures containing up to 600 mM salt resulted in the acquisition of meaningful ESI mass spectra. The technical value of this report was the demonstration that protein solutions could be desalted on a microfludic device, and therefore future developments along these lines could eventually allow complex protein solutions to be examined in a gel-free environment.

As we have stated elsewhere, protein digestion prior to MS analysis is becoming a central step in modem proteome analysis. Interestingly, microfluidic devices can provide very high surface-to-volume ratios. The immobilization of one or more enzymes within the microfluidic channels can provide an efficient digestion. Ekström et al. (2000) described a device in which trypsin was immobilized in a microfluidic channel to provide a high surface-to-volume ratio; the same device was also equipped with a sample pretreatment robot and a microdispenser for transferring the digested proteins to a MALDI target plate for MS analysis. To enhance the surface area and hence the digestion efficiency, the authors used anodic etching in an hydrofluoric acid–ethanol solution, which increased the porosity of the surface and consequently its surface area. In yet another variation of this type of device, Wang et al. (2000) have incorporated beads of immobilized trypsin within a channel for flow-through digestion of proteins. In such a device the protein of interest is pressure driven through a packed-digestion chamber; the flow-through from such a chamber exits into a reservoir which can then be injected into a second channel either separation or infusion into the mass spectrometer.

Most microfabricated devices are based on photolithographic techniques to create various patterns of reservoirs, channels, and reaction chambers on a planner substrate

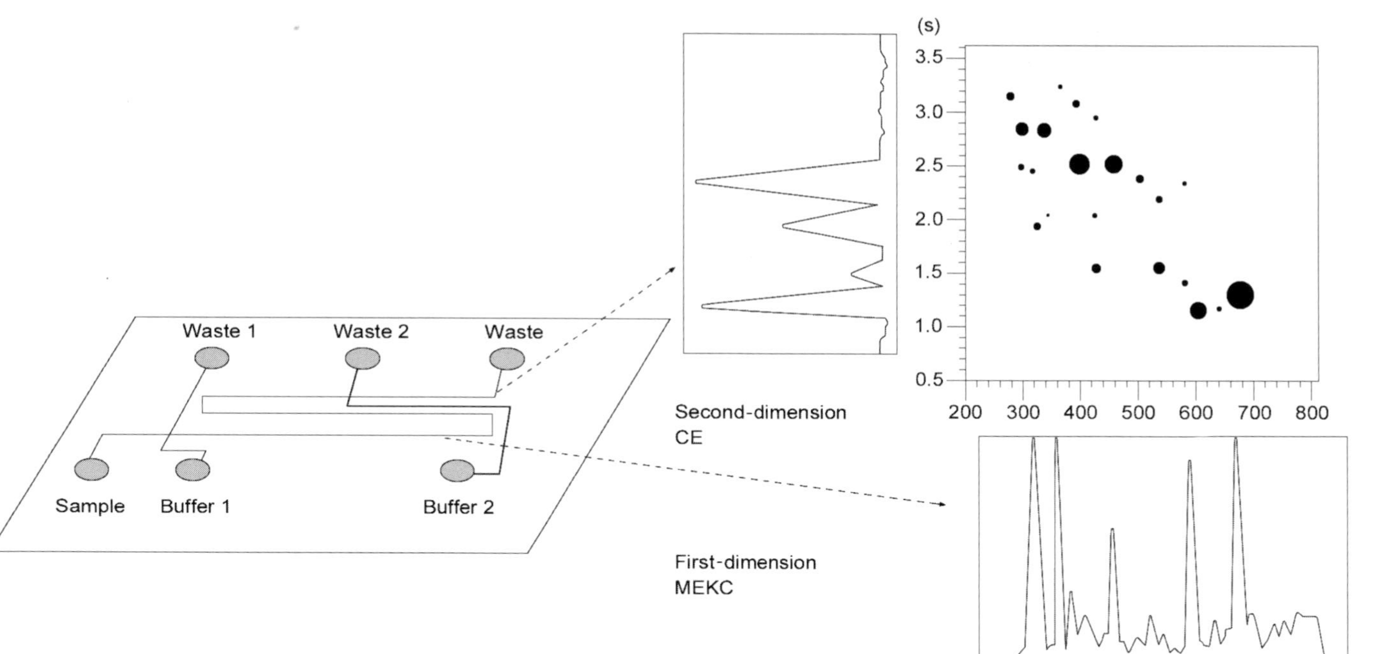

FIGURE 3.11. Illustration of microfluidic system for 2D electrophoretic separation of labeled peptides derived from tryptic digest of cytochrome *c* by MEKC and CE. (Adapted from Figeys, 2002, with permission.)

such as glass or polymer (Figeys, 2002). The coupling of these devices to a mass spectrometer can have different forms depending on the device and the type of mass spectrometer, in particular the type of ionization method. There are two ways of coupling these devices to MS: The first is through continuous fusion of the analyte and the second involves separation of the various components prior to their introduction into the ion source. An example on the first mode of sample introduction was illustrated by Figeys et al. (1997) in which Femtomoles per microliter of standard peptides could be infused continuously via a microdevice and detected by MS. The same group (Figeys and Pinto, 2001) demonstrated a fairly complex nine-reservoir device that was fully computer controlled to manage a flow from a microfluidic device and to trigger the mass spectrometer for automated MS/MS analysis of peptides (see Fig. 3.12).

Microfluidic devices that can perform on-line separations of proteins and peptides have also appeared in recent literature. Devices based on electrophoretic separation, which is highly suitable for microelectrospray interfaces, have been reported in works by Zhang et al. (1999), Wang et al. (2000), Figeys et al. (1998), and J. Li et al. (1999). A representative example of such devices has been reported by Li et al. (1999). Basically, protein digests were introduced one at the time into the microfabricated device, followed by the application of a suitable electrostatic field to fill in an injection space with the digest. This step was followed by the application of another electric field between the buffer reservoir and the electrospray needle, thus driving the separation of the various components toward the electrospray interface and then into the mass

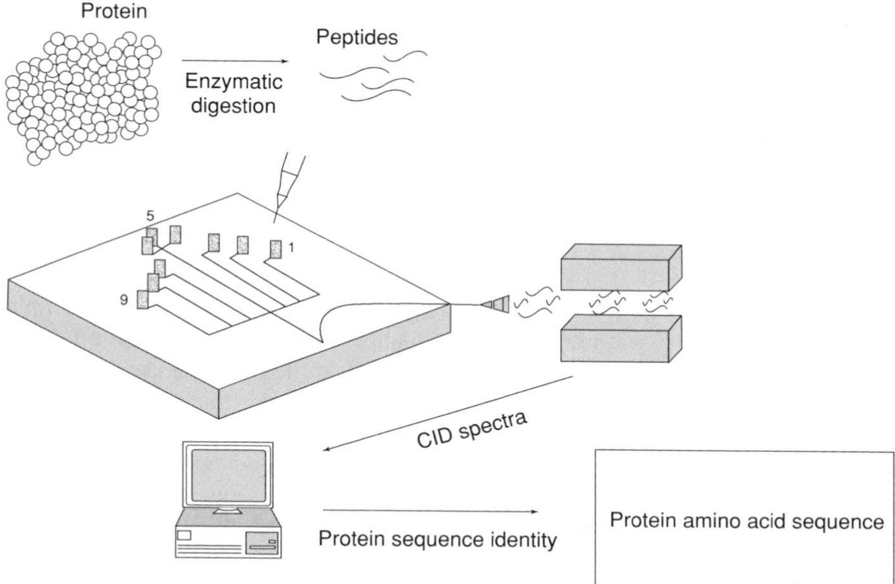

FIGURE 3.12. Schematic representation of microfluidic system for automated identification of proteins. (From Figeys and Pinto, 2001 with permission.)

spectrometer. The authors reported a number of ion chromatograms showing fast and good-quality separation of simple digests (Li et al., 1999). Mature separation methods such as capillary electrophoresis (CE) and various forms of LC and their suitability for coupling with ESI interfaces have shifted the balance in favor of mirofluidic devices adapted for ESI rather than MALDI. Given the higher sensitivity of the latter method, its superior tolerance to contaminants, and its intrinsic potential for high-throughput analysis, we find such imbalance rather surprising. One further justification for more attention to microdevices suitable for MALDI can be sustained by the following observation. Although both ESI and MALDI are destructive ionization methods, that is, the sample is consumed during the analysis, there is an important difference between the two methods. While all the investigated sample is lost once sprayed in the ESI source, only a portion of sample is consumed in the MALDI analysis and the rest can be used for further analysis.

More recently there have been a number of attempts to miniturize and automate sample delivery microdevices for MALDI ion sources. These attempts include off-line fraction collection and on-line coupling. In an attempt to mimic the continuous analysis mode typical of ESI interface, a frit or aerosol sample delivery microdevice was proposed. The idea of continuous-flow probe for MALDI (Li et al., 1993; Nagra and Li, 1995; Whittal et al., 1998) was inspired by the continuous-flow FAB probe operated at a few microliters per minute. Organic solvents such as methanol and ethelene glycol were added to prevent sample freezing within the stream, the frit was directly irradiated with the laser beam, and excess liquid was removed with a paper strip wrapped around the end of the probe. Another variation for delivering continuous sample to a MALDI source has been proposed by Zhang and Caprioli (1996). In such an arrangement zones exiting from CE capillary were deposited on a wet cellulose target coated with a suitable laser matrix. The outlet of the separation capillary was kept in contact with the wet membrane located on a metal plate, which served as one of the CE electrodes. At the end of the separation run, the dried target was inserted into the mass spectrometer. The authors observed that preparation of fresh membrane target was necessary to obtain satisfactory results. It was also observed that the diffusion of the sample into the wet membrane resulted in a substantial decrease in the final separation.

In the rotating-ball design (Orsnes et al., 1998a,b, 1999, 2000) the sample was deposited at atmospheric pressure onto a rotating ball which carried a thin sample film through a vacuum seal into the source chamber of a MALDI source. In this arrangement, which is also called a "Robin interface," volatile components of a solution vaporized outside and inside the source. In the vacuum deposition method (Preisler et al., 1998, 2000; Rejtar et al., 2001) the sample was infused via a fused silica capillary into an evacuated chamber and deposited on a moving quartz, steel, or plastic surface. This method was demonstrated in both on-line and off-line modes. In the first mode the sample was deposited on a rotating wheel or a Mylar tape directly in the ion source. The advantage of a disposable tape cartridge was the possibility of 24 h of uninterrupted deposition of the sample at a tape speed of 1 mm/s.

In a rather curious, yet interesting method for MALDI sample deposition, electrospray was employed to create a homogeneous sample layer (Kahr and Wilkins

1993; Yao et al., 1995). Experiments comparing traditional methods of sample deposition with that using electrospray indicated a substantial improvement in signal reproducibility due to the superior sample homogeneity produced by electrospraying. Although 2DE continues to be wrongly considered a low-throughput approach, there are still efforts to reintegrate this separation approach within the family of automated techniques. A fairly recent article (S. Zhang et al., 2003) describes an automated chip-based nanoelectrospray–MS for the identification of proteins separated by 2D gel electrophoresis. The authors used Nanomate (nanoelectrospray robot described by Van Pelt et al., 2002) together with an ESI-based monolithic microchip device. The system was used for protein identification in 2D gel spots of *E. coli* and yeast extract.

At this point it is wise to remember that proteomics still remains a multifaceted, rapidly evolving, and open-ended endeavor. Within this scenario microfluidic and array technologies have the potential of addressing problems related to protein purification, protein digestion, quantification, and sample introduction to mass spectrometers. However, the introduction of these systems into the proteomic world is at an early stage, proof-of-principle systems that address different aspects already exist, yet the assessment of the performance of these microfluidic devices and protein arrays in terms of sensitivity, throughput, and robustness will be realistic once these technologies have matured and we have enough examples of their successful applications in the real world of biological challenges. We can reasonably state that the pressing need for the determination of protein expression, associated sequences, and their functional role is becoming one of the most demanding challenges in modem bioscience. Such an enormous task is forcing the scientific community to test and use a host of analytical tools regardless of whether such tool(s) are mature enough. Miniaturization is a representative example. Most of us agree that such technology has a role to play in postgenome proteomics. However, we should not entertain the illusion that such technology will be able to address all the problems which other more matured macroapproaches have failed to address. To underline this statement, let us consider the following simple case: If two or more proteins comigrate in the same spot of a 2D map or coelute in an LC column, would a miniaturized system based on one of these separation techniques and MS detection handle such a situation differently from a macrosystem? The answer is certainly no because such a problem is likely to be associated with the intrinsic nature of the two proteins rather than the dimension of the analytical tool. Another commonly encountered situation in proteome analysis is the detection of poorly soluble proteins such as the classes of membrane and nuclear proteins. If this type of problem is not resolved by macrosystems, then certainly the same type of problem is going to be encountered in the miniaturized version of the same analytical approach. The citation of these and other problems is not meant to discourage scientists from moving toward miniaturization; on the contrary, they are meant to draw attention to the fact that many features and working principles of the miniaturized world are based on principles and experiences associated with techniques and approaches based on nonminiaturized setups. This means that, while we are developing our miniaturized analytical tools, we have to bear in mind that there are a number of problems which might be easier to solve in their natural nonminiaturized environment rather than assuming that, by reducing the dimension of our analytical tools, these problems will disappear. As well as these "inherited" problems, there are also problems inherent to the miniaturized world. For example, one of the biggest advantages of such a world is its capability to deal with minute sample amounts, yet in certain situations such an advantage can represent a drawback. For instance, a therapeutically relevant concentration of certain pathologies can be as low as few tens of particles per milliliter of fluid; working at a pumping rate of 1 nL/s would mean a pumping time of 3–4 h to get 1 mL necessary for a meaningful analysis.

3.8. CONCLUDING REMARKS

We have attempted to review various strategies for protein quantification which have been cited in recent literature. These strategies have a number of common steps, such as the initial labeling, separation, and detection. These steps involve various chemical approaches, diverse separation methods, and different working principles depending on whether they are based on macro- or microtechnology. Most of the listed strategies are less than 5 years of age, yet we have an impressive number of variations, each of which is capable of handling a wide range of aspects associated with protein quantification, aspects which only a few years ago would have been unthinkable. Before underlying certain aspects of the reviewed strategies, we ought to consider the following general observations: (i) The basic principle on which most of these strategies is based tells us that none of them can give a complete quantitative account of all proteins present in an entire proteome. In fact, none of the cited strategies have so far successfully characterized the entire proteome of any species. A simple reason for this limitation is the fact that the absence of the tagged amino acid from certain sequences or the failure to tag it due to some modification or due to inefficient labeling will inevitably exclude such sequences from our analysis. (ii) We should remember that the use of a single platform to quantify all the proteins within a given proteome does not carry more scientific weight than achieving the same objective through the combined use of more than one platform. Of course, solving a problem in a single shot has other attractions, including time saving, higher possibility of automation, and obviously certain economic advantages. Having said that, it is worth recording that the extensive chemical diversity of proteins within the simplest proteome together with over 200 known postranslational modifications and the wide dynamic range of protein concentrations are good reasons for the current absence of a single strategy to tackle protein quantification in complex mixtures. (iii) The question of comprehensive analyses in proteomics is proving to be far more challenging than its genomic counterpart. Sample preparation is a good example to remember. We know that the relatively consistent properties of DNA allow the application of a generic sample preparation protocol; protein isolation, on the other hand, can vary dramatically depending on whether they are membrane bound, free in a specific cellular compartment, or part of a complex that must be dissociated. In addition, another hurdle is associated with the large number of permutations in protein structure due to variation in amino acid contents at specific sites (polymorphism) and postranslational modifications. This can lead to multiplicity in protein and peptide structures with similar physicochemical properties, such as pI, hydrophobicity, and molecular size.

A closer look at the most commonly used strategies reveals that some of their limitations, which have been pointed out in their individual sections, are associated with their implementation under certain experimental conditions rather than with the basic principle on which such strategies are based. To clarify this observation, here are some specific considerations.

1. *Isotope Tagging.* In all (except DIGE) strategies described in this chapter, isotope tagging is a central step and is highly suitable for subsequent MS analyses. Setting aside the influence of tagging specificity and efficiency, we shall only consider a limitation which can only manifest under certain experimental conditions and can interfere with the reliability of the quantification. When dealing, for example, with a complex digest of a total cell lysate, one might want to use different chromatographic arrangements in which isotopically labeled peptides could be separated by ion exchange chromatography followed by reverse-phase separation or reverse-phase separation followed by ion mobility chromatography. Now, if the user opts for a high number of deuterium atoms within the tagging molecule to allow for an easier MS differentiation between labeled and unlabeled peptides under ESI conditions, then the use of reverse-phase separation of these peptides will influence the elution of these peptides. In other words, a well-known isotope effect takes place by which the deuterated peptide elutes earlier than its nondeuterated counterpart. Such an effect causes a variation in the isotope ratio across the two differently eluting profiles, an effect which becomes more pronounced with small peptides. This limitation is attributed not to the strategy which uses isotope labeling but to the user who chooses the conditions under which such a strategy is applied. If the user chooses to apply 2DE or chromatography different from reversed phase, then such an effect is likely to be less influential. What we are trying to say is that it would help if we manage to distinguish between inherent limitations of a given strategy and limitations associated with its application.

2. *Fluorescence Tagging.* It is commonly agreed that the introduction of DIGE represents an important development in the use of gel-based electrophoresis for protein quantification. A central step in this approach is the use of *N*-hydroxy-succinimide ester modified cyanine fluorophores, the most popular being named Cy3, Cy5, and Cy2. The users of gel electrophoresis had two good reasons to be excited by this development. First, fluorescence has higher sensitivity and wider dynamic range compared to existing gel stains. Second, DIGE analysis is conducted on the single gel instead of the number of replicas commonly used in the conventional 2DE. This difference has a number of positive implications, such as improved reproducibility, less work, and faster and more reliable quantification. The users of this approach have pointed out a limitation associated with the extent of labeling. For this approach to yield reliable quantification, it is necessary to apply minimal labeling (ideally one dye molecule per each protein). This requirement renders this approach even less sensitive than the well-known silver stain. On the other hand, the failure to apply minimal labeling will result in detrimental effects on the pI and M_r values of the separated proteins.

3. *Modified Proteins.* It is commonly agreed on that the complexity of the cellular proteome increases substantially if protein posttranslational modifications are also taken into account. We have described a limited number of approches which have the capability to quantify proteins which have experienced two of the most common postranslational modifications, phosphorylation and glycosylation. The number of applications using strategies dealing with both modifications remains limited, yet their potential for future general applications is not in doubt. Given the importance of such strategies, they deserve further consideration: On the phosphorylation side the method by Zhou et al. (2001) worked with a purified tyrosine–phosphorylated protein; under their LC/MS conditions, phosphotyrosine containing proteins within cell extracts was not detectable. Furthermore, the proteins observed in each study were among the most highly expressed in yeast. Oda et al. (2001) used β-elimination, which renders the method inherently compromised by the lower reactivity of phosphothreonine and the nonreactivity of phosphotyrosine. Furthermore, β-elimination of O-glycosylated residues may yield false positives and racemization upon Michael addition, which produces diastereomeric peptide mixtures that in some cases tend to reduce sensitivity in reverse-phase LC/MS analysis. Given the age of these strategies, there is no doubt that more optimized versions of these strategies together with the ongoing development in multidimensional chromatography and isotope-labeling protocols will enable these and other similar strategies to tackle protein phosphorylation in cells, particularly its quantification. The success of this type of strategy will provide valuable information on the regulatory mechanisms at the cellular level, information which cannot be obtained from gene sequencing.

Glycosylation is the most common postranslational modification with many clinical biomarkers and therapeutic targets are glycoproteins. These include Her2/neu in breast cancer, β human chorionic gonadotropin and α-fetoprotein in germ cell tumors, prostate-specific antigen in prostate cancer, and CA125 in ovarian cancer. Besides the conventional analysis of posttranslational modifications in individual proteins following LC or 2DE separation, faster and more comprehensive approaches have recently emerged. The two recent methods described by H. Zhang et al. (2003) and Kaji et al. (2003) have demonstrated that large-scale identification of N-linked glycoproteins can be achieved through close interaction between affinity capture, stable isotope tagging, and automated MS–MS. In both methods glycopeptides are immobilized on a solid support; then the N-linked glycopeptides are released by peptide-*N*-glycosidase. Then MS is used to identify the glycosylation sites and to quantify the relative abundance through the use of isotope-coded tags. The main difference between the two methods lies in the mechanisms by which the glycopeptides are captured from the solid support. As far as we are aware, these are the only published works describing strategies to tackle the question of glycosylation on a large scale. This means that further refinement and simplification of these methods are likely to follow.

Accepting the fact that complete proteome analysis by a single strategy is currently out of reach, an approach termed "divide and conquer" is gaining momentum. Strategies tackling a specific set of proteins like those which experiencied phosphorylation or glycosylation are good examples of such approaches.

REFERENCES

Ahn, N.G., Resing, K.A. (2001) *Nat Biotechnol* **19,** 317.

Anderson, N. L., Seilhamer, J. (1997) *Electrophoresis* **18,** 533.

Annan, R. S., Huddleston, M. J., Verma, R., Deshaies, R. J., Carr, S. T. (2001) *Anal. Chem.* **73,** 393.

Arenkove, P., Kukhtin, A., Gemmell, A., Voloschuk, S., Chupeeva, V., Mirzabekov, A. (2000) *Anal. Biochem* **278,** 123.

Bahr, U., StahlZeng, J., Gleitsmann, E., Karas, M. (1997) *J. Mass Spectrom.* **32,** 1111.

Beardsley, R. L., Karty, J. A., Reilly, J. P. (2000) *Rapid Commun. Mass Spectrom.* **14,** 2147.

Berkenkamp, S., Menzel, C., Karas, M., Hillenkamp, F. (1997) *Rapid Commun. Mass Spectrom.* **11,** 1399.

Bordini, E., Hamdan, M. (1999) *Rapid Commun. Mass Spectrom.* **13,** 1143.

Bordini, E., Hamdan, M., Righetti, P. G. (1999a) *Rapid Commun. Mass Spectrom.* **13,** 1818.

Bordini, E., Hamdan, M., Righetti, P. G. (1999b) *Rapid Commun. Mass Spectrom.* **13,** 2209.

Bordini, E., Hamdan, M., Righetti, P. G. (2000) *Electrophoresis* **21,** 2911.

Brancia, F., Oliver, S. G., Gaskell, S. J. (2000) *Rapid Commun. Mass Spectrom.* **14,** 2070.

Brenner, S., Johnson, M., Bridgham, J., Golda, G., Lloyd, D. H., Johnson, D., Luo, S., McCurdy, S., et al. (2000) *Nat. Biotechnol.* **18,** 630.

Cagney, G., Emili, A. (2002) *Nat. Biotechnol.* **20,** 163.

Cahill, M. A., Wozny, W., Schwall, G., Schroer, K., Hölzer, K., Poznanovic, S., Huzinger, C., Vogt, J.A., Stegmann, W., Matthies, H., Schrattenholz, A. (2003) *Rapid Commun. Mass Spectrom.* **17,** 1283.

Caprioli, R. M., Farmer, T. B., Gile, J. (1997) *Anal. Chem.* **69,** 4751.

Casting, R., Slodzian, G. (1962) *J. Microsc.* **1,** 395.

Chakraborty, A., Regnier, F. E. (2002) *J. Chromatogr. A* **949,** 173.

Chaurand, P., Caprioli, R. M. (2002). *Electrophoresis* **23,** 3125.

Chaurand, P., Fouchécourt, S., DaGue, B. B., Xu, B. J., Reyzer, M. L., Orgebin-Crist, M-C., Caprioli, R. M. (2003) *Proteomics* **3,** 2221.

Chaurand, P., Stoeckli, M., Caprioli, R. M. (1999) *Anal. Chem.* **71,** 5263.

Chen, R., Lowe, L., Wilson, J. D., Crowther, E., et al. (1999) *Clin. Chem.* **45,** 1693.

Chiari, M., Nesi, M., Ottolina, G., Righetti, P. G. (1994) *J. Chromatogr. A* **670,** 215.

Ciechanover, A., Orian, A., Schwartz, A. L. (2000) *Bioassays* **22,** 442.

Clackson, T., Hoogenboom, H. R., Griffiths, A. D., Winter, G. (1991) *Nature* **352,** 624.

Davis, M. T., Beierle, J., Bures, E. T., McGinley, M. D., Mort, J., Robinson, J. H., Spahr, C. S., Yu, W., Luethy, R., Patterson, S. D. (2001) *J. Chromatogr. B Biomed. Sci. Appl.* **752,** 281.

de Carvalho, M. G., McCormack, A. L., Olson, E., Ghomashchi, F., Gelb, M. H., Yates III, J. R., Leslie, C. C. (1996) *J. Biol. Chem.* **271,** 6987.

De Jager, W., Te Velthuis, H., Prakken, B. J., Kuis, W., et al. (2003) *Clin. Diagn. Lab. Immunol.* **10**, 133.

de Wildt, R. M. T., Mundy, C. R., Gorick, B. D., Timlinson, I. M. (2000) *Nat. Biotechnol.* **18,** 989.

Dunbar, B. S., Skinner, S. M. (1990) *Methods Enzymol.* **182,** 670.

Dunbar, S. A., Vander Zee, C. A., Oliver, K. G., Karem, K. L., et al. (2003) *J. Microbiol. Methods*. **53,** 245.

Ekström, S., Önnerfjord, P., Nilsson, J., Bengtsson, M., Laurell, T., Marko-Verga, G. (2000) *Anal. Chem.* **72,** 286.

Eng, J., McCormack, A. L., Yates III, J. R. (1994) *J. Am. Soc. Mass Spectrom.* **5,** 976.

Farah, S. J., Wang, S-W., Chang, W-H., Robertson C. R., Gast, A. P. (2001) *Langmuir* **17,** 5731.

Fenyo, D., Quin, J., Chait, B.T. (1998) *Electrophoresis* **19,** 998.

Ferguson, J. A., Steemers, F. J., Walt, D. R. (2000) *Anal. Chem.* **72,** 5618.

Figeys, D. (2002) *Proteomics* **2,** 273.

Figeys, D., Gygi, S. P., McKinnon, G., Aebersold, R. (1998) *Anal. Chem.* **70,** 3728.

Figeys, D., Ning, Y., Aebersold, R. (1997) *Anal. Chem.* **69,** 3153.

Figeys, D., Pinto, D. (2001) *Electrophoresis* **22,** 208.

Finley, D. (2001) *Nature* **412,** 285.

Friedman, M., Krull, L. H., Cavins, J. F. (1970) *J. Biol. Chem.* **245,** 3868.

Fung, E. T., Enderwick, C. (2002) *Biotechniques Suppl.* **32,** S34.

Galvani, M., Hamdan, M., Herbert, B., Righetti, P. G. (2001a) *Electrophoresis* **22,** 2058.

Galvani, M., Hamdan, M., Righetti, P. G. (2000) *Electrophoresis* **21,** 3684.

Galvani, M., Rovatti, L., Hamdan, M., Herbert, B., Righetti, P. G. (2001b) *Electrophoresis* **22,** 2066.

Gehanne, S., Cecconi, D., Carboni, L., Righetti, P. G., Domenici, E., Hamdan, M. (2002) *Rapid Commun. Mass Spectrom.* **16,** 1692.

Geng, M., Ji, J., Regnier, F. (2000) *J. Chromatogr.* **A870,** 295.

Gevaert, K., Van-Damme, J., Goethals, M., Thomas, G. R., Hoorelbeke, B., Demol, H., Martens, L., Puype, M., Staes, A., Vandekerchove, J. (2002) *Mol. Cell Proteomics* **1,** 896.

Gharbi, S., Gaffney, P., Yang, A., Zvelebil, M. J., Cramer, R., Waterfield, M. D., Timms, J. F. (2002) *Mol. Cell Proteomics* **1.2,** 91.

Godovac-Zimmermann, J., Brown, L. R. (2001) *Mass Spectrom. Rev.* **20,** 1.

Goldknopf I. L., et al. (1975) *J. Biol. Chem.* **250,** 7182.

Goodlett, D. R., Keller, A., Watts, J. D., Newitt, R., Yi, E. C., Purvine, S., Eng, J. K., von Haller, P., Aebersold, R., Kolker, E. (2001) *Rapid Commun. Mass Spectrom.* **15,** 1214.

Goshe, M. B., Veenstra, T. D., Panisko, E. A., Conrads, T. P., Angeli, N. H., Smith, R. D. (2002) *Anal. Chem.* **74,** 607.

Griffin, T. J., Lock, C. M., Ii, X-J., Patel, A., Chervetsova, I., Lee, H., Wright, M. E., Ranish, J. A., Chen, S. S., Aebersold, R. (2003) *Anal. Chem.* **75,** 867

Gygi, S. P., Aebersold, R. (2000) *Current Opin. in Chem. BioL.* **4,** 489.

Gygi, S. P., Rist, B., Gerber, S. A., Turecek, F., Gelb, M. H., Aebersold. R. (1999) *Nature Biotech.* **17,** 994.

Haab, B. B., Dunham, M. J., Brown, P. O. (2001) *Genome Biol.* **2,** 1.

Hale, J. E., Butler, J. P., Knierman, M. D., Becker, G. W. (2000) *Anal. Biochem.* **287,** 110

Hamdan, M., Galvani, M., Righetti, P. G. (2001) *Mass Spectrom. Rev.* **20,** 121.

Hamdan, M., Righetti, P. G. (2002) *Mass Spectrom. Rev.* **21,** 287.

Han, D. K., Eng, J., Zhou, H., Aebersold, R. (2001) *Nat. Biotechnol.* **19,** 946.

Herbert, B., Galvani, M., Hamdan, M., Olivieri, E., MacCarthy, J., Pederson, S., Righettiz P. G., (2001) *Electrophoresis* **22,** 2046.

Hoang, V. M., Conrads, T. P., Veenstra, T. D., Blonder, J., Terunuma, A., Vogel, J. C., Fisher, R. J. (2003) *J. Biomol. Techniques* **14,** 216.

Holmes, C. P., Jones, D. G. (1995) *J. Org. Chem.* **60,** 2318.

Hoogenboom, H. R., deBruine, A. P., Hufton, S. E., Hoet, R. M., Arends, J. W., Roovers, R. C. (1998) *Immunotechnology* **4,** 1.

Houseman, B. T., Huh, J. H., Kron, S. J., Mrksich, M. (2002) *Nat. Biotechnol.* **20,** 270.

Huang, R. P. (2001) *Clin. Chem. Lab. Med.* **39,** 209.

Hudson, P. J. (1998) *Curr. Opin. Biotechnol.* **9,** 395.

Ideker, T., Thorsson, V., Ranish, J. A., Christmas, R., Bahler, J., Eng, J. K., Bumgarner, R., Goodlett, D. R., Aebersold, R., Hood, L. (2001) *Science* **292,** 929.

Kahr, M., Wilkins, C. L. (1993) *J. Am. Soc. Mass Spectrom.* **4,** 453.

Kaji, H., Saito, H., Yamauchi, Y., Shinkawa, T., Taoka, M., Hirabayashi, J., Kasai, K-I., Takahashi, N., Isobe, T. (2003) *Nat. Biotechnol.* **21,** 667.

Kambhampati, D., Du, W., Schafferling, M., Kruschina, M. (2001) *Laborwelt* **3,** 45.

Keller, A., Nesizhskii, A. I., Kolker, E., Aebersold, R. (2002) *Anal. Chem.* **74,** 5383.

Keough, T., Lacey, M. P., Youngquist, R. S. (2000) *Rapid Commun. Mass Spectrom.* **14,** 2348.

Knappik, A., Ge, L. M., Honegger, A., Pack, P., Fischer, M., Welinhover, G., Hoess, A., Wolle, J., Pluckthun, A., Vimekas, B. (2000) *J. Mol. Biol.* **296,** 57.

Knezevic, V., Leethanakul, C., Bichsel, V. E., Worth, J. M. (2001) *Proteomics* **1,** 1271.

Knowles, M., Cervino, S., Skynner, H. A., Hunt, S. P., de Felipe, C., Salim, K., Lorente, G. M., McAllister, G., Guest, P. C. (2003) *Proteomics* **3,** 1162.

Kondo, T., Seike, M., Mori, Y., Fujii, K., Yamada, T., Hirohashi, S. (2003) *Proteomics* **3,** 1758.

Krause, E., Wenschcuh, H., Jungblut, P. R. (1999) *Anal. Chem.* **71,** 4160.

Krishna, R. G., Wold, F. (1993) *Adv. Enz. Relat. Areas Mol. Biol.* **67,** 265.

Kusnezow, W., Jacob, A., Walijew, A., Diehl, F., Hoheisel, J. D. (2003) *Proteomics* **3,** 254.

Kuyama, H., Watanabe, M., Toda, C., Ando, E., Tanaka, K., Nishimura, O. (2003) *Rapid Commun. Mass Spectrom.* **17,** 1642.

Laine, R. A. (1994) *Glycobiology* **4,** 759.

Le Naour, F., Hohenkirk, L., Grolleau, A., Misek, D. E., Lescure, P., Geiger, J. D., Hanash, S., Beretta, L. (2001) *J. Biol. Chem.* **276,** 17920.

Levit-Binnun, N., Lindner, A. B., Zik, O., Eshhar, Z., Moses, E. (2003) *Anal. Chem.* **75,** 1436.

Li, J., Thibault, P., Bings, N. H., Skinner, C. D., Wang, C., Colyer, C., Harrison, J. (1999) *Anal. Chem.* **71,** 3036.

Li, L., Garden, R. W., Romanova, E. V., Sweedler, J. V. (1999) *Anal. Chem.* **71,** 5451.

Li, L., Wang, P. L., Coulson. L. D. (1993) *Anal. Chem,* **65,** 493.

Link, A. J., Eng, J., Schieltz, D. M., Carmack, E., Mize, G. J., Morris, D. R., Garvik, B. M., Yates III, J. R. (1999) *Nat. Biotechnol.* **17,** 676.

MacBeath, G., Koehler, A. N., Schreiber, S. L. (1999) *J. Am. Chem. Soc.* **121,** 7967.

MacBeath, G., Schreiber, S. L. (2000) *Science* **289,** 1760.

MacCoss, M. J., McDonald, W. H., Saraf, A., Sadygov, R., Clark, J. M., Tasto, J. J., Gould, K. L., Wolters, D., Washburn, M., Weiss, A., Clark, J. I., Yates III, J. R. (2002) *Proc. Natl. Acad. Sci. USA* **99,** 7900.

McDonald, W. H., Ohi, R., Miyamoto, D. T., Mitchison, T. J., Yates. III, J. R. (2002) *Int. J. Mass Spectrom.* **219,** 245.

Mandoz-Gurpide, J., Wang, H., Misek, D. E., Brichary, F., et al. (2001) *Proteomics* **1,** 1279.

Marks, T., Sharp, R. (2000) *J. Chem. Technol. Biotechnol.* **75,** 6.

Marotti, Jr., L. A., Newitt, R., Wang, Y., Aebersold, R., Dohlman, H. G. (2002) *Biochemistry* **41,** 5067.

Michael, K. L., Taylor, L. C., Schultz, S. L., Walt, D. R. (1998) *Anal. Chem.* **70,** 1242.

Mirgorodskaya, O. A., Kozmin, Y. P., Titov, M. I., Komer, R., Roepstorff, C. P. (2000) *Rapid Commun. Mass Spectrom.* **14,** 1226.

Molloy, M. P., Phadke, N. D., Maddock, J. R., Andrews, P. C. (2001) *Electrophoresis* **22,** 1686.

Moroz, L. L., Gillette, R., Sweedler, J. V. (1999) *J. Exp. Biol.* **202,** 333.

Murphy, R. C. (1979) *Biomed. Mass Spectrom.* **6,** 309.

Nagra, D., Li, L. (1995) *J. Chromatogr.* **711,** 235.

Nelson, B. P., Grimsrud, T. E., Liles, M. R., Goodman, R. M., Corn, R. M. (2001) *Anal. Chem.* **73,** 1.

Nolan, J. P., Sklar, L. A. (2002) *Trends Biotechnol.* **20,** 9.

Oda, Y., Huang, K., Croos, F. R., Coowbum, D., Chait, B. T. (1999) *Proc. Natl. Acad. Sci. USA* **96,** 6951.

Oda, Y., Nagasu, T., Chait, B. T. (2001) *Nat. Biotechnol.* **19,** 379.

O'Farrel., P. H. (1975) *J. Biol. Chem.* **250,** 4007.

Optiek, G. J., Lewis, K. C., Jorgenson, J. W., Anderegg, R. G. (1997) *Anal. Chem.* **69,** 1518.

Optiek, G. J., Ramirez, S. M., Jorgenson, J. W., Moseley III, M. A. (1998) *Anal. Biochem.* **258,** 349.

Orsnes, H., Grat, T., Bohatka, S., Degn, H. (1998a) *Rapid Commun. Mass Spectrom.* **12,** 11.

Orsnes, H., Grat, T., Degn, H. (1998b) *Anal. Chem.* **70,** 4751.

Orsnes, H., Grat, T., Degn, H. (1999) *Anal. Chim. Acta* **390,** 185.

Orsnes, H., Grat, T., Degn, H., Murry, K. K. (2000) *Anal. Chem.* **72,** 251.

Pacholski, M. L., Winograd, N. (1999) *Chem. Rev.* **99,** 2977.

Patel, N., Saunders, G. H. W., Shakesheff, K. M., Cannizzaro, S. M. Langmuir (1999) **15,** 7252.

Patterson, S. D., Aebersold, R. (1995) *Electrophoresis* **16,** 1791.

Paweletz, C. P., Charboneau, L., Bichsel, V. E., Simone, N. L., et al. (2001) *Oncogene* **20,** 1981.

Pederson, S. K., Harry, J. L., Sebastian, L., Baker, J., Traini, M. D., McCarthy, J. T., Manoharan, A., Wilkins, M. R., Gooly, A. A., Righetti, P. G., Packer, N. H., Wiliiams, K. L., Herbert, B. R. (2003) *J. Proteome Res.* **2,** 303.

Peng, J., Schwartz, D., Elias, J. E., Thoreen, C. C., Cheng, D., Marsischky, G., Roelofs, J., Finley, D., Gygi, S. P. (2003) *Nat. Biotechnol.* **21,** 921.

Peters, E. C., Horn, D. M., Tully, D. C., Brock, A. (2001) *Rapid Commun. Mass Spectrom.* **15,** 2387.

Piehler, J., Brecht, A., Valiokas, R., Uedberg, B., Gauglitz, G. (2000) *Biosens. Bioelectron.* **15,** 473.

Preisler, J., Foret, F., Karger, B. L. (1998) *Anal. Chem.* **70,** 5287.

Preisler, J., Hu, P., Rejtar, T., Karger, B. L. (2000) *Anal. Chem.* **72,** 4785.

Rabilloud, T. (2002) *Proteomics* **2,** 3.

Raiteri, R., Nelles, G., Butt, H. J., Knoll, W., Skladal, P. (1999) *Sens. Actuator B* **61,** 213.

Ramdas, I., Wang, J., Hu, L., Cogdell, D., Taylor, E., Zhang, W. (2001) *Biotechniques* **31,** 546.

Reyzer, M. L., Hsieh, Y., Ng. K., Korfmacher, W. A., Caprioli, R. M. (2003) *J. Mass Spectrom.* **38,** 1081.

Rejtar, T., Hu, P., Preisler, J., Foret, F., Karger, B. (2001) *Proc. 49th ASMS Confrence, Chicago,* USA.

Rocklin, R. D., Ramsey, R. S., Ramsey, J. M. (2000) *Anal. Chem.* **72,** 5244.

Rodi, C. P., Darnhover-Patel, B., Stanssens, P., Zabeau, M., van den Boom, D. (2002) *Biotechniques Suppl.* **32,** S62.

Schena, M., Shalon, D., Heller, R., Chai, A., Brown, P. O., Davis, R. W. (1995) *Science* **270,** 467.

Schnöelzer, M., Jedrzejwski, P., Lehman, W. D. (1996) *Electrophoresis* **17,** 945.

Schwarz, S. A., Reyzer, M. L., Caprioli, R. M. (2003) *J. Mass Spectrom.* **38,** 699.

Schweitzer, B., Roberts, S., Grimwade, B., Shad W., et al. (2002) *Nat. Biotechnol.* **20,** 359.

Sechi, S., (2002) *Rapid Commun. Mass Spectrom.* **16,** 1416.

Sechi, S., Chait, B. T. (1998) *Anal. Chem.* **70,** 5150.

Shaw, J., Rowlinson, R., Nickson, J., Stone, T., Sweet, A., Williams, K., Tonge, R. (2003) *Proteomics* **3,** 1181.

Smith, R. D., Anderson, G. A., Lipton, M. S., Pasa-Tolic, L., Shen, Y., Conrads, T. P., Veenstra, T. D., Udseth, H. R. (2002) *Proteomics* **2,** 513.

Smolka, M., Zhou, H., Purkayastha, S. (2001) *Anal. Biochem.* **297,** 25.

Stoeckli, M., Chaurand, P., Hallahan, D. E., Caprioli, R. M. (2001) *Nat. Med.* **7,** 493.

Stoeckli, M., Staab, D., Staufenbiel, M., Wiederhold, K-H. (2002) *Anal. Biochem.* **311,** 33.

Taylor, J. D., Briley, D., Nguyen, Q., Long, K., Iannone, M. A., Li, M. S., Ye, F., Afshari, A., Lai, E., Wagner, M., Chen, J., Weiner, M. P. (2001) *Biotechniques* **30,** 661.

Thompson, A., Schäfer, J., Kuhn, K., Kienle, S., Schwarz, J., Schmidt, G., Neumann, T., Hamon, C. (2003) *Anal. Chem.* **75,** 1895.

Troendle, F. J., Reddick, C. D., Yost, R. A. (1999) *J. Am. Soc. Mass Spectrom.* **10,** 1315.

Twerenbold, D., Vuilleumier, J-L., Gerber, D., Tadsen, A., Brandt, B. V. D., Gillevert, P. M. (1996) *Appl. Phys. Lett.* **68,** 3503.

Tureček, F. (2002) *J. Mass Spectrom.* **37**, 1.

Ünlü, M., Morgan, M. E., Minden, J. S. (1997) *Electrophoresis* **18,** 2071.

Van der Geer, P., Hunter, T. (1994) *Electrophoresis* **15,** 544.

Van Pelt, C. K., Zhang, S., Heinon, J. D. (2002) *J. Biomol. Tech.* **13,** 72.

Vaughan, T. J., Osboum, J. K., Tempest, P. R. (1998) *Nat. Biotechnol.* **16,** 535.

Veenstra, T. D., Martinovic, S., Anderson, G. A., Pasa-Tolic, L., Smith, R. D. (2000) *J. Am. Soc. Mass Spectrom.* **11,** 78.

von Heller, P. D., Yi, E., Donohoe, S., Vaughn, K., Keller, A., Nesvizhskii, A. I., Eng, J., Li, X-J., Goodlett, D. R., Aebersold, R., Watts, J. D. (2003) *Mol. Cell. Proteomics* **2.7,** 428.

Vuong, G. L., Weiss, S. M., Kammer, W., Priemer, M., Vingrom, M., Nordheim, A., Cahill, M. A. (2000) *Electrophoresis* **21,** 2594.

Walton, I. D., Norton, S. M., Balasingham, A., He, L., Oviso, Jr., D. F., Gupta, D., Raju, P. A., Natan, M. J., Freeman, R. G. (2002) *Anal. Chem.* **74,** 2240.

Wang, C., Oleschuk, R., Ouchen, F., Thibault, J. U. P., Harrison, J. (2000) *Rapid Commun. Mass Spectrom.* **14,** 1377.

Wang, S., Regnier, F. E. (2001) *J. Chromatogr. A* **924,** 345.

Washburn, M. P., Wolters, D., Yates III, J. R. (2001) *Nat. Biotechnol.* **19,** 242.

Watts, J. D., Affolter, M., Krebs, D. L., Wange, R. L., Samelson, L. E., Aebersold, R. (1994) *J. Biol. Chem.* **269,** 29520.

Weinberger, S. R., Viner, R. I., Ho, P. (2002) *Electrophoresis* **23,** 3182.

Whittal, R. M., Russon, L. M., L. Li, (1998) *J. Chromatogr. A* **794,** 367.

Wind, M., Feldmann, I., Jakubowwski, N., Lehmann, W. D. (2003) *Electrophoresis* **24,** 1276.

Winter, G., Griffiths, A. D., Hawkins, R. E., Hoogenboom, H. R. (1994) *Annu. Rev. lmmunol.* **12,** 433.

Xu, N., Un, Y., Hofstadler, S., Matson, D., Call, C. J., Smith, R. D. (1998) *Anal. Chem.* **70,** 3553.

Yan, J. X., Sanchez, J. C., Rouge, V., Williams, K. L., Hochstrasser, D. F. (1999) *Electrophoresis* **20,** 723.

Yao, J., Dey, M., Pastor, S. J., Wilkins, C. L. (1995) *Anal. Chem.* **67,** 3638.

Yao, X., Freas, A., Ramirez, J., Demirev, P. A., Fenselau, C. (2001) *Anal. Chem.* **73,** 2836.

Zhang, B., Liu, H., Karger, B. L., Foret, F. (1999) *Anal. Chem.* **71,** 3258.

Zhang, H.Y., Caprioli, R.M. (1996) *J. Mass Spectrom.* **31,** 1039.

Zhang, H., Li, X-J., Martin, D. B., Aebersold, R. (2003) *Nat. Biotechnol.* **21,** 660.

Zhang, H., Yan, W., Aebersold, R. (2004) *Curr. Opin. Chem. Biol.* **8**, 1.

Zhang, R., Regnier, F. J. (2002) *J. Proteome Res.* **1,** 139.

Zhang, S., Van Pelt, C. K., Heinon, J. D. (2003) *Electrophoresis* **24,** 3620.

Zhou, G., Li, H., DeChamp, D., Chen, S., Shu, H., Gong, Y., Flaig, M., Gillespie, J. W., Hu, N., Taylor, P. R., Emmert-Buck, M. R., Liotta, L. A., Petricoin III, E. F., Zhao, Y. (2002) *Mol. Cell. Proteomics* **1.2,** 117.

Zhou, H., Ranish, J. A., Watts, J. D., Aebersold, R. (2002) *Nat. Biotecnol.* **19,** 512.

Zhou, H., Watts, J. D., Aebersold, R. (2001) *Nat. Biotechnol.* **19,** 375.

Zhu, H., Bilgin, M., Bangham, R., Hall, D., Casamayor, A., Bertone, P., Lan, N., Jansen, R., Bildingmaier, S., Houfek, T., Mitchell, T., Miller, P., Dean, R. A., Gerstein, M., Snyder, M. (2001) *Science* **293,** 2101.

Zuo, X., Echan, L., Hembach, P., Tang, H. Y., Speicher, K. D., Santoli, D., Speicher, D. W. (2001) *Electrophorsis* **22,** 1603.

II

PROTEOMICS TODAY: SEPARATION SCIENCE AT WORK

INTRODUCTION

Three major events (in fact four) helped shape the current state of the art on two-dimensional map analysis. The first one occurred in the late 1950s and early 1960s when Harry Svensson, a freelance researcher at the Karolinska Institute in Stockholm, developed the theory of isoelectric focusing (IEF) and laid down its theoretical foundations in a few, by now, classic papers (Svensson 1961a,b; Svensson, 1962). Much earlier, during World War II, he had been forced by his maestro, Tiselius, to work on the creation of stationary pH gradients in the presence of an electric field with nonamphoteric compounds, namely by steady-state electrolysis of a salt in free solution. Of course the ion components would vacate the grounds, collect at the opposite electrodes, and leave an empty trail on their wake with no soldiers to guard the battlefield. Svensson understood that a multitude of buffers was required, that all of them had to be amphoteric and, in addition, they had to have decent buffering power as ensured by not too large ΔpK values. Only in this way would the zones of isoelectric "carrier ampholytes" form a continuous chain as the electric field would tie them to their isoelectric zones while diffusion would cause them to broaden just enough to penetrate the neighboring ampholyte zones, thus simultaneously ensuring buffer capacity and conductivity. It was too bad that this army was barely a handful of soldiers, hardly able to cover the grounds in the pH 3–10 interval. Nonetheless, Svensson published, in 1964, some remarkable color pictures of unique hemoglobin (Hb) separations in his 110-mL focusing column stabilized by a sucrose density gradient. In these

Proteomics Today. By Mahmoud Hamdan and Pier Giorgio Righetti
ISBN 0-471-64817-5

separations, he used protein lysates, notably of casein, albumin, Hb, and whole blood, as background carrier ampholytes. Peptides rich in this provided the much needed buffering power and conductivity in the pH 6–7 gap.

Aware of these limitations, Svensson hired a medical student, Olof Vesterberg, to help him devise a proper synthesis of the much needed carrier ampholytes. After three years of slow progress, Svensson became professor of Physical Chemistry at Gothenburg and the IEF team broke up. Vesterberg continued on this project in Stockholm and, in the spring of 1964, Svensson received a phone call from an excited Vesterberg, who appeared to have solved the problem. Well, the chap had been moonlighting and pouring over textbooks of organic chemistry and surfaced with a remarkable synthesis of the much wanted "carrier ampholytes": a chaotic synthesis, to be sure, as chaotic as a medical student could possibly devise. A most ingenious chaotic process, in fact, by which concoctions of oligoamines (from tetra- to hexa-amino groups) were reacted with limiting amounts of an α-β-unsaturated acid, acrylic acid (Vesterberg, 1969). Chaos generated order! In a steep voltage gradient, this army of synthetic amphoteres join arms in an orderly fashion, with each assuming a (quasi) Gaussian distribution about its respective isoelectric point (pI) value (Rilbe, 1973).

Was everything under control then? Well, with such a superb technique (one of the few able to counteract entropy's tendency to dissipate peaks via diffusion in the surrounding medium by producing sharply focused zones no matter how carelessly the experiment is carried out), one would think yes, all quiet on the Western Front! Svensson and Vesterberg were indeed convinced that a breakthrough had been achieved, and they approached Arne Tiselius for his approval. Poor Arne, who had been witnessing the steady erosion of his U-tube method, scolded them, infuriated at the notion that they would dare to bring forward a technique surely bound for failure, considering that macro-ions, on their approach to pI values, would likely aggregate and precipitate.

The IEF technique was launched anyway, as a preparative method, requiring 110 mL and 440 mL columns for operation. An entire experiment, including column set-up, focusing, elution, and analysis of hundreds of fractions, required a minimum of one week of hard labor! Although during the 1960s the growth of IEF was painfully slow, by the beginning of the 1970s, especially due to the introduction of the analytical counterpart in polyacrylamide gels (Righetti and Drysdale, 1971). IEF enjoyed such a marked growth as to soon become a leading separation technique in all fields of biological sciences.

The second event occurred later in the same decade, 1960s. A Carioca trio, very fond of carnivals in Rio, decided they would carry the masquerade all year around, for the merriment of all those poor scientists chained to the bench in desperate search of fresh data aimed at getting fresh money for their research, in an endless turning of the wheel. Just as IEF would screen proteins solely on the basis of surface charge, it would have been nice to be able to sort them out on the basis of their pure mass value. On these premises, Shapiro et al. (1967) came up with the notion of disguising: Let proteins swim in a solution of an anionic surfactant, above the critical micellar concentration (sodium dodecyl sulphate, SDS, seemed well suited for this task). They would surely be coated by micelles of this chemical, to the point of swamping the original amphoteric charge of polypeptide chains and forcing them to behave just like nucleic acids, odd macromolecules in which the charge to mass ratio becomes nearly constant above *ca.* 400 bp in length (Stellwagen, Gelfi, and Righetti, 1997). Most proteins would just adsorb SDS to a magic ratio of 1.4 mg SDS/mg protein (Pitt-Rivers and Impiombato, 1968): if one would then drive them into a porosity gradient gel (Margolis and Kenrick, 1968), may be even by exploiting discontinuous buffers (Laemmli, 1970), one would end up with razor-blade-sharp zones, whose mobility, when plotted against the log of molecular mass, would result in a linear relationship (Chrambach and Rodbard, 1971).

By mid 1970s, another explosive concept became a reality: two-dimensional map analysis. Somebody out there realized that, as we now had two electrokinetic methodologies perfectly orthogonal (in the Gidding's sense) (Giddings, 1991), combining them at right angle would offer a unique map in which the protein spots would be maximally spread in the two-dimensional space having as coordinates charge (pI) and mass (in the SDS-PAGE second run). Three labs reported this 2D technique simultaneously and independently in 1975, although most of the credit went just to O'Farrell (O'Farrell, 1975; Klose, 1975; Scheete, 1975). Perhaps because his system was the most elaborate: In fact, he was able to resolve and detect about 1100 different proteins from lysed *E. coli* cells on a single 2D map and suggested that the maximum resolution capability might have been as high as 5000 different proteins. O'Farrell had elaborated the technique to its utmost sophistication, nullifying the attempts of the second and third wave of "discoverers" to get credit for it; even today 2D maps represent the most popular technique in proteome analysis.

There is no way, though, to escape the crippling disease of aging on spaceship Earth. By the end of 1970s, it was clear that a number of ailments was besieging Svensson-Rilbe's creature, namely uneven buffering capacity and conductivity, irreproducibility of CA synthesis, the most-dreaded cathodic drift impeding proper focusing conditions, to name just a few. It was with these problems in mind that we teamed up with LKB scientists to work out a totally new concept, immobilized pH gradients, that seemed to solve all those problems at once (Bjellqvist et al., 1982, 1983). This new work of art was unveiled on April 22, 1982, at the electrophoresis meeting organized by Stathakos in Athens and acclaimed with standing ovations. At least that is what we had hoped. In reality, we presented these data to an almost empty room, since most of the delegates had never been to Athens before and opted to enjoy lovely spring weather on the Acropolis, on the Licabetto, on the Plaka, strolling just about around any corner in the capital except at the Hellenic Academy of Science, where the meeting was held.

It would be unfair to end this Introduction without mentioning one of the latest developments in the field of proteome analysis, namely "multidimensional chromatography coupled to mass spectrometry." Although this approach is amply treated in the following chapters, I have some reservations in dealing with it *in extenso* here, for at least two reasons. First of all, because of the pompous term "multidimensional" given to this technique, which in fact is just two-dimensional as our glorious 2D maps (the two dimensions being, in general, the coupling of an ion exchange with a reversed-phase column, ensuring orthogonality of the system). The term "multidimensional" would call for at least >4 dimensions, since, even if it were only three (which is absolutely not the case!) or four-dimensions, the English dictionary would suggest more humble terms such as tridimensional or tetradimensional. The other reason that brings strong reservations to this technique is the most unfortunate abbreviation introduced by Yates III and his group, who have called it "MudPit" (Yates III et al., 1995). Since I am not an English native, I consulted my *Webster's New World Dictionary* (which I suppose applies to the "Old World" as well), and here is what I find:

- Mud: "wet, soft, sticky earth", or even "defamatory remarks, libel or slander";
- Pit: "a hole or cavity in the ground"; "an abyss"; "hell"; "any concealed danger, trap, snare" and much more.

Well, I would like to send to my friend, Yates III, this warning coming from my Latin ancestors: "*nomen omen*," which in current English slang translates as: "the label indicates the content!" I would urge him to abjure this credo and give a better name to his creature.

REFERENCES

Bjellqvist, B., Ek, K., Righetti, P.G., Gianazza, E., Görg, A., Westermeier, R. and Postel, W. (1982). *J. Biochem. Biophys. Methods* **6**, 317–339.

Bjellqvist, B., Ek, K., Righetti, P.G., Gianazza, E., Görg, A., and Postel, W. (1983). *In* "Electrophoresis '82" (D. Stathakos, Ed.), pp. 61–74. de Gruyter, Berlin.

Chrambach, A., Rodbard, D. (1971) *Science* **172,** 440–445.

Giddings, J.C. (1991) *Unified Separation Science*, Wiley, New York, pp. 112–131.

Klose, J. (1975) *Humangenetik* **26,** 231–243.

Laemmli, U.K. (1970) *Nature* **227,** 680–681.

Margolis, J, Kenrick, K.G. (1968) *Anal. Biochem.* **25,** 347–355.

O'Farrell, P.H. (1975) *J. Biol. Chem.* **250,** 4007–4021.

Pitt-Rivers, R., Impiombato, F.S.A. (1968) *Biochem. J.* **109,** 825–831.

Righetti, P.G., and Drysdale, J.W. (1971). *Biochim. Biophys. Acta* **236**, 17–24.

Rilbe, H. (1973). *Ann. N. Y. Acad. Sci.* **209**, 11–22.

Scheele, G.A. (1975) *J. Biol. Chem.* **250,** 5375–5385.

Shapiro, A.L., Viñuela, E., Maizel, J.V. (1967) *Biochem. Biophys. Research Commun.* **28,** 815–822.

Stellwagen, N., Gelfi, C., Righetti, P.G. (1997) *Biopolymers* **42,** 687–703.

Svensson, H. (1961). *Acta Chem. Scand.* **15**, 325–341.

Svensson, H.(1961). *Acta Chem. Scand.* **16**, 456–466.

Svensson, H. (1962). *Arch. Biochem. Biophys.*, **Suppl. 1**, 132–140.

Vesterberg, O. (1969). *Acta Chem. Scand.* **23**, 2653–2666.

Yates III, J.R., McCormack, A.L., Schieltz, D., Carmack, E., Link, A. (1995) *Anal. Chem.* **67,** 3202–3210.

4

CONVENTIONAL ISOELECTRIC FOCUSING IN GEL MATRICES AND CAPILLARIES AND IMMOBILIZED pH GRADIENTS

4.1. INTRODUCTION

This chapter is organized as follows: We will first treat conventional isoelectric focusing in soluble, carrier ampholyte (CA) buffers as originally envisaged by Svensson-Rilbe (Righetti, 1983). In this section, we will give a brief historical survey of its evolution, from a preparative-scale approach to the extremely popular gel-slab version. This chapter will also contain an ample description of the properties of CA buffers, especially in regard to their buffering capacity (β) and conductivity, since these basic concepts have been found to be fundamental in capillary zone electrophoresis (CZE). We will then proceed to a description of immobilized pH gradients (IPGs), the novel version of focusing launched in 1982 (Righetti, 1990). In the third section, we will describe isoelectric focusing (IEF) in a capillary format, the rising star in separation science (Righetti and Gelfi, 1994; Righetti et al., 1996b; Righetti and Bossi, 1998). Finally, we will describe CZE separations exploiting isoelectric buffers as a background electrolyte. Although this is not a focusing technique per se, it is the natural evolution of the know-how developed in IEF, and it appears to hold a unique separation potential in CZE. Since this zonal electrophoresis in isoelectric buffers relies heavily on the properties of CAs, the importance of the theoretical treatment offered at the beginning on the physicochemical properties of such buffers will be here appreciated.

Proteomics Today. By Mahmoud Hamdan and Pier Giorgio Righetti
ISBN 0-471-64817-5

4.1.1. Brief Historical Survey

In 1960–1962, a major breakthrough came with the discovery of conventional IEF in CA (soluble, amphoteric buffers) reported by Svensson–Rilbe in a series of now classical articles (Svensson, 1961, 1962a,b). At just about the same time, Meselson et al. (1957) described isopycnic centrifugation (IPC), a related, high-resolution technique, another member of a family that Kolin (1977) called isoperichoric focusing. Unlike conventional chromatographic and electrophoretic techniques available up to that time, which could not provide any means for avoiding peak decay during the transport process, IEF and IPC had built-in mechanisms opposing entropic forces trying to dissipate the zone. As the analyte reached an environment (the "perichoron" in Greek) in which its physicochemical parameters were equal (iso) with those of the surroundings, it focused, or condensed, in an ultrathin zone kept stable and sharp in time by two opposing force fields: diffusion (tending to dissipate the zone) and external fields (voltage gradients in IEF or centrifugal fields in IPC) forcing the "escaping" analyte back into its "focusing" zone. It was truly a unique event in separation science by which all of us, general practitioners in the field and humble Clark Kents in the laboratory, felt (and acted) like Superman.

According to the original idea of Svensson–Rilbe, CA–IEF was born as a preparative technique in vertical, hollow columns exploiting two colinear gradients: a sucrose density gradient acting as an anticonvective medium and a pH gradient generated by the current during the focusing step (Vesterberg et al., 1967). A typical experiment took several days to complete and the subsequent analyses of fractions eluted from the gradients were tedious and laborious. It was the advent of gel IEF (Awdeh et al., 1968; Righetti and Drysdale, 1971, 1973) that rendered the technique so popular. The evolution of CA–IEF can be followed in a series of meetings, starting with the cornerstone symposium in New York (Catsimpoolas, 1973) and soon followed by Glasgow (Arbuthnott and Beeley, 1975), Milan (Righetti, 1975), Hamburg (Radola and Graesslin, 1977), and Cambridge (Massachusetts) (Catsimpoolas, 1978). With the latter meeting, the specific interest in IEF and isotachophoresis (ITP, note that ITP is also a steady-state technique with this proviso: in IEF all zones have identical mobilities all equal to zero, whereas in ITP all zones acquire the same, nonzero velocity) moved to the general field of electrophoresis (in which, however, IEF still played the role of Prima Donna); there followed a series of meetings in Munich (Radola, 1980a), Charleston (Allen and Arnaud, 1981), Athens (Stathakos, 1983), Tokyo (Hirai, 1984), Göttingen (Neuhoff, 1984), London (Dunn, 1986), and Copenhagen (Schafer-Nielsen, 1988). With the latter meeting, the publication of proceedings ceased, since the official journal of the International Electrophoresis Society was launched, that is, *Electrophoresis*, which appeared to be the proper forum for such events.

It was in 1975 that IEF took another, important turn: In that year O'Farrell (1975) introduced 2DE and demonstrated that, by sequentially running IEF in the first dimension followed by SDS electrophoresis at right angles, more than 1100 individual polypeptides could be resolved in an *Escherichia coli* lysate. These 2D maps became the nightmare of chromatographers: They still had to accept the hard reality that, even in the best gas chromatographic separation, it was hard to resolve barely 100 distinct peaks (Giddings, 1991). Soon 2DE excited grandiose projects, like the Human Protein Index System of Anderson and Anderson (1982), who started the far-reaching goal of mapping all possible phenotypes expressed by any and all different cells in our organism—an Herculean task, indeed. This started a series of meetings of the 2DE group, particularly strong in the field of clinical chemistry (Young and Anderson, 1982; King, 1984). A number of books were devoted to this 2DE issue (Celis and Bravo, 1984; Dunbar, 1987) and a series of meetings were started on this special topic (Dunn, 1991a). Today, 2DE has found a proper forum in *Electrophoresis*, which began hosting individual papers dealing with variegate topics in 2D maps. Starting in 1988, *Electrophoresis* launched special issues devoted to 2D maps not only in clinical chemistry and human molecular anatomy but also in every possible living organism and tissue, the first one dedicated to plant proteins (Damerval and de Vienne, 1988). Soon, a host of such "paper symposia" appeared, collecting databases on any new spots of which a sequence and a function could be elucidated (Celis, 1989, 1990a,b, 1991, 1992, 1994a,b, 1996, 1999; Dunn, 1991b, 1995, 1997, 1999; Lottspeich, 1996; Tümmler, 1998; Williams, 1998; Humphery-Smith, 1997; Appel et al., 1997, 1999). Today, 2D map analysis is properly hosted in a number of new journals, such as *Proteomics*, *Journal of Proteome Research*, and *Molecular and Cellular Proteomics*.

4.2. CONVENTIONAL ISOELECTRIC FOCUSING IN AMPHOTERIC BUFFERS

4.2.1. General Considerations

All fractionations which rely on differential rates of migration of sample molecules, for example along the axis of a chromatographic column or along the electric field lines in electrophoresis, generally lead to concentration bands or zones which are essentially always out of equilibrium. The narrower the band or zone, the steeper the concentration gradients and the greater the tendency of these gradients to dissipate spontaneously. This dissipative transport is thermodynamically driven; it relates to the tendency of entropy to break down all gradients, to maximize dilution, and, during this process, to thoroughly mix all components (Giddings, 1991). Most frequently, entropy exerts its effects via diffusion, which causes molecules to move down concentration gradients and so produces band broadening and component intermixing.

The process of IEF in CA (Svensson, 1961, 1962a,b) and in IPGs (Bjellqvist et al., 1982) provides an additional force which counteracts CA diffusion and so maximizes the ratio of separative to dissipative transports. This substantially increases the resolution of the fractionation method. The sample focuses on its isoelectric point (pI) driven by the voltage gradient and by the shape of the pH gradient along the separation axis (Fig. 4.1). The separation can be optimized by using thin or ultrathin matrices (0.5 mm or less in thickness; Righetti and Bello, 1992; Radola, 1980b) and by applying very low sample loads (as permitted by high-sensitivity detection techniques such as silver and gold staining, radioactive labeling, and immunoprecipitation

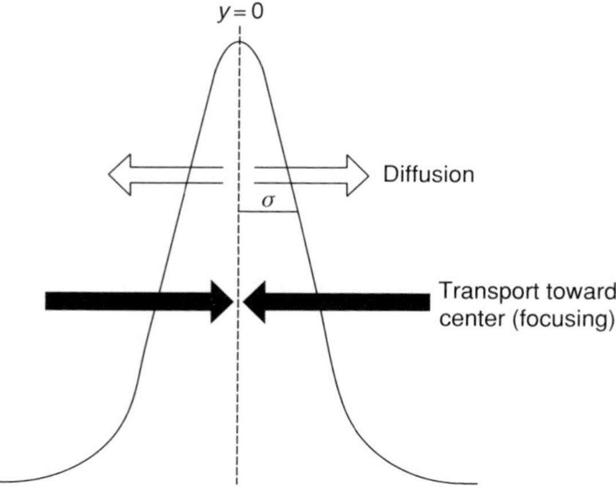

FIGURE 4.1. Forces acting on condensed zone in IEF. The focused zone is represented as a symmetric Gaussian peak about its focusing point (pI; $y = 0$). Migration of sample toward the pI position is driven by the voltage gradient and by the slope of the pH gradient. σ is the standard deviation of the peak. (Courtesy of Dr. O. Vesterberg.)

followed by amplification with peroxidase- or alkaline phosphatase–linked secondary antibodies).

4.2.1.1. Basic Method Isoelectric focusing is an electrophoretic technique by which amphoteric compounds are fractionated according to their pI values along a continuous pH gradient (Rilbe, 1973). Contrary to zone electrophoresis, where the constant (buffered) pH of the separation medium establishes a constant charge density at the surface of the molecule and causes it to migrate with constant velocity (in the absence of molecular sieving), the surface charge of an amphoteric compound in IEF keeps changing and decreasing, according to its titration curve, as it moves along a pH gradient until it reaches its equilibrium position, that is, the region where the pH matches its pI. There, its mobility equals zero and the molecule comes to a stop.

The gradient is created and maintained by the passage of an electric current through a solution of amphoteric compounds which have closely spaced pI values, encompassing a given pH range. The electrophoretic transport causes these CAs to stack according to their pI values, and a pH gradient, increasing from anode to cathode, is established. At the beginning of the run, the medium has a uniform pH, which equals the average pI of the CAs. Thus most ampholytes have a net charge and a net mobility. The most acidic CA moves toward the anode, where it concentrates in a zone whose pH equals its pI, while the more basic CAs are driven toward the cathode. A less acidic ampholyte migrates adjacent and just cathodal to the previous one, and so on, until all the components of the system reach a steady state. After this stacking process is completed, some CAs still enter, by diffusion, zones of higher or lower pH where they are no longer in isoelectric equilibrium. But as soon as they enter these zones, the CAs become charged and the applied voltage forces them back to their steady-state

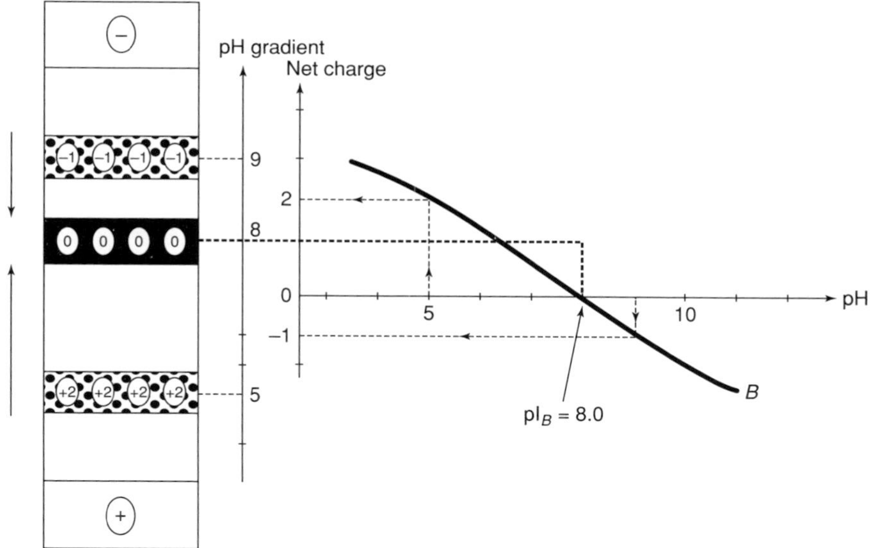

FIGURE 4.2. Principle of IEF in CA buffers. When a protein macroion is applied to a prefocused gel slab, both above and below its pI (here pI = 8.0), it acquires negative and positive charges, according to its titration curve (shown on the right). The two fronts then merge at pH = pI, where the net charge is zero. (Courtesy of LKb Produkter AB.)

position. This pendulum movement, diffusion versus electrophoresis, is the primary cause of the residual current observed under isoelectric steady-state conditions. Finally, as time progresses, the sample protein molecules also reach their pI. Figure 4.2 shows this focusing process for a protein with a pI = 8.0. If this macroion is applied simultaneously above and below its pI, its negatively and positively charged species (whose net charge is defined by the titration curve shown at the right), respectively, migrate toward each other until they fuse, or merge (or focus, tout court), at the pI zone, having zero net charge.

4.2.1.2. Applications and Limitations The technique only applies to amphoteric compounds and more precisely to good ampholytes with a steep titration curve around their pI, *conditio sine qua non* for any compound to focus in a narrow band. This is very seldom a problem with proteins, but it may be so for short peptides that need to contain at least one acidic or basic amino acid residue in addition to the $-NH_2$ and –COOH termini. Peptides which have only these terminal charges are isoelectric over the entire range of approximately pH 4 and pH 8 and so do not focus. Another limitation with short peptides is encountered at the level of the detection methods: CAs are reactive to most peptide stains. This problem may be circumvented by using specific stains, when appropriate (Gianazza et al., 1979, 1980), or by resorting to IPGs which do not give background reactivity to nihydrin and other common stains for primary amino groups (e.g., dansyl chloride, fluorescamine) (Gianazza et al., 1984b).

In practice, notwithstanding the availability of CAs covering the pH 2.5–11 range, the practical limit of CA–IEF is in the pH 3.5–10 interval. Since most protein pI values cluster between pH 4 and 6 (Gianazza and Righetti, 1980), this may pose a major problem only for specific applications.

When a restrictive support like polyacrylamide (PAA) is used, a size limit is also imposed for sample proteins. This can be defined as the size of the largest molecules which retain an acceptable mobility through the gel. A conservative evaluation sets an upper molecular mass limit of about 750,000 when using standard techniques. The molecular form in which the proteins are separated strongly depends upon the presence of additives, such as urea and/or detergents. Moreover, supramolecular aggregates or complexes with charged ligands can be focused only if their K_d is lower than 1 μM and if the complex is stable at pH = pI (Krishnamoorthy et al., 1978). An aggregate with a higher K_d is easily split by the pulling force of the current.

4.2.1.3. Specific Advantages

(a) Isoelectric focusing is an equilibrium technique; therefore the results do not depend (within reasonable limits) upon the mode of sample application, the total protein load, or the time of operation.

(b) An intrinsic physicochemical parameter of the protein (its pI) may be measured.

(c) Isoelectric focusing requires only a limited number of chemicals, is completed within a few hours, and is less sensitive than most other techniques to the skill (or lack of it) of the operator.

(d) Isoelectric focusing allows resolution of proteins whose pI differs by only 0.01 pH unit (with immobilized pH gradients, up to about 0.001 pH unit); the protein bands are very sharp due to the focusing effect.

4.2.1.4. Carrier Ampholytes Table 4.1 lists the general properties of carrier ampholytes, that is, of the amphoteric buffers used to generate and stabilize the pH gradient in IEF. The fundamental and performance properties listed in this table are usually required for a well-behaved IEF system, whereas the "phenomena" properties are in fact the drawbacks or failures inherent to the technique. For instance, the "plateau effect" or "cathodic drift" is a slow decay of the pH gradient with time whereby, upon prolonged focusing at high voltages, the pH gradient with the focused proteins drifts toward the cathode and is eventually lost in the cathodic compartment. There seems to be no remedy to this problem (except from abandoning CA–IEF in favor of the IPG technique), since there are complex physicochemical causes underlying it, including a strong electroosmotic flow (EOF) generated by the covalently bound negative charges of the matrix (carboxyls and sulfate in both polyacrylamide and agarose) (as reviewed in Righetti and Tonani, 1992). In addition, it appears that some basic CAs may bind to hydrophobic proteins, such as membrane proteins, by

TABLE 4.1. Properties of Carrier Ampholytes

Fundamental classical properties
1. Buffering ion has mobility of zero at pI
2. Good conductance
3. Good buffering capacity

Performance properties
1. Good solubility
2. No influence on detection systems
3. No influence on sample
4. Separable from sample

Phenomena properties
1. Plateau effect (i.e., drift of pH gradient)
2. Chemical change in sample
3. Complex formation

hydrophobic interaction. This cannot be prevented during electrophoresis, whereas ionic CA–protein complexes are easily split by the voltage gradient (Sinha and Righetti, 1986).

In chemical terms, CAs are oligoamino, oligo-carboxylic acids available from different suppliers under different trade names. There are three basic synthetic approaches: Vesterberg's (1973) approach, which involves reacting different oligoamines (tetra-, penta-, and hexa-amines) with acrylic acid; the Söderberg et al. (1980) synthetic process, which involves the copolymerization of amines, amino acids, and dipeptides with epichlorohydrin; and the Pogacar–Grubhofer approach, which utilizes ethyleneimine and propylenediamine for subsequent reaction with propanesultone, sodium vinylsulfonate, and sodium chloromethyl phosphonate (Pogacar and Jarecki, 1974; Grubhofer and Borja, 1977). Accordingly, there are three types of products on the market: the Amersham-Pharmacia Biotech Ampholines (formerly from LKB-Produkter AB) and Bio-Rad Biolytes, which belong to the first class; the Amersham-Pharmacia Biotech Pharmalytes, which should be listed in the second class; and the Novex-Servalyt Ampholytes and Genomic Solutions pH 3–10 Ampholytes, which should be classified in the third category.

The wide-range synthetic mixtures (pH 3–10) contain hundreds, possibly thousands, of different amphoteric chemicals having pI values evenly distributed along the pH scale. Since they are obtained by different synthetic approaches, CAs from different manufacturers are bound to have somewhat different pI values. Thus, if higher resolution is needed, particularly for 2D maps of complex samples, we suggest using blends of the different commercially available CAs. A useful blend is 50% Pharmalyte, 30% Ampholine, and 20% Biolyte by volume).

Carrier ampholytes from any source should have an average molecular mass (M_r) of about 750 (size interval 600–900, the higher M_r referring to the more acidic CA species) (Bianchi-Bosisio et al., 1981). Thus CAs should be readily separable (unless they are hydrophobically complexed to proteins) from macromolecules by gel filtration. Dialysis is not recommended due to the tendency of CAs to aggregate. Salting out of proteins with ammonium sulfate seems to completely eliminate any contaminating CAs.

A further complication arises from the chelating effect of acidic CAs, especially towards Cu^{2+} ions, which may inactivate some metallo-enzymes (Galante et al., 1975). In addition, focused CAs represent a medium of very low ionic strength (<1 mEq/L at the steady state) (Righetti, 1980). Since the isoelectric state involves a minimum of solvation and thus of solubility for the protein macroion, there is a tendency for some proteins (e.g., globulins) to precipitate during the IEF run near their pI position. This is a severe problem in preparative runs. In analytical procedures it can be minimized by reducing the total amount of sample applied.

The hallmark of a "carrier ampholyte" is the absolute value of pI $-$ pK_{prox} (or ½ ΔpK): The smaller is this value, the higher the conductivity and buffering capacity (at pH $=$ pI) of the amphotere. A ΔpK $=$ log 4 (i.e., pI $-$ pK $= 0.3$) would provide an incredible molar buffering power (ß) at the pI 2.0 (unfortunately, such compounds do not exist in nature). A ΔpK $=$ log 16 (i.e., pI $-$ pK $= 0.6$) offers a ß value of 1.35 at pH $=$ pI. Let us take a practical example: Lys and His, two amino acids that can be considered good carrier ampholytes for IEF. For Lys, the pI value (9.74) is nested on a high saddle between two neighboring protolytic groups (the ε- and α-amino; pI $-$ pK $= 0.79$), thus providing an excellent ß power ($\sim$1.0). Conversely, the situation is not so brilliant with His: The pI value (7.57) is located in a valley with a ß value of only 0.24 (due to a pI $-$ pK of 1.5). When plotting the molar ß power of a weak protolyte along the pH axis, one reaches a maximum of ß $= 1.0$ at pH $=$ pK. If one accepts as a still reasonable ß a value of one-third of this maximum, this is located at pH $-$ pK $= \pm 0.996$. It is thus seen that even His, generally considered as a good carrier ampholyte, is in fact barely acceptable and falls just below this one-third limit of acceptance (Rilbe, 1996).

4.2.2. Equipment

4.2.2.1. Electrophoretic Equipment Three major items of apparatus are required: an electrophoretic chamber, a power supply, and a thermostatting unit. The optimal configuration of the electrophoretic chamber is for the lid to contain movable platinum wires (e.g., in the Multiphor 2, in the Pharmacia FBE3000, or in the Bio-Rad chambers models 1045 and 1415). This allows the use of gels of various sizes and the application of high field strengths across just a portion of the separation path. A typical chamber is shown in Figure 4.3. The most suitable power supplies for IEF are those with automatic constant-power operation and with voltage maxima as high as 5000–6000 V. The minimal requirements for good resolution are a limiting voltage of 1000 V and a reliable amperometer with a full scale not exceeding 50 mA. Lower field strengths (E) cause the protein bands to spread (resolution is proportional to $\sqrt{E}$). The amperometer monitors the conductivity and so allows periodic manual adjustment of the electrophoretic conditions to keep the delivered power as close as possible to a constant value.

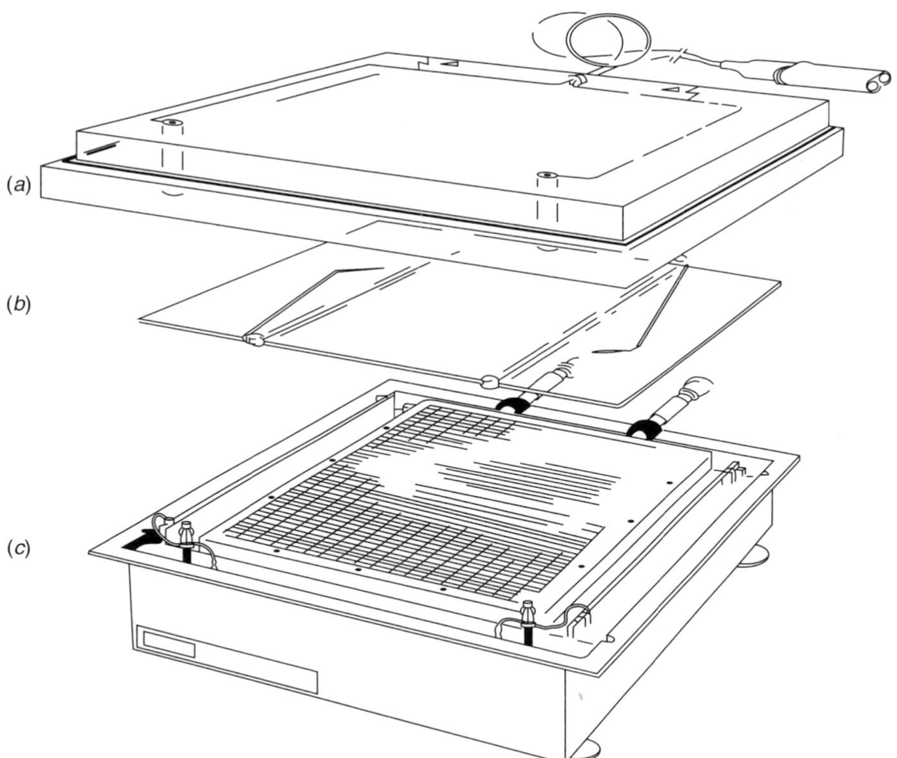

FIGURE 4.3. Drawing of LKB Multiphor II chamber: (*a*) cover lid; (*b*) cover plate with movable platinum electrodes; (*c*) base chamber with ceramic cooling block for supporting gel slab. (Courtesy of LKB Produkter AB.)

Efficient cooling is important for IEF because it allows high field strengths to be applied without overheating. Tap water circulation is adequate for 8 M urea gels but not acceptable for gels lacking urea. Placing the electrophoretic apparatus in a cold room may be beneficial to prevent water condensation around the unit in very humid climates, but it is inadequate as a substitute for coolant circulation.

4.2.2.2. Polymerization Cassette The polymerization cassette is the chamber that is used to form the gel for IEF. It is assembled from the following elements: a gel-supporting plate, a spacer, a cover (molding) plate, and some clamps (see Fig. 4.4). A plain glass plate is sufficient to support the gel when detection of the separated proteins does not require processing through several solutions (e.g., when the sandwich technique for zymograms or immunoblotting is to be applied) or when the polyacrylamide matrix is sturdy (gels >1 mm thick, gel concentration $T > 5\%$). However, for thin, soft gels a permanent support is required. Glass coated with

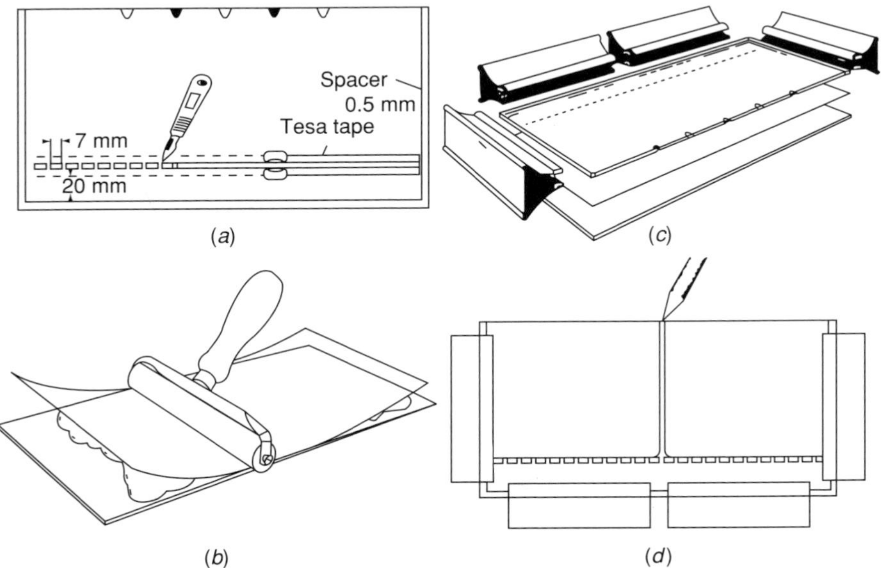

FIGURE 4.4. Preparation of gel cassette: (*a*) preparation of slot former: onto cover plate (bearing rubber gasket U-frame) is glued a strip of tesa tape out of which rectangular tabs are cut with a scalpel; (*b*) application of Gel Bond PAG film to supporting glass plate; (*c*) assembling gel cassette; (*d*) pouring gelling solution in vertically standing cassette using pipette. (Courtesy of LKB Produkter AB.)

γ-methacryl-oxypropyl-trimethoxy-silane is the most reliable reactive substratum and is the most suitable for autoradiographic procedures (Bianchi-Bosisio et al., 1980) (see Protocol 4.1 for the procedure). It is also the cheapest of such supports: Dried-out gels can be removed with a blade and then a brush with some scrubbing powder. Unreacted silane may be hydrolyzed by keeping the plates in Clorox for a few days. This step is unnecessary, however, if they have to go through successive cycles of silanization. The glass plates used as a support should not be thicker than 1–1.2 mm. On the other hand, thin plastic sheets designed to bind polyacrylamide gels firmly (e.g., Gel Bond PAG by Marine Colloids, PAG foils by Amersham-Pharmacia, Gel Fix by Serva) are more practical if the records of a large number of experiments have to be filed or when different parts of the gel need to be processed independently (e.g., the first step of a 2D separation or a comparison between different stains). The plastic sheet is applied to a supporting glass plate and the gel is cast onto this. The binding of the polyacrylamide matrix to these substrata, however, is not always stable and so care should be taken in using them, especially for detergent-containing gels and when using aqueous staining solutions. For good adherence, the best procedure is to cast "empty" gels (i.e., polyacrylamide gel lacking CAs), wash and dry them, and then reswell with the solvent of choice (see Section 4.3.2.2).

U-gaskets of any thickness between 0.2 and 5 mm can be cut from rubber sheets (para-, silicone-, or nitrile-rubber) and used as spacers. For thin gels, a few layers

PROTOCOL 4.1. SILICONIZING GLASS PLATES

EQUIPMENT AND REAGENTS

- Binding silane (γ-methacryloxy-propyl-trimethoxy-silane) prepared by Union Carbide (available through Pharmacia, Serva, or LKB) or Repel-silane (dimethyl-dichloro-silane) available from Merck, Serva, etc.
- Leaf gel-supporting plates

METHOD

Binding silane

Two alternative procedures are available:

Either

1. Add 4 mL of silane to 1 L of distilled water adjusted to pH 3.5 with acetic acid.
2. Leave the plates in this solution for 30 min.
3. Rinse with distilled water and dry in air.

or

4. Dip the plates for 30 s in a 0.2% solution of silane in anhydrous acetone.
5. Thoroughly evaporate the solvent using a hair drier.
6. Rinse with ethanol if required.

In either case, store the siliconized plates away from untreated glass.

Repel-silane

1. Swab the glass plates with a wad impregnated with a 2% (w/v) solution of Repel-silane in 1,1,1-trichloroethane.
2. Dry the plates in a stream of air and rinse with distilled water.

of Parafilm (each about 120 μm thick) can be stacked and cut with a razor blade. The width of such U-gaskets should be about 4 mm. In addition, cover plates with a permanent frame are commercially available. The same applies to a plastic tray with two lateral ridges for horizontal polymerization (see Fig. 4.5*a*). A similar device may be home made using Dymo tape strips which are 250 μm thick to form the permanent spacer frame. Mylar foil strips or self-adhesive tape may be used as spacers for 50–100-μm-thick gels. Rubber- or tape-gaskets should never be left to soak in soap (which they absorb) but just rinsed and dried promptly.

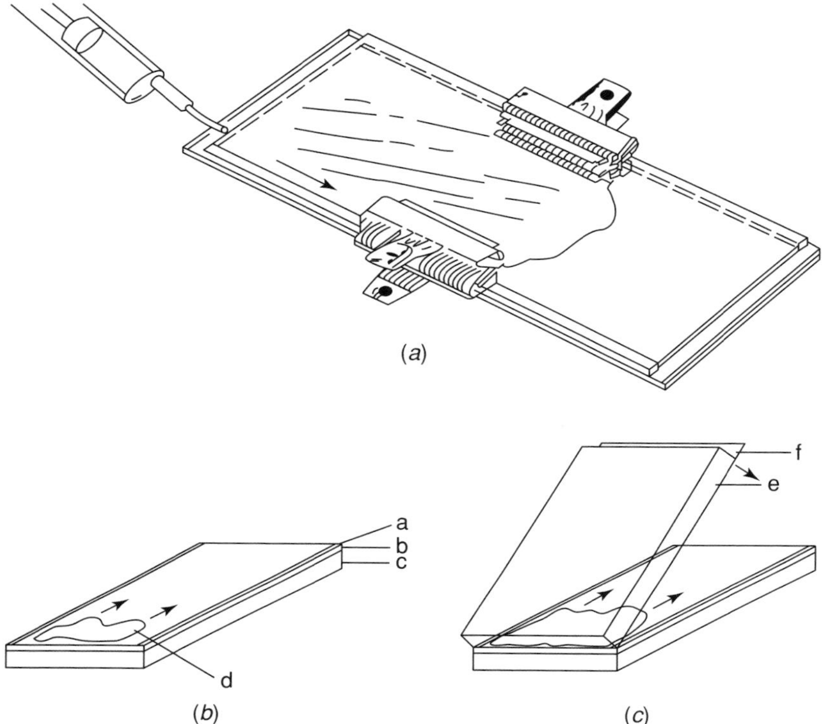

FIGURE 4.5. Casting of thin gel plates: (*a*) by capillary filling of horizontally placed cassette; (*b*, *c*) by "flap" technique: a, spacer strips; b, silanized glass plate or polyester film; c, glass base plate; d, polymerization mixture; e, glass cover plate; f, cover film. [(*a*) Courtesy of LKB. (*b*, *c*) By permission from Radola, 1980b.]

Clean glass coated with dimethyl dichloro silane (Repel silane; see Protocol 4.1) and a thick Perspex sheet are suitable materials for the cover plate. If you wish to mold sample application pockets into the gel slab during preparation, attach Dymo tape pieces to the plate or glue small Perspex blocks to the plate with drops of chloroform. Perspex should never be exposed unevenly to high temperatures (for example, by being rinsed in running hot water) because it bends even if cut in thick slabs.

4.2.3. Polyacrylamide Gel Matrix

4.2.3.1. Reagents Stocks of dry chemicals (acrylamide, Bis, ammonium persulfate) may be kept at room temperature provided they are protected from moisture by being stored in air-tight containers. Very large stocks are better sealed into plastic bags together with Drierite (Mercks) and stored in a freezer. TEMED stocks should also ideally be kept in a freezer in an air-tight bottle or, better, under nitrogen. Avoid contaminating acrylamide solutions with heavy metals, which can initiate its polymerization.

Acrylamide and Bis for IEF must be of the highest purity to avoid poor polymerization and strong electroosmosis resulting from acrylic acid. Bis is more hydrophobic and more difficult to dissolve than acrylamide, so start by stirring it in a little amount of lukewarm distilled water (the solution process is endothermic), then add acrylamide and water as required. Recently, novel monomers endowed with extreme resistance to alkaline hydrolysis and higher hydrophilicity have been reported: they are *N*-acryloylamino ethoxy ethanol (Chiari et al., 1994) and *N*-acryloylamino propanol (Simò-Alfonso et al., 1996). Table 4.2 lists the most commonly used additives in IEF.

TABLE 4.2. Common Additives for IEF

Additive	Purpose	Concentration	Limitations
Sucrose, glycerol	To improve mechanical properties of low %*T* gels and to reduce water transport and drift	5–20%	Increased viscosity slightly slows focusing process
Glycine, taurine	To increase dielectric constant of medium; increases solubility of some proteins (e.g., globulins) and reduces ionic interactions	0.1–0.5 M	Glycine is zwitterionic between pH 4 and 8, taurine between pH 3 and 7; their presence somewhat slows focusing process and shifts resulting gradient
Urea	Disaggregation of supramolecular complexes	2–4 M	Unstable in solution, especially at alkaline pH
	Solubilization of water-insoluble proteins, denaturation of hydrophilic proteins	6–8 M	Urea soluble at $\geq$10°C; it accelerates polyacrylamide polymerization, so reduces amount of TEMED added
Nonionic and zwitterionic detergents	Solubilization of amphiphilic proteins.	0.1–1%	To be added to polymerizing solutions just before catalysts to avoid foaming; they interfere with polyacrylamide binding to reactive substrata; they are precipitated by trichloroacetic acid (TCA) and require specific staining protocol

4.2.3.2. Gel Formulations To allow all the sample components to reach their equilibrium position at essentially the same rate and the experiment to be terminated before the pH gradient decay process adversely affects the quality of the separation, it is best to choose a nonrestrictive anticonvective support. There are virtually no theoretical but only practical lower limits for the gel concentration (the minimum being about $T = 2.2\%, C = 2\%$). Large pore sizes can be obtained both by decreasing $\%T$ and by either decreasing or increasing $\%C$ from the critical value of 5%. Although the pore size of polyacrylamide can be enormously enlarged by increasing the percentage of crosslinker, two undesirable effects also occur in parallel, namely increasing gel turbidity and proneness to syneresis (Righetti et al., 1981; Gelfi and Righetti, 1981). In this respect, N, N'-(1,2-dihydroxyethylene)bisacrylamide (DHEBA), with its superior hydrophilic properties, appears superior to bisacrylamide. In contrast, N, N'-diallyltartardiamide (DATD) inhibits the polymerization process and so gives porous gels just by reducing the actual $\%T$ of the matrix. Because unpolymerized acryloyl monomers may react with $-NH_2$ and react readily with –SH groups on proteins (Chiari et al., 1992; Bordini et al., 1999a,b) and, once absorbed through the skin, act as neurotoxins, the use of DATD should be avoided altogether.

4.2.3.3. Choice of Carrier Ampholytes A simple way to extend and stabilize the extremes of a wide (pH 3–10) gradient is to add acidic and basic (natural) amino acids. Thus lysine, arginine, aspartic acid, and glutamic acid are prepared as individual stock solutions containing 0.004% sodium azide and stored at 0–4°C. They are added in volumes sufficient to give 2–5 mM final concentration. To cover ranges spanning between 3 and 5 pH units, a few narrow cuts of CA need to be blended, with the proviso that the resulting slope of the gradient will be (over each segment of the pH interval) inversely proportional to the amount of ampholytes isoelectric in that region.

Shallow pH gradients are often used to increase the resolution of sample components. However, longer focusing times and more diffuse bands will result unless the gels are electrophoresed at higher field strengths. Shallow pH gradients (shallower than the commercial 2-pH-unit cuts) can be obtained in different ways:

(a) By subfractionating the relevant commercial carrier ampholyte blend. This can be done by focusing the CAs at high concentration in a multicompartment electrolyzer (Bossi and Righetti, 1995).
(b) By allowing trace amounts of acrylic acid to induce a controlled cathodic drift during prolonged runs. This is effective in the acidic pH region but it is rather difficult to obtain reproducible results from run to run.
(c) By preparing gels containing different concentrations of carrier ampholytes in adjacent strips (Låås and Olsson, 1981) or with different thicknesses along the separation path (Altland and Kaempfer, 1980).
(d) By adding specific amphoteric compounds (spacers) at high concentration (Caspers et al., 1977).

TABLE 4.3. Good Carrier Ampholytes Acting as Spacers

Carrier Ampholyte	pI	Carrier Ampholyte	pI
Aspartic acid	2.77	β-Aspartyl-histidine	4.94
Glutathione	2.82	Cysteinyl-cysteine	4.96
Aspartyl-tyrosine	2.85	Pentaglycine	5.32
o-Aminophenylarsonic acid	3.00	Tetraglycine	5.40
Aspartyl-aspartic acid	3.04	Triglycine	5.59
p-Aminophenylarsonic acid	3.15	Tyrosyl-tyrosine	5.60
Picolinic acid	3.16	Isoglutamine	5.85
Glutamic acid	3.22	Lysyl-glutamic acid	6.10
β-Hydroxyglutamic acid	3.29	Histidyl-glycine	6.81
Aspartyl-glycine	3.31	Histidyl-histidine	7.30
Isonicotinic acid	3.44	Histidine	7.47
Nicotinic acid	3.44	L-Methylhistidine	7.67
Anthranilic acid	3.51	Carnosine	8.17
p-Aminobenzoic acid	3.62	α,β-Diaminopropionic acid	8.20
Glycyl-aspartic acid	3.63	Anserine	8.27
m-Aminobenzoic acid	3.93	Tyrosyl-arginine	8.38–8.68
Diiodotyrosine	4.29	L-Ornithine	9.70
Cysteinyl-diglycine	4.74	Lysine	9.74
α-Hydroxyasparagine	4.74	Lysyl-lysine	10.04
α-Aspartyl-histidine	4.92	Arginine	10.76

In the last case, two kinds of ampholytes may be used for locally flattening the pH gradient: "good" and "poor." Good CAs, those with a small pI–pK_1 (i.e., possessing good conductivity and buffering capacity at the pI) are able to focus in narrow zones. Low concentrations (5–50 mM) are sufficient to induce a pronounced flattening of the pH curve around their pI. A list of these CAs is given in Table 4.3. Poor carrier ampholytes, on the other hand, form broad plateaux in the region of their pI and should be used at high concentrations (0.2–1.0 M). Their presence usually slows down the focusing process. Some of them are listed in Table 4.4.

A note of caution: In an IEF system, the distribution of acids and bases is according to their dissociation curve, in a pattern that may be defined as protonation (or deprotonation) stacking. If large amounts of these compounds originate from the samples (in the form of buffers), the limits of the pH gradient shift from the expected values. For example, ß-mercaptoethanol, as added to denatured samples for 2D PAGE analysis, lowers the alkaline end from pH 10 to approximately pH 7.5 (Righetti et al., 1982). This effect, however, may sometimes be usefully exploited. For example, high levels of TEMED in the gel mixture appear to stabilize alkaline pH gradients (Yao and Bishop, 1982). In addition, TEMED is utilized in cIEF for blocking the capillary region after the detection point, so that basic proteins would focus in the region prior to the detector (Zhu et al., 1991).

TABLE 4.4. Poor Carrier Ampholytes Acting as Spacers

Carrier Ampholyte	pK_1	pK_2	pI
Carrier ampholytes with pI 7–8			
β-Alanine	3.55	10.24	6.90
γ-Aminobutyric acid	4.03	10.56	7.30
δ-Aminovaleric acid	4.26	10.77	7.52
ε-Aminocaproic acid	4.42	11.66	8.04
Good's buffers with acidic pI			
MES [2-(N-morpholino)ethanesulfonic acid]	1.3	6.1	3.70
PIPES [Piperazine-1,4-bis(2-ethanesulfonic acid)]	1.3	6.8	4.05
ACES [N-(2-acetamido)-2-aminoethane sulfonic acid]	1.3	6.8	4.05
BES [N, N-Bis(2-hydroxyethyl)-2-aminoethanesulfonic acid]	1.3	7.1	4.20
MOPS [3-(N-morpholino)propane sulfonic acid]	1.3	7.2	4.25
TES [N-[Tris(hydroxymethyl)methyl]-2-aminoethanesulfonic acid]	1.3	7.5	4.40
HEPES [N-(2-hydroxyethyl)piperazine-N′-2-ethane sulphonic acid]	1.3	7.5	4.40
EPPS [N-(2-Hydroxyethyl)piperazine-N′-(3-propanesulfonic acid)]	1.3	8.0	4.65
TAPS [N-[Tris(hydroxymethyl)methyl]-3-aminopropanesulfonic acid]	1.3	8.4	4.85

pI is Isoelectric point.

4.2.4. Gel Preparation and Electrophoresis

Protocol 4.2 outlines the series of steps required for an IEF run. The key steps are described in more detail below.

4.2.4.1. Assembling Gel Mold This procedure is illustrated in Figures 4.4*a*–*c*. Figure 4.4*a* shows the preparation of the slot former which will give 20 sample application slots in the final gel. In the method shown, the slot former is prepared by gluing a strip of embossing tape onto the cover plate and cutting rectangular tabs, with the dimensions shown, using a scalpel. A rubber U-gasket covering three edges of the cover plate is also glued to the plate. In Figure 4.4*b*, a sheet of Gel Bond PAG film is applied to the supporting glass plate in a thin film of water. When employing reactive polyester foils as the gel backing, use glycerol rather than water. Avoid leaving air pockets behind the gel-backing sheet since this will produce gels of uneven thickness, which will create distortions in the pH gradient. Also ensure that the backing sheet is cut flush with the glass support since any overhang easily bends. Finally, the gel mold is assembled using clamps (Fig. 4.4*c*). Assembly is usually made easier by wetting and blotting the rubber gasket just before use. Note that the cover plate has three V-shaped indentations on one edge, which allow insertion of a pipette or syringe tip into the narrow gap between the two plates of the mold to facilitate opening the gel.

PROTOCOL 4.2. CA–IEF FLOW SHEET

1. Assemble the gel mold.
2. Mix all components of the polymerizing mixture, except ammonium persulfate.
3. Degas the mixture for a few minutes and reequilibrate (if possible) with nitrogen.
4. Add the required amount of ammonium persulfate stock solution and mix.
5. Transfer the mixture to the gel mold and overlay it with water.
6. Leave the mixture to polymerize (at least 1 h at room temperature or 30 min at 37°C).
7. Open the gel mold and blot any moisture from the gel edges and surface.
8. Lay the gel on the cooling block of the electrophoretic chamber.
9. Apply the electrode strips.
10. Prerun the gel, if appropriate.
11. Apply the samples.
12. Run the gel.
13. Measure the pH gradient.
14. Reveal the protein bands using a suitable detection procedure.

4.2.4.2. Filling Mold One of three methods may be chosen: gravity, capillarity, or the flap technique. The gravity procedure uses a vertical cassette with a rubber gasket U-frame glued to the cover plate (Fig. 4.4*a*). If the cover plate has V-indentations along its free edge as shown in Fig. 4.4*a*, the gel mixture can be transferred simply by using a pipette or a syringe with its tip resting on one of these indentations (Fig. 4.4*d*). Avoid filling the mold too fast, which will create turbulence and trap air bubbles. If an air bubble appears, stop pouring the solution and try to remove the bubble by tilting and knocking the mold. If this maneuver is unsuccessful, displace the bubble with a 1-cm-wide strip of polyester foil.

The capillary method is mainly used for casting reasonably thin gels (0.2–0.5 mm). It requires a horizontal sandwich with two lateral spacers (see Fig. 4.5*a*). The solution is fed either from a pipette or from a syringe fitted with a short piece of fine-bore tubing. It is essential that the solution flow evenly across the whole width of the mold during casting. If an air bubble appears, do not stop pumping in the solution or you will produce more bubbles. Remove all the air bubbles at the end using a strip of polyester foil. A level table is not mandatory, but the mold should be left laying flat until the gel is completely polymerized.

The flap technique is mainly used for preparing ultrathin gels. A 20% to 50% excess of gel mixture is poured along one edge of the cover (with spacers on both sides) (Fig. 4.5*b*) and the support plate is slowly lowered on it (Fig. 4.5*c*) (Radola, 1980b). Air bubbles can be avoided by using clean plates. If bubbles do get trapped, they can also be removed by lifting and lowering the cover plate once more. Since this method may lead to spilled unpolymerized acrylamide, take precautions for its containment (wear gloves and use absorbent towels to mop up excess).

4.2.4.3. Gel Polymerization Protocol 4.3 gives a step-by-step procedure for casting a CA–IEF gel which might include, if needed, additives such as urea and surfactants. As a variant of this protocol, one can polymerize "empty" gels (i.e., devoid of CAs and of any leachable additive), wash and dry them, and reswell them in the appropriate CA solution (including any additive, as needed). This is a direct application of the IPG technology (Righetti, 1990). After polymerization (as above, but in the absence of CAs), wash the gel three times in 300 mL distilled water each time to remove catalysts and unreacted monomers. Equilibrate the washed gel (20 min with shaking) in 1.5% glycerol and finally dry it onto Gel Bond PAG foil. It is essential that the gel does not bend so, before drying, the foil should be made to adhere to a supporting glass plate (taking care to remove all air bubbles in between and fastening it in position with clean, rust-proof clamps). Drying must be at room temperature in front of a fan. Finally, mount the dried gel back in the polymerization cassette and allow it to reswell in the appropriate CA (and suitable additive) solution. This can be conveniently done by using a reswelling cassette for dry IPG gels.

In ultrathin gels, pH gradients are sensitive to the presence of salts, including TEMED and persulfate. Moreover, unreacted monomers are toxic and noxious to proteins. The preparation, washing, and reequilibration of empty gels removes these components from the gel and so eliminates these problems.

4.2.4.4. Sample Loading and Electrophoresis The electrophoresis procedure is described in Protocol 4.4. The gel is placed on the cooling plate of the electrophoresis chamber. It is necessary to perform the electrophoresis at a constant temperature and with well-defined conditions, since the temperature influences the pH gradient and consequently the separation positions of the proteins. Also, the presence of additives (e.g., urea) strongly affects the separation positions by changing the physicochemical parameters of the solution trapped inside the gel matrix. Electrode strips filled with electrode solutions are placed on the surface of the gel (at anodic and cathodic extremes). Whenever possible (and compatible with the width of the pH gradient to be developed), strong acids and bases as anolytes and catholytes, respectively, should be avoided. For example, 1 M NaOH at the cathode could hydrolyze the amido groups in the underlying polyacrylamide gel, thus inducing a strong EOF (100 mM Tris free base is to be preferred). Also, amphoteric compounds could be used for this purpose: for example, at the anode aspartic acid (pI 2.77 at 50 mM concentration) or iminodiacetic acid (pI 2.33 for a 100 mM solution) and at the cathode lysine freebase (pI 9.74).

The electrophoretic procedure is divided into two steps. The first is a preelectrophoresis (characterized by the application of low voltages) which allows prefocusing of carrier ampholytes and the elimination of all contaminants or unreacted

PROTOCOL 4.3. POLYMERIZATION OF GEL FOR IEF

EQUIPMENT AND REAGENTS

- 30% T acrylamide–bisacrylamide solution (see Table 4.5)
- 2% carrier ampholytes (Ampholine, Servalyte, Resolyte, or Pharmalyte)
- Urea
- TEMED
- 40% ammonium persulfate (prepared fresh)
- Detergents (as desired, for gel mixture)
- Graduated glass measuring cylinder sufficient to hold gel mixture during preparation
- Vacuum flask
- Mechanical vacuum pump
- Gel mold

METHOD

1. Mix all the components of the gel formulation (from Table 4.5, except TEMED, ammonium persulfate, and detergents, when used) in a cylinder, add distilled water to the required volume, and transfer the mixture to a vacuum flask.
2. Degas the solution using the suction from a water pump; the operation should be continued as long as gas bubbles form. Manually swirl the mixture or use a magnetic stirrer during degassing. The use of a mechanical vacuum pump is desirable for the preparation of very soft gels but is unnecessarily cumbersome for urea gels (urea would crystallize). At the beginning of the degassing step the solution should be at or above room temperature to decrease oxygen solubility. Its cooling during degassing is then useful in slowing down the onset of polymerization.
3. If possible, it is beneficial to reequilibrate the degassed solution against nitrogen instead of air (even better with argon, which is denser than air). From this step on, the processing of the gel mixture should be as prompt as possible.
4. Add detergents, if required. If they are viscous liquids, prepare a stock solution beforehand (e.g., 30%) but do not allow detergent to take up more than 5% of the total volume. Mix briefly with a magnetic stirrer.
5. Add the volumes of TEMED and then ammonium persulfate as specified in Table 4.5. Immediately mix by swirling, and then transfer the mixture to the gel mold. Carefully overlay with water or butanol.

Continued

PROTOCOL 4.3. *Continued*

6. Leave to polymerize for at least 1 h at room temperature or 30 min at 37°C. Never tilt the mold to check whether the gel has polymerized; if it has not polymerized when you tilt it, then the top never will, because of the mixing caused by tilting. Instead, the differential refractive index between liquid (at the top and usually around the gasket) and gel phase is an effective and safe index of polymerization and is shown by the appearance of a distinct line after polymerization.
7. The gels may be stored in their molds for a couple of days in a refrigerator. For longer storage (up to 2 weeks for neutral and acidic pH ranges), it is better to disassemble the molds and (after covering the gels with Parafilm and wrapping with Saran foil) store them in a moist box.
8. Before opening the mold, allow the gel to cool at room temperature if polymerized at a higher temperature. Laying one face of the mold onto the cooling block of the electrophoresis unit may facilitate opening the cassette. Carefully remove the overlay by blotting and also remove the clamps.
9. (a) If the gel is cast against a plastic foil, remove its glass support; then carefully peel the gel from the cover plate.
   (b) If the gel is polymerized on silanized glass, simply force the two plates apart with a spatula or a blade.
   (c) If the gel is not bound to its support, lay the mold on the bench, with the plate to be removed uppermost and one side protruding a few centimeters from the edge of the table. Insert a spatula at one corner and force against the upper plate; Turn the spatula gently until a few air bubbles form between the gel and the plate; then use the spatula as a lever to open the mold. Wipe any liquid from the gel surface with a moistened Kleenex, but be careful; keep moving the swab to avoid it sticking to the surface.

   If bubbles form on both gel sides of the gel, try at the next corner of the mold. If the gel separates from both glass plates and folds up, you may still be able to salvage it, provided the gel thickness is at least 400 μm and the gel concentration T is at least 5%. Using a microsyringe, force a small volume of water below the gel; then carefully make the gel lay flat again by maneuvering it with a gloved hand. Remove any remaining air bubbles using the needle of the microsyringe. Cover the gel with Parafilm, and gently roll it flat with a rubber roller (take care not to damage it with excess pressure). Carefully remove any residual liquid by blotting.

catalysts from the gel matrix. Then the protein sample is loaded. In IEF gels, the sample should be applied along the whole pH gradient to determine its optimum application point. The second electrophoretic step is then started. This is carried out at constant power; low voltages are used for the sample entrance followed by higher voltages during the separation and band sharpening. The actual voltage used in each step depends on gel thickness.

PROTOCOL 4.4. ELECTROPHORETIC PROCEDURE

EQUIPMENT AND REAGENTS

- Electrophoresis apparatus (see Section 4.2.2.1)
- Gel polymerized in gel mold (from Protocol 4.3)
- Whatmann No. 17 filter paper
- Electrode solutions (see Table 4.8)
- Protein samples
- Filter paper (e.g., Whatmann No. 1, Whatmann 3 mm, Paratex II) if sample is to be aplied via filter paper tabs (see step 6 below)
- 10 mM KCl
- Narrow-bore combination pH electrode (e.g., from Radiometer)

METHOD

1. Set the cooling unit at 2–4°C for normal gels, at 8–10°C for 6 M urea, and at 10–12°C for 8 M urea gels.
2. To enable rapid heat transfer, pour a few milimeters of a nonconductive liquid (distilled water, 1% nonionic detergent, or light kerosene) onto the cooling block of the electrophoretic chamber. Form a continuous liquid layer between the gel support and the apparatus and gently lower the plate into place, avoiding trapping air bubbles and splashing water onto or around the gel. Should this happen, remove all liquid by careful blotting. When the gel is narrower than the cooling block, apply the plate on its middle. If this is not possible or if the electrode lid is too heavy to be supported by just a strip of gel, insert a wedge (e.g., several layers of Parafilm) between the electrodes and cooling plate.
3. Cut electrode strips (e.g., from Whatmann No. 17 filter paper); note that most paper exposed to alkaline solutions becomes swollen and fragile, so we use only Whatmann No. 17 approx. 5 mm wide and about 3 mm shorter than the gel width. Saturate them with electrode solutions (see Table 4.8). However, they should not be dripping; blot them on paper towels if required.
4. Wearing disposable gloves, transfer the electrode strips onto the gel. They must be parallel and aligned with the electrodes. The cathodic strip firmly adheres to the gel: Do not try to change its position once applied. Avoid cross contaminating the electrode strips (including with your fingers). If any electrode solution spills over the gel, blot it off immediately, rinse with a few drops of water, and blot again. Check that the wet electrode strips do not exceed the size of the gel and cut away any excess pieces. Be sure to apply the most alkaline solution at the cathode and the most acidic at the anode (if

Continued

PROTOCOL 4.4. ***Continued***

you fear you have misplaced them: The color of the NaOH-soaked paper is yellowish but, of course, you may also check this with litmus paper). If you discover a mistake at this point, simply turn your plate around or change the electrode polarity. However, there is no remedy after the current has been on for a while.

5. The salt content of the samples should be kept as low as possible. If necessary, dialyze them against glycine (any suitable concentration) or dissolve in diluted CAs. When buffers are required, use low-molarity buffers composed of weak acids and bases.
6. The samples are best loaded into precast pockets. These should not be deeper than 50% of the gel thickness. If they are longer than a couple of millimeters and it is necessary to preelectrophorese the gel (see step 7), the pockets should be filled with dilute CAs. After the preelectrophoresis (see Table 4.9 for conditions) remove this solution by blotting. Then apply the samples.

   The amount of sample applied should fill the pockets. Try to equalize the volumes and the salt content among different samples. After about 30 min of electrophoresis at high voltage (see Table 4.9 for conditions), the content of the pockets can be removed by blotting and new aliquots of the same samples loaded. The procedure can be repeated a third time. Alternatively, the samples can be applied to the gel surface and absorbed into pieces of filter paper. Different sizes and material of different absorbing power are used to accommodate various volumes of liquid (e.g., a 5×10-mm tab of Whatman No. 1 can retain about 5 μL, of Whatman 3 m about 10 μL, of Paratex II more than 15 μL). Up to three layers of paper may be stacked. If the exact amount of sample loaded has little importance, the simplest procedure is to dip the tabs into it, then blot them to remove excess sample. Otherwise, align dry tabs on the gel and apply measured volumes of each sample with a micropipette. For stacks of paper pieces, feed the solution slowly from one side rather than from the top. Do not allow a pool of sample to drag around its pad, but stop feeding liquid to add an extra tab when required. This method of application of samples is not suitable for samples containing alcohol.
7. Most samples may be applied to the gel near the cathode without preelectrophoresis. However, pre-running is advisable if the proteins are sensitive to oxidation or unstable at the average pH of the gel before the run. Prerunning is not suitable for those proteins with a tendency to aggregate upon concentration or whose solubility is increased by high ionic strength and dielectric constant or which are very sensitive to pH extremes. Anodic application should be excluded for high-salt samples and for proteins (e.g., a host of serum components) denatured at acidic pH. Besides this guideline, however, as a rule the optimal conditions for sample loading should be experimentally determined, together with the minimum focusing time, with a pilot run in which the sample of interest is applied in different positions of the gel and at different times.

PROTOCOL 4.4. ***Continued***

Smears, or lack of confluence of the bands after long focusing time, denote improper sample handling and protein alteration.

8. After electrophoresis the pH gradient in the focused gel may be read using a contact electrode. The most general approach, however, is to cut a strip along the focusing path (0.5–2 cm wide, with an inverse relation to the gel thickness).Then cut segments between 3 and 10 mm long from this and elute each for about 15 min in 0.3–0.5 mL of 10 mM KCl (or with the same urea concentration as present in the gel). For alkaline pH gradients, this processing of the gel should be carried out as quickly as possible, the elution medium should be thoroughly degassed, and air-tight vials flushed with nitrogen should be used. Measure the pH of the eluted medium using a narrow-bore combination electrode. For most purposes, it is sufficient to note just the temperature of the coolant and the pH measurement in order to define an operational pH. For a proper physicochemical characterization, the temperature differences should be corrected for, as suggested in Table 4.10 (which gives also corrections for the presence of urea).

4.2.5. General Protein Staining

Table 4.5 lists, with pertinent references, some of the most common protein stains used in IEF. Detailed recipes are given below. Extensive reviews covering general staining methods in gel electrophoresis have recently appeared (Wirth and Romano, 1995; Merril and Washart, 1998a). Merril and Washart (1998b) have also given a very

TABLE 4.5. Protein-Staining Methods

Stains	Application	Sensitivity	Reference
Coomassie blue G-250 (micellar)	General use	Low	Righetti and Chillemi, 1978
Coomassie blue R-250/$CuSO_4$	General use	Medium	Righetti and Drysdale, 1974
Coomassie blue R-250/ sulfosalicilic acid	General use	High	Neuhoff et al., 1985
Silver stain	General use	Very high	Merril et al., 1981
Coomassie blue R-250/$CuSO_4$	In presence of detergents	Medium	Righetti and Drysdale, 1974
Light green SF	General use	Medium	Radola, 1973
Fast green FCF	General use	Low	Riley and Coleman, 1968
Coomassie blue R-250 at 60°C	In presence of detergents	High	Vesterberg, 1972

extensive bibliography list covering all aspects of polypeptide detection methods, including enzyme localization protocols (over 350 citations). Additionally, although not specifically reported for IEF, some recent developments include use of eosin B dye (Waheed and Gupta, 1996); a mixed-dye technique comprising Coomassie blue R-250 and Bismark brown R (Choi et al., 1996); Stains-All for highly acidic molecules (Myers et al., 1996); fluorescent dyes for proteins, such as SYPRO orange and SYPRO red (Steinberg et al., 1996a,b); and a 2-min Nile red staining (Javier Alba et al., 1996).

Before entering in detail in all methods listed below, a general comment seems appropriate. This stems from a recent review by Rabilloud (2000), who evaluated all possible stains adopted for IEF and 2D maps according to key issues such as sensitivity (detection threshold), linearity, homogeneity (i.e., variation from one protein to another), and reproducibility. Accordingly, four steps have been considered: (a) affixing the label before IEF; (b) labeling in between IEF and SDS–PAGE; (c) tagging after SDS–PAGE, and (d) labeling after blotting (i.e., transfer of the separated protein onto an inert membrane). We will thus briefly review here these four steps, with the understanding that they apply not only to plain IEF and IPG, as in this chapter, but also to the various electrophoretic steps reported in the following chapters.

(a) Affixing Label before IEF. In principle, this approach would be very convenient since it could be done in a minimal volume with a high concentration of reactants; however, any noncovalently bound tag will be disrupted. On the other hand, any prelabeling method that alters the charge state of the protein, either by removing a charged group or by adding a spurious one, is not compatible with 2D maps. Thus, amine or thiol alkylation with reactive acidic dyes such as Remazol blue (Bosshard and Datyner, 1977) must be avoided because the proteins will gain negative charges. Covalent fluorescent tagging would appear to be superior provided that fluorophors with intense light absorption, high quantum yield, large Stokes shifts, and limited fading can be found. Since fluorescein derivatives carry negative charges, pre-IEF labeling has been limited for a long time to thiol alkylation with electrically neutral probes (Urwin and Jackson, 1993), although these species offer limited sensitivity. Use of cyanine-based probes (Unlii et al., 1997), with attachment via amine acylation, improves the process and does not alter the protein charge (except for rather basic proteins) since the positive charge lost upon covalent bonding is replaced by the quaternary ammonium on the cyanine fluorophore. Radioactive labeling followed by fluorography (Bonner and Laskey, 1974) or by phosphor storage (Johnston et al., 1990) appears to be the most sensitive and IEF- and 2D-compatible methods. The latter technique, in which the β-radiation induces an energy change in an europium salt, which, via laser excitation, is converted into visible light, appears to offer a 20- to 100-fold higher sensitivity than autoradiography, coupled to a linear dynamic range or four orders of magnitude. Another modern detection method is based on amplification detectors similar to those used in high-energy physics and is compatible with dual-isotope detection (Charpak et al., 1989). The conclusion: Radioactive

detection probably offers the best signal-to-noise (S/N) ratio and the best ultimate sensitivity.

(b) Labeling between IEF and SDS–PAGE. This stage offers much greater flexibility because the charge state of the proteins can be altered provided that the apparent M_r is kept constant. Covalent labeling dominates at this point, and thus fluorescent tags are preferred. However, since protein nucleophiles (amino and thiol groups) are the preferred targets for such a grafting, care to remove carrier ampholytes should be exerted, since these compounds could consume most of the label. This process is usually achieved with a few aqueous acid alcohol baths or with 10% TCA (Urwin and Jackson, 1991). When tagging with fluorophores, it should be remembered that detection is an important point. In 2D maps, a scanning device should be used for acquiring all the spots in the gel. Since laser-induced fluorescence offers the highest sensitivity, fluorophors having an excitation maximum close to an available laser source (e.g., fluorescein and rhodamine) are preferred. In another setup, illumination is done in the gel plane, preferably by UV excitation, and the fluorescent light is collected by a charge-coupled device (CCD) camera (Jackson et al., 1988).

(c) Tagging after SDS–PAGE. After this step, which can be either mono- or bidimensional (2D maps), any solution can be adopted, according to the sensitivity desired. Noncovalent detection with organic dyes, the most common being the Coomassie brilliant blue G and R, is quite popular due to a simple protocol, although the sensitivity is of the order of ~1 μg (Fazekas de St. Groth et al., 1963). These kinds of stains are of the regressive type; that is, the gel is first saturated with dye and then destained to remove the dye unbound to the protein. Alternatively, in dilute dye solutions, the gel can be stained to the endpoint, which however requires very long staining times (Chen et al., 1993). As an additional option, as described below, micellar or colloidal stainings can be adopted with sensitivities of about 100 ng. Noncovalent, fluorescent probes are also very popular after this stage. Generally, such probes are nonfluorescent in water but highly fluorescent in apolar media such as detergent; thus they take advantage of SDS binding to proteins to create a microenvironment promoting fluorescence at the protein spot. Typical labels of this type are naphthalene derivatives (Horowitz and Bowman, 1987), SYPRO dyes (Steinberg et al., 1996a), and Nile red (Alba et al., 1996). Detection by differential salt binding is also an attractive procedure, although negative stains are problematic if photographic documentation is required. All negative stains use divalent cations (Cu, Zn) to form a precipitate with SDS. Submicrogram sensitivities are claimed, the staining is very rapid, and destaining is achieved by simple metal chelators such as ethylenediaminetetraacetic acid (EDTA). Increased sensitivities are obtained when the precipitate in the gel is no longer Zn–SDS, but a complex salt of zinc and imidazole (Ortiz et al., 1992; Fernandez-Patron et al., 1998). In this last case, nanogram sensitivities can be obtained. Perhaps one of the most popular staining methods, however, is metal ion reduction, that is, silver staining, although it is one of the most complex detection procedures. What makes a silver stain highly sensitive is the strong autocatalytic character of silver reduction (Rabilloud et al., 1994). This condition is achieved by using a very weak developer (e.g., dilute formaldehyde) and sensitizers between fixation and silver impregnation.

Silver-staining protocols are divided into two families. In one, the silvering agent is silver nitrate and the developer is formaldehyde in an alkaline carbonate solution. In the second, the silvering agent is a silver–ammonia complex and the developer is formaldehyde in dilute citric acid. The sensitivity with silver staining in modern protocols is in the low-nanogram range for both procedures. This is 100-fold better than classical Coomassie brilliant blue staining, 10-fold better than colloidal brilliant Coomassie blue tags, and about 2-fold better than zinc staining, with the extra benefit of a much better contrast.

(d) Labeling after Blotting. Staining on blots with organic dyes is done almost exclusively by regressive protocols. Because most dyes bind weakly to neutral membranes [nitrocellulose and polyvinylidene difluoride (PVDF)], and due to the concentration effect afforded by blotting, high-nanogram sensitivities are easily reached (Christiansen and Houen, 1992). Noncovalent fluorescence detection is also feasible on blots. In this case, lanthanide complexes are used (Lim et al., 1997), since they have the advantage of providing time-lapse detection with a very good S/N ratio. In the case of PVDF membranes, not wettable with water, fluorescein or rhodamine can be used, since they will bind only to the wetted protein spots (Coull and Pappin, 1990). Under these conditions, the detection limit on blots appears to be around 10 ng protein.

At the end of this long excursus, some interesting conclusions have been drawn by Rabilloud (2000). According to this author, most staining methods have reached a plateau close to their theoretical maximum. For organic dye staining, the maximum occurs when the dye is bound to all available sites on the protein, as is the case with colloidal Coomassie brilliant blue staining. A maximum also seems to have been reached for silver staining, since no increase of sensitivity has been reported in the last 5 years. The only method left with most potential for improvement is fluorescence. Recently, in fact, an ultrasensitive protocol utilizing SYPRO ruby IEF protein gel stain (Molecular Probes, Eugene, OR) has ben reported. This luminescent dye can be excited with 302- or 470-nm light and it has been optimized for protein detection in IEF (and IPG) gels. Proteins are stained in a ruthenium-containing metal complex overnight and then simply rinsed in distilled water for 2 h. Stained proteins can be excited by UV light at ~302 nm (UV-B transilluminator) or with visible wavelengths at ~470 nm. Fluorescent emission of the dye is maximal at 610 nm. The sensitivity of SYPRO ruby is superior to colloidal Coomassie stains and the best silver-staining protocols by a factor of 3–30 times. SYPRO ruby is suitable for staining proteins in both nondenaturing and denaturing carrier ampholyte IEF and IPGs. The unique advantage of this stain is that it does not contain extraneous chemicals (formaldehyde, glutaraldehyde, Tween-20) that frequently interfere with peptide identification in MS. In fact, successful identification of stained proteins by peptide mass profiling was demonstrated, rendering this stain procedure a most promising tool for 2D mapping (Steinberg et al., 2000). It was demonstrated that SYPRO ruby interacts strongly with Lys, Arg, and His residues in proteins, and as such it closely resembles Coomassie stains. On the contrary, acidic silver nitrate stain primarily interacts only with Lys. Although it has been reported that Nile red (Bermudez et al., 1994) as well as SYPRO red and SYPRO orange can also stain IEF gels, the gels must be first incubated in SDS, since all three of these lipophilic dyes bind to proteins indirectly through the anionic detergent. Moreover, with these three dyes, the sensitivity is often poorer than with standard Coomassie staining; on all these accounts, it would appear that this novel SYPRO ruby stain could represent a major revolution in detection techniques (more on fluorescent stains can be found in Chapter 5). Examples of some staining procedures will be given below.

4.2.5.1. Micellar Coomassie Blue G-250 The advantages of this procedure are that only one step is required (i.e., no protein fixation, no destain), peptides down to ~1500 molecular mass can be detected, there is little interference from CAs, and, finally, the staining mixture has a long shelf life (Righetti and Chillemi, 1978; Neuhoff et al., 1985, 1988). The small amount of dye that may precipitate with time can be removed by filtration or washed from the surface of the gels with liquid soap. Note that the dye, as prepared, is in a leuco form (i.e., it is almost colorless, pale greenish instead of deep blue) and in a micellar state. When adsorbed by the peptide/protein zone, it stains the latter in the usual blue color while leaving the gel slab colorless. The process is rather slow, since the dye has to be extracted from the micelle and slowly be adsorbed by the protein zone; it is best to leave the gel in this solution for at least a day under gentle agitation. A more sensitive colloidal Coomassie G-250 (dubbed blue silver, on account of its sensitivity being quite close to that of silvering protocols) has been recently described by Candiano et al. (2004).

4.2.5.2. Coomassie Blue R-250/$CuSO_4$ This staining procedure, by which copper sulfate is admixed to the Coomassie solution, is easily carried out and has good sensitivity (Righetti and Drysdale, 1974). Apparently, divalent copper becomes coordinated around the peptide bond, thus enhancing the color yield.

4.2.5.3. Coomassie Blue R-250/Sulfosalicylic Acid Heat the gels for 15 min at 60°C in a solution of 1 g Coomassie R-250 in 280 mL methanol and 730 mL water, containing 110 g TCA and 35 g sulfosalicylic acid (SSA). Destain at 60°C in 500 mL ethanol, 160 mL acetic acid, and 1340 mL water. Precipitation of the dye at the gel surface is a common problem (remove it with alkaline liquid soap) (Neuhoff et al., 1985). In a milder approach, Neuhoff et al. (1988) substituted strong acids such as sulfosalicylic acid with an acidic medium containing ammonium sulfate. The microprecipitates of Coomassie G act as a reservoir of dye molecules, so that enough dye is available to occupy all the binding sites on all the proteins, provided the staining is long enough to reach steady state (48 h). The presence of ammonium sulfate in the colloidal dye suspension increases the strength of the hydrophobic interaction with the protein moiety, resulting in better sensitivity.

Warning: Often present in many colloidal Coomassie formulations, TCA, often leads to esterification of glutamic acid residues, which might hinder identification of proteins using MALDI–TOF–MS, as often done after 2D mapping (Haebel et al., 1998; Scheler et al., 1998; Patterson and Aebersold, 1995). In addition, the aspartate–proline bond is susceptible to hydrolysis in acidic staining and destaining solutions, resulting in cleavage of some proteins (Hunkapillar et al., 1983).

4.2.5.4. Coomassie Blue G-250/Urea/Perchloric Acid This staining procedure has been proposed by Vesterberg (1972). The procedure is here briefly outlined:

- Fixative: Dissolve 147 g TCA and 44 g sulfosalicilic acid in 910 mL of water.
- Staining solution: Dissolve 0.4 g Coomassie blue G-250 and 39 g urea in ~800 mL of water. Immediatly before use, add 29 mL of 70% $HClO_4$ with vigorous stirring. Bring the volume to 1 L with distilled water.
- Destaining solution: Prepare this by mixing 100 mL acetic acid, 140 mL ethanol, 200 mL ethyl acetate, and 1560 mL water.

1. Fix the protein bands for 30 min at 60°C in fixative.
2. Stain the gel for 30 min at 60°C in staining solution.
3. Destain for 4–22 h in destaining solution.

4.2.5.5. Silver Stain A typical method for silver staining is described in Protocol 4.5 (Merril et al., 1981). A myriad of silvering procedures exist. A good review with guidelines can be found in Rabilloud (1990).

Warning: All silver-staining protocols which utilize aldehydes are in general not compatible with subsequent MS analysis. Formaldehyde, in general, leads to alkylation of α- and ε-amino groups (Haebel et al., 1998; Scheler et al., 1998; Patterson and Aebersold, 1995), which hinders identification of proteins using MS analysis. When fixation is avoided, good results are often obtained with methods in which the proteins are digested, such as in peptide mass fingerprinting. In this last case, protocols that include silver nitrate (Shevchenko et al., 1996) or silver ammonia (Rabilloud et al., 1998) have been reported to be successful in peptide mass identification. Also, destaining silver-stained gels with ferricyanide and thiosulfate seems to greatly improve subsequent analysis by MS (Gharahdaghi et al., 1999). Although aldehyde-free, MS-compatible silver protocols have been recently described, their sensitivity is much reduced, to the point at which "blue silver" might be preferred (Sinha et al., 2001; Mortz et al., 2001; Yan et al., 2000).

4.2.6. Specific Protein Detection Methods

Table 4.6 lists some of the most common specific detection techniques: It is given as an example and with appropriate references but is not meant to be exhaustive.

4.2.7. Quantitation of Focused Bands

We refer the readers to Westermeier's book (1997) for an in-depth treatise. Basically, the instrumentation for image acquisition consists of (a) a video camera (mostly

PROTOCOL 4.5. SILVER-STAINING PROCEDURE

REAGENTS

The reagents required are listed below.

METHOD

The silver-staining procedure involves exposing the gel to a series of reagents in a strict sequence, as described in the following steps.

Step	Reagent	Volume	Time
1.	12% TCA	500 mL	30 min
2.	50% methanol, 12% acetic acid	1 L	30 min
3.	1% HIO_4	250 mL	30 min
4.	10% ethanol, 5% acetic acid	200 mL	10 min (repeat 3×)
5.	3.4 mM potassium dichromate, 3.2 mM nitric acid	200 mL	5 min
6.	Water	200 mL	30 s (repeat 4×)
7.	12 mM silver nitrate	200 mL	5 min in light, 25 min in dark
8.	0.28 M sodium carbonate, 0.05% formaldehyde	300 mL 300 mL	30 s (repeat 2×) Several minutes
9.	10% acetic acid	100 mL	2 min
10.	Photographic fixative (e.g., Kodak Rapid Fix)	200 mL	10 min
11.	Water	Several liters	Extensive washes

12. Remove any silver precipitate on the gel surface using a swab.

TABLE 4.6. Specific Protein Detection Techniques

Protein detected	Technique	Reference
Glycoproteins	PAS (periodic acid-Schiff) stain	Hebert and Strobbel, 1974
Lipoproteins	Sudan black stain	Godolphin and Stinson, 1974
Radioactive proteins	Autoradiography	Laskey and Mills, 1975
	Fluorography	Laskey, 1980
Enzymes	Zymograms[a]	Harris and Hopkinson, 1976
Antigens	Immunoprecipitation in situ	Richtie and Smith, 1976
	Print immunofixation	Arnaud et al., 1977
	Blotting	Towbin et al., 1979

[a] The concentration of the buffer in the assay medium usually needs to be increased in comparison with zone electrophoresis to counteract the buffering action by carrier ampholytes.

CCD cameras, since their digital signals can be fed directyl into a computer); (b) desktop scanners, usually equipped with visible light sources, able to scan in general in both reflectance and transmittance; and (c) densitometers, equipped in general with white light or laser light sources. The quantitation of bands is only possible when the absorptions (optical density, OD) are linear. With white light this is possible up to 2.5 OD at most; with laser light up to 4 OD.

Some general rules about densitometry are worth repeating:

(a) The scanning photometer should have a spatial resolution of the same order as the fractionation technique (for correct analysis, 50–100 μm resolution is required).
(b) The stoichiometry of the protein–dye complex varies among different proteins.
(c) This same relationship is linear only over a limited range of protein concentrations.
(d) At high absorbance, the limitations of the spectrophotometer interfere with the photometric measurement.

4.2.8. Troubleshooting

A number of problems, with their potential solutions, are listed here. This troubleshooting guide is the result of some 20 years of use, of innumerable workshops conducted among the general practitioners in the field, and of listening to a medley of complaints among users.

4.2.8.1. Waviness of Bands Near Anode This may be caused by:

(a) Carbonation of the catholyte: In this case, prepare fresh NaOH with degassed distilled water and store properly.
(b) Excess catalysts: Reduce the amount of ammonium persulfate.
(c) Too long sample slots: Fill them with dilute CAs.
(d) Too low concentrations of CAs: Check the gel formulation.

To alleviate the problem, it is usually beneficial to add low concentrations of sucrose, glycine, or urea and to apply the sample near the cathode. To salvage a gel during the run, as soon as the waves appear, apply a new anodic strip soaked with a weaker acid (e.g., acetic acid vs. phosphoric acid) inside the original one and move the electrodes closer to one another.

4.2.8.2. Burning along Cathodic Strip This may be caused by:

(a) The formation of a zone of pure water at pH 7: Add to your acidic pH range a 10% solution of either the 3–10 or the 6–8 range ampholytes.
(b) The hydrolysis of the acrylamide matrix after prolonged exposure to alkaline pH: Choose a weaker base, if adequate, and, unless a prerun of the gel is strictly required, apply the electrodic strips after loading the samples.

4.2.8.3. pH Gradients Different from Expected

(a) For acidic and alkaline pH ranges, the problem is alleviated by the choice of anolytes and catholytes, whose pH is close to the extremes of the pH gradient.

(b) Alkaline pH ranges should be protected from carbon dioxide by flushing the electrophoretic chamber with moisture-saturated N_2 (or better with argon) and by surrounding the plate with pads soaked in NaOH. It is worth remembering that pH readings on unprotected alkaline solutions become meaningless within half an hour or so.

(c) A large amount of a weak acid or base, supplied as sample buffer, may shift the pH range (ß-mercaptoethanol is one such base). The typical effect of the addition of urea is to increase the apparent pI of the CAs.

(d) It may be due to the cathodic drift; to counteract this:

Reduce the running time to the required minimum (as experimentally determined for the protein of interest or for a colored marker of similar M_r).

Increase the viscosity of the medium (with sucrose, glycerol, or urea).

Reduce the amount of ammonium persulfate.

Remove acrylic acid impurities by recrystallizing acrylamide and Bis and by treating the monomer solution with mixed-bed ion exchange resin.

For a final cure, incorporate into the gel matrix a reactive base, such as 2-dimethylamino propyl methacrylamide (Polyscience) (Righetti and Macelloni, 1982): Its optimal concentration (of the order of 1 μM) should be experimentally determined for the system being used.

4.2.8.4. Sample Precipitation at Application Point If large amounts of material precipitate at the application point, even when the M_r of the sample proteins is well below the sieving limits of the gel, the trouble is usually caused by protein aggregation. Some remedies:

(a) Try applying the sample in different positions on the gel with and without prerunning: Some proteins might be altered only by a given pH.

(b) If you have evidence that the sample contains high-M_r components, reduce the value of $\%T$ of the polyacrylamide gel.

(c) If you suspect protein aggregation brought about by the high concentration of the sample (e.g., when the problem is reproduced by disc electrophoresis runs), do not prerun and set a low voltage (100–200 V) for several hours to avoid the concentrating effect of an established pH gradient at the beginning of the run. Also consider decreasing the protein load and switching to a more sensitive detection technique. The addition of surfactants and/or urea is usually beneficial.

(d) If the proteins are only sensitive to the ionic strength and/or the dielectric constant of the medium (in this case they perform well in disc electrophoresis and are precipitated if dialysed against distilled water), increasing the CA concentration, adding glycine or taurine, and sample application without prerunning may overcome the problem.

(e) The direct choice of denaturing conditions (8 M urea, detergents, ß-mercaptoethanol) very often minimizes these solubility problems, dissociating proteins (and macromolecular aggregates) to polypeptide chains.

4.2.9. Some Typical Applications of IEF

Isoelectric focusing is a fine-tuned analytical tool for investigating posttranslational processing and chemical modification of proteins. Protein processing may be grouped into two kinds. The first covers changes in primary structure, with proteolysis at peptide bonds, for example, for removal of intervening sequences or of leader peptides, as in intracellular processing. Such modifications are typically identified by size fractionation techniques, such as SDS–PAGE. A second class covers another mode of processing, in which the size is only marginally affected and no peptide bond is cleaved but the surface charge is modified. Such posttranslational processing is the typical realm for IEF analysis. It includes attachment of oligosaccharide chains, such as in glycoproteins, or of individual sugar moieties, such as in glycated proteins (e.g., hemoglobin A_{1c}), or phosphorylation, methylation, adenosine 5′-diphosphate (ADP) ribosylation, and ubiquitination, to name just a few. But there are also a number of modifications occurring at the NH_2 groups (e.g., acetylation), at the COOH groups (e.g., deamidation), and at the SH groups as well as a number of chemical modifications occurring in vitro, such as accidental carbamylation due to the presence of urea. All of these modifications can be properly investigated by IEF, as amply discussed in a broad review dedicated to this topic (Gianazza, 1995). Of course, IEF can only give a pI shift and cannot tell much more than that in assessing the modification. However, IEF as a fractionation tool coupled to MS can be a formidable hyphenated method for assessing the extent of modification involved, the stoichiometry of the reaction, and the reaction site too. From this point of view, MALDI–TOF–MS appears to be a versatile tool. Isoelectric focusing can also be used as a probe for interacting systems, such as protein ligand interactions. If the interacting species are stable in the electric field, IEF enables determination of intrinsic ligand-binding constants for statistical binding of a charged ligand, binding to heterogeneous sites, and cooperative binding. The interacting systems can be a protein with a small ligand or a protein with a macromolecular ligand. In extreme cases, such interactions could be engendered by isomerization (pH-dependent conformational transition) or by interaction with the carrier ampholyte buffers. An excellent review on all these aspects is now available (Cann, 1998).

4.2.10. Examples of Some Fractionations

After so many methodological tips, it is refreshing to look at some results obtained with this technique. Since its inception, especially in the gel format (preparative IEF in sucrose density gradients was a quite demanding technique and never became popular), it was immediately realized that IEF represented a turning point in separation techniques, unrivaled by any other fractionation method available in the 1960s, since it offered exquisite resolution. A case in point was the hemoglobin (Hb) separation shown in Figure 4.6. It is known that Hb in vitro can slowly undergo spontaneous autooxidation; in vivo, however, in healthy individuals, Hb is usually in its reduced form, since, in order to fulfil its physiological role, the heme iron must remain in a ferrous state. Heme that has oxidized to the ferric form (met Hb) is unable to bind oxygen. Despite the significant difference in charge between ferro Hb ($\alpha_2\beta_2$) and its oxidized forms (${\alpha^+}_2\beta_2$, $\alpha_2{\beta^+}_2$, ${\alpha^+}_2{\beta^+}_2$), these species had never been isolated before by conventional electrophoresis. In contrast, gel IEF permitted clearcut separations of fully and partially oxidized intermediates. After Hb A was partially oxidized with an agent such as ferricyanide, three new bands could be visualized (Fig. 4.6). The fully oxidized Hb (${\alpha^+}_2{\beta^+}_2$) appears as a brown band with a pI of 7.20, 0.25 pH units above that of Hb A (Bunn and Drysdale, 1971). Equidistant between the fully oxidized Hb and reduced Hb ($\alpha_2\beta_2$), two closely spaced bands could be detected, which later were shown by Park (1973) to be $\alpha_2{\beta^+}_2$ and ${\alpha^+}_2\beta_2$, respectively. The fact that these two components could be separated in this high-resolution system added to the large body of evidence indicating differences in the heme environments of the α- and β-chains, inducing a slightly different net surface charge (conformational transitions) in otherwise identical total charge proteins. As it turned out, this story of the valence hybrids was even more complex. In theory, there should be 10 valence

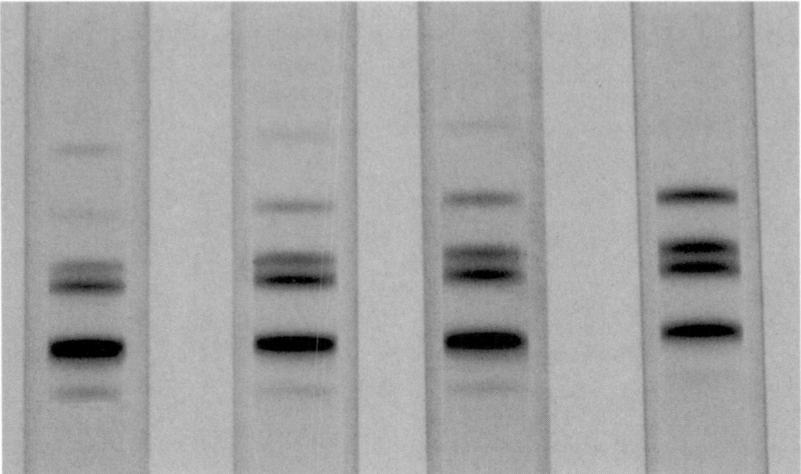

FIGURE 4.6. Separation of oxidized forms of human hemoglobin. Isoelectric fractionations were performed on hemolysates on 0.05 M phosphate buffer (pH 7.1) in presence of 2% carrier ampholytes pH 6–8. Samples (from left to right): control, no oxidant added and Hbs with increasing amounts of oxidant up to addition of a half equivalent amount of ferricyanide. Zones (from bottom to top): oxyhemoglobin (red, $\alpha_2\beta_2$); $\alpha_2{\beta_2}^+$; $\alpha_2{\beta_2}^+$; ferrihemoglobin (brown, ${\alpha_2}^+{\beta_2}^+$). (Modified from Righetti and Drysdale, 1973.)

hybrids obtainable by partial oxidation of $\alpha_2\beta_2$. By quenching a Hb solution partially saturated with CO into a hydro-organic solvent containing ferricyanide, Perrella et al. (1981) were able to produce a population of partially oxidized and CO-bound Hb molecules. When these valence hybrids were analyzed by IEF at temperatures as low as –23°C, all the intermediate species were resolved. This was an excellent example of the use of IEF as a structural probe.

In a gel-slab format, IEF, was soon adopted also in clinical chemistry. A nice example is the screening of thalassemia syndromes, which had to be performed on a large scale in regions of high incidence, such as the island of Sardinia. To assess the extent of the syndrome, the amount of γ-globin (belonging to fetal Hb) chains in newborns had to be evaluated and the ratio β/γ determined. In the past, this was accomplished by the three-step procedure of Clegg et al. (1966): incubation of red blood cells with [^{3}H] leucine, removal of heme, and chromatographic separation of globin chains in CM-cellulose. This procedure was very complex and placed a heavy burden on laboratories where large populations at risk had to be screened. Aware of these shortcomings, we proposed a novel method for globin chain separation consisting of an IEF step in conventionally thick (Righetti et al., 1979; Saglio et al., 1979) or in ultrathin gel layers (Valkonen et al., 1980) in the presence of urea and detergents, the latter greatly improving the β/γ separation and even inducing the splitting of two γ-chain phenotypes, called Aγ and Gγ, due to a replacement of a Ala with a Gly residue in position 136 of the γ-chains. Even this remarkable improvement fell short of expectations, since it was still labor intensive, due to the need of preparing heme-free globin chains and to the lengthy staining/destaining steps caused by the high levels of surfactant in the gel. A breakthrough finally came with the work of Cossu et al. (1982), as shown in Figure 4.7, who proposed focusing the

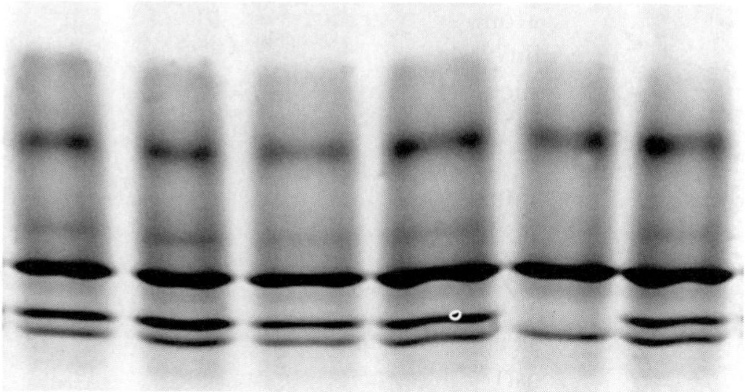

FIGURE 4.7. Isoelectric focusing of cord blood hemolysates in pH 6–8 gradient added with 0.2 M β-Ala and 0.2 M 6-amino caproic acid. Experimental: $T = 6\%, C = 4\%$ gel; run: 15 min at 400 V, 90 min at 15 W–1500 V_{max}; protein load: 20 μg/track. Closeup of a few sample tracks in a gel slab. Note that the second sample from right is a homozygous β-thalassemic, since there is a complete absence of adult hemoglobin. The three major zones are, from bottom to top: acetylated fetal (Hb F_{ac}), adult (Hb A), and fetal (Hb F) hemoglobins. (From Cossu et al., 1982, by permission.)

intact components of umbilical cord blood: Hb F, Hb A, and Hb F_{ac} (acetylated Hb F). Nobody had attempted that before due to the very poor resolution between the Hb A and Hb F_{ac} bands caused by minute pI differences. The authors succeeded in that by using nonlinear pH 6–8 gradients, flattened in the region around pH 7 by addition of 0.2 M β-Ala and 0.2 M 6-amino caproic acids, a technique called "separator IEF." It can be seen, by examination of a close-up of the gel, that the resolution is now excellent. The technique was user friendly, did not require any sample manipulation steps prior to IEF, was fast (ca. 2 h focusing time), and was adaptable to large sample numbers (up to 30 tracks could be run simultaneously). Even the staining was rapid: It consisted in dipping the gels into 0.2% bromophenol blue in 50% ethanol and 5% acetic acid and in a quick destaining step in 30% ethanol and 5% acetic acid, a method first reported by Awdeh (1969), producing Hb bands stained in deep red-purple color. As a result of this massive screening (which soon spread to other regions at risk, such as Sicily, Calabria etc.) and of genetic counseling to couples at risk, homozygous thalassemic conditions in Italy have been essentially eradicated today.

Another interesting aspect of the power of IEF comes from Figure 4.8, which shows the analysis of active fragments of the human growth hormone (hGH), synthesized by the solid-phase method of Merrifield (1969). In the early days of IEF it was thought that the IEF of peptides would not be feasible, first, because proteins have a higher diffusion coefficient than proteins and, second, because they would not be precipitated and fixed into the gel matrix by the common protein stains, like alcoholic solutions of Coomassie blue. Figure 4.8 dispelled both myths and proved that the IEF of peptides was highly feasible and would indeed produce very sharp bands (Righetti and Chillemi, 1978). The problem of fixation was solved by adopting a leuco-stain, a micellar suspension of Coomassie brilliant blue G-250 (G stays for its greenish hue, which cannot be appreciated in monomeric solutions but is clearly visible in the micellar state) in TCA: As the focused gel is bathed directly in this solution, the peptides adsorb the dye from the micelle and are fixed by the dye molecules, which probably act as crosslinks over the different peptide chains, thus forming a macromolecular aggregate which is trapped in the gel matrix fibers. This experiment was very important for peptide chemists since it helped in redirecting their synthetic strategies. As seen from Figure 4.8, when medium-length peptides were produced via the Merrifield approach, the amounts of failed and truncated sequences were in large excess over the desired product, to the point that the latter could not be recognized any longer. Although this had been theoretically predicted, it had not been experimentally verified up to that time due to lack of high-resolution techniques. As news spread, in vitro synthesis of peptides took an important turn: It was done only in short sequences (five to six amino acids at most) which were at the end joined together by splicing.

Another example of the unique performance of IEF is the zymogram of Figure 4.9: We were following the purification of a dehydrogenase from spinach leaves. The progress in this process could be monitored by staining gel tubes either for general proteins (blue gels) or by a specific zymogram in situ (purple bands; Righetti and Zanetti, ca. 1976, unpublished). Note that sharp bands could be maintained for quite long staining times, after extruding the gels from the glass tubes, due to the lower diffusion ability of focused zones.

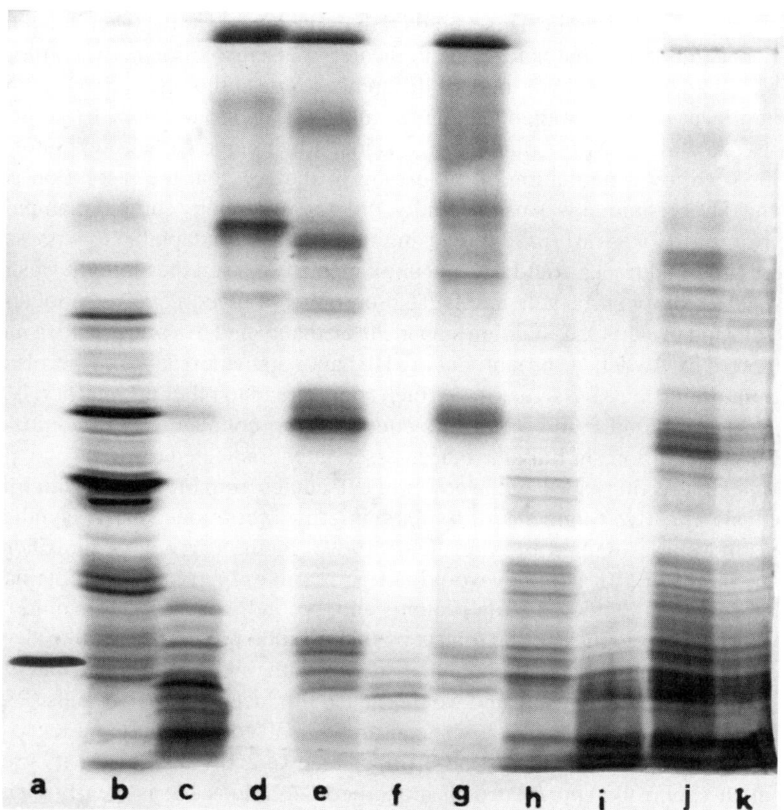

FIGURE 4.8. Isoelectric focusing of peptides in length range 8–54 amino acids. The gel slab was 0.7 mm thick and contained 7% acrylamide, 2% Ampholine pH 3.5–10, and 8 M urea. Ten to 15 μL of sample (10 mg/mL) was applied in filter paper strips at the anode after 1 h prefocusing. Total running time 4 h at 10 W (1000 V at equilibrium). The gel was then dipped in a colloidal dispersion of Coomassie brilliant blue G-250 in 12% TCA. The samples were the following synthetic fragments of the human growth hormone (hGH): a = hGH 31–44; b = hGH 15–36; c = hGH 111–134; d = hGH 1–24; e = hGH 166–191; f = hGH 25–51; g = hGH 157–191; h = hGH 1–36; i = hGH 96–134; j = hGH 115–156; k = hGH 103–156. Shorter fragments (octa-, nona-, and decapeptides) were neither fixed in the gel nor stained. (From Righetti and Chillemi, 1978, by permission.)

4.2.11. Artifacts or Not?

This question was the object of hot debate at the inception of the technique, and there were just as many reports on artifacts as additional ones denying them. The field was so controversial that in 1983 Righetti wrote a chapter of his book on IEF by the title "Artefacts: A Unified View." The final consensus was that artifacts, when reported, occurred only in extreme cases and with "extreme" structures. One of the most glorious examples was the suspicious report that heparin gave a large number (up to 21) of

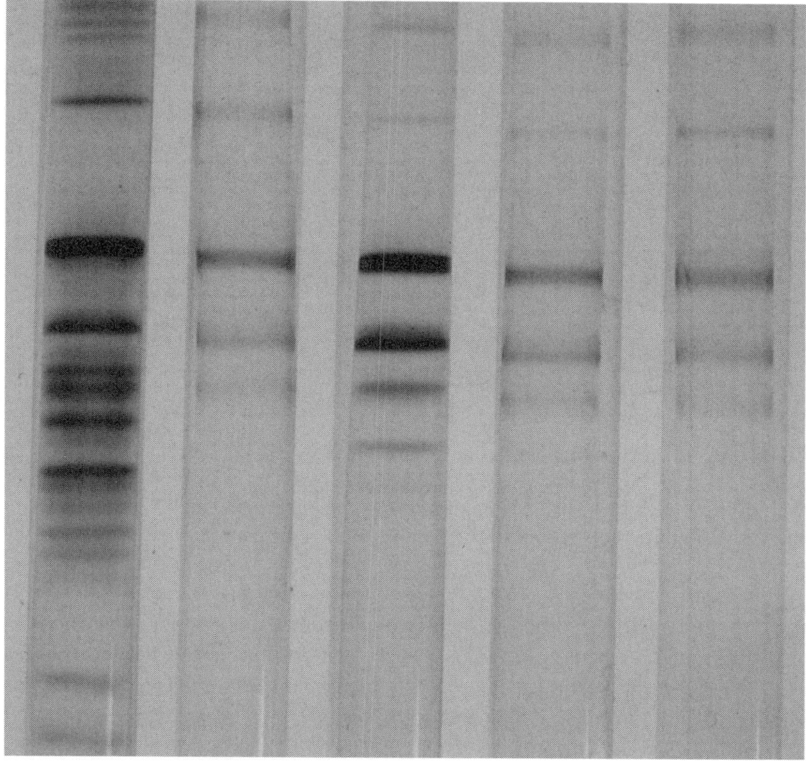

FIGURE 4.9. Purification of dehydrogenase from spinach leaves. Blue gels: staining with general protein stain (Coomassie blue); purple bands: specific staining with an in situ zymogramming for dehydrogenases (Righetti and Zanetti, ca. 1976, unpublished).

bands focusing in the pH 4–5 range (Johnson and Mulloy, 1976). Now, how could a pure polyanion, bearing no positive counterions on its polymeric backbone, exhibit an "isoelectric point" remained a mystery to anybody with a minimum knowledge of chemistry. It took the work of Righetti and Gianazza (1978) and Righetti et al. (1978) to find out that these 21 fractions indeed represented 21 different complexes of the same heparin polymer with 21 different Ampholine molecules (see Fig. 4.10). This was the "catch 22": In those days, when binding of small molecules to macromolecules was suspected, all theoreticians had worked out a bimodal distribution, that is, the bound versus the unbound species! After our report, it was clear that multimodal distribution could be the real trend in IEF, so theoreticians like Cann (1979, 1998) hurried to change their models. It then became apparent that, in a specular fashion, also polycations would produce such an artifactual binding pattern (Gianazza and Righetti, 1980). Except for these artifacts generated by peculiar structures, there was no practical evidence that carrier ampholytes would elicit the same multimodal distribution by interacting with proteins, so general users widely accepted the technique. While that

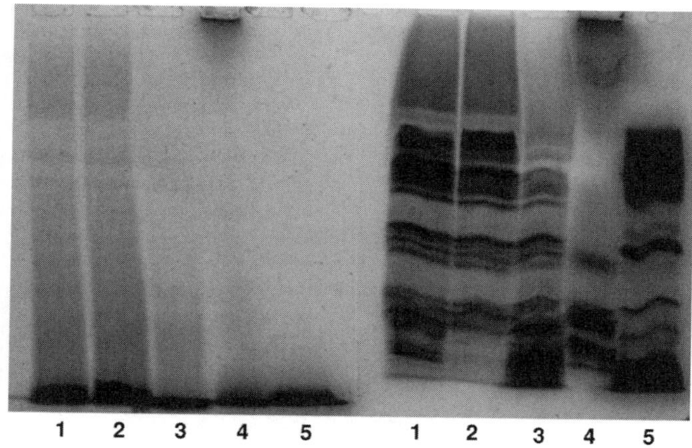

FIGURE 4.10. Isoelectric focusing of polyanions in $T = 5\%$ polyacrylamide gels, 2% Ampholine pH 2.5–5 in the absence (right side) and presence (left side) of 8 M urea; 1, heparin A; 2, heparin B; 3, carboxyl-reduced heparin B; 4, polygalacturonic acid; 5, polyglutamic acid. Staining with toluidine blue. (From Righetti et al., 1978, by permission.)

might be generally true, some artifacts were recently reported in the focusing of diaspirin crosslinked hemoglobin (Bossi et al., 1999b) and in the IEF purification of basic fatty acid–binding protein, where it would appear that some CAs could adhere to the protein, possibly in the hydrophobic ligand-binding pocket (Perduca et al., 2000).

Even in the case of Immobilines, though, which are surely immune from the drawbacks of CA–IEF, some disturbing artifacts have been recently reported (Berven et al., 2003). Some trains of spots belonging to outer membrane proteins from the gram-negative bacterium *Methylococcus capsulatus* run in 2D maps were found upon excision and rererunning of individual bands to reproduce the whole family of spots in the original track. Since spots were regenerated on both sides of the reapplied polypeptide spot and no difference in the apparent M_r was observed, truncation/degradation as the origin for this phenomenon had to be discarded. Multiple equilibria with CA buffers could not be accounted for, since no CAs were present in the Immobiline gel. It was concluded that this family of proteins belonged to a class of stable polypeptides able to generate conformational equilibria representing both native and denatured polypeptide isoforms (although the experiments were run in 7 M urea). Such folding isomers, all originated from a single polypeptide chain, would then give rise to pI heterogeneity and polypeptide spot trains. Other proteins giving rise to such trains of spots were found to be OMPA-related proteins from *P. gingivalis* as well as mistletoe lectin and trasthyretin (Guruprasad et al., 1990). Nevertheless, it would appear that these few examples are rather rare and are limited to peculiar classes of extremely stable (thus denaturing-resistant) proteins, tending to coexist with partially refolded species even in strong denaturing milieus. Given all the problems of CA–IEF, as illustrated in these examples and in the following section, it is not surprising that the natural evolution of CA–IEF would be the Immobiline technology.

4.3. IMMOBILIZED pH GRADIENTS

4.3.1. General Considerations

As illustrated below, IPGs represent perhaps the ultimate development in all focusing techniques, a big revolution in the field in fact. Due to the possibility of engineering the pH gradient at whim, from the narrowest (which, for practical purposes, has been set at 0.1 pH unit over a 10-cm distance) to the widest possible one (a pH 2.5–12 gradient), IPGs permit the highest possible resolving power on the one hand and the widest possible collection of spots (in 2D maps) on the other hand. The chemistry is precise and amply developed; so are all algorithms for implementing any possible width and shape of the pH gradient. Due to their unique performance, IPGs now represent the best possible first dimension for 2D maps and are increasingly adopted for this purpose.

4.3.1.1. Problems of Conventional IEF Table 4.7 lists some of the major problems associated with conventional IEF using amphoteric buffers. Some of them are quite severe; for example (point 1), low ionic strength often induces near-isoelectric precipitation and smearing of proteins, even in analytical runs at low protein loads. The problem of uneven conductivity is magnified in poor ampholyte mixtures, like the Poly Sep 47 (a mixture of 47 amphoteric and nonamphoteric buffers, claimed to be superior to CAs) (Cuono and Chapo, 1982). Due to their poor composition, huge conductivity gaps form along the migration path, against which proteins of different pI values are stacked. The results are simply disastrous (Righetti et al., 1988c). Cathodic drift (point 6) is also a major unsolved problem of CA–IEF, resulting in extensive loss of proteins at the gel cathodic extremity upon prolonged runs and characterized by a gradual isotachophoretic decay from the gradient ends (Mosher and Thormann, 2002). The dynamics of CA–IEF are illustrated in the simulation of Figure 4.11. This simulator is quite realistic, since it accepts 150 components and voltage gradients (300 V/cm) employed in laboratory practice. Before passage of the current, the column is at constant pH and the multitude of amphoteric buffers are randomly distributed, resulting in reciprocal neutralization. After starting the experiment, the focusing process appears to proceed in two phases, a relative fast separation phase followed by a slow stabilizing phase during which a steady state should be reached. As shown in the bottom tracing, a pH 3–10 gradient appears to be fully established within 10–12 min, with the CAs forming Gaussian-like, overlapping zones of high concentration

TABLE 4.7. Problems with Carrier Ampholyte Focusing

1. Medium of very low and unknown ionic strength
2. Uneven buffering capacity
3. Uneven conductivity
4. Unknown chemical environment
5. Not amenable to pH gradient engineering
6. Cathodic drift (plateau phenomenon)

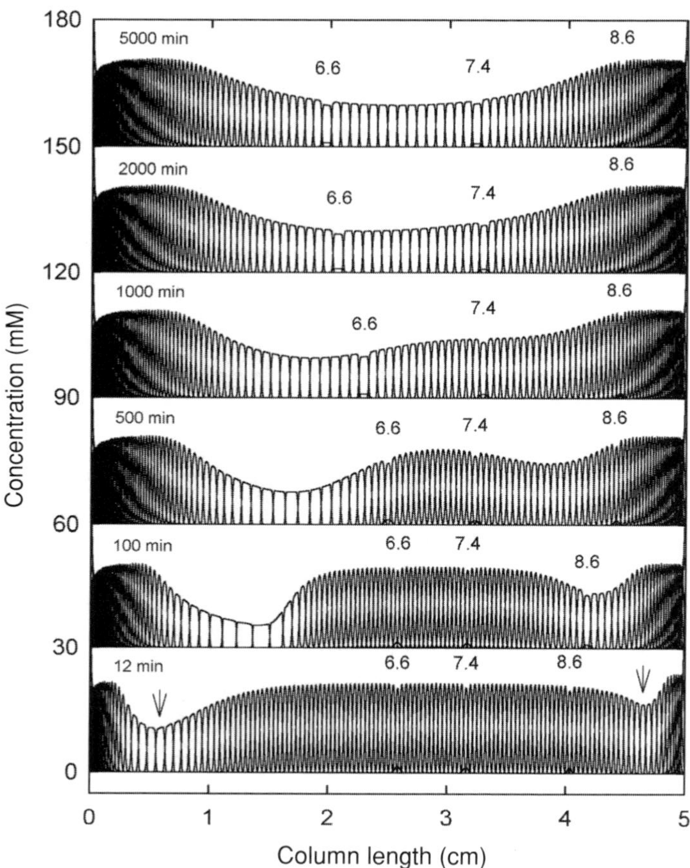

FIGURE 4.11. Computer-simulated distributions of 140 carrier ampholytes and three dyes after 12, 100, 500, 1000, 2000, and 5000 min of current flow. The numbers refer to the pI values of the dyes and the arrowheads point to their locations. Successive graphs are presented with a y-axis offset of 30 mM. The arrows at the bottom graph mark the two transient concentration valleys that are characteristic for the stabilizing phase. The cathode is to the right. (From Mosher and Thormann, 2002, by permission.)

(>100-fold increase compared to the initial stage) with the three samples (dyes with pI values of 6.6, 7.4, and 8.6) well focused too. Close to the column ends, large changes of carrier concentrations (valleys) are marked with arrows. This initial CA distribution generates an almost linear pH profile. This simulation, however, shows also the major drawback of all CA-based focusing processes: Steady state can never be achieved. The two valleys at the extremes migrate inward, flattening the focused CA from Gaussian into square-shaped peaks and displacing them toward the terminal electrodes (see the simulated profiles from 100 to 5000 min). As a result, the pH gradient keeps changing slopes until it assumes a sigmoidal profile with marked flattening of the central

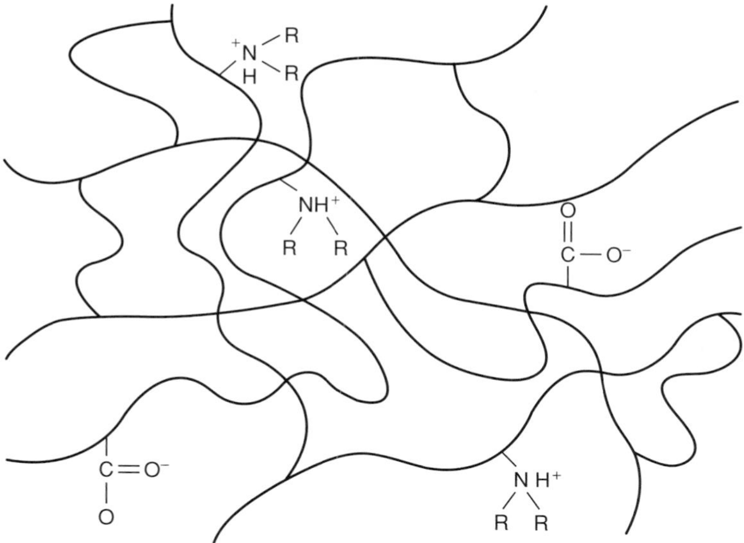

FIGURE 4.12. Isoelectric focusing in IPGs. A hypothetical gel structure is depicted, where the strings represent the neutral acrylamide residues, the crossover points the Bis crosslinking, and the positive and negative charges the grafted Immobiline molecules. (Courtesy of LKB Produkter AB.)

region with the inflection point around neutrality (called plateau phenomenon). This results in a continuous shift of the foci of the focused samples along the column length (see the diverging positions of the three dyes from bottom to top in Fig. 4.3). For all these reasons, in 1982 Bjellqvist et al. launched the technique of IPGs.

4.3.1.2. Immobiline Matrix Immobilized pH gradients are based on the principle that the pH gradient, which exists prior to the IEF run itself, is copolymerized, and thus insolubilized within the fibers of a polyacrylamide matrix (see Fig. 4.12 for a pictorial representation). This is achieved by using, as buffers, a set of six nonamphoteric, weak acids and bases having the following general chemical composition: CH_2=CH–CO–NH–R, where R denotes either two different weak carboxyl groups with p*K* 3.6 and 4.6 or four tertiary amino groups with p*K* 6.2, 7.0, 8.5, and 9.3 (available under the trade name Immobiline from Pharmacia-Upjohn and now also from Bio-Rad). Their synthesis has been described by Chiari et al. (1989a,b). A more extensive set comprising 10 chemicals (a p*K* 3.1 acidic buffer, a p*K* 10.3 basic buffer and two strong titrants, a p*K* 1 acid and a p*K* $>$ 12 quaternary base) is available as "pI select" from Fluka AG, Buchs, Switzerland (see Tables 4.8 and 4.9 for their formulas) (Righetti et al., 1996c). All of the above chemicals have been reported by our group; for example, the synthesis of the p*K* 3.1 buffer was utilized by Righetti et al. (1988b) for separation of isoforms of very acidic proteins (pepsin) and the p*K* 10.3 species was first adopted by Sinha and Righetti (1987) for creating alkaline gradients for

TABLE 4.8. Acidic Acrylamido Buffers

p*K*	Formula	Name	M_r
1.2	CH_2=CH-CO-NH-C(CH_3)(CH_2-SO_3H)-CH_3	2-Acrylamido-2-methylpropane sulfonic acid	207
3.1	CH_2=CH-CO-NH-CH(OH)-COOH	2-Acrylamido-glycolic acid	145
3.6	CH_2=CH-CO-NH-CH_2-COOH	N-Acryloyl-glycine	129
4.6	CH_2=CH-CO-NH-$(CH_2)_3$-COOH	4-Acrylamido-butyric acid	157

separation of elastase isoforms. Over the years, we have reported the synthesis of a number of other buffering ions produced with the aim of closing some gaps between the available Immobilines, especially in the pH 7.0–8.5 interval. These are a p*K* 6.6 2-thiomorpholinoethylacrylamide and a p*K* 7.4 3-thiomorpholinopropylacrylamide (Chiari et al., 1990b); a p*K* 6.85 1-acryloyl-4-methylpiperazine (Chiari et al., 1991a); an alternative p*K* 7.0 2-(4-imidazolyl)ethylamine-2-acrylamide (Chiari et al., 1991b), and a p*K* 8.05 *N*,*N*-bis(2-hydroxyethyl)-*N′*-acryloyl-1,3-diaminopropane (Chiari et al., 1990a). These compounds are listed in Table 4.10. Additional species have recently been described by Bellini and Manchester (1998).

TABLE 4.9. Basic Acrylamido Buffers

p*K*	Formula	Name	M_r
6.2	CH_2=CH-CO-NH-$(CH_2)_2$-N(morpholine ring)O	2-Morpholino ethylacrylamide	184
7.0	CH_2=CH-CO-NH-$(CH_2)_3$-N(morpholine ring)O	3-Morpholino propylacrylamide	199
8.5	CH_2=CH-CO-NH-$(CH_2)_2$-N(CH_3)-CH_3	*N*, *N*-Dimethyl aminoethyl acrylamide	142
9.3	CH_2=CH-CO-NH-$(CH_2)_3$-N(CH_3)-CH_3	*N*, *N*-Dimethyl aminoethyl acrylamide	156
10.3	CH_2=CH-CO-NH-$(CH_2)_3$-N(C_2H_5)-C_2H_5	*N*, *N*-Diethyl aminopropyl acrylamide	184
>12	CH_2=CH-CO-NH-$(CH_2)_2$-N^+(C_2H_5)(C_2H_5)-C_2H_5	*N*, *N*, *N*-Triethyl aminoethyl acrylamide	198

TABLE 4.10. New Basic Acrylamido Buffers

pK^a	Formula	Name	M_r	Ref.
6.6	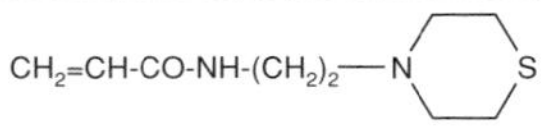 CH_2=CH-CO-NH-$(CH_2)_2$—N(thiomorpholine ring)S	2-Thiomorpholinoethylacrylamide	200	*b*
6.85	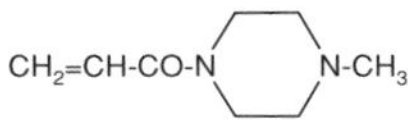 CH_2=CH-CO-N(piperazine ring)N-CH_3	1-Acryloyl-4-methylpiperazine	154	*c*
7.0	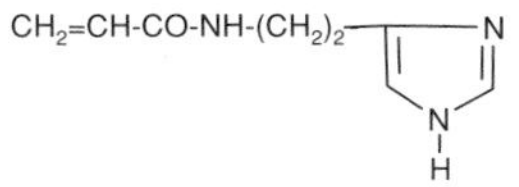 CH_2=CH-CO-NH-$(CH_2)_2$—(imidazole ring, N, N-H)	2-(4-Imidazolyl)ethylamine-2-acrylamide	165	*d*
7.4	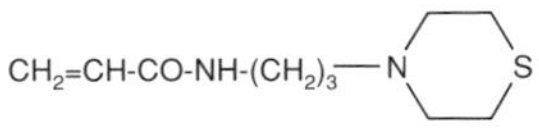 CH_2=CH-CO-NH-$(CH_2)_3$—N(thiomorpholine ring)S	3-Thiomorpholinopropylacrylamide	214	*b*
8.05	CH_2=CH-CO-NH-$(CH_2)_3$-$N(CH_2CH_2OH)_2$	*N*,*N*-Bis(2-hydroxyethyl)-*N*′-acryloyl-1,3-diaminopropane	200	*e*

[a] All pK values measured at 25°C.
[b] From Chiari et al., 1990b.
[c] From Chiari et al., 1991a.
[d] From Chiari et al., 1991b.
[e] From Chiari et al., 1990a.

During gel polymerization, these buffering species are efficiently incorporated into the gel (84% to 86% conversion efficiency at 50°C for 1 h) (Righetti et al., 1984). Immobiline-based pH gradients can be cast in the same way as conventional polyacrylamide gradient gels, using a density gradient to stabilize the Immobiline concentration gradient, with the aid of a standard, two-vessel gradient mixer. As shown in their formulas, these buffers are no longer amphoteric, as in conventional IEF, but are bifunctional. At one end of the molecule is located the buffering (or titrant) group, and at the other end is an acrylic double bond which disappears during immobilization of the buffer on the gel matrix. The three carboxyl Immobilines have rather small temperature coefficients (dpK/dT) in the 10–25°C range due to their small standard heats of ionization (≈1 kcal/mol) and thus exhibit negligible pK variations in this temperature interval. On the other hand, the five basic Immobilines exhibit rather large ΔpK (as much as ΔpK = 0.44 for the pK 8.5 species) due to their larger heats of ionization (6–12 kcal/mol). Therefore, for reproducible runs and pH gradient calculations, all the experimental parameters have been fixed at 10°C.

Temperature is not the only variable that affects Immobiline pK (and therefore the actual pH gradient generated). Additives in the gel that change the water structure (chaotropic agents, e.g., urea) or lower its dielectric constant and the ionic strength of the solution alter their pK values. The largest changes, in fact, are due to the presence of urea: Acidic Immobilines increase their pK in 8 M urea by as much as 0.9 pH unit, while the basic Immobilines increase their pK by only 0.45 pH unit (Gianazza et al., 1983). Detergents in the gel (2%) do not alter the Immobiline pK, suggesting that they are not incorporated into the surfactant micelle. For generating extended pH gradients, we use two additional chemicals which are strong titrants having pK well outside the desired pH range. One is QAE (quaternary amino ethyl) acrylamide (pK > 12) and the other is AMPS (2-acrylamido-2-methyl propane sulfonic acid, pK 1.0) (see Tables 4.8 and 4.9).

As shown in Figure 4.12, the proteins are placed on a gel with a preformed, immobilized pH gradient (represented by carboxyl and tertiary amino groups grafted to the polyacrylamide chains). When the field is applied, only the sample molecules (and any ungrafted ions) migrate in the electric field. Upon termination of electrophoresis, the proteins are separated into stationary, isoelectric zones. Due to the possibility of designing stable pH gradients at will, separations have been reported in only 0.1 pH unitwide gradients over the entire separation axis, leading to an extremely high resolving power (ΔpI = 0.001 pH unit).

4.3.1.3. Narrow and Ultranarrow pH Gradients We define the gradients from 0.1 to 1 pH unit as narrow (toward the 1 pH unit limit) and ultranarrow (close to the 0.1 pH unit limit) gradients. Within these limits we work on a tandem principle; that is, we choose a buffering Immobiline, either a base or an acid, with its pK within the pH interval we want to generate and a nonbuffering Immobiline, then an acid or a base, respectively, with its pK at least 2 pH units removed from either the minimum or maximum of our pH range. The titrant will provide equivalents of acid or base to titrate the buffering group but will not itself buffer in the desired pH interval. For these calculations, we used to resort to modified Henderson–Hasselbalch equations and to rather complex nomograms found in the LKB Application Note No. 321. A list of fifty-eight gradients each 1 pH unit wide, starting with the pH 3.8–4.8 interval and ending with the pH 9.5–10.5 range, separated by 0.1-pH-unit increments can be found in Righetti (1990). For 1-pH-unit gradients between the limits pH 4.6–6.1 and pH 7.2–8.4 there are wide gaps in the pK of neighboring Immobilines and so more than two Immobilines need to be used to generate the desired pH_{min} and pH_{max} values. As an example, take the pH interval 4.6–5.6. There are no available Immobilines with pK values within this pH region, so the nearest species, pK 4.6 and 6.2, will act as both partial buffers and partial titrants. A third Immobiline is needed in each vessel, a true titrant that will bring the pH to the desired value. As titrant for the acidic solution, for pH_{min} we use pK 3.6 Immobiline and for pH_{max} we use pK 9.3 Immobiline. If a narrower pH gradient is needed, it can be derived from any of the fifty-eight 1-pH intervals by a simple linear interpolation of intermediate Immobiline molarities. This process can be repeated for any desired pH interval down to ranges as narrow as 0.1 pH unit.

4.3.1.4. Extended pH Gradients: General Rules for Their Generation and Optimization Linear pH gradients are obtained by arranging for an even buffering power throughout. The latter could be ensured only by ideal buffers spaced apart by $\Delta pK = 1$. In practice, there are only eight buffering Immobilines with some wider gaps in ΔpK; therefore other approaches must be used to solve this problem. Two methods are possible. In one approach (constant buffer concentration), the concentration of each buffer is kept constant throughout the span of the pH gradient and "holes" of buffering power are filled by increasing the amounts of the buffering species bordering the largest ΔpK. In the other approach (varying buffer concentration) the variation in concentration of different buffers along the width of the desired pH gradient results in a shift in each buffer's apparent p*K* together with the Δp*K* values evening out. The second approach is by far preferred, since it gives much higher flexibility in the computational approach. In a series of papers (Gianazza et al., 1984a; 1985a,b, 1989; Mosher et al., 1986; Righetti et al., 1986a, 1988c; Celentano et al., 1987, 1988; Tonani and Righetti 1991; Righetti and Tonani 1991), we have described a computer approach able to calculate and optimize any such pH interval up to the most extended one (which can cover a span of pH 2.5–11). Tables for these recipes can be found in the book by Righetti (1990) and in many of the above references. We prefer not to give here such recipes, since anyone can easily calculate any desired pH interval with the user-friendly computer program of Giaffreda et al. (1993). However, we will give here general guidelines for the use of such a program and optimization of various recipes:

(a) When calculating recipes up to 4 pH units, in the pH 4–9 interval, there is no need to use strong titrants. As most acidic and basic titrants, the p*K* 3.1 and 10.3 Immobilines can be used, respectively.

(b) When optimizing recipes >4 pH units (or close to the pH 3 or pH 11 extremes), strong titrants are preferably used; otherwise it will be quite difficult to obtain linear pH gradients, since weak titrants will act, per force, as buffering ions as well.

(c) When calculating recipes of 4 pH units, it is best to insert in the recipe all the eight weak buffering Immobilines. The computer program will automatically exclude the ones not needed for optimization.

(d) The program of Giaffreda et al. (1993) can calculate not only linear but also concave or convex exponential gradients (including sigmoidal ones). To limit consumption of Immobilines (at high concentration in the gel they could give rise to ominous reswelling and interact with the macromolecule separand via ion exchange mechanisms), limit the total Immobiline molarity (e.g., to only 15–20 mM) and the average buffering power (ß) (these two bits of information are specifically asked for when preparing any recipe). In particular, please note that recipes with an average ß value of only 2–3 mEq/L pH are quite adequate in IPGs. The separand macroions, even at concentration >10 mg/mL, rarely have ß values >1 μEg/L pH.

(e) When working at acidic and alkaline pH extremes, however, note that the average ß power of the recipe should be progressively higher, so as to counteract the ß value of bulk water. Additionally, at such pH extremes, the matrix acquires a net positive or negative charge, and this gives rise to strong EOF. To quench EOF, the washed and dried matrix should be reswollen against a gradient of viscous polymers (e.g., liquid linear polyacrylamide, hydroxyethyl cellulose) (Bianchi-Bosisio et al., 1986).

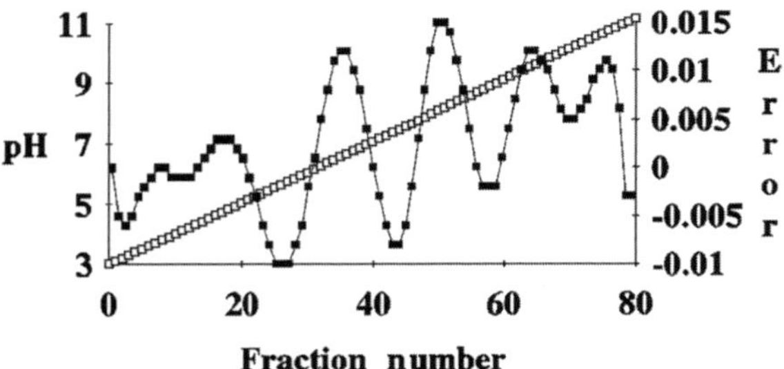

FIGURE 4.13. Optimization of linear IPG interval spanning pH 3–11 range. Notice the minute deviation from linearity. Open squares: pH gradient; solid squares: deviation. (From Giaffreda et al., 1993, by permission.)

It might be asked how good such wide gradients could be made. Figure 4.13 gives an example of a linear pH gradient (as calculated with the program of Giaffreda et al., 1993) covering the pH 3–11 interval: It can be appreciated that the gradient is indeed highly linear, the maximum deviation from the target function being at most ±0.01 pH units but often considerably smaller than that (see also Goerg et al., 1998, 1999, 2000).

4.3.1.5. Nonlinear, Extended pH Gradients Although originally most IPG formulations for extended pH intervals had been given only in terms of rigorously linear pH gradients, this might not be the optimal solution in some cases. The pH slope might need to be altered in pH regions that are overcrowded with proteins. This is particularly important in the general case involving the separation of proteins in a complex mixture, such as cell lysates, and thus is imperative when performing 2D maps. We have computed the statistical distribution of the pI values of water-soluble proteins and plotted them in the histogram of Figure 4.14. From the histogram, given the relative abundance of different species, it is clear that an optimally resolving pH gradient should have a gentler slope in the acidic portion and a steeper profile in the alkaline region. Such a course has been calculated by assigning to each 0.5-pH-unit interval in the pH 3.5–10 region a slope inversely proportional to the relative abundance of proteins in that interval. This generated the ideal curve (dotted line) in Figure 4.14. Here, too, it might be asked how good is the gradient compared with the target function. Although that gradient had been optimized with somewhat more primitive algorithms (it was in 1985), we have recalculated it with the program of Giaffreda et al. (1993); as shown in Figure 4.15, this nonlinear, concave gradient can be made with extreme accuracy, the error, at least in the acidic region of the pH interval, being close to zero. What is also important, in the present case, is the establishment of a new principle in IPG technology, namely that the pH gradient and the density gradient stabilizing it need not be colinear because the pH can be adjusted by localized flattening for increased resolution while leaving the density gradient unaltered. Although nonlinear gradients, of just about any shape, are much in vogue in chromatography, it should be emphasized that the gradients displayed

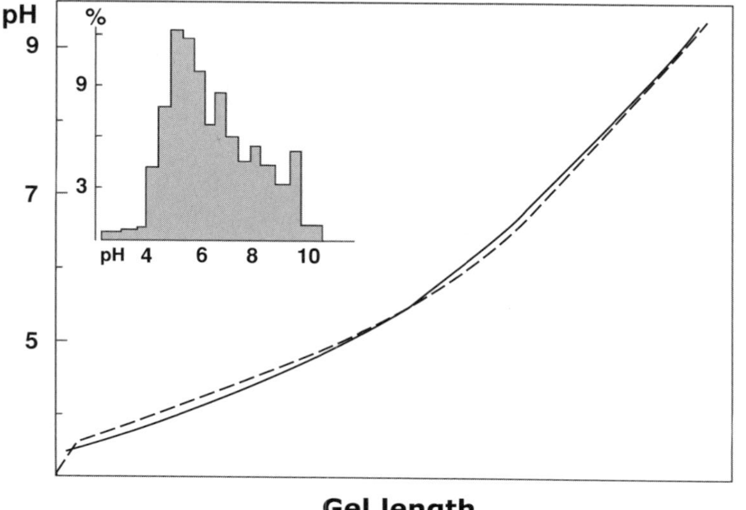

FIGURE 4.14. Nonlinear pH 4–10 gradient: "ideal" (solid line) and actual (dashed line) courses. The shape of the "ideal" profile was computed from data on statistical distribution of protein pI values. The relevant histogram is redrawn in the inset. (From Gianazza et al., 1985b, by permission).

here are somewhat unique in that they are created and optimized simply with the aid of a two-vessel gradient mixer by manipulating the composition (in Immobilines) of the two limiting solutions. On the contrary, in chromatography, this is achieved by drawing solutions of different compositions from several vessels and/or altering the pumping speed of said solutions to be then mixed in a mixing chamber. Conversely, our setup for casting Immobiline gels of any pH shape is gadget free! Though we have considered only examples of extended pH gradients, narrower pH intervals can be treated in the same fashion.

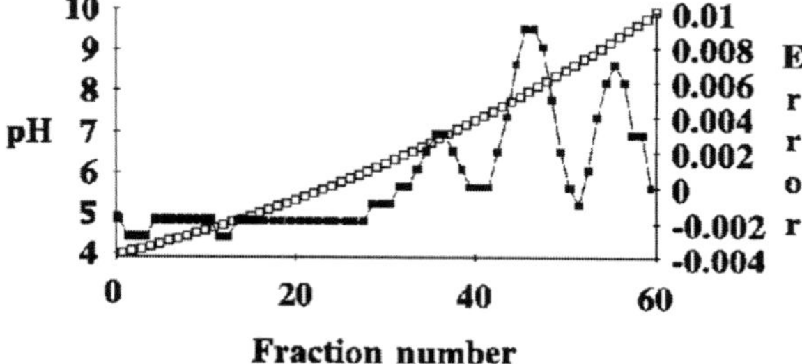

FIGURE 4.15. Calculation of nonlinear, concave IPG interval spanning pH 4–10 range. Notice the minute deviation from the target shape. Open squares: pH gradient; solid squares: deviation. (From Giaffreda et al., 1993, by permission.)

4.3.1.6. Extremely Alkaline pH Gradients We have recently optimized a recipe for producing an extremely alkaline immobilized pH gradient covering nonlinearly the pH 10–12 interval for separation of very alkaline proteins such as subtilisins and histones (Bossi et al., 1994a,b). Successful separations were obtained in $T = 6\%$, $C = 4\%$ polyacrylamide matrices, reswollen in 8 M urea, 1.5% Tween 20, 1.5% Nonidet P-40, and 0.5% Ampholine pH 9–11. Additionally, to quench the very high conductivity of the gel region on the cathodic side, the reswelling solution contained a 0–10% (anode-to-cathode) sorbitol gradient [or an equivalent 0–1% HEC (hydroxyethyl cellulose) gradient]. The best focusing was obtained by running the gel at 17°C, instead of the customary 10°C temperature. In the case of histones, all their major components had pI values between pH 11 and 12 and only minor components (possibly acetylated and phosphorylated forms) focused below pH 11. By summing all bands in Arg- and Lys-rich fractions, 8–10 major components and at least 12 minor zones were clearly resolved (Fig. 4.16). This same recipe could be used as a first-dimension run for a 2D separation of histones (Righetti et al., 1996a) and of ribosomal and nuclear proteins (Goerg et al., 1997).

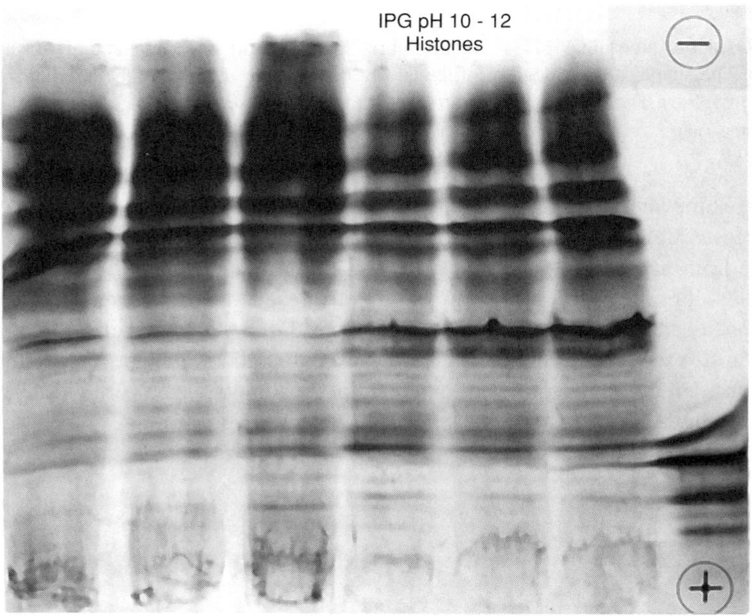

FIGURE 4.16. Focusing of histones in pH 10–12 interval. Gel: $T = 6\%$, $C = 4\%$ polyacrylamide matrix, containing an IPG 10–12 gradient, reswollen in 7 M urea, 1.5% Nonidet P-40, and 0.5% Ampholine pH 9–11. The gel was run at 10°C under a layer of light paraffin oil at 500 V for the first h followed by increasing voltage gradients, after sample penetration, up to 1300 V for a total of 4 h. The samples (2 mg/mL; 50 μL seeded) were loaded in plastic wells at the anodic gel surface. Staining with Coomassie brilliant blue R-250 in Cu^{2+}. Samples (from left to right): tracks 1–3: VIII-S histones; 4–6: II-AS histones; 7: cytochrome *c* (the main upper band has a pI of 10.6). (From Bossi et al., 1994, by permission.)

PROTOCOL 4.6. IPG FLOW SHEET

1. Assemble the gel mould (Protocol 4.7) and mark the polarity of the pH gradient on the back of the supporting plate.
2. Mix the required amounts of Immobilines. Fill to one-half of the final volume with distilled water.
3. Check the pH of the solution and adjust as required.
4. Add the correct volume of $T = 30\%$ acrylamide–bisacrylamide monomer, glycerol (0.2–0.3 mL/mL of the "dense" solution only), and TEMED and bring to final volume with distilled water.
5. For ranges removed from neutrality, titrate to about pH 7.5 using Tris base for acidic solution and acetic acid for alkaline solutions.
6. Transfer the denser solution to the mixing chamber and the lighter solution to the other reservoir of the gradient mixer. Center the mixer on a magnetic stirrer and check for the absence of air bubbles in the connecting duct.
7. Add the required amount of ammonium persulfate to the solutions.
8. Allow the gradient to pour into the mold from the gradient mixer.
9. Allow the gel to polymerize for 1 h at 50°C.
10. Disassemble the mold and weigh the gel.
10. Wash the gel for 1 h three times (20 min each) with 200 mL of distilled water with gentle shaking.
11. Reduce the gel back to its original weight using a nonheating fan.
12. Transfer the gel to the electrophoresis chamber (at 10°C) and apply the electrodic strips.
13. Load the samples and start the run.
14. After the electrophoresis, stain the gel to detect the separated proteins.

4.3.2. IPG Methodology

The overall procedure is outlined in Protocol 4.6. Note that the basic equipment required is the same as for conventional CA–IEF gels. Thus the reader should consult Sections 4.2.2.1 and 4.2.2.2. In addition, as we essentially use the same polyacrylamide matrix, the reader is referred to Section 4.2.3 for a description of its general properties.

4.3.2.1. Casting an Immobiline Gel When preparing for an IPG experiment, two pieces of information are required: the total liquid volume needed to fill the gel cassette and the required pH interval. Once the first is known, this volume is divided into two halves: One half is titrated to one extreme of the pH interval, the other to the opposite extreme. As the analytical cassette usually has a thickness of 0.5 mm and for the standard 12×25-cm size (see Fig. 4.4) contains 15 mL of liquid to be gelled, in principle, two solutions, each of 7.5 mL, should be prepared. However, because the volume of some Immobilines to be added to 7.5 mL might sometimes be rather small (i.e., $<50\ \mu L$), we prefer to prepare a double volume, which will be enough for casting two gel slabs. The Immobiline solutions (mostly the basic ones) tend to leave droplets on the plastic disposable tips of micropipettes. For accurate dispensing, therefore, we suggest rinsing the tips once or twice with distilled water after each measurement. The polymerization cassette is filled with the aid of a two-vessel gradient mixer, and thus the liquid elements which fill the vertically standing cassette have to be stabilized against remixing by a density gradient. These two solutions are called "acidic dense" and "basic light" solutions. This choice is, however, a purely conventional one and can be reversed provided one marks the bottom of the mold as the cathodic side. To understand the sequence of steps needed, the reader is referred to Protocols 4.6–4.8 and Figure 4.17 for the final gel assembly.

(i) Preparation of Gel Mold. Figure 4.17 gives the final assembly for the cassette and gradient mixer. The cassette is assembled as described in Protocol 4.7. The capillary tubing conveying the solution from the mixer into the cassette is inserted in the central V-shaped indentation. As for the gradient mixer, one chamber contains a magnetic stirrer, while in the reservoir is inserted a plastic cylinder having the same volume that is held by a trapezoidal rod. The latter, in reality, is a "compensating cone" needed to raise the liquid level to such an extent that the two solutions (in the mixing chamber and in the reservoir) will be hydrostatically equilibrated. In addition, this plastic rod can also be utilized for manually stirring the reservoir after addition of TEMED and persulfate.

(ii) Polymerization of Linear pH Gradient. It is preferable to use "soft" gels, that is, with a low $\%T$. Originally, all recipes were given for $T = 5\%$ matrices, but today we prefer gels down to as low as $T = 3\%$ (Candiano et al., 2002; Bruschi et al., 2003). These soft gels can be easily dried without cracking and allow better entry of larger proteins. In addition, the local ionic strength along the polymer coil is increased, and this permits sharper protein bands due to increased solubility at the pI. A linear pH gradient is generated by mixing equal volumes of the two starting solutions in a gradient mixer. It is a must for any gel formulation removed from neutrality (pH 6.5–7.5) to titrate the two solutions to neutral pH, so as to ensure reproducible polymerization conditions and avoid hydrolysis of the five alkaline buffering Immobilines. If the pH interval used is acidic, add Tris; if it is basic, add formic acid. We recommend that a minimum of 15 mL of each solution (enough for two gels) is prepared and that the volumes of Immobiline needed are measured with a well-calibrated microsyringe to ensure high accuracy. Prepare the acidic, dense solution and the basic, light solution for the pH gradient as described in Protocol 4.6. If the same gradient is to be prepared repeatedly, the buffering and nonbuffering Immobiline and water mixtures can be prepared as stock solutions and stored according to the recommendations for Immobiline. Prepared gel solutions must not be stored. However, gels with a pH less than 8 can be stored in a humidity chamber for up to 1 week after polymerization. An example of a preparation of a linear pH gradient is given in Protocol 4.8.

PROTOCOL 4.7. ASSEMBLING MOLD

EQUIPMENT

- Gel mold (Pharmacia)
- Dymo tape
- Repel-silane (ready to be used from Sigma or as described in Protocol 4.1)
- Gel Bond PAG film (Pharmacia)

METHOD

1. Wash the glass plate bearing the U-frame with detergent and rinse with distilled water.
2. Dry with paper tissue.
3. To mold sample application slots in the gel, apply suitably sized pieces of Dymo tape to the glass plate with the U-frame; a 5×3-mm slot can be used for sample volumes between 5 and 20 μL (this step is only necessary when preparing a new mold or rearranging an old one; see Fig. 4.5*a*). To prevent the gel from sticking to the glass plates with U-frame and slot former, coat them with Repel-silane according to Table 4.2. Make sure that no dust or fragments of gel from previous experiments remain on the surface of the gasket, since this can cause the mold to leak.
4. Use a drop of water on the Gel Bond PAG film to determine the hydrophilic side. Apply a few drops of water to the plain glass plate and carefully lay the sheet of Gel Bond PAG film on top with the hydrophobic side down (see Fig. 4.5*b*). Avoid touching the surface of the film with fingers. Allow the film to protrude 1 mm over one of the long sides of the plate as a support for the tubing from the gradient mixer when filling the cassette with gel solution (but only if using a cover plate without any V-indentations). Roll the film flat to remove air bubbles and to ensure good contact with the glass plate.
5. Clamp the glass plates together with the Gel Bond PAG film and slot former on the inside by means of clamps placed all along the U-frame, opposite to the protruding film. To avoid leakage, the clamps must be positioned so that the maximum possible pressure is applied (see Fig. 4.4*c*).

4.3.2.2. Reswelling Dry Immobiline Gels Precast, dried Immobiline gels encompassing a number of ranges are now available from Amersham Pharmacia Biotech and Bio-Rad. They all contain $T = 4\%$ and they span the following pH narrow ranges: 3.5–4.5, 4.0–5.0 (e.g., for α_1-antitrypsin analysis), 4.5–5.5 (e.g., for Gc screening), 5.0–6.0 (e.g., for transferrin analysis), and 5.5–6.7 (e.g., for phosphoglucomutase screening). In addition, there are a number of wide pH ranges: 4–7L, 6–9, 6–11,

PROTOCOL 4.8. POLYMERIZATION OF LINEAR pH GRADIENT GEL

EQUIPMENT AND REAGENTS

- Gradient maker (the Pharmacia-Hoefer catalogs offer a large choice of gradient mixers, in this case the model having vessels of 15 or 30 mL volume is very suitable)
- Magnetic stirrer and stirring bar
- Gel mold with V-indentations (see text and Fig. 4.4*a*)
- Oven at 50°C
- Basic light and acidic dense gel mixtures (see Section 4.3.2.1); 7.5 mL of each
- Nonheating fan

METHOD

1. Check that the valve in the gradient mixer and the clamp on the outlet tubing are both closed.
2. Transfer 7.5 mL of the basic, light solution to the reservoir chamber.
3. Slowly open the valve just enough to fill the connecting channel with the solution and quickly close it again. Then transfer 7.5 mL of the acidic dense solution to the mixing chamber.
4. Place the prepared mold upright on a leveled surface. The optimum flow rate is obtained when the outlet of the gradient mixer is 5 cm above the top of the mold. Open the clamp of the outlet tubing, fill the tubing halfway with the dense solution, and close the clamp again.
5. Switch on the stirrer and set to a speed of about 500 rpm.
6. Add the catalysts to each chamber.
7. Insert the free end of the tubing between the glass plates of the mold at the central V-indentation (Fig. 4.17).
8. Open the clamp on the outlet tubing; then immediately open the valve between the dense and light solutions so that the gradient solution starts to flow down into the mold by gravity. Make sure that the levels of liquid in the two chambers fall at the same rate. The mold will be filled within 5 min. To assist the mold to fill uniformly across its width, the tubing from the mixer may be substituted with a two- or three-way outlet assembled from small glass or plastic connectors (e.g., spare parts of chromatographic equipment) and butterfly needles.
9. When the gradient mixer is empty, carefully remove the tubing from the mold. After leaving the cassette to rest for 5 min, place it on a leveled surface in

an oven at 50°C. Polymerization is allowed to continue for 1 h. Meanwhile, wash and dry the mixer and tubing.

10. When polymerization is complete, remove the clamps and carefully take the mold apart. Start by removing the glass plate from the supporting foil. Then hold the remaining part so that the glass surface is on top and the supporting foil underneath. Gently peel the gel away from the slot former, taking special care not to tear the gel around the slots.
11. Weigh the gel and then place it in 300 mL of distilled water for 1 h to wash out any remaining ammonium persulfate, TEMED, and unreacted monomers and Immobilines. Change the water three times (changes are every 20 min).
12. After washing the gel, carefully remove any excess water from the surface with a moist paper tissue. To remove the water absorbed by the gel during the washing step, leave it at room temperature until the weight has returned to within 5% of the original weight. To shorten the drying time, use a nonheating fan placed at about 50 cm from the gel to increase the rate of evaporation. Check the weight of the gel after 5 min and, from this, estimate the total drying time. The drying step is essential, as a gel containing too much water will "sweat" during the electrofocusing run and droplets of water will form on the surface. However, if the gel dries too much, the value of $\%T$ will be increased, resulting in longer focusing times and a greater sieving effect.

3–10L, and 3–10NL (where L = linear, NL = nonlinear). Some of them are available in 7, 11, 18, and even 24 cm in length, whereas all the narrow ranges are cast only as long (18-cm) strips. All of them are 3 mm wide and, when reswollen, 0.5 mm thick (gel layer). Precast, dried IPG gels in alkaline narrow ranges should be handled with care because at high pH the hydrolysis of both the gel matrix and the Immobiline chemicals bound to it is much more pronounced.

It has been found that the diffusion of water through Immobiline gels does not follow a simple Fick's law of passive transport from high (the water phase) to zero (the dried gel phase) concentration regions, but it is an active phenomenon: even under isoionic conditions, acidic ranges cause swelling 4–5 times faster than alkaline ones (Gelfi and Righetti, 1984). Given these findings, it is preferable to reswell dried Immobiline gels in a cassette similar to the one for casting the IPG gel. Figure 4.18 shows the reswelling system: The dried gel is inserted in the cassette, which is clamped and allowed to stand on the short side. The reswelling solution is gently injected into the chamber via a small hole in the lower right side using a cannula and a syringe until the cassette is completely filled. As the system is volume controlled, it can be left to reswell overnight, if needed. Gel drying and reswelling are preferred procedure when an IPG gel containing additives is needed. In this case it is always best to cast an "empty" gel, wash it, dry it, and then reconstitute it in presence of the desired additive (e.g., urea, alkyl ureas, detergents, carrier ampholytes, and mixtures thereof).

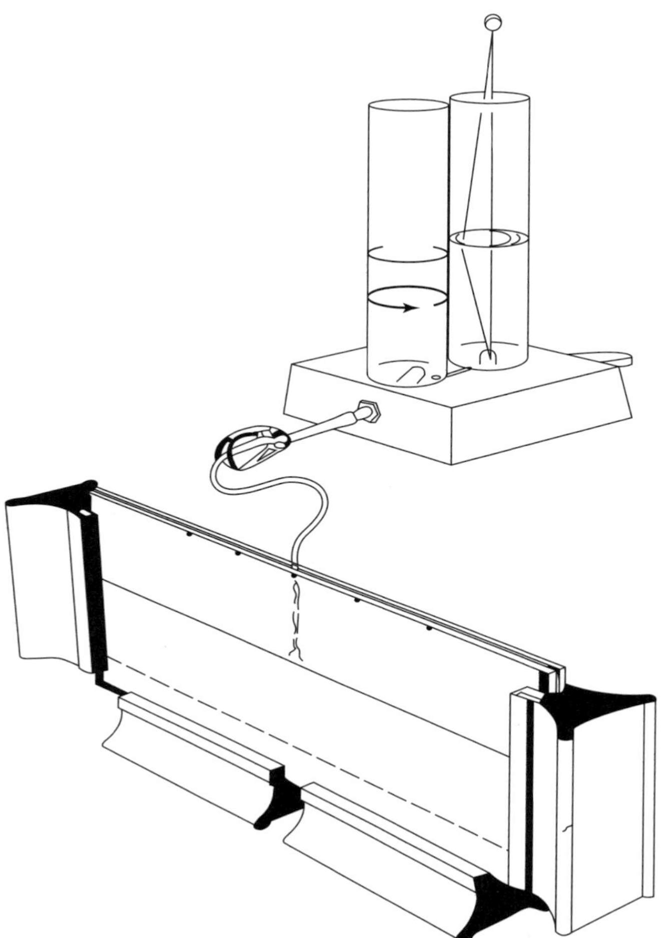

FIGURE 4.17. Setup for casting IPG gel. A linear pH gradient is generated by mixing equal volumes of a dense and light solution, titrated to the extremes of the desired pH interval. Note the "compensating" rod in the reservoir, used as a stirrer after addition of catalysts and for hydrostatically equilibrating the two solutions. Insertion of the capillary conveying the solution from the mixer to the cassette is greatly facilitated by using modern cover plates bearing 3 V-shaped indentations. (Courtesy of LKB Produkter AB.)

4.3.2.3. Electrophoresis A common electrophoresis protocol consists of an initial voltage setting of 500 V for 1–2 h followed by an overnight run at 2–2500 V. Ultranarrow gradients are further subjected to a couple of hours at 5000 V, or better at about 1000 V/cm across the region containing the bands of interest.

4.3.2.4. Staining and pH Measurements Immobilized pH gradients tend to bind strongly to dyes, so the gels are better stained for a relatively short time

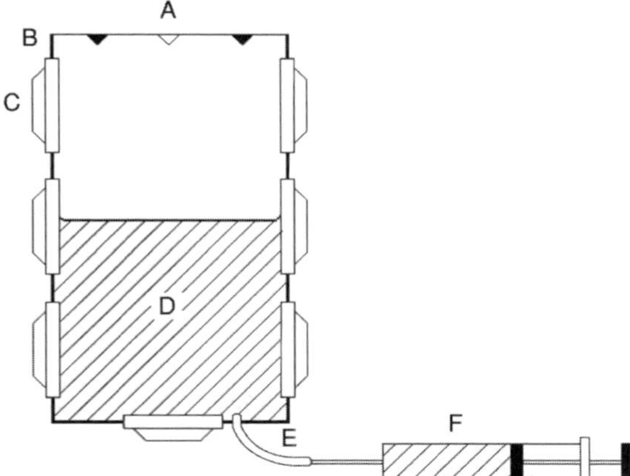

FIGURE 4.18. Reswelling cassette for dry IPG gels. The dried IPG gel (on its plastic backing) is inserted in the cassette, which is then gently filled with any desired reswelling solution via a bottom hole with the help of tubing and a syringe. A: gel cassette; B: gasket; C: clamps; D: gel solution; E: tubing; F: syringe.

(30–60 min) with a stain of medium intensity. For silver staining, a novel recipe optimized for IPG gels has been published (Rabilloud et al., 1992). A new fluorescent staining protocol exploiting the dye SYPRO ruby was reported in Section 4.2.5.6 and is valid for both conventional IEF and IPGs. It appears to have the highest possible sensitivity so far reported for any staining procedure.

Accurate pH measurements are virtually impossible by equilibration between a gel slice and excess water and not very reliable with a contact electrode, although in mixed Immobiline–Ampholine gels this is feasible due to elution of the soluble CA buffers into the supernatant (Righetti et al., 1986b; Gelfi et al., 1986; Rovida et al., 1986). One can preferably either refer to the banding pattern of a set of marker proteins or elute CAs from a mixed-bed gel, when applicable.

4.3.2.5. Storage of Immobiline Chemicals There are two major problems with the Immobiline chemicals, especially with the alkaline ones: hydrolysis and spontaneous autopolymerization. Hydrolysis is quite a nuisance because then only acrylic acid is incorporated into the IPG matrix, with a strong acidification of the calculated pH gradient. Hydrolysis is an autocatalyzed process for the basic Immobilines, since it is pH dependent. For the p*K* 8.5 and 9.3 species, such a cleavage reaction on the amido bond can occur, even in the frozen state, at a rate of about 10% per year (Astrua-Testori et al., 1986). Autopolymerization is also quite deleterious for the IPG technique. Again, this reaction occurs particularly with alkaline Immobilines and is purely autocatalytic, as it is greatly accelerated by deprotonated amino groups. Oligomers and *n-mers* are formed which stay in solution and can even be incorporated into the IPG gel. These products of autopolymerization, when added to proteins in solution, are able to bridge them via two unlike binding surfaces. A lattice is formed and the proteins (especially larger ones, such as ferritin, α_2-macroglobulin, a thyroglobulin) are precipitated out of solution. This precipitation effect is quite strong and begins even at the level of short oligomers (>decamer) (Rabilloud et al., 1987; Righetti et al., 1987 a,b). These problems of autopolymerization and oligomer formation, although discovered only in 1987, originated in 1986 when a few scientists lamented loss of large serum proteins from 2D maps (Hochstrasser et al., 1986). It probably originated from the "Chernobyl cloud" since gamma rays induce polymerization of even the most stubborn double bonds. Today, in fact, this problem has rarely surfaced (Esteve-Romero et al., 1996). One clear fact has emerged: The species most prone to autopolymerization is the p*K* 7.0 Immobiline and, in present-day Immobilines, we could detect, by MS, only traces (<0.5%) of barely dimers.

These problems with the basic Immobilines have been solved by dissolving them in anhydrous *n*-propanol (containing a maximum of 60 ppm water), which stabilizes them against both hydrolysis and autopolymerization for a virtually unlimited period of time (less than 1% degradation per year even when stored at +4°C). Thus, present-day alkaline Immobiline bottles are now supplied as 0.2 M solutions in *n*-propanol. The acidic Immobilines, being much more stable, are available as water solutions laced with 10 ppm of an inhibitor (Gåveby et al., 1988).

4.3.2.6. Mixed-Bed CA–IPG Gels In CA–IPG gels the primary, immobilized pH gradient is admixed with a secondary, soluble carrier ampholyte–driven pH gradient. It sounds strange that, given the problems connected with the CA buffers (discontinuities along the electrophoretic path, pH gradient decay, etc.), which the IPG technique was supposed to solve, one should resurrect this past methodology. In fact, when working with membrane proteins (Rimpilainen and Righetti, 1985) and microvillar hydrolases partly embedded in biological membranes (Sinha and Righetti, 1986), we found that the addition of CAs to the sample and IPG gel would increase protein solubility, possibly by forming mixed micelles with the detergent used for membrane solubilization or by directly complexing with the protein itself. It is a fact that, in the absence of CAs, these same proteins essentially fail to enter the gel and mostly precipitate or give elongated smears around the application site (in general, cathodic sample loading). More recently, it was found that, on a relative hydrophobicity scale, the five basic Immobilines (p*K* 6.2, 7.0, 8.5, 9.3, and 10.3) are decidedly more hydrophobic than their acidic counterparts (p*K* 3.1, 3.6, and 4.6). Upon incorporation in the gel matrix, the phenomenon becomes cooperative and could lead to the formation of hydrophobic patches on the surface of such a hydrophilic gel as polyacrylamide. As the strength of a hydrophobic interaction is directly proportional to the product of the cavity area times its surface tension, it is clear that experimental conditions which lead to a decrement of molecular contact area axiomatically weaken such interactions. Thus CAs might quench the direct hydrophobic protein–IPG matrix interaction, effectively detaching the protein from the surrounding polymer coils and allowing good focusing into sharp bands. For this to happen, the CA-shielding species should already be impregnated in the Immobiline gel and be present in the sample solution as well. In other words, CAs can only prevent the phenomenon and cannot cure it a posteriori. It has been additionally found that addition of CAs to the sample protects it from strongly acidic and alkaline boundaries originating from the presence of salts in the sample zone (especially "strong" salts such as NaCl, phosphates, etc.) (Righetti et al., 1988).

A note of caution should be mentioned concerning the indiscriminate use of the CA–IPG technique: At high CA levels (>1%) and high voltages (>100 V/cm) these gels start exuding water with dissolved carrier ampholytes, with severe risks of short circuits, sparks, and burning on the gel surface. The phenomenon is minimized by chaotropes (e.g., 8 M urea), by polyols (e.g., 20% sucrose), and by lowering the CA molarity in the gel (Astrua-Testori and Righetti, 1987). As an answer to the basic question of when and how much CAs to add, we suggest the following guidelines:

(a) If your sample focuses well as such, ignore the mixed-bed technique (which presumably will be mostly needed with hydrophobic proteins and in alkaline pH ranges).
(b) Add only the minimum amount of CAs (in general, ~0.5% to 1%) needed for avoiding sample precipitation in the pocket and for producing sharply focused bands.

4.3.3. Troubleshooting

One could cover pages with a description of all the troubles and possible remedies in any methodology. However, here we will list only the major ones encountered with the IPG technique and we refer the readers to Table 4.11 for all the possible causes and remedies suggested. We highlight the following points:

(a) When the gel is gluey and there is poor incorporation of Immobilines, the biggest offenders are generally the catalysts (e.g., too old persulfate, crystals wet due to adsorbed humidity; wrong amounts of catalysts added to the gel mix); check in addition the polymerization temperature and the pH of the gelling solutions.
(b) Bear in mind the last point in Table 4.11: if you have done everything right and still you do not see any focused protein, you might have simply positioned the platinum wires on the gel with the wrong polarity. Unlike conventional IEF gels, in IPGs the anode has to be positioned at the acidic (or less alkaline) gel extremity while the cathode has to be placed at the alkaline (or less acidic) gel end.

4.3.4. Some Analytical Results with IPGs

We will limit this section to some examples of separations in ultranarrow-pH intervals, where the tremendous resolving power (ΔpI) of IPGs can be fully appreciated. The

TABLE 4.11. Troubleshooting Guide for IPGs

Symptom	Cause	Remedy
Drifting of pH during measurement of basic starting solution	Inaccuracy of glass pH electrodes (alkaline error)	Consult information supplied by electrode manufacturer
Leaking mould	Dust or gel fragment on gasket	Carefully clean gel plate and gasket
Gel consistency is not firm, gel does not hold its shape after removal from mold	Inefficient polymerization	Prepare fresh ammonium persulfate and check that recommended polymerization conditions are being used
Plateau visible in anodic and/or cathodic section of gel during electrofocusing, no focusing proteins seen in that part of gel	High concentration of salts in system	Check that correct amounts of ammonium persulfate and TEMED are used
Overheating of gel near sample application when beginning electrofocusing	High salt content in sample	Reduce salt concentration by dialysis or gel filtration
Nonlinear pH gradient	Backflow in gradient mixer	Find and mark optimal position for gradient mixer on stirrer
Refractive line at pH 6.2 in gel after focusing	Unincorporated polymers	Wash gel in 2 L distilled water; change water once and wash overnight
Curved protein zones in that portion of gel which was at top of mold during polymerization	Too rapid polymerization	Decrease rate of polymerization by putting mold at −20°C for 10 min before filling it with gel solution or place solutions at 4°C for 15 min before casting gel
Uneven protein distribution across a zone	Slot or sample application not perpendicular to running direction	Place slot or sample application pieces perfectly perpendicular to running direction
Diffuse zones with unstained spots or drops of water on gel surface during electrofocusing	Incomplete drying of gel after washing step	Dry gel until it is within 5% of its original weight
No zones detected	Gel focused with wrong polarity	Mark polarity on gel when removing it from mold

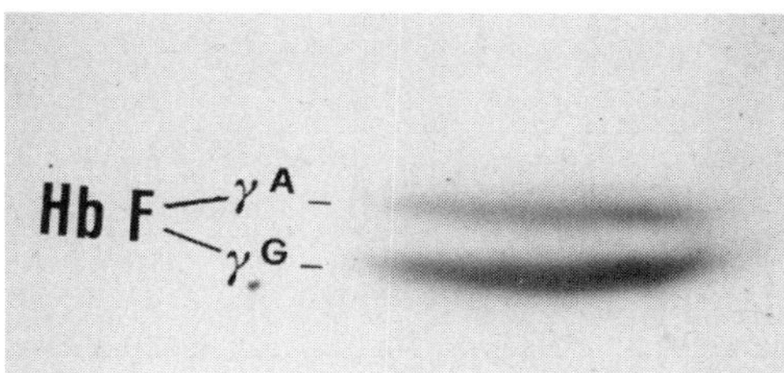

FIGURE 4.19. Focusing of umbilical cord lysates from an individual heterozygous for fetal hemoglobin (Hb F) Sardinia (only the Hb F bands are shown, since the two other major components of cord blood, i.e., Hb A and Hb F_{ac}, are lost in very shallow pH gradients). In an IPG gel spanning 0.1 pH unit, over a standard 10 cm gel length, two bands are resolved, identified as the Gly vs. Ala gene products. The resolved Aγ/Gγ bands are in a 20 : 80 ratio, as theoretically predicted from gene expression. Their identity was established by eluting the two zones and fingerprinting the γ-chains. (From Cossu and Righetti, 1987, by permission.)

ΔpI is the difference, in surface charge, in pI units, between two barely resolved protein species. Rilbe (1973) has defined ΔpI as

$$\Delta\text{pI} = 3.07\sqrt{\frac{D(d\text{pH}/dx)}{E(-du/d\text{pH})}}$$

where D and du/dpH are the diffusion coefficient and titration curve of proteins, E is the voltage gradient applied, and dpH/dx is the slope of the pH gradient over the separation distance. Experimental conditions that minimize ΔpI will maximize the resolving power. Ideally, this can be achieved by simultaneously increasing E and decreasing dpH/dx, an operation for which IPGs seem well suited. As stated previously, with conventional IEF it is very difficult to engineer pH gradients that are narrower than 1 pH unit. One can push the ΔpI, in IPGs, to the limit of 0.001 pH unit, the corresponding limit in CA–IEF being only 0.01 pH unit. We began to investigate the possibility of resolving neutral mutants, which carry a point mutation involving amino acids with nonionizable side chains and are, in fact, described as "electrophoretically silent" because they cannot be distinguished by conventional electrophoretic techniques. The results were quite exciting. As shown in Figure 4.19, fetal hemoglobin (Hb F), which is an envelope of two components, called Aγ and Gγ, carrying a Gly → Ala mutation in γ-136, could be resolved into two main bands in a very shallow pH gradient spanning 0.1 pH unit (over a standard 10-cm migration length). These two tetramers, normal components during fetal life, are found in approximately an 80 : 20 ratio, as they should, according to gene expression. The resolution here is close to the practical limit of ΔpI = 0.001 (Cossu and Righetti, 1987).

Another interesting example is given in Figure 4.20. Here a preparation of bovine tubulin is resolved into at least 21 isoforms in the IPG pH 4–7 interval (Williams et al., 1999). The advantage of the IPG technique is that, unlike in CA–IEF, true steady-state patterns are obtained, rendering the method highly reproducible. The very low

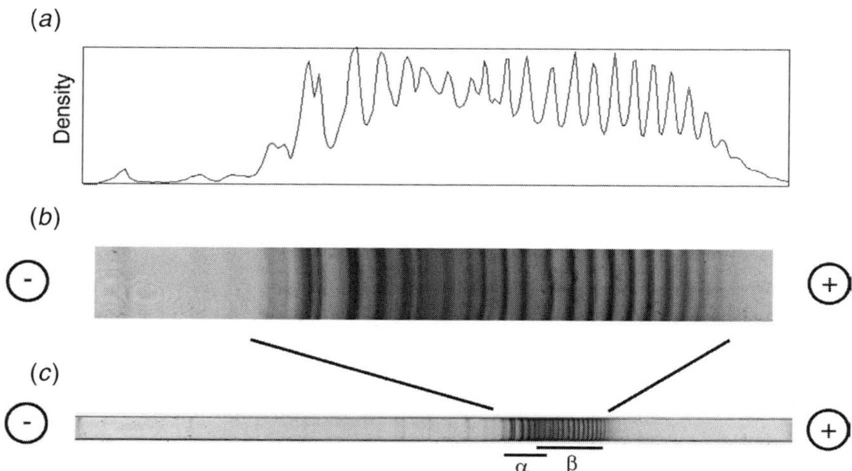

FIGURE 4.20. Isoelectric focusing of bovine tubulin in an IPG pH 4–7 gradient. (*a*): IEF of 25 μg of tubulin. Bottom: entire IPG strip. Middle: enlarged detail of principal protein-containing region. Top: scan of this region. At least 21 bands are visible. Zones in which α- and β-tubulin are known to focus are indicated in (c). The 11-cm-long IPG strip was rehydrated for 14 h at 22°C. Focusing was for 1 h at 500 V, 1 h at 1000 V, and 3 h at 5000 V, all at 20°C. Current was limited to 50 μA per strip. The positions of positive and negative electrodes are indicated. (From Williams et al., 1999, by permission.)

conductivity of the IPG strips, in addition, allows much higher voltage gradients to be delivered along the focusing axis, thus producing much sharper protein zones and ultimately leading to much higher resolution.

Another interesting aspect of IPGs which renders them so appealing as a first dimension in 2D mapping is their unique load ability, even when used as analytical gels (typically 0.5 mm thick). A case in point is shown in Figure 4.21, which displays the focusing pattern of a commercial preparation of conalbumin in the pH 4.5–6.5 interval (Ek et al., 1983). It is seen that, as the protein load is increased from 20 up to 500 μg, there is no disruption of the focusing pattern, except from some minor overloading effect on the major component: The bands are still very sharp and highly resolved, so that minor components not visible at lower loads can be appreciated at higher loads. It is this unique property of IPGs that renders them so attractive for the IEF first dimension in 2D maps. By exploiting this unique load ability, some authors have even reported loading up to 10 mg total protein in the IPG strips, a truly impressive amount. This load capability is strongly favored by adopting dilute gels ($T = 4\%$ and lower), since the gel matrix stretches and swells considerably in thickness in the highly packed, focused protein zone, thus permitting such high loads (Gelfi and Righetti, 1983; Righetti and Gelfi, 1984).

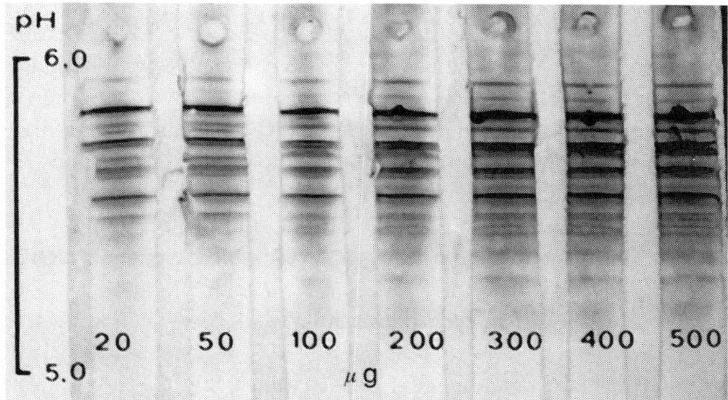

FIGURE 4.21. Isoelectric focusing of conalbumin in IPG pH 4.5–6.5 interval. Gel: 11-cm-long, 0.5-mm-thick strip. Focusing: overnight, 2000 V, 10°C. The samples were loaded in holes bored in the gel matrix at the cathodic gel extremity. Staining with Coomassie brilliant blue. (From Ek et al., 1983, by permission.)

4.4. CAPILLARY ISOELECTRIC FOCUSING

Although capillary isoelectric focusing (cIEF) still is beleaguered by some problems (notably how to cohabit with the ever-present hazard of EOF), the technique is gaining momentum and is becoming quite popular, especially in the analysis of rDNA products and heterogeneity due to differential glycosylation patterns and "protein aging," that is, Asn and Gln deamidation in vitro. A spin-off of cIEF is zone electrophoresis in zwitterionic, isoelectric buffers, a technique that exploits all the basic concepts of IEF and offers unrivaled resolution due to fast analysis in high voltage gradients.

4.4.1. General Considerations

In addition to the reviews suggested at the beginning of this chapter (Righetti and Gelfi, 1994; Righetti et al., 1996b; Righetti and Bossi, 1998), we recommend, as further readings, the following ones: Mazzeo and Krull (1993), Righetti and Chiari (1993), Hjertèn (1992), Pritchett (1996), Kilár (1994, 2003), Wehr et al. (1996), Rodriguez-Diaz et al. (1997a,b), Righetti et al. (1997), Taverna et al. (1998), Shimura (2002), Jin et al. (2001), Shen and Smith (2002), and Jensen et al. (2000).

Capillary electrophoresis offers some unique advantages over conventional gel slab techniques: The amount of sample required is truly minute (a few microliters at the injection port, but only a few nanoliters in the moving zone); the analysis time is in general very short (often just a few minutes) due to the very high voltages applicable; and analyte detection is online and is coupled to a fully instrumental approach (with automatic storage of electropherograms on a magnetic support).

A principal difference between IEF in a gel and in a capillary is that, in the latter, mobilization of the focused proteins past the detector has to be carried out if an

TABLE 4.12. cIEF Separation of Proteins in Coated Capillaries

1. Ampholytes: commercially available ampholyte solutions (e.g., Pharmalyte, Biolyte, Servalyte) in desired pH range
2. Sample: 1–2 mg/mL Protein solution, mixed with 3–4% carrier ampholytes (final concentration) in desired pH range; sample solution should be desalted or equilibrated in weak buffer-counterion system
3. Capillary: 25–30 cm long, 50–75 μm ID, coated, filled with sample–CA solution
4. Anolyte: 10 mM phosphoric acid (or any other suitable weak acid, such as acetic, formic acids) or low-pI zwitterions
5. Catholyte: 20 mM NaOH (or any other suitable weak base, such as Tris, ethanolamine) or high-pI zwitterions
6. Focusing: 8–10 kV, constant voltage, for 5–10 min
7. Mobilizer: 50–80 mM NaCl, or 20 mM NaOH (cathodic) or 20 mM NaOH (anodic)
8. Mobilization: 5–6 kV, constant voltage
9. Detection: 280 nm, near mobilizer (or any appropriate visible wavelength for colored proteins)
10. Washing: after each run, with 1% neutral detergent (such as Nonidet P-40, Triton X-100) followed by distilled water

online imaging detection system is not being used. Mainly three techniques are used: chemical and hydrodynamic flow mobilization (in coated capillaries) and mobilization utilizing EOF (in uncoated or partially coated capillaries). We do not particularly encourage the last approach (Thormann et al., 1992), since the transit times of the focused zones change severely from run to run; thus we will only describe the cIEF approach in well-coated capillaries, where EOF is completely suppressed.

4.4.2. cIEF Methodology

Table 4.12 gives a typical methodology for cIEF in a coated capillary. Since many procedures for silanol deactivation have been reported (Chiari et al., 1996) and good coating practice is very difficult to achieve in a general biochemical laboratory, we recommend buying precoated capillaries (e.g., from Beckman and Bio-Rad).

The following general guidelines are additionally suggested:

(a) All solutions should be degassed.

(b) The ionic strength of the sample may dramatically influence the length of the focusing step and also completely ruin the separation; therefore, sample desalting prior to focusing or a low buffer concentration (ideally made of weak buffering ion and weak counterion) is preferable. Easy sample desalting can be achieved via centrifugation through Centricon membranes (Amicon).

(c) The hydrolytic stability of the coating is poor at alkaline pH; therefore mobilization with NaOH may destroy such coatings after a few runs.

(d) Ideally, nonbuffering ions should be excluded in all compartments for cIEF. This means that in the electrodic reservoirs one should use weak acids (at the anode) and weak bases (at the cathode) instead of phosphoric acid and NAOH, as adopted today by most cIEF users. This includes the use of zwitterions (e.g., Asp, pI = 2.77, or Glu, pI = 3.25, at the anode and Lys, pI = 9.74, or Arg, pI = 10.76, at the cathode).

(e) When eluting the focused bands past the detector, we have found that resolution is maintained better by a combination of salt elution (e.g., adding 20 mM NaCl or Na phosphate to the appropriate compartment) and a siphoning effect, obtained by having a higher liquid level in one compartment and a lower level in the other. The volumes to use will depend on the apparatus. For the BioFocus 2000 apparatus from Bio-Rad, the volumes used are 650 and 450 μL, respectively.

4.4.2.1. Increasing Resolution by Altering Slope of pH Gradient Methods have not yet been devised for casting IPGs in a capillary format, and so it is difficult in cIEF to achieve the resolution typical of IPGs, namely $\Delta pI = 0.001$. Nevertheless, one efficient way for incrementing the resolving power of CA–IEF is to add "spacers" to a regular 2-pH-unit interval. In fact, Cossu et al. (1982) reported a modified gel slab spacer IEF technique for screening for β-thalassemia by measuring the relative proportions of three major Hbs present in cord blood of newborns: adult (A), fetal (F), and acetylated fetal (F_{ac}). We have applied this to cIEF of cord blood. Whereas a standard cIEF run in a pH 6–8 gradient (the best for focusing any Hb variant) could hardly separate the fetal from the adult Hb (F/A), a simple addition of 100 mM β-Ala (Fig. 4.22*a*) brought about excellent resolution among the three species, thus allowing proper quantitation and correct diagnosis of thalassemic conditions (Conti et al., 1995).

Another case in point is the separation of Hb from its glycated form, Hb A_{1c}. Determination of the percentage of Hb A_{1c} in adult blood is important for evaluation of some pathological alterations of the glycosidic metabolic pathways. In particular, the percentage of Hb A_{1c} is routinely used for assessing the degree of diabetes by providing an integrated measurement of blood glucose according to the red blood cell life span. Hemoglobin A and A_{1c} have minute pI differences, of the order of 0.01 and less. In conventional IEF in gel slabs, Cossu et al. (1984) again solved this problem by resorting to the same combination of 0.2 M β-alanine and 0.2 M 6-amino caproic acid, which provided baseline resolution between Hb A and Hb A_{1c}. Conti et al. (1996) have thus reapplied this pH gradient manipulation procedure to cIEF with very good results: As shown in Figure 4.22*b*, an artificial mixture of 65% A and 35% A_{1c}, as purified by IPGs, gave the expected pattern with full baseline resolution.

Another interesting application of cIEF in combination with CZE is the study of protein–protein interaction. This is usually done by immobilizing in the capillary one of the reactants and letting the interacting species to move through the immobilized zone by electrophoretic transport. But there is another, efficient and simpler procedure: One of the two reactants could be immobilized "as a temporal" event in a pH gradient. The ligand could then be swept through the stationary zone and the stoichiometric complex, provided its pI value is outside the bounds of the pH gradient created in the

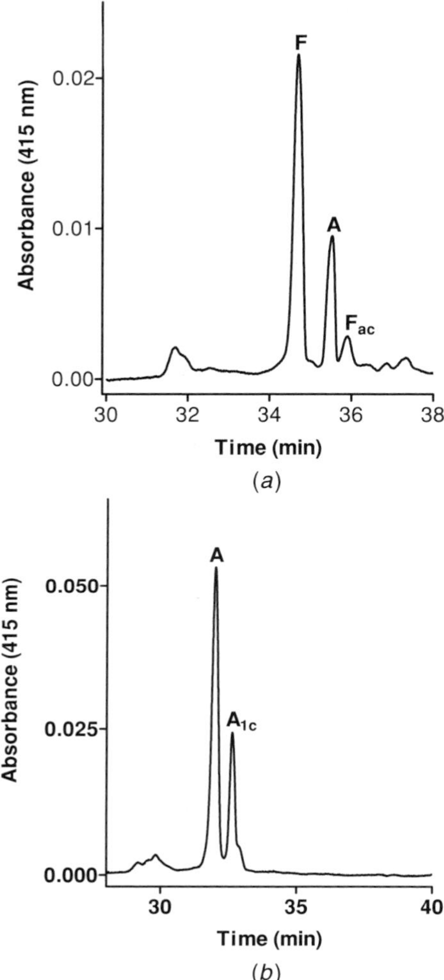

FIGURE 4.22. Capillary IEF of hemoglobins. (*a*) separation of Hb F, A, and F_{ac}. Background electrolyte: 5% Ampholine, pH 6–8, added with 3% short-chain polyacrylamide and 100 mM β-Ala. (*b*) Separation of an artifical mixture of 65% Hb A and 35% Hb A_{1c} in 5% Ampholine, pH 6–8, added with an equimolar mixture of separators, 0.33 M β-Ala and 0.33 M 6-amino caproic acid. Anolyte: 20 mM Asp; catholyte: 40 mM Tris. Sample loading: by pressure for 60 s. Focusing run: 20 kV constant at 7 μA (initial) to 1 μA (final current), 20°C in a Bio-Rad Bio Focus 2000 unit. Capillary: coated with poly(*N*-acryloyl amino ethoxy ethanol), 25 μm ID, 23.6/19.1 total/effective length. Mobilization conditions: with 200 mM NaCl added to anolyte, 22 kV. Detection at 415 nm. (From Conti et al., 1995 and 1996, by permission.)

capillary, emerge at the detector window and thus be quantified. This concept was applied to the study of haptoglobin (Hp)–hemoglobin (Hb) complex formation by Righetti et al. (1997a). A known amount of Hb is focused in a capillary in a pH 6–8 range (pI of Hb is 7.0) and thus kept temporarily "immobilized" in the electrophoretic chamber. Subsequently, increasing amounts of ligand (Hp) are loaded cathodically and allowed to sweep past the focused Hb zone. As the complex formed has a pI value well outside the bounds of such a pH gradient (the 1 : 1 molar Hb–Hp complex has a pI of 5.5; the $1 : \frac{1}{2}$ molar Hp–Hb complex has a pI of 5.0), it escapes immobilization and moves past the detector window, where it is monitored and quantified, thus permitting the analysis of the stoichiometry of the Hp–Hb complex.

4.4.2.2. On Problem of Protein Solubility at Their pI One of the most severe shortcomings of all IEF techniques (whether in gel slabs or capillaries, in soluble buffers or IPGs) is protein precipitation at the pI value. This problem is aggravated by increasing sample concentrations (overloading is often necessary in order to reveal minor components) and by decreasing the ionic strength (I) of the background electrolyte. In this last case, it has been calculated that a 1% carrier ampholyte solution, once focused, would exhibit a remarkably low I value (Righetti, 1980), of the order of 0.5 mEq/L. As demonstrated by Grönwall (1942), the solubility of an isoionic protein plotted against pH near the isoionic point is a parabola with a fairly narrow minimum at relatively high I but with progressively wider minima on the pH axis at decreasing I values. This means that, in unfavorable conditions, protein precipitation will not simply occur at a precise point of the pH scale (the pI), but it will occur in the form of smears covering as much as 0.5 pH units.

In the past we used, with some success, glycerol, ethylene, and propylene glycols (Ettori et al., 1992) when purifying proteins by preparative IPGs. However, we recently found proteins completely insensitive to these solubilizers. Of course, one could always use the classical mixture of 8 M urea and 2% detergents, as routinely adopted for solubilizing entire cell lysates in 2D maps. Yet, we were looking for nondenaturing solubilizers, so that proteins could be recovered in a native form or enzyme assays could be performed in the capillary. Particularly encouraging results were obtained with the use of zwitterions in cIEF, especially at high concentrations (~1 M) (Conti et al., 1997). When attempting to focus the flavoprotein LASPO (L-aspartate oxydase), only massive precipitates were obtained. The only mixture that could restore full solubility was a combination of nondetergent sulfobetaines (NDSBs), 0.5 M of the $M_r = 195$ and 0.5 M of the $M_r = 256$ compounds. Interestingly, a very similar pattern was obtained by using, as additive, 1 M bicine, that is, one of the Good's buffers. Another difficult case was the analysis of thermamylase, an endoamylase that catalyzes the hydrolysis of α-D-(1,4) glycosidic linkages in starch components, which produced only smears and precipitates upon IEF. Excellent resolution and focusing patterns were finally obtained in mixtures of neutral additives (typically 20% sucrose, but also sorbitol and, to a lesser extent, sorbose) and zwitterions, in particular 0.1 M taurine (Esteve-Romero et al., 1996b).

The use of zwitterions had been advocated long ago by Alper et al. (1975). Although this use had fallen into oblivion, recent reports by Vuillard et al. (1994, 1995a,b) suggest that this was indeed an avenue worth exploring, as their results with this novel class of zwitterions synthesized by them (NDSBs) have been encouraging not only in focusing mildly hydrophobic membrane proteins but also in improving protein crystal growth. An explanation for this solubilizing power could come from the work of Timasheff and Arakawa (1989) on stabilization of protein structure by solvents. They explored the physical basis of the stabilization of native protein structures in aqueous solution by the addition of cosolvents at high ($\sim$1 M) concentration. According to them, class I stabilizers (such as sucrose and zwitterions, e.g., amino acids, taurine) act by increasing the surface tension of water and by being preferentially excluded from the hydration shell of the protein. In fact, all these chemicals show a negative value of the binding parameters, signifying that there is an excess of water in the domain of the protein, that is, that the macromolecule is preferentially hydrated. All these phenomena occur at high concentrations of the cosolvents, typically above 1 M, as found in the present report. As a result, the protein is in a state of "superhydration," which might prevent binding to Immobilines in the gel matrix and might markedly improve solubility at the pI value in cIEF. It goes without saying that these additives fully maintain enzyme activity throughout the IEF process.

It should finally be noted that, often, protein precipitation and denaturation could be induced by the presence of high salt levels in the sample, as typically occurring in biological fluids [e.g., urine, cerebrospinal fluid (CSF)]. Thus, desalting prior to the IEF step is often a must. Manabe et al. (1999) reported a method for microdialysis of CSF able to process al little as a 20–30 μL volume coupled to a conductivity device. With this pre-cIEF step, these authors were able to successfully fractionate CSF and resolve as many as 70 peaks.

4.4.2.3. Assessment of pH Gradients and pI Values in cIEF In conventional IEF in gel slabs, protein pI markers are commonly offered by a number of suppliers (e.g., Amersham Pharmacia Biotech, Bio-Rad). In its simplest approach, unknown pI values can be assessed also in cIEF by plotting the pI values of a set of markers cofocused with the proteins under investigation versus their relative mobility upon elution; this is the procedure adopted by Chen and Wiktorowicz (1992). In another, more intriguing approach, in the focusing of transferrin, Kilàr (1991) has proposed a novel method for pI assessments: Monitoring the current in the mobilization step. If one records simultaneously the peaks of the mobilized stack of proteins and the rising current due to passage of the salt wave in the capillary, one can correlate a given pI value (which should be known from the literature a priori) with a given current associated with the transit of a given peak at the detector port. The system can thus be standardized and used for constructing a calibration graph to be adopted in further work without resorting to "internal standards." One such graph correlating current with pI values is shown in Figure 4.23: This appears to be a precise method since the error is given as only about $\pm$0.03 pH units.

The use of protein markers has problems, though, since they are not only difficult to be obtained as single pI components (they are often a family of closely related species) but also subjected to aging, due to hydrolysis of side-chain amides in Asn

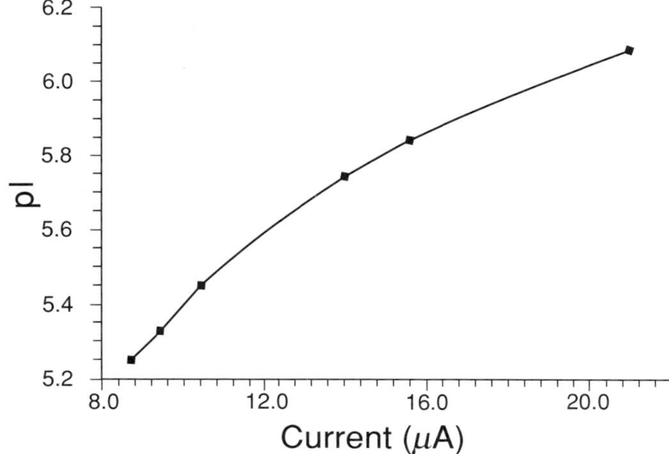

FIGURE 4.23. Calibration graph for pI determination using current during mobilization step as parameter in cIEF. The six experimental points represent six forms of transferrin, containing different amounts of sialic acid and iron. (From Kilàr, 1991, by permission.)

and Gln residues. As alternatives to such proteinaceous material, a number of other markers have been proposed, such as amphoteric dyes (Conway-Jacobs and Lewin, 1971) and phenanthroline iron complexes (Nahkleh et al., 1972).

In another approach, Slais and Friedl (1994) proposed the synthesis of 10 different dyes of the aminomethylnitro phenol and the aminomethylated sulfonaphthalein families. These dyes covered the grounds from pH 3.9 up to pH 10.3 and appeared to be quite evenly distributed along the pH scale. They had the following advantages: They would be more stable than protein pI markers (known to be subjected to aging, a process leading to deamidation and thus to strings of spots situated on the acidic side of the parental protein) and could be detected by both UV absorption (in the near-UV region) and fluorimetric detection, since they could be excited in the same UV region to fluoresce. While this approach is certainly most interesting, there is a caveat, though: Already at its inception, Molteni et al. (1994), who had tested them prior to their official release, reported that these dyes could be easily adsorbed by the silica wall, producing additional EOF and unstable drifting of the pH gradient. In another approach, Shimura and Kasai (1995) suggested stable, fluorescence-labeled peptides for assessing protein pI values. This approach was problem prone too, since, due to the fact that the markers could only be detected by fluorescence, the protein, whose pI had to be measured, had to be derivatized with fluorescence markers as well. This posed a serious problem, since it is known that the derivatization of proteins at all possible reacting sites is an impossible task. For that reason, Kobayashi et al. (1997) proposed peptide digests from protamine (this relatively basic protein being selected for the specific purpose of obtaining also alkaline pI markers) and two additional synthetic peptides (Gly–Gly–Gly and Gly–Gly–His), all of them labeled with dansyl chloride and subsequently purified by preparative IEF in the Rotofor as pI markers in cIEF. Since dansyl absorbs closes to 300 nm, with protein maxima of absorption being at 280 nm, they reasoned that measuring absorption at both wavelengths would allow to distinguish protein peaks from those of the pI markers. According to these authors, resolution (and precision on pI assessments) of 0.1 pH would be routinely achieved, whereas a pI 0.01 resolution could be possible only under certain conditions.

In yet another approach, Mohan and Lee (2002) proposed a hybrid technique consisting in using a set of 10 commercially available pI marker proteins combined with 7 UV-absorbing, tryptic peptides isolated from cytochrome *c*. By using this approach (allowing to map pH values high in the alkaline pH region) combined with admixing to the pH 3–10 Pharmalytes the pH 9–11 Ampholine interval and TEMED, they could resolve and measure the pI (pH 12) of even the very alkaline peptide bradykinin.

Perhaps the best approach, though, could be the one of Shimura et al. (2000), who proposed a set of 16 synthetic oligopeptides (trimers to hexamers) as pI markers for cIEF, fully compatible with UV absorption detection. Each peptide was made to contain one Trp residue for detection by UV absorption and other residues having ionic side chains, responsible for giving sharply focusing peaks during cIEF. To obtain this set of 16 pI markers with a fairly even distribution along the pH scale, some rules were followed: The basic ones were made to contain mostly Lys and Arg residues; the neutral ones had to be made with His residues (it is the only amino acid able to buffer along neutrality!); and the acidic ones were made to contain progressively higher levels of Glu and, finally, Asp residues. The pI values of these peptides were determined by slab gel IEF by using commercial carrier ampholytes. The pI values of the peptides range from as low as 3.38 up to 10.17. The measured values agreed well with the predicted ones, based on amino acid composition, with root-mean-square differences of 0.15 pH units. The sharp focusing, stability, high purity, and high solubility of these synthetic pI markers should facilitate the profiling of a pH gradient in cIEF and the determination of pI values of proteins. The quality of these ampholytes is given by the value of dz/dpH, that is, the slope of the charge over the pH axis at the pI value: The higher this value, the better is the ampholyte, since it will exhibit good conductivity and good buffering capacity in the proximity of its pI. These figures are very high when the pI is rather close to the pK of the ionizable side chains (accordingly, these values are above unity at the extremes of the pH scale, that is, for the most acidic and most basic peptides). Figure 4.24 shows the remarkable mapping of the pH course in cIEF obtainable with this set of 16 markers. Figure 4.24*b* is an enlargement of the time portion from 9 to 11 min, showing the still good resolution of the high-pI markers, the fusion between peptides 28 and 29 being surely due to the lack of extension of the pH gradient toward more alkaline pH values. The authors correctly stress one important point, often overlooked by experimenters: pI values of samples should be estimated by assuming a linear relationship for pH against detection time only between two flanking marker peptides!

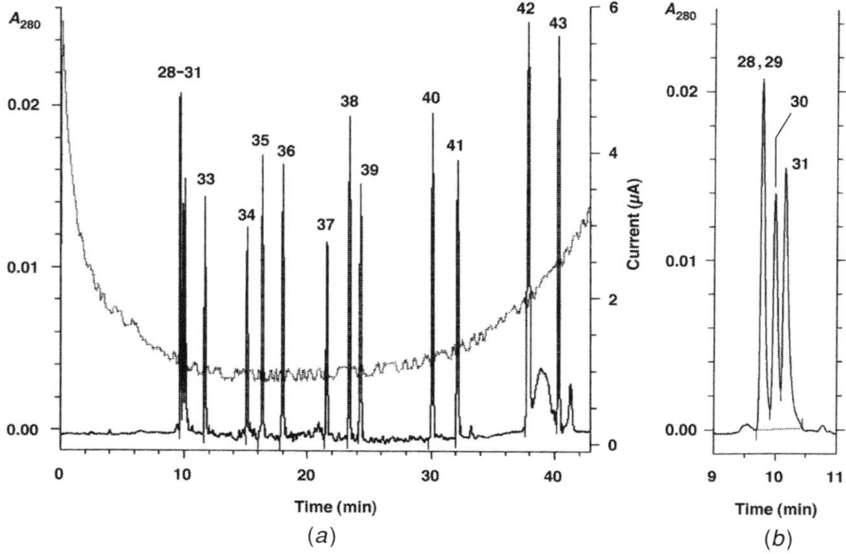

FIGURE 4.24. (*a*) Capillary IEF separation of 16 pI markers. The capillary (50 mm ID × 27 cm) was filled with the ampholyte solution and the mixture of the peptide marker solution (0.25 mM of each peptide) was injected from the anodic side for 30 s. Focusing was carried out at 500 V/cm for 2 min at 25°C. The focused peptide zones were mobilized by a low-pressure rinse mode while maintaining a field strength of 500 V/cm. The peptides were detected via their absorption at 280 nm. (*b*) Enlargement of the portion from 9 to 11 min of (*a*). The bow-shaped profile in (*a*) is the monitoring of the current (in μA). (From Shimura et al., 2000, by permission.)

4.5. SEPARATION OF PEPTIDES AND PROTEINS BY CZE IN ISOELECTRIC BUFFERS

This is an interesting development of cIEF whereby zone electrophoresis can be performed in isoelectric, very low conductivity buffers, allowing the highest possible voltage gradients and thus much improved resolution of peptides and proteins due to reduced diffusion in short analysis times. Although originally described in alkaline pH values (notably in Lys, pI = 9.87) and at neutral pH (His buffers, pH = pI = 7.6, albeit in this last case mostly for DNA and oligonucleotides; Gelfi et al., 1996, 1998; Magnusdottir et al., 1999), we have found that acidic zwitterions offer an extra bonus: They allow the use of uncoated capillaries, due to protonation of silanols at the prevailing pH of the background electrolyte.

4.5.1. General Properties of Amphoteric, Isoelectric Buffers

Although not strictly related to IEF per se, the use of isoelectric buffers stems from IEF know-how and is having a unique impact in CZE; thus we feel it is appropriate to end this review with a glimpse at this field. The physicochemical properties of such buffers (especially in regard to their buffering power) have been discussed in Section 4.2.2. Moreover, Stoyanov et al. (1997) introduced a new parameter for evaluating the performance of amphoteric buffers: the β/λ ratio, that is, the ratio between the molar buffering power and conductivity. Ideal buffers are those with the highest possible β/λ ratio (for nonzero l values), since they allow delivering very high voltage gradients with minimal joule effects. In the field of proteins and for other small M_r compounds, Hjertèn et al. (1995) have explored a number of different amphoteric compounds and given proper guidelines for their use. These authors reported separations at voltage gradients as high as 2000 V/cm. Their results with protein separations have been modeled also by Blanco et al. (1996), who obtained simulated protein separations in close agreement with the experimental ones of Hjertèn et al. (1995). More recently, Righetti and Nembri (1997) have generated peptide maps in isoelectric aspartic acid and shown that such maps could be developed in only 8–10 min, as opposed to 70–80 min in the standard pH 2.0 phosphate buffer, with much superior resolution. Isoelectric Asp, at 50 mM concentration, produces a pH in solution almost identical to its pI value (pI $=$ 2.77 at 25°C). At this pH value, some of the large peptides (tryptic digests of casein) analyzed were strongly adsorbed by the uncoated capillary wall. Generation of peptide maps in isoelectric Asp would be great if one could use plain, uncoated capillaries, since the technique would be extremely simple (no sample derivatization, due to reading at 214 nm, where the adsorption of the peptide bond is quite strong, use of unmodified capillaries, very short analysis times, and very high resolution). After many attempts, a buffer mixture was optimized comprising 50 mM Asp, pH 2.77; 0.5% hydroxyethyl cellulose (HEC, for dynamic coating of the silanols); and 5% 2,2,2-trifluoroethanol (TFE, for modulating peptide mobility). We have now applied this system to the routine analysis of tryptic digests of α- and β-globin chains so as to identify point mutations producing amino acid substitutions (Capelli et al., 1997). Yet, this system was able to resolve only 11 out of 13 fragments present in the β-chain digest. In attempts at ameliorating the system, we have finally adopted the following buffer mixture: 30 mM Asp, pH 2.97; 0.5% HEC; 10% TFE; and 50 mM NDSB-195. This buffer mixture performed extremely well, fully resolving 13 of 13 peptides, in a total time window of 15–16 min in a 75-μm-ID capillary at 600 V/cm (see Fig. 4.25. The inset shows, on an enlarged area, the spectrum of the first eight peptides eluted in the 3–6-min time window. But one could do better than that: By adopting a 50-μm-ID capillary and at 900 V/cm (and doubling the injection time from 15 to 30 s, which corresponds to a sample plug of 17 nL), one can accomplish the analysis in only 9 min with excellent resolution. One can appreciate the remarkable speed, resolution, and sensitivity offered by such a simple technique. One last, very important remark: It is not true that, by adopting a single isoelectric buffer, one has to necessarily work at a fixed pH value (which would greatly limit the usefulness of the technique). One basic rule that should be remembered is that the pH produced in solution by an ampholyte has two limits: on one side (the upper bounds for an acidic, the lower bounds for an alkaline amphotere) is located the pI of water (i.e., pH 7.0); on the opposite side is located the true pI of the ampholyte (Stoyanov and Righetti, 1997; Bossi and Righetti, 1997). These two limits can be reached by modulating the concentration of the amphotere in solution. At extreme dilutions (practically useless, of course, since the β power would

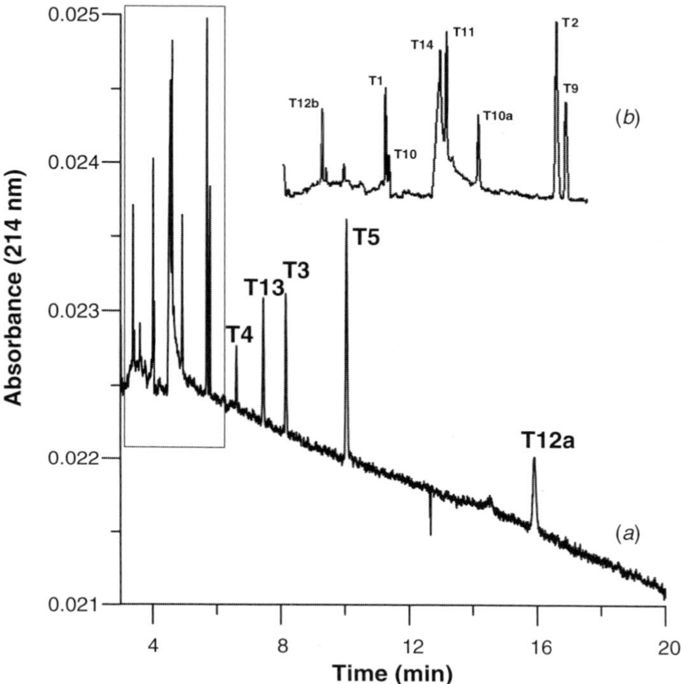

FIGURE 4.25. Separation of 13 peptides of tryptic digest of human α-globin chains in isoelectric buffers. Background electrolyte: 30 mM Asp, pH 2.97, 0.5% HEC, 10% TFE, and 50 mM NDSB-195. Capillary: 30 cm long (23 cm effective length), 75 μm ID. Run at 600 V/cm in a Bio-Rad Bio Focus 2000 unit. Sample injection: 30 s, corresponding to a plug of 17 nL. (*a*) Entire electropherogram with 13 eluted peaks. (*b*) Expanded-scale representation of boxed area in (*a*). Note that the third eluting peak from left (4 min transit time), almost invisible in (*a*), is in fact completely resolved from its neighbor (T_1/T_{10} bands). (From Capelli et al., 1997, by permission.)

approach zero!) the pI of the ampholyte will be that of water, that is, pH 7.0. At the correct concentration (which is not a universal value; it depends on ΔpK, i.e., on how "good" or "bad" is a carrier ampholyte), the pH in solution will approach the pI of the amphotere. We have exploited this subtle rule for implementing the separation shown in Figure 4.25: Note that here, by lowering the concentration of Asp from 50 mM (pH = 2.77) to 30 mM (pH = 2.97), we have in fact moved the pH of the background electrolyte by as much as 0.2 pH units. With this simple modification, we could move along the pH/mobility curves of the 13 peptides and find a pH window where no nodal (or cross-over) points existed among all the curves, thus ensuring separation of all 13 peptides. Soon after the report on Asp, we described another amphoteric buffer, imino diacetic acid (IDA), whose physicochemical parameters were found, by theoretically modeling and experimental verification, to be pI 2.23 (at 100 mM concentration), $pK_1 = 1.73$ and $pK_2 = 2.73$. The IDA was found to be compatible with most hydro-organic solvents, including trifluoroethanol (TFE), up to at least 40% (v/v), typically used for modulating peptide mobility. In naked capillaries, a buffer comprising 50 mM IDA, 10% TFE (or 6–8 M urea), and 0.5% hydroxyethyl cellulose (HEC) allows generation of peptide maps with high resolution, reduced transit times, and no interaction of even large peptides with the wall. Thus IDA appears to be another valid isoelectric buffer system operating in a different pH window (pH 2.33 in 50 mM IDA) as compared to the amphotere previously adopted (50 mM Asp, pH 2.77) for the same kind of analysis (Bossi and Righetti, 1997). Another amphoteric, isoelectric, acidic buffer for separation of oligo- and polypeptides by CZE is cysteic acid (Cys-A). At 200 mM concentration, Cys-A exhibited a pI of 1.80; given $\Delta pK = 0{,}6$, the pK of the carboxyl was assessed as 2.1 and the pK of the sulfate group as 1.50. At 100 mM concentration, this buffer provided an extraordinary buffering power: 140×10^{-3} Eq/L pH. In the presence of 30% (v/v) hexafluoro-2-propanol (HFP), this buffer did not change its apparent pI value but drastically reduced its conductivity. In Cys-A/HFP buffer, small peptides exhibited a mobility closely following the Offord equation, that is, proportional to the ratio $m^{2/3}/z$). When added with 4–5 M urea, there was an inversion in the mobility of some peptides, suggesting strong pK changes as an effect of urea addition. It was found that the minimum mass increment for proper peptide separation was $\Delta M_r = \sim 1\%$. For simultaneous M_r and pK changes, the minimum ΔM_r is reduced to only 0.6% provided that a concomitant minimum $\Delta pK = 0.08$ took place. When separating large peptides (human globin chains) in 100 mM Cys-A, baseline separations among α-, β-, and γ-chains could be obtained (Bossi and Righetti, 1999). The properties of four acidic, isoelectric buffers were summarized: cysteic acid (Cys-A, pI = 1.85), IDA (pI = 2.23), aspartic acid (Asp, pI = 2.77), and glutamic acid (Glu, pI = 3.22). These four buffers allow us to explore an acidic portion of the titration curves of macroions, covering about 1.6 pH units (from pH 1.85 to ~3.45), thus permitting resolution of compounds having coincident titration curves at a given pH value. It was additionally shown that the acidic buffers are not quite stationary in the electric field but can be transported at progressively higher rates (according to the pI value) from the cathodic to the anodic vessel. This is due to the fact that at their respective pI values a fraction of the amphotere has to be negatively charged in order to provide counterions to the excess of protons due to bulk water dissociation (Bossi et al., 1999a). A number of applications of CZE in isoelectric, acidic buffers are here briefly mentioned: screening of wheat cultivars via analysis of the endosperm protein gliadins (Capelli et al., 1998); analysis of maize lines via CZE profiling of zeins (the prolamins or seed storage proteins in maize) (Righetti et al., 1998a; Olivieri et al., 1999); determination of cow's milk in non-bovine cheese via CZE of whey proteins (Herrero-Martinez et al., 2000a,b); and separation of globin chains in umbilical cord blood and screening for point mutations in α and β human globin chains (Righetti et al., 1998b; Saccomani et al., 1999).

4.5.2. Troubleshooting for CZE in Isoelectric Buffers

Although the technique seems to be working quite well, one should be aware of the following problems:

(a) As stated above, these rather acidic buffers might not be very stable in solution. Some of them are amino acids (Asp, Glu) and could be good pabulum for bacterial growth. Thus, unless these solutions are made sterile and kept that way, they should not be used for more than a week and made fresh after that.

(b) Although zwitterionic buffers used at pH = pI should in theory be stationary, this is less and less valid the more acidic is their pI value, since a fraction of these species must be negatively charged to act as counterions to the excess of protons in solution. This would result in net migration of such "isoelectric buffers" from cathode to anode. In addition, because the buffer chambers in CZE are rather small (often less than 1 mL in volume), depletion of the buffering ions could ensue after only a few runs. Thus, the electrode reservoirs might have to be replenished after only three or four runs. An easy check for that is to measure the pH and conductivity of both vessels and change the solutions when these values vary by a given amount (e.g., 15% of the original value).

(c) Although in rather acidic zwitterions (Cys-A, IDA) addition of 0.5% HEC seems to be quite effective in preventing binding of polypeptide chains to the silica wall, this will not hold true in Asp and Glu, especially in 7–8 M urea solutions, due to the considerably higher pH values and the concomitant dissociation of silanols. Thus, other remedies for preventing such binding have to be sought. The best remedy is to add minute amounts (1 mM) of oligoamines to the background electrolyte, the best ones being spermine and TETA (tetraethylene pentamine) (Verzola et al., 2000a).

(d) Even under the most controlled conditions, very small amounts of proteinaceous material could stick to the wall, carpeting the silica surface after an adequate number of runs. Although washes in strong acid (0.1 M HCl) and strong base (0.1 M NaOH) are usually recommended, we have found that the best cleansing method consists in an electrophoretic desorption brought about by micellar SDS (60 mM) in 30 mM phosphate buffer, pH 7. The SDS is placed into the anodic compartment only and driven electrophoretically inside the capillary lumen, where it efficiently sweeps any residue of protein bound to the silica (Verzola et al., 2000a,b).

4.5.3. Novel EOF Modulators

Electroosmotic flow is an ever-present hazard in protein separations, to the point at which countless procedures have been described for masking ionized silanols (as reviewed by Chiari et al., 1996; Regnier and Lin, 1998; Schomburg, 1998). Most of these procedures are very complex and cumbersome, so they are rarely applied in daily practice. Our group has developed different classes of novel quaternarized piperazine compounds able to modulate and reverse the EOF flow in a most peculiar manner. The first class comprises mono-cyclo compounds, with ω-iodoalkyl chains of different lengths (typically C_4 to C_8), able to be adsorbed by silicas, at alkaline pH, and spontaneously alkylate ionized silanols, thus becoming covalently affixed to it. The

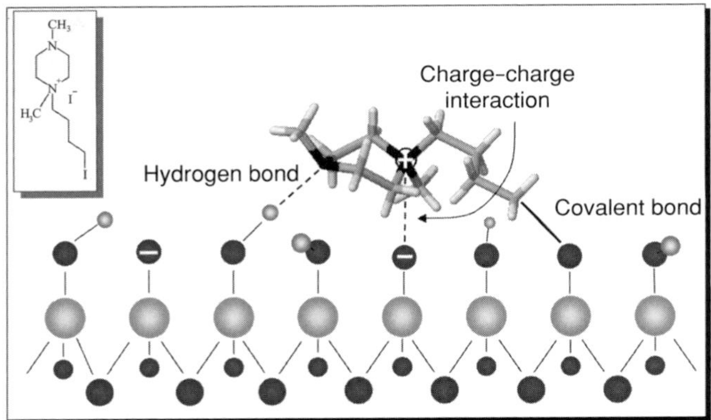

FIGURE 4.26. Proposed mechanism of binding of quaternarized piperazine M1C4I (*N*-methyl-*N*-4-iodobutyl)-*N′*-methylpiperazine; formula in upper left insert) to silica wall. (From Sebastiano et al., 2000, by permission.)

second class is constituted by bicyclic compounds attached at the termini of an alkyl chain of variable lengths (here, too, typically C_4 to C_8). This second class is unable to covalently bind silica surfaces, although in thin-layer chromatography it exhibits an extraordinary affinity for silica beads. On the basis of the strikingly different behavior, structural rules are derived for the minimum requirements for general classes of amines to bind to silica walls and modify EOF. For compounds unable to bind covalently to the wall, the most important structural motif is two quaternary nitrogens spaced apart by a C_4 chain: This seems to be the average distance (i.e., 0.8 nm) between two adjacent, ionized silanols for a snug fit. The other structural binding motif is the "hydrophobic decoration," that is, the ratio of charged groups to alkane groups in the various amines; amines with high levels of such alkane groups (i.e., with higher hydrophobicity) seem to bind more tenaciously to the wall, probably due to hydrophobic interaction not to the wall but among the amine derivatives themselves, when carpeting the silica. Their synthesis, general properties, and applications to peptide–protein separations can be found in Sebastiano et al. (2000, 2001, 2003), Galvani et al. (2001), Gelfi et al. (2001), and Verzola et al. (2003). Figure 4.26 gives the chemical formula of one of them with the mechanisms for docking onto the silica wall and forming a covalent bond via alkylation of ionized silanols.

4.6. CONCLUSIONS

As we hope we have demonstrated in this chapter, modern IEF techniques, both in soluble and immobilized buffers, have much to offer to users. We feel that now adequate solutions exist to the two most noxious impediments to a well-functioning technique, namely lack of flexibility in modulating the slope of the pH gradient and protein precipitation at (and in proximity of) the pI value. The solutions we have discussed (use of spacers and novel mixtures of solubilizers, comprising sugars and high molarities of zwitterions) seem to be working quite satisfactorily. In addition, an important spin-off of IEF know-how seems to be gaining importance in zone electrophoretic separations: the use of isoelectric buffers. Such buffers allow delivery of extremely high voltage gradients, permitting separations of the order of a few minutes, thus favoring very high resolution due to minimum, diffusion-driven peak spreading. As an extra bonus, by properly modulating the molarity of the isoelectric buffer in solution, it is possible to move along the pH scale by as much as 0.3–0.4 pH units, thus optimizing the pH window for separation. The new rising star is cIEF, which has a lot to offer for future users. Particularly appealing is the fact that cIEF provides a fully instrumental approach to electrophoresis, thus lessening dramatically the experimental burden and the labor-intensive approach of gel slab operations. While capillary electrophoresis equipment is currently available mainly as single-channel units, the new generation of equipment will offer multichannel capabilities in batteries from 20 to 96 capillary arrays (but several hundreds have also been envisioned). An interesting development of CZE and cIEF is its interfacing with mass spectrometry, thus opening a new window in proteome analysis. Since excellent reviews have recently appearead from Hille et al. (2001), Shen and Smith (2002), Moini (2002), Shimura 2002), and Manabe (2003), this aspect will not be treated here, also because it will be dealt with in the next chapter.

Thus rapid growth is expected in this field. Last, but not least, we mention the latest evolution of CZE, that is, integrated, chip-based capillary electrophoresis (ICCE). This is emerging as a new analytical tool allowing fast, automated, miniaturized and multiplexed assays, thus meeting the needs of the pharmaceutical industry in drug development. It already allows pre- and postcolumn derivatization, DNA sequencing, online PCR analysis, on-chip enzymatic sample digestion, fraction isolation, and immunoassays, all in picoliter sample volumes injected and electrophoretic time scales of the order of a few to a few hundred seconds (Effenhauser et al., 1997). More specifically, miniaturized cIEF in plastic microfluidic devices (Tan et al., 2002) or in very short (1.2-cm) fused-silica capillaries (Wu et al., 2001) has been recently reported. In the last case, whole column imaging was obtained by using a LED (light-emitting diode) as a light source and via detection with a CCD camera.

REFERENCES

Alba, F. J., Bermudez, A., Bartolome, S., Daban, J. R. (1996) *BioTechniques* **21,** 625–626.

Allen, R. C., Arnaud, P. (Eds.) (1981) *Electrophoresis '81: Advanced Methods, Biochemical and Chemical Applications*, W. de Gruyter, Berlin.

Alper, C. A., Hobart, M. J., Lachmann, P. J. (1975), in Arbuthnott, J. P., Beeley, J. A. (Eds.), *Isoelectric Focusing*, London, Butterworths, pp. 306–312.

Altland, K., Kaempfer, M. (1980) *Electrophoresis* **1**, 57–62.

Anderson, N. G., Anderson, L. (1982) *Clin. Chem.* **28,** 739–748.

Appel, R. D., Dunn, M. J., Hochstrasser, D. F. (Guest Eds.) (1997) Paper Symposium: Biomedicine and Bioinformatics, *Electrophoresis* **18,** 2703–2842.

Appel, R. D., Dunn, M. J., Hochstrasser, D. F. (Guest Eds.) (1999) Paper Symposium: Biomedicine and Bioinformatics, *Electrophoresis* **20,** 3481–3686.

Arbuthnott, J. P., Beeley, J. A. (Eds.) (1975) *Isoelectric Focusing*, Butterworths, London.

Arnaud, P., Wilson, G. B., Koistinen, J., Fudenberg, H. H. (1977) *J. Immunol. Methods* **16,** 221–231.

Astrua-Testori, S., Pernelle, J. J., Wahrmann, J. P., Righetti, P. G. (1986) *Electrophoresis* **7,** 527–529.

Astrua-Testori, S., Righetti, P. G. (1987) *J. Chromatogr.* **387,** 121–127.

Awdeh, Z. L. (1969) *Sci. Tools* **16,** 42–43.

Awdeh, Z. L., Williamson, A. R., Askonas, B. A. (1968) *Nature* **219,** 66–67.

Bellini, M. P., Manchester, K. L. (1998) *Electrophoresis* **19,** 1590–1595.

Bermudez, A., Daban, J., Garcia, J., Mendez, E. (1994) *BioTechniques* **16,** 621–624.

Berven, F. S., Karlsen, O. A., Murrell, J. C., Jensen, H. B. (2003) *Electrophoresis* **24,** 757–761.

Bianchi-Bosisio, A., Loehrlein, C., Snyder, R. S., Righetti, P. G. (1980) *J. Chromatogr.* **189,** 317–330.

Bianchi-Bosisio, A., Righetti, P. G., Egen, N. B., Bier, M. (1986) *Electrophoresis* **7,** 128–133.

Bianchi-Bosisio, A., Snyder, R. S., Righetti, P. G. (1981) *J. Chromatogr.* **209**, 265–272.

Bjellqvist, B., Ek, K, Righetti, P. G., Gianazza, E., Görg, A., Postel, W., Westermeier, R. (1982) *J. Biochem. Biophys. Methods* **6,** 317–339.

Blanco, S., Clifton, J. M., Loly, J. L., Peltre, G. (1996) *Electrophoresis* **17,** 1126–1133.

Bonner, W. M., Laskey, R. A. (1974) *Eur. J. Biochem.* **46,** 83–88.

Bordini, E., Hamdan, M., Righetti, P. G. (1999a) *Rapid Commun. Mass Spectrom.* **13,** 1818–1827.

Bordini, E., Hamdan, M., Righetti, P. G. (1999b) *Rapid Commun. Mass Spectrom.* **13,** 2209–2215.

Bosshard, H. F., Datyner, A. (1977) *Anal. Biochem.* **82,** 327–333.

Bossi, A., Gelfi, C., Orsi, A., Righetti, P. G. (1994a) *J. Chromatogr. A* **686,** 121–128.

Bossi, A., Olivieri, E., Castelletti, L., Gelfi, C., Hamdan, M., Righetti, P. G. (1999a) *J. Chromatogr. A* **853,** 71–82.

Bossi, A., Patel, M. J., Webb, E. J., Baldwin, M. A., Jacob, R. J., Burlingame, A. L., Righetti, P. G. (1999b) *Electrophoresis* **20,** 2810–2817.

Bossi, A., Righetti, P. G. (1995) *Electrophoresis* **16,** 1930–1934.

Bossi, A., Righetti, P. G. (1997) *Electrophoresis* **18,** 2012–2018.

Bossi, A., Righetti, P. G. (1999) *J. Chromatogr. A* **840,** 117–129.

Bossi, A., Righetti, P. G., Vecchio, G., Severinsen, S. (1994b) *Electrophoresis* **15,** 1535–1540.

Bruschi, M., Musante, L., Candiano, G., Herbert, B., Antonucci, F., Righetti, P. G. (2003) *Proteomics* **3,** 821–825.

Bunn, H. F., Drysdale, J. W. (1971) *Biochim. Biophys. Acta* **229,** 51–57.

Candiano, G., Bruschi, M., Musante, L., Santucci, L., Ghiggeri, G. M., Carnemolla, B., Orecchia, P., Zardi, L., Righetti, P. G. (2004) *Electrophoresis*, **25** 1327–1333.

Candiano, G., Musante, L., Bruschi, M., Ghiggeri, G. M., Herbert, B., Antonucci, F., Righetti, P. G. (2002) *Electrophoresis* **23,** 292–297.

Cann, J. R. (1979) in Righetti, P. G., Van Oss, C. J., Vanderhoff, J. W. (Eds.), *Electrokinetic Separation Methods*, Elsevier, Amsterdam, pp. 369–388.

Cann, J. R. (1998) *Electrophoresis* **19,** 1577–1585.

Capelli, L., Forlani, F., Perini, F., Guerrieri, N., Cerletti, P., Righetti, P. G. (1998) *Electrophoresis* **19,** 311–318.

Capelli, L., Stoyanov, A. V., Wajcman, H., Righetti, P. G. (1997) *J. Chromatogr. A* **791,** 313–322.

Cash, P. (Guest Ed.) (1999) Paper Symposium: Microbial Proteomes, *Electrophoresis* **20,** 2149–2310.

Caspers, M. L., Posey, Y., Brown, R. K. (1977) *Anal. Biochem.* **79,** 166–180.

Catsimpoolas, N. (Ed.) (1973) *Ann. N. Y. Acad. Sci.* **209,** 1–529.

Catsimpoolas, N. (Ed.) (1978) *Electrophoresis '78*, Elsevier, Amsterdam.

Celentano, F., Gianazza, E., Dossi, G., Righetti, P. G. (1987) *Chemometr. Intell. Lab. Systems* **1,** 349–358.

Celentano, F. C., Tonani, C., Fazio, M., Gianazza, E., Righetti, P. G. (1988) *J. Biochem. Biophys. Methods* **16,** 109–128.

Celis, J. E. (Guest Ed.) (1989) Paper Symposium: Protein Databases in Two Dimensional Electrophoresis, *Electrophoresis* **10,** 71–164.

Celis, J. E. (Guest Ed.) (1990a) Paper Symposium: Cell Biology, *Electrophoresis* **11,** 189–280.

Celis, J. E. (Guest Ed.) (1990b) Paper Symposium: Two Dimensional Gel Protein Databases, *Electrophoresis* **11,** 987–1168.

Celis, J. E. (Guest Ed.) (1991) Paper Symposium: Two Dimensional Gel Protein Databases, *Electrophoresis* **12,** 763–996.

Celis, J. E. (Guest Ed.) (1992) Paper Symposium: Two Dimensional Gel Protein Databases, *Electrophoresis* **13,** 891–1062.

Celis, J. E. (Guest Ed.) (1994a) Paper Symposium: Electrophoresis in Cancer Research, *Electrophoresis* **15,** 307–556.

Celis, J. E. (Guest Ed.) (1994b) Paper Symposium: Two Dimensional Gel Protein Databases, *Electrophoresis* **15,** 1347–1492.

Celis, J. E. (Guest Ed.) (1995) Paper Symposium: Two Dimensional Gel Protein Databases, *Electrophoresis* **16,** 2175–2264.

Celis, J. E. (Guest Ed.) (1996) Paper Symposium: Two Dimensional Gel Protein Databases, *Electrophoresis* **17,** 1653–1798.

Celis, J. E. (Guest Ed.) (1999) Genomics and Proteomics of Cancer, *Electrophoresis* **20,** 223–429.

Celis, J. E., Bravo, R. (Eds.) (1984) *Two Dimensional Gel Electrophoresis of Proteins*, Academic, Orlando, FL.

Charpak, G., Dominik, W., Zaganidis, N. (1989) *Proc. Natl. Acad. Sci. USA* **86,** 1741–1745.

Chen, H., Cheng, H., Bjerknes, M. (1993) *Anal. Biochem.* **212,** 295–296.

Chen, S. M., Wiktorowicz, J. E. (1992) *Anal. Biochem.* **206,** 84–90.

Chiari, M., Casale, E., Santaniello, E., Righetti, P. G. (1989a) *Theor. Appl. Electr.* **1,** 99–102.

Chiari, M., Casale, E., Santaniello, E., Righetti, P. G. (1989b) *Theor. Appl. Electr.* **1,** 103–107.

Chiari, M., Ettori, C., Manzocchi, A., Righetti, P. G. (1991a) *J. Chromatogr.* **548,** 381–392.

Chiari, M., Giacomini, M., Micheletti, C., Righetti, P. G. (1991b) *J. Chromatogr.* **558,** 285–295.

Chiari, M., Micheletti, C., Nesi, M., Fazio, M., Righetti, P. G. (1994) *Electrophoresis* **15,** 177–186.

Chiari, M., Nesi, M., Righetti, P. G. (1996) in Righetti, P. G. (Ed.), *Capillary Electrophoresis in Analytical Biotechnology*, CRC Press, Boca Raton, FL, pp. 1–36.

Chiari, M., Pagani, L., Righetti, P. G., Jain, T., Shor, R., Rabilloud, T. (1990a) *J. Biochem. Biophys. Methods* **21,** 165–172.

Chiari, M., Righetti, P. G., Ferraboschi, P., Jain, T., Shorr, R. (1990b) *Electrophoresis* **11,** 617–620.

Chiari, M., Righetti, P. G., Negri, A., Ceciliani, F., Ronchi, S. (1992) *Electrophoresis* **13,** 882–884.

Choi, J. K., Yoon, S. H., Hong, H. Y., Choi, D. K., Yoo, G. S. (1996) *Anal. Biochem.* **236,** 82–84.

Christiansen, J., Houen, G. (1992) *Electrophoresis* **13,** 179–183.

Clegg, J. B., Naughton, M. A., Weatherall, D. J. (1966) *J. Mol. Biol.* **19,** 91–100.

Conti, M., Galassi, M., Bossi, A., Righetti, P. G. (1997) *J. Chromatogr. A* **757,** 237–245.

Conti, M., Gelfi, C., Bianchi-Bosisio, A., Righetti, P. G. (1996) *Electrophoresis* **17,** 1590–1596.

Conti, M., Gelfi, C., Righetti, P. G. (1995) *Electrophoresis* **16,** 1485–1491.

Conway-Jacobs, A., Lewin, L. A. (1971) *Anal. Biochem.* **43,** 394–400.

Cossu, G., Manca, M., Gavina, P. M., Bullitta, R., Bianchi-Bosisio, A., Gianazza, E., Righetti, P. G. (1982) *Am. J. Haematol.* **13,** 149–157.

Cossu, G., Manca, M., Pirastru, M. G., Bullitta, R., Bianchi-Bosisio, A., Righetti, P. G. (1984) *J. Chromatogr.* **307,** 103–110.

Cossu, G., Righetti, P. G. (1987) *J. Chromatogr.* **398,** 211–216.

Coull, J. M., Pappin, D. J. C. (1990) *J. Prot. Chem.* **9,** 259–260.

Cuono, C. B., Chapo, G.A. (1982) *Electrophoresis* **3,** 65–70.

Damerval, C., de Vienne, D. (Guest Eds.) (1988) Paper Symposium: Two Dimensional Electrophoresis of Plant Proteins, *Electrophoresis* **9,** 679–796.

Dunbar, B. S. (1987) *Two-Dimensional Electrophoresis and Immunological Techniques*, Plenum, New York.

Dunn, M. J. (Ed.) (1986) *Electrophoresis '86: Proceedings of the fifth Meeting of the International Electrophoresis Society*. Verlag Chemie, Weinheim.

Dunn, M. J. (Ed.) (1991a) *2-D PAGE '91: Proceedings of the International Meeting. On Two-Dimensional Electrophoresis*, Zebra Printing, Perivale.

Dunn, M. J. (Guest Ed.) (1991b) Paper Symposium: Biomedical Applications of Two-Dimensional Gel Electrophoresis, *Electrophoresis* **12,** 459–606.

Dunn, M. J. (Guest Ed.) (1995) 2D Electrophoresis: From Protein Maps to Genomes, *Electrophoresis* **16,** 1077–1326.

Dunn, M. J. (Guest Ed.) (1997) From Protein Maps to Genomes, Proceedings of the Second Siena Two-Dimensional Electrophoresis Meeting, *Electrophoresis* **18,** 305–662.

Dunn, M. J. (Guest Ed.) (1999) From Genome to Proteome: Proceedings of the Third Siena Two-Dimensional Electrophoresis Meeting, *Electrophoresis* **20,** 643–1122.

Effenhauser, C. S., Bruin, G. J. M., Paulus, A. (1997) *Electrophoresis* **18,** 2203–2213.

Ek, K., Bjellqvist, B., Righetti, P. G. (1983) *J. Biochem. Biophys. Methods* **8,** 134–155.

Esteve-Romero, J., Simò-Alfonso, E., Bossi, A., Bresciani, F., Righetti, P.G. (1996a) *Electrophoresis* **17,** 704–708.

Esteve-Romero, J. S., Bossi, A., Righetti, P. G. (1996b) *Electrophoresis* **17,** 1242–1247.

Ettori, C., Righetti, P. G., Chiesa, C., Frigerio, F., Galli, G., Grandi, G. (1992) *J. Biotechnol.* **25,** 307–318.

Fazekas de St. Groth, S., Webster, R. G., Datyner, A. (1963) *Biochim. Biophys. Acta* **71,** 377–391.

Fernandez-Patron, C. Castellanos-Serra, L., Hardy, E., Guerra, M., Estevez, E., Mehl, E., Frank, R. W. (1998) *Electrophoresis* **19,** 2398–2406.

Galante, E., Caravaggio, T., Righetti, P. G. (1975) in Righetti, P. G. (Ed.), *Progress in Isoelectric Focusing and Isotachophoresis*, Elsevier, Amsterdam, pp. 3–12.

Galvani, M., Hamdan, M., Righetti, P. G., Gelfi, C., Sebastiano, R., Citterio, A. (2001) *Rapid Commun. Mass Spectrom.* **15,** 210–216.

Gåveby, B. M., Pettersson, P., Andrasko, J., Ineva-Flygare, L., Johannesson, U., Görg, A., Postel, W., Domscheit, A., Mauri, P. L., Pietta, P., Gianazza, E., Righetti, P. G. (1988) *J. Biochem. Biophys. Methods* **16,** 141–164.

Gelfi, C., Perego, M., Righetti, P. G. (1996) *Electrophoresis* **17,** 1470–1475.

Gelfi, C., Perego, M., Righetti, P. G., Cainarca, S., Firpo, S., Ferrari, M., Cremonesi, L. (1998) *Clin. Chem.* **44,** 906–913.

Gelfi, C., Righetti, P. G. (1981) *Electrophoresis* **2,** 213–219.

Gelfi, C., Righetti, P. G. (1983) *J. Biochem. Biophys. Methods* **8,** 156–171.

Gelfi, C., Righetti, P. G. (1984) *Electrophoresis* **5,** 257–262.

Gelfi, C., Morelli, A., Rovida, E., Righetti, P. G. (1986) *J. Biochem. Biophys. Methods* **13,** 113–124.

Gelfi, C., Viganò, A., Ripamonti, M., Righetti, P. G., Sebastiano, R., Citterio, A. (2001) *Anal. Chem.* **73,** 3862–3868.

Gharahdaghi, F., Weinberg, C. R., Meagher, D. A., Imai, B. S., Mische, S. M. (1999) *Electrophoresis* **20,** 601–605.

Giaffreda, E., Tonani, C., Righetti, P. G. (1993) *J. Chromatogr.* **630,** 313–327.

Gianazza, E. (1995) *J. Chromatogr. A* **705,** 67–87.

Gianazza, E., Artoni, G., Righetti, P. G. (1983) *Electrophoresis* **4,** 321–324.

Gianazza, E., Astrua-Testori, S., Righetti, P. G. (1985a) *Electrophoresis* **6,** 113–117.

Gianazza, E., Celentano, F., Dossi, G., Bjellqvist, B., Righetti, P. G. (1984a) *Electrophoresis* **5,** 88–97.

Gianazza, E., Celentano, F. C., Magenes, S., Ettori, C., Righetti, P. G. (1989) *Electrophoresis* **10,** 806–808.

Gianazza, E., Chillemi, F. Duranti, M., Righetti, P. G. (1984b) *J. Biochem. Biophys. Methods* **8,** 339–351.

Gianazza, E., Chillemi, F., Gelfi, C., Righetti, P. G. (1979) *J. Biochem. Biophys. Methods* **1,** 237–251.

Gianazza, E., Chillemi, F., Righetti, P. G. (1980) *J. Biochem. Biophys. Methods* **3,** 135–141.

Gianazza, E., Giacon, P., Sahlin, B., Righetti, P. G. (1985b) *Electrophoresis* **6,** 53–56.

Gianazza, E., Righetti, P. G. (1980) *J. Chromatogr.* **193,** 1–8.

Giddings, J. C. (1991) *Unified Separation Science*, Wiley, New York, pp. 86–109.

Godolphin, W. J., Stinson, R. A. (1974) *Clin. Chim. Acta* **56,** 97–103.

Goerg, A., Boguth, G., Obermaier, C., Harder, A., Weiss, W. (1998) *Electrophoresis* **19,** 1516–1519.

Goerg, A., Obermaier, C., Boguth, G., Csordas, A., Diaz, J. J., Madjar, J. J. (1997) *Electrophoresis* **18,** 328–337.

Goerg, A., Obermaier, C., Boguth, G., Harder, A., Scheibe, B., Wildgruber, R., Weiss, W. (2000) *Electrophoresis* **21,** 1037–1053.

Goerg, A., Obermaier, C., Boguth, G., Weiss, W. (1999) *Electrophoresis* **20,** 712–717.

Grönwall, A. (1942) *Com. Rend. Lab. Carlsberg, Ser. Chem.* **24,** 185–195.

Grubhofer, N., Borja, C. (1977) in Radola, B. J., Graesslin, D. (Eds.), *Electrofocusing and Isotachophoresis*, de Gruyter, Berlin, 1974, pp. 111–120.

Guruprasad, K, Reddy, B. V., Pandit, M. W. (1990) *Protein Eng.* **4,** 155–161.

Haebel, S., Albrecth, T., Sparbier, K., Walden, P., Korner, R., Steup, M. (1998) *Electrophoresis* **19,** 679–686.

Harris, H., Hopkinson, D. A. (1976) *Handbook of Enzyme Electrophoresis in Human Genetics*, Elsevier, Amsterdam.

Hebert, J. P., Strobbel, B. (1974) LKB Application Note 151 Bromma, Sweden.

Herrero-Martinez, J., Simò-Alfonso, E., Ramis-Ramos, G., Gelfi, C., Righetti, P. G. (2000a) *Electrophoresis* **21,** 633–640.

Herrero-Martinez, J., Simò-Alfonso, E., Ramis-Ramos, G., Gelfi, C., Righetti, P. G. (2000b) *J. Chromatogr. A*, **878,** 261–271.

Hille, J. M., Freed, A. L., Wätzig, H. (2001) *Electrophoresis* **22,** 4035–4052.

Hirai, H. (Ed.) (1984) *Electrophoresis '83: Advanced Methods, Biochemical and Chemical Applications*, W. de Gruyter, Berlin.

Hjertèn, S. (1992) in Grossman, P. D., Colburn, J. C. (Eds.), *Capillary Electrophoresis: Theory and Practice*, Academic, San Diego, pp. 191–214.

Hjertèn, S., Valtcheva, L., Elenbring, K., Liao, J. L. (1995) *Electrophoresis* **16,** 584–594.

Hochstrasser, D., Augsburger, V., Funk, M., Appel, R., Pellegrini, C., Muller, A. F. (1986) in Dunn, M. J. (Ed.), *Electrophoresis '86: Proceedings of the Fifth Meeting of the International Electrophoresis Society*, VCH, Weinheim, pp. 566–568.

Horowitz, P. M., Bowman, S. (1987) *Anal. Biochem.* **165,** 430–434.

Humphery-Smith (Guest Ed.) (1997) Paper Symposium: Microbial Proteomes, *Electrophoresis* **18,** 1207–1497.

Hunkapillar, M., Lujan, E., Ostrander, F., Hood, L. (1983) *Methods Enzymol.* **91,** 227–236.

Jackson, P., Urwin, V. E., Mackay, C. D. (1988) *Electrophoresis* **9,** 330–339.

Javier Alba, F., Bermudez, A., Bartolome S., Daban, J. R. (1996) *BioTechniques* **21,** 625–626.

Jensen, P. K., Pasa-Tolic, L., Peden, K. K., Martinovic, S., Lipton, M. S., Anderson, G. A., Tolic, N., Wong, K. K., Smith, R. D. (2000) *Electrophoresis* **23,** 1372–1380.

Jin, L. J., Ferrance, J., Landers, J. P. (2001) *BioTechniques* **31,** 1332–1353.

Johnson, E. A., Mulloy, B. (1976) *Carbohydr. Res.* **51,** 119–127.

Johnston, R. F., Pickett, S. C., Barker, D. L. (1990) *Electrophoresis* **11,** 355–360.

Kilár, F. (1991) *J. Chromatogr.* **545,** 403–410.

Kilár, F. (1994) in Landers, J. P. (Ed.), *Handbook of Capillary Electrophoresis*, CRC Press, Boca Raton, FL, pp. 95–109.

Kilár, F. (2003) *Electrophoresis* **24,** 3908–3916.

King, J. S. (Guest Ed.) (1984) Special Issue in Two Dimensional Electrophoresis, *Clin. Chem.* **30,** 1897–2108.

Kobayashi, H., Aoki, M., Suzuki, M., Yanagisawa, A., Arai, E. (1997) *J. Chromatogr. A* **772,** 137–144.

Kolin, A. (1977) in Radola, B. J., Graesslin, D. (Eds.), *Electrofocusing and Isotachophoresis*, de Gruyter, Berlin, pp. 3–33.

Krishnamoorthy, R., Bianchi-Bosisio, A., Labie, D., Righetti, P. G. (1978) *FEBS Lett.* **94,** 319–323.

Låås, T., Olsson, I. (1981) *Anal. Biochem.* **114,** 167–172.

Laskey, R. A. (1980) *Methods Enzymol.* **65,** 363–371.

Laskey, R. A., Mills, A. D. (1975) *Eur. J. Biochem.* **56,** 335–341.

Lim, M. J., Patton, W. F., Lopez, M. F., Spofford, K. H., Shojaee, N., Shepro, D. (1997) *Anal. Biochem.* **245,** 184–195.

Lottspeich, F. (Guest Ed.) (1996) Paper Symposium: Electrophoresis and Amino Acid Sequencing, *Electrophoresis* **17,** 811–966.

Magnusdottir, S., Gelfi, C., Hamdan, M., Righetti, P. G. (1999) *J. Chromatogr. A* 859, 87–98.

Manabe, T. (2003) *J. Chromatogr. B* **787,** 29–41.

Manabe, T., Miyamoto, H., Inoue, K., Nakatsu, M., Arai, M. (1999) *Electrophoresis* **20,** 3677–3683.

Mazzeo, J. R., Krull, I. S. (1993) in Guzman, A. N. (Ed.), *Capillary Electrophoresis Technology*, Dekker, New York, pp. 795–818.

Merrifield, R. B. (1969) *Adv. Enzymol.* **32,** 221–241.

Merril, C. R., Goldman, D., Sedman, S. A., Ebert, M. H. (1981) *Science* **211,** 1438–1440.

Merril, C. R., Washart, K. M. (1998a) in Hames, B. D. (Ed.), *Gel Electrophoresis of Proteins: A Practical Approach* Oxford University Press, Oxford, pp. 53–91.

Merril, C. R., Washart, K. M. (1998b) in Hames, B. D. (Ed.), *Gel Electrophoresis of Proteins: A Practical Approach* Oxford University Press, Oxford, pp. 319–343.

Meselson, M., Stahl, F. W., Vinogradov, J. (1957) *Proc. Natl. Acad. Sci. USA* **43,** 581–585.

Mohan, D., Lee, D. S. (2002) *J. Chromatogr. A* **979,** 271–276.

Moini, M. (2002) *Anal. Bioanal. Chem.* **373,** 466–480.

Molteni, S., Frischknecht, H., Thormann, W. (1994) *Electrophoresis* **15,** 23–30.

Mortz, E., Krogh, T. N., Vorum, H., Goerg, A. (2001) *Proteomics* **1,** 1359–1363.

Mosher, R. A., Bier, M., Righetti, P. G. (1986) *Electrophoresis* **7,** 59–66.

Mosher, R. A., Thormann, W. (2002) *Electrophoresis* **23,** 1803–1814.

Myers J. M., Veis, A., Sabsay, B., Wheeler, A. P. (1996) *Anal. Biochem.* **240,** 300–302.

Nahkleh, E. T., Samra, S. A., Awdeh, Z. L. (1972) *Anal. Biochem.* **49,** 218–224.

Neuhoff, V. (Ed.) (1984) *Electrophoresis '84*, Verlag Chemie, Weinheim.

Neuhoff, V., Arold, N., Taube, D., Ehrhardt, W. (1988) *Electrophoresis* **9,** 255–262.

Neuhoff, V., Stamm, R., Eibl, H. (1985) *Electrophoresis* **6,** 427–437.

O'Farrell, P. (1975) *J. Biol. Chem.* **250,** 4007–4021.

Olivieri, E., Viotti, A., Lauria, M., Simò-Alfonso, E., Righetti, P. G. (1999) *Electrophoresis* **20,** 1595–1604.

Ortiz, M. L., Calero, M., Fernandez-Patron, C., Castellanos, L., Mendez, E. (1992) *FEBS Lett.* **296,** 300–304.

Park, C. M. (1973) *Ann. N. Y. Acad. Sci.* **209,** 237–256.

Patterson, S., Aebersold, R. (1995) *Electrophoresis* **16,** 1791–1814.

Perduca, M., Bossi, A., Goldoni, L., Monaco, H., Righetti, P. G. (2000) *Electrophoresis* **21,** 2316–2320.

Perrella, M., Cremonesi, L., Benazzi, L., Rossi-Bernardi, L. (1981) *J. Biol. Chem.* **256,** 11098–11103.

Pogacar, P., Jarecki, R. (1974) in Allen, R., Maurer, H. (Eds.), *Electrophoresis and Isoelectric Focusing in Polyacrylamide Gels*, de Gruyter, Berlin, pp. 153–158.

Pritchett, T. J. (1996) *Electrophoresis* **17,** 1195–1201.

Rabilloud, T. (1990), *Electrophoresis* **11,** 785–794.

Rabilloud. T. (2000) *Anal. Chem.* **72,** 48A–55A.

Rabilloud, T., Brodard, V., Peltre, G., Righetti, P. G., Ettori, C. (1992) *Electrophoresis* **13,** 264–266.

Rabilloud, T., Gelfi, C., Bossi, M. L., Righetti, P. G. (1987) *Electrophoresis* **8,** 305–312.

Rabilloud, T., Kieffer, S., Procaccio, V., Louwagie, M., Courchesne, P. L., Patterson, S. D., Martinez, P., Garin, J., Lunardi, J. (1998) *Electrophoresis* **19,** 1006–1014.

Rabilloud, T., Vuillard, L., Gilly, C., Lawrence, J. J. (1994) *Cell. Mol. Biol.* **40,** 57–75.

Radola, B. J. (1973) *Biochim. Biophys. Acta* **295,** 412–428.

Radola, B. J. (Ed.) (1980a) *Electrophoresis '79: Advanced Methods, Biochemical and Chemical Applications*, W. de Gruyter, Berlin.

Radola, B. J. (1980b) *Electrophoresis* **1,** 43–56.

Radola, B. J., Graesslin, D. (Eds.) (1977) *Electrofocusing and Isotachophoresis*, W. de Gruyter, Berlin.

Regnier, F. E., Lin, S. (1998) in Khaledi, M. G. (Ed.), *High Performance Capillary Electrophoresis*, Wiley, New York, pp. 683–728.

Richtie, R. F., Smith, R. (1976) *Clin. Chem.* **22,** 497–499.

Righetti, P. G. (1980) *J. Chromatogr.* **190,** 275–282.

Righetti, P. G. (1983) *Isoelectric Focusing: Theory, Methodology and Applications*, Elsevier, Amsterdam.

Righetti, P. G. (1990) *Immobilized pH Gradients: Theory and Methodology*, Elsevier, Amsterdam.

Righetti, P. G. (Ed.) (1975) *Progress in Isoelectric Focusing and Isotachophoresis* Elsevier, Amsterdam, pp. 1–395.

Righetti, P. G., Bello, M. (1992) *Electrophoresis* **13,** 275–279.

Righetti, P. G., Bianchi-Bosisio, A. (1981) *Electrophoresis* **2,** 65–75.

Righetti, P. G., Bossi, A. (1998) *Anal. Chim. Acta* **372,** 1–19.

Righetti, P. G., Bossi, A., Görg, A., Obermaier, C., Boguth, G. (1996a) *J. Biochem. Biophys. Methods* **31,** 81–91.

Righetti, P. G., Brost, B. C., Snyder, R. S. (1981) *J. Biochem. Biophys. Methods* **4,** 347–363.

Righetti, P. G., Brown, R., Stone, A. L. (1978) *Biochim. Biophys. Acta* **542,** 222–231.

Righetti, P. G., Chiari, M. (1993) in Guzman, A. N. (Ed.), *Capillary Electrophoresis Technology*, Dekker, New York, pp. 89–116.

Righetti, P. G., Chiari, M., Gelfi, C. (1988a) *Electrophoresis* **9,** 65–73.

Righetti, P. G., Chiari, M., Sinha, P. K., Santaniello, E. (1988b) *J. Biochem. Biophys. Methods* **16,** 185–192.

Righetti, P. G., Chillemi, F. (1978) *J. Chromatogr.* **157,** 243–251.

Righetti, P. G., Conti, M., Gelfi, C. (1997a) *J. Chromatogr. A* **767,** 255–262.

Righetti, P. G., Drysdale, J. W. (1971) *Biochim. Biophys. Acta* **236,** 17–28.

Righetti, P. G., Drysdale, J. W. (1973) *Ann. N. Y. Acad. Sci.* **209,** 163–186.

Righetti, P. G., Drysdale, J. W. (1974) *J. Chromatogr.* **98,** 271–321.

Righetti, P. G., Ek, K., Bjellqvist, B. (1984) *J. Chromatogr.* **291,** 31–42.

Righetti, P. G., Fazio, M., Tonani, C. (1988c) *J. Chromatogr.* **440,** 367–377.

Righetti, P. G., Fazio, M., Tonani, C., Gianazza, E., Celentano, F. C. (1988d) *J. Biochem. Biophys. Methods* **16,** 129–140.

Righetti, P. G., Gelfi, C. (1984) *J. Biochem. Biophys. Methods* **9,** 103–119.

Righetti, P. G., Gelfi, C. (1994) *J. Cap. Elec.* **1,** 27–35.

Righetti, P. G., Gelfi, C., Bossi, M. L. (1987a) *J. Chromatogr.* **392,** 123–132.

Righetti, P. G., Gelfi, C., Bossi, M.L., Boschetti, E. (1987b) *Electrophoresis* **8,** 62–70.

Righetti, P. G., Gelfi, C., Chiari, M. (1996b) in Righetti, P. G. (Ed.), *Capillary Electrophoresis in Analytical Biotechnology* CRC Press, Boca Raton, FL, pp. 509–538.

Righetti, P. G., Gelfi, C., Chiari, M. (1996c) in B. L. Karger and W. S. Hancock (Eds.), *Methods in Enzymology: High Resolution Separation and Analysis of Biological Macromolecules, Part A: Fundamentals*, Vol. 270, Academic, San Diego, pp. 235–255.

Righetti, P. G., Gelfi, C., Conti, M. (1997b) *J. Chromatogr. B* **699,** 91–104.

Righetti, P. G., Gianazza, E. (1978) *Biochim. Biophys. Acta* **532,** 137–146.

Righetti, P. G., Gianazza, E., Celentano, F. (1986a) *J. Chromatogr.* **356,** 9–14.

Righetti, P. G., Gianazza, E., Gianni, A. M., Comi, P., Giglioni, B., Ottolenghi, S., Secchi, C., Rossi-Bernardi, L. (1979) *J. Biochem. Biophys. Methods* **1,** 47–59.

Righetti, P. G., Macelloni, C. (1982) *J. Biochem. Biophys. Methods* **6,** 1–15.

Righetti, P. G., Morelli, A., Gelfi, C., Westermeier, R. (1986b) *J. Biochem. Biophys. Methods* **13,** 151–159.

Righetti, P. G., Nembri, F. (1997) *J. Chromatogr. A* **772,** 203–211.

Righetti, P. G., Olivieri, E., Viotti, A. (1998a) *Electrophoresis* **19,** 1738–1741.

Righetti, P. G., Saccomani, A., Stoyanov, A. V., Gelfi, C. (1998b) *Electrophoresis* **19,** 1733–1737.

Righetti, P. G., Tonani, C. (1991) *Electrophoresis* **12,** 1021–1027.

Righetti, P. G., Tonani, C. (1992) in Dondi, F., Guiochon, G. (Eds.), *Theoretical Advancement in Chromatography and Related Separation Techniques NATO ASI Series C: Mathematical and Physical Sciences*, Vol. 383, Kluwer Academic, Dordrecht, pp. 581–605.

Righetti, P. G., Tudor, G., Gianazza, E. (1982) *J. Biochem. Biophys. Methods* **6,** 219–227.

Rilbe, H. (1973) *Ann. N.Y. Acad. Sci.* **209,** 11–22.

Rilbe, H. (1996) *pH and Buffer Theory. A New Approach.* Wiley, Chichester.

Riley, R. F., Coleman, M. K. (1968) *J. Lab. Clin. Med.* **72,** 714–720.

Rimpilainen, M., Righetti, P. G. (1985) *Electrophoresis* **6,** 419–422.

Rodriguez-Diaz, R., Zhu, M., Wehr, T. (1997a) *J. Chromatogr. A* **772,** 145–160.

Rodriguez-Diaz, R., Zhu, M., Wehr, T. (1997b) *Electrophoresis* **18,** 2134–2144.

Rovida, E., Gelfi, C., Morelli, A., Righetti, P. G. (1986) *J. Chromatogr.* **353,** 159–171.

Saccomani, A., Gelfi, C., Wajcman, H., Righetti, P. G. (1999) *J. Chromatogr. A* 832, 225–238.

Saglio, G., Ricco, G., Mazza, U., Camaschella, C., Pich, P. G., Gianni, A. M., Gianazza, E., Righetti, P. G., Giglioni, B., Comi, P., Gusmeroli, M., Ottolenghi, S. (1979) *Proc. Natl. Acad. Sci. USA* **76,** 3420–3424.

Schafer-Nielsen, C. (Ed.) (1988) *Electrophoresis '88: Sixth Meeting of the International Electrophoresis Society*, Verlag Chemie, Weinheim.

Scheler, C., Lamer, S., Pan, Z., Li, X., Salnikov, J., Jungblut, P. (1998) *Electrophoresis* **19,** 918–927.

Schomburg, G. (1998) in Khaledi, M. G. (Ed.), *High Performance Capillary Electrophoresis*, Wiley, New York, pp. 481–524.

Sebastiano, R., Gelfi, C., Righetti, P. G., Citterio, A. (2000) *J. Chromatogr. A* **894,** 53–61.

Sebastiano, R., Gelfi, C., Righetti, P. G., Citterio, A. (2001) *J. Chromatogr. A* **924,** 71–81.

Sebastiano, R., Lapadula, M., Righetti, P. G., Gelfi, C., Citterio, A. (2003) *Electrophoresis* **24,** 4189–4196.

Shen, Y., Smith, R. D. (2002) *Electrophoresis* **23,** 3106–3124.

Shevchenko, A., Wilm, M., Mann, M. (1996) *Anal. Chem.* **68,** 850–858.

Shimura, H., Kasai, K. (1995), *Electrophoresis* **16,** 1479–1484.

Shimura, K. (2002) *Electrophoresis* **23,** 3847–3857.

Shimura, K., Wang, Z., Matsumoto, H., Kasai, K. I. (2000) *Electrophoresis* **21,** 603–610.

Simò-Alfonso, E., Gelfi, C., Sebastiano, R., Citterio, A., Righetti, P. G. (1996) *Electrophoresis* **17,** 723–731, 732–737.

Sinha, P. K., Righetti, P. G. (1986) *J. Biochem. Biophys. Methods* **12,** 289–297.

Sinha, P. K., Righetti, P. G. (1987) *J. Biochem. Biophys. Methods* **15,** 199–206.

Sinha, P. K., Poland, J., Schnolzer, M., Rabilloud, T. (2001) *Proteomics* **1,** 835–840.

Slais, K., Friedl, Z. (1994) *J. Chromatogr. A* **661,** 249–256.

Söderberg, L., Buckley, D., Hagström, G., Bergström, J. (1980) *Prot. Biol. Fluids* **27,** 687–691.

Stathakos, D. (Ed.) (1983) *Electrophoresis '82: Advanced Methods, Biochemical and Chemical Applications*, W. de Gruyter, Berlin, 1983.

Steinberg, T. H., Chernokalskaya, E., Berggren, K., Lopez, M. F., Diwu, Z., Haugland, R. P., Patton, W. F. (2000) *Electrophoresis* **21,** 8–496.

Steinberg, T. H., Haugland, R. P., Singer, V. L. (1996a) *Anal. Biochem.* **239,** 238–245.

Steinberg, T. H., Jones, L. J., Haugland, R. P., Singer, V. L. (1996b) *Anal. Biochem.* **239,** 223–237.

Stoyanov, A. V., Gelfi, C., Righetti, P. G. (1997) *Electrophoresis* **18,** 717–723.

Stoyanov, A. V., Righetti, P. G. (1997) *J. Chromatogr. A* **790,** 169–176.

Svensson, H. (1961) *Acta Chem. Scand.* **15,** 325–341.

Svensson, H. (1962a) *Acta Chem. Scand.* **16,** 456–466.

Svensson, H. (1962b) *Arch. Biochem. Biophys.* (*Suppl. 1*), 132–140.

Tan, W., Fan, Z. H., Qiu, C. X., Ricco, A. J., Gibbons, I. (2002) *Electrophoresis* **23,** 3638–3645.

Taverna, M., Tran, N. T., Merry, T., Horvath, E., Ferrier, D. (1998) *Electrophoresis* **19,** 2527–2594.

Thormann, W., Caslavska, J., Molteni, S., Chmelik, J. (1992) *J. Chromatogr*. **589,** 321–327.

Timasheff, S. N., Arakawa, T. (1989) in Creighton, T. E. (Ed.), *Protein Structure, a Practical Approach*, IRL Press, Oxford, pp. 331–345.

Tonani, C., Righetti, P. G. (1991) *Electrophoresis* **12,** 1011–1021.

Towbin, H., Staehelin, T., Gordon, J. (1979) *Proc. Natl. Acad. Sci. USA* **76,** 4350–4352.

Tümmler, B. (Guest Ed.) (1998) Microbial Genomes: Biology and Technology, *Electrophoresis* **19,** 467–624.

Unlii, M., Morgan, E. M., Minden, J. S. (1997) *Electrophoresis* **18,** 2071–2077.

Urwin, V. E., Jackson, P. (1991) *Anal. Biochem*. **195,** 30–37.

Urwin, V. E., and Jackson, P. (1993) *Anal. Biochem*. **209,** 57–62.

Valkonen, K., Gianazza, E., Righetti, P. G. (1980) *Clin. Chim. Acta* **107,** 223–229.

Verzola, B., Gelfi, C., Righetti, P. G. (2000a) *J. Chromatogr. A*, **868,** 85–99.

Verzola, B., Gelfi, C., Righetti, P. G. (2000b) *J. Chromatogr. A*, **874,** 293–303.

Verzola, B., Sebastiano, R., Righetti, P. G., Gelfi, C., Lapadula, M., Citterio, A. (2003) *Electrophoresis* **24,** 121–129.

Vesterberg, O. (1972) *Biochim. Biophys. Acta* **257,** 11–22.

Vesterberg, O. (1973) *Ann. N. Y. Acad. Sci*. **209,** 23–33.

Vesterberg, O., Wadstrom, T., Vesterberg, K., Svensson, H., Malmgren, B. (1967) *Biochim. Biophys. Acta*, **133,** 435–445.

Vuillard, L., Braun-Breton, C., Rabilloud, T. (1995a) *Biochem. J*. **305,** 337–343.

Vuillard, L., Marret, N., Rabilloud, T. (1995b) *Electrophoresis* **16,** 295–297.

Vuillard, L., Rabilloud, T., Leberman, R., Berthet-Colominas, C., Cusack., S. (1994) *FEBS Lett*. **353,** 294–296.

Waheed A. A., Gupta, P. D. (1996) *Anal. Biochem*. **233,** 249–152.

Wehr, T., Zhu, M. D., Rodriguez-Diaz, R. (1996) in Karger, B. L., Hancock, W. S. (Eds.), *Methods in Enzymology: High Resolution Separation and Analysis of Biological Macromolecules, Part A: Fundamentals*. Vol. 270, Academic, San Diego, pp. 358–374.

Westermeier, R. (1997) *Electrophoresis in Practice*, VCH, Weinheim, pp. 81–90.

Williams, K. L. (Guest Ed.) (1998) Strategies in Proteome Research, *Electrophoresis* **19,** 1853–2050.

Williams, Jr., R. C., Shah, C., Sackett, D. (1999) *Anal. Biochem*. **275,** 265–267.

Wirth, P. J. Romano, A. (1995) *J. Chromatogr. A* **698,** 123–143.

Wu, X. Z., Sze, N. S. K., Pawliszyn, J. (2001) *Electrophoresis* **22,** 3968–3971.

Yan, J. X., Wait, R., Berkelman, T., Harry, R. A., Westbrook, J. A., Wheeler, C. H., Dunn, M. J. (2000) *Electrophoresis* **21,** 3666–3672.

Yao, J. G., Bishop, R. (1982) *J. Chromatogr*. **234,** 459–462.

Young, D. S., Anderson, N. G. (Guest Eds.) (1982) Special Issue in Two Dimensional Electrophoresis, *Clin. Chem*. **28,** 737–1092.

Zhu, M. D., Rodriguez, R., Wehr, T. (1991) *J. Chromatogr*. **559,** 479–485.

5

SODIUM DODECYL SULFATE–POLYACRYLAMIDE GEL ELECTROPHORESIS

5.1. INTRODUCTION

Sodium dodecyl sulfate–polyacrylamide gel electrophoresis is perhaps the most popular and direct method for assessing, in a fast and reproducible manner, the M_r of denatured polypeptide chains and the purity of a protein preparation. Accounts on SDS–PAGE can be found in, for example, Andrews (1986), Rothe and Maurer (1986), Rothe (1991), Westermeier (1997), and Shi and Jackowski (1998). In small or large (or very large) gel slab formats, it is the standard second dimension of 2D maps, and that is why it will be treated in detail in this book. Typically, in disc electrophoresis, proteins migrate according to both surface charge and mass, so that discriminating the two contributions is not an easy task [although it can be done in a number of mathematical treatments, such as the Ferguson (1964) plots]. A possible approach to M_r measurements of proteins would be to cancel out differences in molecular charge by chemical means, so that migration would then occur solely according to size. Sodium dodecyl sulfate, an amphipatic molecule, is known to form complexes with both nonpolar side chains and charged groups of amino acid residues in polypeptides of all possible sizes and shapes without rupturing polypeptide bonds. Surprisingly, large amounts of SDS can be bound to proteins, assessed at about 1.4 g SDS/g protein by a number of authors (Pitt-Rivers and Impiombato, 1968; Fish et al., 1970; Reynolds and Tanford, 1970a,b). This means that the number of SDS molecules is of the order of half the number of amino acid residues in the polypeptide chain. This amount of highly charged surfactant molecules is sufficient to overwhelm effectively the intrinsic charges on the polymer chain, so that the net charge per unit mass becomes

Proteomics Today. By Mahmoud Hamdan and Pier Giorgio Righetti
ISBN 0-471-64817-5

approximately constant. Electrophoretic migration is then proportional to the effective molecular radius, that is, to the molecular mass of the polypeptide chain (Shapiro et al., 1967). Although it is an oversimplification to which a number of exceptions are to be found, the relationship does in fact hold true for a very large number of proteins (Weber and Osborn, 1969; Dunker and Rueckert, 1969) and the method has become one of the most widely used for measurement of protein molecular mass. The unique properties of SDS are due to its long, flexible alkyl tail, which is able to establish hydrophobic interactions with all combinations of amino acids, which leads to massive unfolding of proteins. The action of SDS is also due to its ionic head, which can break ionic interactions between proteins and drive an important electrostatic repulsion between SDS–protein complexes. This prevents reassociation of such complexes, even at the very high concentrations encountered in gel electrophoresis. Another important property of SDS lies in the fact that the ionic head is a strong electrolyte, so that it is fully ionized in the pH 2–12 interval, where most of the biochemical separations take place. According to Reynolds and Tanford (1970a), a concentration of SDS above 0.5 mM is sufficient for binding 1.4 g SDS/g protein in a primarily hydrophobic way, provided the disulfide bridges are reduced and the polymer chain is thus in an extended conformation. Under the influence of SDS, proteins assume the shape of rodlike particles the length of which varies uniquely with the molecular mass of the protein moiety, occupying 0.074 nm (1.4 g SDS bound) per amino acid residue. Certain proteins, such as papain, pepsin, and glucose oxidase (Nelson, 1971), and two different classes of proteins, the glycoproteins (Segrest et al., 1971) and the histones (Panyim and Chalkley, 1969), show an "anomalous" behavior toward SDS. Either they bind a relatively low amount of detergent or SDS cannot compensate for their intrinsic charges, such as in the case of histones. But these examples are not sufficient to cast severe doubts on the validity of M_r estimation by SDS (Waehneldt, 1975; Lambin, 1978), and in fact SDS–PAGE has become by far the most popular method for M_r assessments of denatured polypeptide chains.

The reasons for this rapid acceptance are not difficult to see. The apparatus required is readily available in most laboratories and is inexpensive. Once learned, the procedure is straightforward and highly reproducible, and results can be obtained within a few hours using only a few micrograms of material. As with other PAGE methods, in many cases the samples need not to be totally pure. The degree of purity required depends largely upon the sample being studied and the ease with which the component of interest can be identified on the final gel pattern.

In summary, SDS–PAGE is used mainly for the following purposes:

- Estimation of protein size
- Assessment of protein purity
- Protein quantitation
- Monitoring protein integrity
- Comparison of the protein composition of different samples
- Analysis of the number and size of polypeptide subunits
- When using, for example, Western blotting
- As a second dimension of 2D maps

5.2. SDS–PROTEIN COMPLEXES: A REFINEMENT OF THE MODEL

Over the years, some models for the structure of complexes between proteins and SDS have been proposed:

- A *rodlike particle model*, proposed on the basis of hydrodynamic measurements. This was one of the earliest models suggested (Reynolds and Tanford, 1970a), and it hypothesized that, upon binding of SDS, the polypeptide would form "rodlike" structures about 3.6 nm in diameter and 0.074 nm/amino acid residue in length. In a more recent reinterpretation, the particle was described as a "short, rigid, rod-like segment, with intervening regions possessing some flexibility" (Tanford, 1980).
- A *necklace model*, in contradiction with the rodlike particle above. In this second model, "the polymer chain is flexible and micelle-like clusters of SDS are scattered along the chain" (Shirahama et al., 1974; Takagi et al., 1975). In an improved version, SDS binds to the protein in the form of spherical micelles and the polypeptide forms α-helices mostly in the hydrophobic region of the micelles.
- A flexible *α-helix/random-coil* structure, suggested on the basis of circular dichroism changes of proteins upon SDS binding (Mattice et al., 1976).
- A flexible *helix model*, in which the polypeptide chain is helically coiled around an SDS micelle, attached by hydrogen bonds between sulfate group oxygens and peptide bond nitrogens (Lundahl et al., 1986).

Over the years, Lundahl's group has refined this last model (Mascher and Lundahl, 1989; Wallstén and Lundahl, 1990; Ibel et al., 1990, 1994; Lundahl et al., 1992), which has now become the "*protein-decorated, micelle model*. Based on small-angle neutron scattering data, this model proposes that adjacent, protein-decorated, spherical micelles are formed, rather than cylindrical structures, as previously suggested by the same group (Lundahl et al., 1986). These authors have studied the formation of SDS–protein complexes at SDS concentrations in the proximity of its critical micellar concentration (CMC; 1.8 mM SDS in 100 mM buffer). They propose the following series of events in complex formation. First the dodecyl chains of the detergent penetrate the surface of the protein and come into contact with the hydrophobic interior of the protein or of its domains. As a consequence, polypeptide segments from the interior of the protein become displaced toward the surface of the complex, since they are less hydrophobic than the dodecyl chain. Many SDS molecules become locked in the inserted position by ion pair formation and hydrogen bonding. Additional SDS molecules then become included in the complex until a spherical SDS micelle is completed, around which the polypeptide is wound. Any length of polypeptide that, for steric reasons, cannot be accommodated in direct contact with the micelles forms the core for growth of another protein-covered micelle. This process is repeated until the whole polypeptide is coiled around adjacent SDS micelles that are linked with short polypeptide segments. For a protein of average size (50,000 Da) it is believed that this complex could be composed of three protein-decorated SDS micelles not necessarily

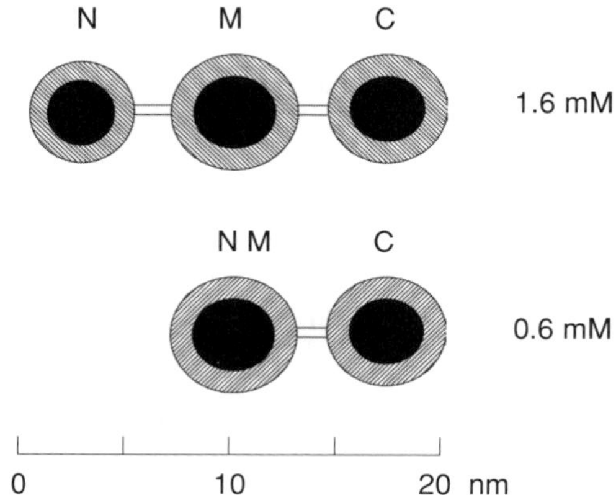

FIGURE 5.1. Schematic scale models of mutual disposition of three-protein decorated micelles of enzyme *N*-5′-phosphoribosylanthranilate isomerase/indole-3-glycerol-phosphate synthase in buffers with SDS concentrations approaching CMC (1.6 mM SDS, top) and of two protein-decorated micelles in buffers below CMC (0.6 mM SDS, bottom). The independent micelles N (N-terminal), M (middle), and C (C-terminal) of the three-micelle complex are connected, at saturation, by flexible oligopeptide linkers about five or six amino acid residues long; with decreasing SDS concentrations to subsaturating levels, the small micelle N coalesces with M and micelle C approaches M. The hydrophobic micelle core is filled in black, whereas the outer hydrophilic shell, formed by hydrophilic stretches of the polypeptide chain and the SDS sulfate groups, is not shaded (From Ibel et al., 1994, by permission.)

of equal size. For example, in the case of the enzyme *N*-5′-phosphoribosylanthranilate isomerase/indole-3-glycerol-phosphate synthase ($M_r = 49{,}484$ Da) the SDS–protein complex contained the dodecyl hydrocarbon moieties in three globular cores, of which the central one was the largest (see Fig. 5.1; notice, though, that at SDS levels below the CMC the micelles can be reduced to only two). Each core was surrounded by a hydrophilic shell, formed by the hydrophilic and amphiphilic stretches of the polypeptide chain, and by the sulfate groups of the detergent, whereas, presumably, most or all nonpolar side groups of the polypeptide chain penetrate partly or completely into the hydrophobic micelle cores (Ibel et al., 1990). The model has received support from a recent paper by Westerhuis et al. (2000) reevaluating the migration behavior of SDS–protein complexes. These authors have detected at least two independent electrophoretic migration mechanisms for SDS–protein micelles: (i) for proteins in the 14–65-kDa range in a $T < 15\%$ polyacrylamide matrix, linear Ferguson plots suggested that they migrated ideally and that their effective radii could be estimated in this manner; (ii) concave plots at higher gel concentrations and for complexes derived for larger proteins indicated that migration in these cases could be described by reptation theory. Migration of the large proteins at lower gel concentrations and small proteins at higher gel concentrations was not well described by either theory, representing intermediate behavior not contemplated by these mechanisms. Such data support the model in which all but the smallest SDS–protein complexes adopt a necklacelike structure in which spherical micelles are distributed along the unfolded polypeptide chains, as proposed by Lundahl's group and also by Samsò et al. (1995).

5.3. THEORETICAL BACKGROUND OF M_r MEASUREMENT BY SDS–PAGE

Methods for measurements of molecular size by electrophoresis fall into two main categories, namely those using a relationship between mobility and various gel concentrations (Ferguson, 1964) and those for which a single concentration is used. The latter can only be applied to families of molecules with the same m/z ratio, such as nucleic acids, or to molecules in which uniform charge densities have been produced by binding large amounts of a charge ligand, such as in the case of SDS.

Ferguson plots try to discriminate the contributions of charge and mass, to the electrophoretic mobility of proteins, according to the following relationship:

$$\log R_f = \log Y_0 - K_R T$$

where R_f is the relative protein mobility (corrected to the mobility of a fast migrating dye), Y_0 the extrapolated free mobility on the y axis, T the polyacrylamide gel concentration, and K_R (coefficient of retardation) the slope of the curve obtained in a plot of log R_f versus $\%T$ and is proportional to the mass of the macromolecule subjected to electrophoresis in a series of gels at various $\%T$. When a series of Ferguson plots of R_f versus $\%T$ were constructed for a number of SDS-treated proteins (Neville, 1971), it was found that the intercept at $T = 0$, which is the apparent free mobility (Y_0), was almost identical for all the species examined. This demonstrates that, for SDS-laden proteins, the effective m/z ratio is, to a first approximation, constant. To determine M_r from K_R values, all that is necessary is to show that there is a uniform dependence of M_r on K_R. Rodbard and Chrambach (1970, 1971) and Chrambach and Rodbard (1971) have shown that K_R is dependent on the effective molecular radius. Since Reynolds and Tanford (1970b) and Fish et al. (1970) have shown that the hydrodynamic properties of protein–SDS complexes are a unique function of polypeptide chain length, K_R must clearly be dependent on this also. In practice, plots of log K_R versus log M_r are generally linear with slopes proportional to the relationship between M_r and the Stokes radii. It must be emphasized that plots of log M_r versus R_f (in gels of constant $\%T$) are linear only over a certain range of mass values, typically between 15,000 and 70,000 Da (see Fig. 5.2); however, if porosity gradients are used (e.g., gels ranging in concentration between $T = 5\%$ and $T = 20\%$), plots of log M_r versus log $\%T$ are linear over a much broader M_r range (Fig. 5.3) and are thus to be preferred over the previous ones when highly dispersed samples are to be analyzed. In this last case, the relationship found (Rothe, 1982) has been

$$\log_{10} M_r = a \log_{10} T + b$$

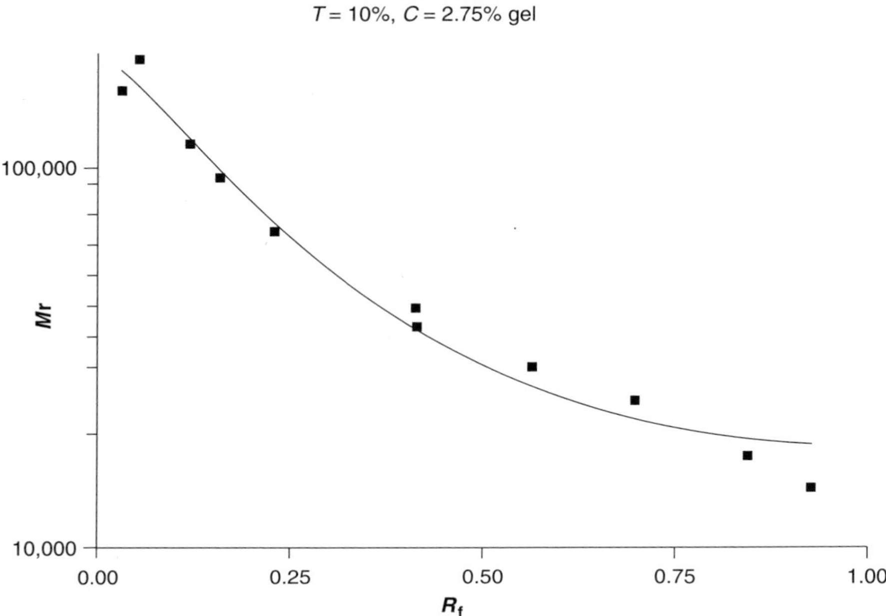

FIGURE 5.2. Calibration curve of log M_r vs. R_f plotted with series of markers in 10,000–200,000-Da range separated on constant $\%T$ ($T = 10\%$ at $C = 2.75\%$) SDS–PAGE slab. Note the deviation from linearity. The markers are (M_r in kDa): myosin (194); RNA polymerase (β-subunit) (160); β-galactosidase (116); phosphorylase B (94); RNA polymerase (σ-subunit) (95); bovine serum albumin (68); ovalbumin (43); RNA polymerase (α-subunit) (38.4); carbonic anhydrase (30); trypsinogen (24.5); β-lactoglobulin (17.5); lysozyme (14.5).

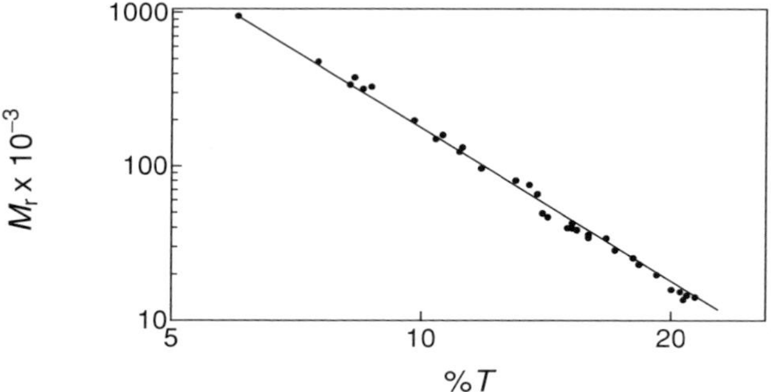

FIGURE 5.3. Calibration curve of log M_r vs. log $\%T$ for series of 34 standard proteins covering M_r 13,000–95,000 range, separated on $T = 3\%$ to $T = 30\%$ (at constant $C = 8.4\%$) linear gradient SDS–PAGE slab gel. (From data of Rothe, 1982.)

where a and b are the slope and intercept, respectively, of the linear regression line, which is established from measurements of $\%T$ for a number of standard proteins of known M_r run at the same time on the same SDS–PAGE gel. This relationship holds true and results in linear plots of $\log_{10} M_r$ versus $\log_{10} T$ for gels with any shape of acrylamide concentration gradient (Lambin, 1978; Poduslo and Rodbard, 1980). The practical difficulty with the use of gradient gels for M_r determination is that the accuracy of the method depends on a knowledge of the precise shape of the gradient, such that $\%T$ can be accurately estimated for each protein band. This is usually the case in linear gradients, since $\%T$ will be easily estimated by measurements of the distance migrated by the protein zone. When using nonlinear porosity gradients, this is more difficult. The range of M_r values over which there is a linear relationship between $\log_{10} M_r$ and $\log_{10} T$ depends on the acrylamide gradient employed. Small polypeptides (<13,000 Da) can be analyzed if high concentrations of urea are added to the gradient gel system (Hashimoto et al., 1983). Alternatively, it has been found that small polypeptides can be effectively separated using a tricine, rather than a glycine, discontinuous buffer for SDS–PAGE (Schägger and von Jagow, 1987). This last procedure is able to separate peptides with M_r values as low as 1000 Da. An alternative approach for determining M_r using linear gradient SDS–PAGE gels has been developed by Rothe and Maurer (1986). A straight-line plot is produced from the relationship

$$\log_{10} M_r = a\sqrt{D} + b$$

where D is the migration distance (mm) and a and b the slope and intercept of the straight line (see Fig. 5.4). Linearity is independent of the buffer system, the concentration of the crosslinker in the range $C = 1\%$ to $C = 8\%$, and the concentration range of the gradient within the range $T = 3\%$ to $T = 30\%$ at a gel length between 8 and 15 cm. However, the values of a and b are altered by changes in these parameters.

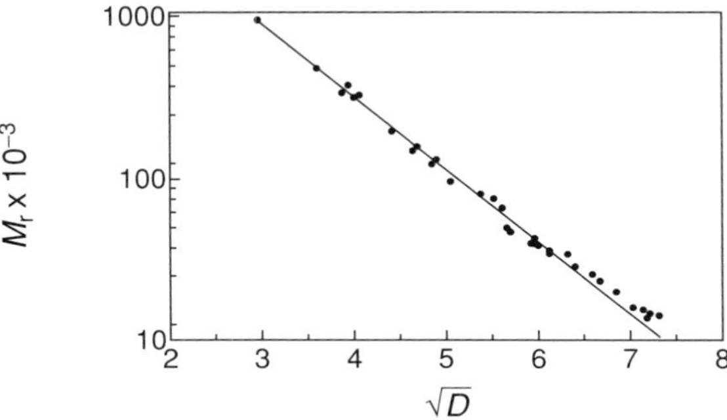

FIGURE 5.4. Method of Rothe and Maurer (1986) for determining M_r using linear gradient SDS–PAGE. Plot shows the linear relationship between log M_r and $\sqrt{D}$ for a series of 34 standard proteins covering the M_r 13,000–95,000 range, separated on $T = 3\%$ to $T = 30\%$ (at constant $C = 8.4\%$) linear gradient SDS–PAGE slab gel. (From data of Rothe, 1982.)

The linear relationship between $\log_{10} M_r$ and $\sqrt{D}$ is, in practice, time independent, so that estimations of M_r can be made when the optimal resolution for the particular protein sample being analyzed has been obtained.

The typical standard error in assessment of the polypeptide subunit M_r by SDS–PAGE could be as high as $\pm 10\%$, definitely much too high for today's standards but acceptable in the 1970s, when the technique was adopted in most laboratories. Today, of course, with the availability of MS measurements, the M_r of macromolecules subjected to SDS–PAGE can be determined with absolute precision, the error being typically of the order of ± 1 Da over 10,000 Da (Burlingame et al., 1998). This error could be much larger for proteins exhibiting anomalous behavior; Neville (1971) has pointed out that analytes with abnormal migration in SDS–PAGE can be readily observed by using plots of log K_R versus log M_r, instead of the common semilog plot of log R_f versus M_r, as proposed by the discoverers of the technique (Shapiro et al., 1967). The major sources of error in the above are the assumptions that Y_0 is the same for all the standards and unknown proteins and also that there is a constant relationship between the effective molecular radius and the M_r values. The assumption of a constant value for Y_0 implies that the free mobility of protein–SDS complexes is independent of size and charge. Independence of charge, as stated above, requires the binding of a large amount of SDS, so that the intrinsic charge inherent to the polypeptide chain makes an insignificant contribution to the net charge of the complex; the other requirement is that the weight of SDS bound to a given weight of protein is the same for both the standard proteins and the unknowns. To achieve complete binding of SDS, it is important the polypeptide chains are not conformationally constrained, which implies that disulfide bridges should always be reduced. As an example, it was reported (Pitt-Rivers and Impiombato, 1968; Reynolds and Tanford, 1970b) that BSA and ribonuclease bound only 0.9 g of SDS per gram of protein without reduction but bound the usual value of about 1.4 g upon reduction of disulfide bonds. In addition to conformation effects, also the amino acid composition of proteins may give rise to anomalous behavior on electrophoresis. This is either caused by an atypical degree of SDS binding per gram of protein or because the intrinsic charge of the polypeptide chain makes a significant contribution to the net charge of the protein–SDS complex, so that Y_0 deviates markedly from the average value. According to Panyim (1971), the latter case occurs with histones, which migrate slower than would be expected on the basis of their known M_r's. Also, many glycoproteins behave anomalously even when SDS and thiol reagent are in excess, probably because they bind SDS only to the proteinaceous part of the molecule. The reduced net charge resulting from decreased SDS binding lowers the polypeptide mobility during electrophoresis, yielding artifactually high M_r estimates. However, Segrest and Jackson (1972) have found that, with increasing polyacrylamide gel concentration, molecular sieving predominates over the charge effect and the apparent M_r's of glycoproteins decrease and approach their real M_r's. Thus, one could perform a series of SDS runs at increasing $\%T$ and extrapolate an asymptotic minimum M_r or, as suggested by Lambin (1978) and Lambin and Fine (1979), use directly a pore gradient gel. An alternative way for assessment of glycoprotein M_r's by SDS–PAGE is to use Tris–borate–EDTA buffer systems (Poduslo, 1981). At alkaline pH, borate ions can form complexes with neutral sugars, converting them into charged species. The formation of such borate complexes could increase the net negative charge, which would offset the decreased binding of SDS to glycoproteins, resulting in a charge density producing migration rates in SDS gels that now correlate with their molecular sizes. Finally, polypeptides with M_r's below ~10,000 are not well resolved on uniform concentration polyacrylamide gels. Their separation can be improved in pore gradient gels or in 8 M urea–SDS gels containing a high percentage of crosslinker (Swank and Munkres, 1971).

5.4. METHODOLOGY

5.4.1. Purity and Detection of SDS

Purity of commercial preparations of SDS is fundamental for reproducible results. Swaney et al. (1974) reported anomalous behavior in the banding patterns upon SDS–electrophoresis of foot-and-mouth disease virus polypeptides, which they attributed to the fact that SDS contains a contamination in excess of 10% with chains longer than 12 carbon atoms (e.g., C_{14} to C_{22}). Such contaminants could perturb the binding ratio of SDS to proteins away from the expected value of 1.4 g SDS/g protein. That this could be the case was also demonstrated by Dohnal and Garvin (1979). They found that the major contaminant of SDS is STS (sodium tetradecyl sulfate), which is often present at levels from 10 to 30% in SDS. The STS has much greater affinity for proteins than SDS, so that its removal from the protein moiety is extremely difficult. Apparently, it is the presence of STS in the SDS–polypeptide micelle which is responsible for the staining of proteins by pinacryptol yellow, as reported by Stoklosa and Latz (1974). The purity of SDS preparations can be checked by gas chromatography after hydrolysis to 1-dodecanol in 4 N HCl for 2 h at 100°C (Sigrist, 1974). Alternatively, SDS as such can be injected into the gas chromatograph, where it undergoes quantitative in situ conversion to the corresponding alcohol (Malin and Chapoteau, 1981). After an SDS fractionation, if the protein has to be recovered free of SDS, it is important to have a sensitive analytical method for its determination in the presence of protein. This can be done in a simple way by complexing this surfactant with cationic dyes, such as *p*-rosaniline (Jarush and Sonenberg, 1950) or methylene blue (Mukerjee, 1956). A spectrophotometric assay for SDS using acridine orange has also been described (Sokoloff and Frigon, 1981). By today's standards, it is doubtful that such highly contaminated preparations of SDS should still be around and sold commercially; nevertheless, if something were to go wrong, it would pay to remember that these problems are associated with the early days of the technique.

5.4.2. Molecular Mass Markers

It is important to have at hand a wide range of M_r markers so as to explore any polypeptide size. Those listed in Table 5.1 should suffice for most purposes since they span an interval of about 1000-fold M_r increments (although markers smaller than 10 kDa are rarely used). Kits covering limited M_r intervals are in general available

TABLE 5.1. Molecular Mass Markers for SDS Electrophoresis

Protein	M_r	Protein	M_r
Bacitracin	1,480	Enolase (muscle)	41,000
Glucacon	3,500	Ovalbumin	43,000
Insulin (reduced)	6,600	Fumarase (muscle)	49,000
Trypsin inhibitor (lima bean)	9,000	IgG heavy chains	50,000
Cytochrome c (muscle)	11,700	Glutamate dehydrogenase (liver)	53,000
α-Lactalbumin	14,176	Pyruvate kinase (muscle)	57,000
Lysozyme	14,314	Catalase (liver)	57,500
Ribonuclease B	14,700	Bovine serum albumin	66,290
Hemoglobin (β-chains)	15,500	Transferrin	76,000
Avidin	1,600	Plasminogen	81,000
Myoglobin	17,200	Lactoperoxidase	93,000
β-Lactoglobulin B	18,363	Phosphorylase a (muscle)	100,000
Soybean trypsin inhibitor	20,095	Ceruloplasmin	124,000
Trypsin	23,300	β-Lactosidase (*Escherichia coli*)	130,000
Chymotrypsinogen	25,666	Serum albumin (dimer)	132,580
Carbonic anhydrase B	28,739	Immunoglobulin G (unreduced)	150,000
Carboxypeptidase A	34,409	Immunoglobulin A (unreduced)	160,000
Pepsin	34,700	α_2-Macroglobulin (reduced)	190,000
Glyceraldehyde-3-phosphate dehydrogenase (muscle)	35,700	Myosin (heavy chain)	220,000
		Thyroglobulin	335,000
Lactate dehydrogenase (muscle)	36,180	α_2-Macroglobulin (unreduced)	380,000
Aldolase (muscle)	38,994	Immunoglobulin M (unreduced)	950,000
Alcohol dehydrogenase (liver)	39,805		

from several commercial sources (e.g., Bio-Rad, Pharmacia, Serva, B.D.H.). For example, Bio-Rad offers three kits, called low, high, and broad range, to be used as standards for separations in the 14,000–90,000, 45,000–200,000, or 6500–200,000 M_r intervals, respectively. In addition, a polymeric series can be prepared by crosslinking the polypeptide chains of some small proteins [e.g., lysozyme, hemoglobin, bovine serum albumin (BSA)]. Payne (1973) has used glutaraldehyde in this way for preparing soluble polymers with M_r values from 3×10^4 up to 2×10^7; Wolf et al. (1970) have obtained similar polymer families by crosslinking with diethylpyrocarbonate. However, these crosslinked oligomers have come under some criticism (Steele and Nielsen, 1978), since they tend to migrate faster than regular proteins, thus leading to overestimation of apparent M_r's by 5% to 15% when they are used as calibration markers. A variety of M_r standards are also commercially available, as listed below:

(a) Prestained protein standards prepared by covalent attachment of chromophores. However, a possible drawback of this type of standard is that the M_r values are not predictable after modification. Indeed, each batch lot of the same proteins may have different apparent sizes by SDS–PAGE;

(b) Biotin- or ^{14}C-labeled SDS–PAGE standards allowing accurate M_r assessments directly on Western blots.

(c) Protein ladders (e.g., from Gibco BRL) consisting of equally spaced (in mass terms) proteins prepared by controlled oligomerization with suitable crosslinkers.
(d) SDS–PAGE standards prepared to give even band intensities with no extraneous bands when detected by silver staining.

5.4.3. Prelabeling with Dyes or Fluorescent Markers

It is in SDS–PAGE that protein prelabeling with a dye or fluorescent marker has found a major application. In general, introduction of these markers changes the net molecular charge by reactions involving the terminal α-NH_2 and ε-NH_2 groups of Lys residues. In disc electrophoresis or in IEF and IPG this would be anathema, as it would generate a series of bands of slightly changed mobility or pI from otherwise homogeneous proteins. In SDS–PAGE, the intrinsic charge on the polypeptide chain is not important (to a given extent), so labeled and unlabeled molecules migrate together according to their size only. The small increment in size caused by the introduction of the label cannot usually be detected by SDS–PAGE (given the fact that, in order to resolve adjacent bands, one would need a mass increase of a minimum of 2000 Da) and would typically result, at worst, in a more diffuse analyte zone (of course, such size increments would be readily detectable by modern MS techniques, such as MALDI–TOF) (Burlingame et al., 1998; Ogorzalek Loo et al., 1996; Jeannot et al., 1999; Liang et al., 1996; Strupat et al., 1994).

Griffith (1972) proposed prelabeling with Remazol dye (which reacts with –SH groups, primary and secondary amines, and alcoholic –OH), while Bosshard and Datyner (1977) used Drimarene brilliant blue or Uniblue A. Inouye (1971) has incorporated into proteins the fluorescent label 1-dimethylaminonaphthalene-5-sulfonyl chloride (dansyl chloride). The detection limit was as low as 8 ng protein/band. Other fluorescent tags are 4-phenylspiro [furan-2 (3H), 1′-phthalan-3,3′-dione] (fluorescamine), as suggested by Ragland et al. (1974) or MDPF [2-methoxy-2,4-diphenyl-3(2*H*)furanone] (Barger et al., 1976) or *o*-phthaldialdehyde (OPA, Weidekamm et al., 1973). The MDPF was reported to be about 2.5 times more sensitive than fluorescamine while OPA may be as much as one order of magnitude or more so. Use of these dyes is summarized in Table 5.2. Another labeling agent is *N*-iodoacetyl-*N*′-(5-sulfo-1-naphthyl)ethylenediamine (IAEDANS), a modified iodoacetic acid reagent formed by linking it to an amino–naphtholsulfonic acid fluorophor. The reagent specifically modifies Cys residues and exhibits an excitation maximum at 340 nm and an emission maximum at 418 nm (Ursitti et al., 1995). Of course, if the SDS step is used as the second dimension of 2D maps and the protein has to be eluted and its mass assessed by MALDI–TOF for elucidating its identity by interrogation of databases, prelabeling would be disastrous, since often the extent of reaction cannot be ascertained with precision and this would lead to substantial errors in mass measurements. Thus one would have to resort to postlabeling (see below) or to other means for protein in-run detection without covalent attachment of a label. Chen and Chrambach (1996) and Yarmola et al. (1998) have recently proposed an in situ detection step relying on placing a fluorescing paper under the plate supporting the gel in an

TABLE 5.2. Prelabeling of Proteins Prior to SDS–PAGE

Method	Excitation Wavelength λ_{exc}	Emission Wavelength λ_{emis}
Remazol Dye		
1. Mix 0.2 mL protein (2–10 mg/mL) in 0.15 M NaCl with 0.05 mL 1 M Na phosphate (pH 7.2–9.2).		
2. Add 0.05 mL Remazol BBR (10 mg/mL) in 10% SDS.		
3. Heat at 100°C for 5 min or 56°C for 10 min; add 1% β-mercaptoethanol and repeat the heating cycle.		
4. Apply 4-μL samples to gels.		
Dansylation		
1. Mix protein solution (4 mg/mL) in 2% $NaHCO_3$ with an equal volume of dansyl chloride in acetone (2 mg/mL).	340	520
2. Incubate at 37°C for 2 h in the dark; shake intermittently.		
3. Precipitate protein with 2 vol acetone.		
4. Centrifuge; wash precipitate with 2 vol acetone.		
5. Dissolve precipitate in buffer and pretreat with SDS as usual.		
Fluorescamine		
1. Dissolve 50–100 μg protein in 100 μL 15 mM Na phosphate buffer (pH 8.5) in 5% SDS and 5% sucrose.	390	475
2. Heat at 100°C for 5 min.		
3. Cool; add 5 μL fluorescamine (1 mg/mL) in acetone and shake.		
4. Add 5 μL bromophenol blue tracking dye (5 mg/mL) if needed.		
5. Apply 10-μL sample to gel.		
2-Methoxy-2,4-diphenyl-3 (2H)-furanone (MDPF)		
As for fluorescamine, but at 3 use MDPF (2 mg/mL) in dimethyl sulfoxide.	390	480
O-Phthaldialdehyde		
1. Mix 1-mL protein solution (0.1–50 μg) in 0.05 M Na phosphate, pH 8.5, with 2.5 μL β-mercaptoethanol.	340	460
2. Let stand for 10 min.		
3. Add 2.5 μL of 1% *o*-phthaldialdehyde in methanol.		
4. Place in the dark for 2 h.		
5. Pretreat with SDS in the usual way and apply to gel.		

automatic scanning instrument called HPGE-1000 (LabIntelligence, Belmont, CA). The sensitivity, however, was pretty low (being based on fluorescence quenching, it was 20-fold less sensitive than direct fluorescence staining). More recently, Yefimov et al. (2000) reported a novel fluorescent-labeling procedure based on Cascade blue acetyl azide (Molecular Probes, Eugene, OR). Curiously, this label does not react covalently with the protein, yet it stays bound to the SDS–protein micelle (apparently only when using a triethylammonium–barbiturate buffer!), so that the protein can be monitored in situ, yet when eluted and analyzed by MS it would give the correct mass value. The proteins were also found not to be subjected to other mass modifications so commonly encountered when running (and eluting) proteins in polyacrylamide gels because the SDS runs occurred in 3% agarose gels. It remains to be seen if this procedure will give the fine-tuned mass resolution so typical of SDS–PAGE and if it will really be an improvement over existing protocols, since here too sensitivity is very low (10 μg/lane is loaded, which would result in a sensitivity even lower than with Coomassie blue).

5.4.4. Postelectrophoretic Detection

Although the stain protocols described in the previous chapter (isoelectric focusing) in general apply, it should be remembered that the presence of SDS might interfere, so that some procedures might have to be adapted to this technique. For example, the standard Coomassie brilliant blue R-250 procedure would work here, but it would be preferable to perform the staining step by incubating at 50°C, since at this temperature the SDS micelle will disaggregate more quickly and diffuse more rapidly outside the gel matrix. Of course, silver-staining protocols fully apply (a number of them are described by Merril and Washart, 1998), with the proviso that, if the protein has to be eluted for further analysis, such staining protocols are in general incompatible with subsequent MS analysis. Below, a classical silver nitrate staining procedure is described.

5.4.4.1. Nondiamine, Silver Nitrate Stain In this type of stain, silver ions released from silver nitrate under acidic conditions are reduced to metallic silver when the gels are placed in an alkaline solution containing a reducing agent. In the reduction reaction formaldehyde is oxidized to formic acid, which is buffered by sodium carbonate (Merril and Washart, 1998). The procedure below is optimized for gels of <1 mm in thickness and the volumes given apply for gels of 16×16 cm in size. All steps should be performed in glass trays with only one gel per tray. The trays should be placed on a shaker at a gentle speed. All steps are run at room temperature.

- After electrophoresis, place each gel into 500 mL of fixing solution (50% v/v methanol, 10% v/v acetic acid). Soak the gels for 1 h. Staining can occur just after 1 h; however, the gels may remain in this solution overnight, if needed.
- Transfer each gel from the fixing solution into 500 mL of rehydration solution (10% v/v methanol, 5% v/v acetic acid). Soak the gels in this solution for 10 min.

- Discard the rehydration solution and add 200 mL of glutaraldehyde solution (1% w/v) to each tray. Equilibrate the gels in this solution for 30 min.
- Discard the glutaraldehyde solution and wash each gel with 500 mL of deionized water for 15 min.
- Eliminate the wash and add to each gel 200 mL of dichromate solution (34 mM potassium dichromate, 32 mM nitric acid). Equilibrate the gels in this solution for 5 min.
- Discard the dichromate solution and add to each gel 200 mL of staining solution (0.118 mM silver nitrate).
- Eliminate the staining solution.
- Wash each gel briefly (approximately 5 s) with 50 mL of reducing solution (0.283 M sodium carbonate, 7 mM formaldehyde) and then discard this wash. Follow this initial rinse with another 500 mL of the same reducing solution. If colloidal particles collect in the solution prior to full image development, replace the reducing solution with fresh solution. Continue the development in this solution until the proteins are sufficiently stained. Proteins should become visible within 1–3 min.
- To block the image development, dispense with the reducing solution and add 500 mL of stop-bath solution (3% v/v acetic acid). Equilibrate the gels in this solution for 5 min.
- Dispose of the stop-bath solution. Wash each gel twice with 500 mL of distilled water with 10% methanol or ethanol for 10 min each wash.
- Store the gels moist in sealed, clear plastic bags.

Caution: Artifactual bands in SDS–PAGE gels are often observed when using highly sensitive silver-staining protocols. Skin keratins contaminating the protein samples or the buffers are believed to be a major cause of these bands, appearing in the 50–68-kDa region (Ochs, 1983; Bérubé et al., 1994; Shapiro, 1987). A recent report (Yokota et al., 2000) suggests some remedies to it based on the observation that such spurious bands seem to originate from the polyacrylamide gels themselves rather than from the protein samples, sample buffer, or electrode buffer. In addition, the gel region responsible for such spurious keratin bands seems to be located at or near the sample application wells in the polyacrylamide gels (contaminated combs?). On these grounds, the remedy proposed is quite simple: Prior to the actual run and to real sample application, fill the wells with only sample buffer, devoid of protein sample. Conduct electrophoresis with reversed polarity (5 min with a constant current of 40 mA in a standard Laemmli buffer); remove the sample buffer (into which the keratins will have collected). At this point resume the normal gel polarity, add your sample the wells, and run your SDS–PAGE separation. Upon silver staining, the spurious keratin bands will have almost completely disappeared (Yokota et al., 2000). A number of silvering protocols compatible with subsequent protein analysis by MS have been recently described (Sinha et al., 2001; Mortz et al., 2001; Yan et al., 2000b; Gharahdaghi et al., 1999).

5.4.4.2. Colloidal Staining In the following sections, methods will be presented exploiting staining with the Coomassie dyes, a family of dyes extremely popular in electrophoresis due to their relatively high sensitivity and simplicity of use. Today, they are particularly exploited in their micellar form, which imparts to them a much higher sensitivity. Figure 5.5 gives the formulas of the two most common ones, the R-250 (R refers to its reddish hue) and G-250 (G for its greenish hue) forms. The latter has been exploited, in a colloidal form, by Neuhoff et al. (1985, 1988) for a staining method quite popular today in the second dimension of a 2D gel. This method has a high sensitivity (~30 ng protein/band) but requires overnight for development. The procedure is as follows:

- Slowly add 100 g of ammonium sulfate to 980 mL of a 2% phosphoric acid solution until it has completely dissolved. Bring it to 1 L volume by adding 20 mL of a solution of 1 g of Coomassie brilliant blue G-250 in water. Shake before use. This staining solution can be used several times.
- Fix the gel for 1 h in 12% (w/v) TCA.
- Stain overnight with 160 mL of staining solution (0.1% w/v Coomassie G-250 in 2% phosphoric acid, 10% ammonium sulfate) plus 40 mL of methanol (added during staining).
- Wash for 1–3 min in 0.1 mol/L Tris–phosphate buffer, pH 6.5.
- Rinse briefly in 25% (v/v) aqueous methanol.
- Stabilize the protein–dye complex in 20% ammonium sulfate.

A modified Neuhoff's colloidal Coomassie blue G-250 stain has been recently reported, dubbed "blue silver" on account of its considerably higher sensitivity, approaching the one of conventional silver staining (Candiano et al., 2004). The main modifications, as compared to the Neuhoff protocol, were a 20% increment in dye concentration (from 0.1% up to 0.12%) and a much higher level of phosphoric acid in the recipe (from 2% up to 10%). The "blue silver" exhibits a much faster dye uptake

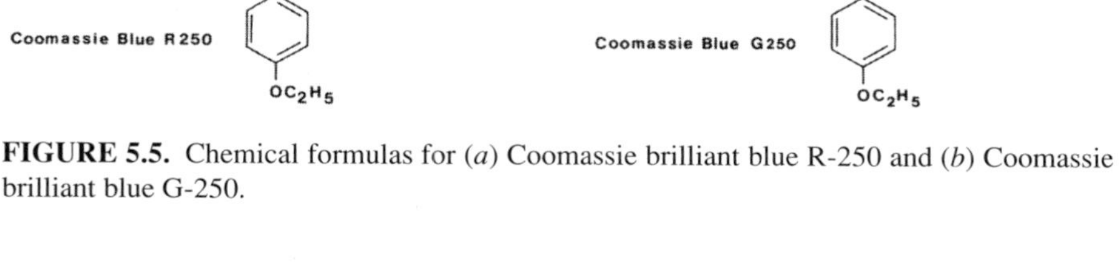

FIGURE 5.5. Chemical formulas for (*a*) Coomassie brilliant blue R-250 and (*b*) Coomassie brilliant blue G-250.

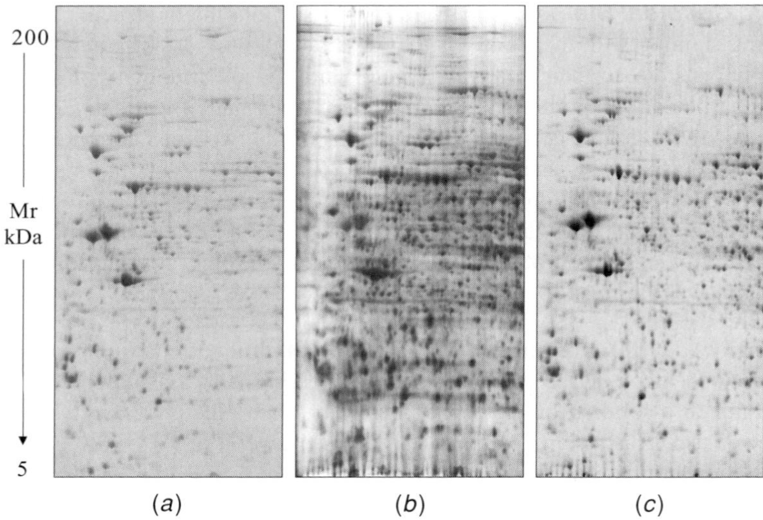

FIGURE 5.6. Comparison of stain sensitivity after 2D map of human kidney tubular epithelial cells of three stains tested: (*a*) Neuhoff; (*b*) silver stained; (*c*) blue silver. The three panels show an identical portion of the 2D map. In all cases the total protein load was 100 μg. (From Candiano et al., 2004, by permission.)

(80% during the first hour of coloration, versus none with a commercial preparation from Sigma). Even at equilibrium (24 h staining), the blue silver exhibits a much higher sensitivity than all other recipes, approaching (but lower than) the one of the classical silver stain. Measurements of stain sensitivity after SDS–PAGE of BSA gave a detection limit (signal-to-noise ratio >3) of 1 ng in a single zone. The somewhat lower sensitivity of blue silver as compared to classical silvering protocols in the presence of aldehydes is amply compensated for by its full compatibility with MS of eluted polypeptide chains after a 2D map analysis, thus confirming that no dye is covalently bound to (or permanently modifies) any residue in the proteinaceous material. It is believed that the higher level of phosphoric acid in the recipe, and thus its lower final pH, helps in protonating the last dissociated residues of Asp and Glu in the polypeptide coils, thus greatly favoring ionic anchoring of dye molecules to the protein moiety. Such a binding, though, must be followed by considerable hydrophobic association with the aromatic and hydrophobic residues along the polypeptide backbone. An example of the sensitivity achieved is shown in Figure 5.6.

5.4.4.3. "Hot" Coomassie Staining A number of recent reports have appeared recommending high temperatures for quick staining processes while guaranteeing high sensitivity. The Kurien and Scofield (1998) method involves a 5-min staining step at 70°C followed by a 20-min destaining step at the same temperature and a rinsing step with distilled water at room temperature for a further 20 min (total processing time of 45 min; sensitivity no greater than 100 ng/band). In another approach

(Wu and Welsh, 1995), 2.5% bleach at 55°C followed by rinsing of the gel in distilled water has been recommended as a method for speeding up the destaining process (90 min total processing time; 100 ng protein/band as upper detection limit). As a latest report, the evolution of a protocol published by Fairbanks et al. (1971), who suggested three staining solutions of progressively lower concentration of Coomassie and isopropanol for increased sensitivity, requiring up to two working days has been proposed. This method (Wong et al., 2000) comprises the following steps:

- After SDS–PAGE, place the gels in plastic boxes (Nalgene, Tupperware) with a hole on the lid. Add 100 mL of staining solution (0.05% Coomassie, 25% isopropanol, 10% acetic acid).
- Heat the gel in staining solution in a conventional 100-W-output microwave oven on full power until the boiling point is reached (~2 min).
- Cool at room temperature for 5 min with gentle shaking. After this step, bands with ~100 ng protein could be visualized despite the blue background.
- Discard the staining solution, rinse with distilled water, and immediately discard.
- Add 100 mL of a new staining solution (0.005% Coomassie, 10% isopropanol, 10% acetic acid) and microwave again to the boiling point (~1 min and 20 s).
- Discard the hot staining solution, rinse with distilled water, and discard. At this step, bands with ~50 ng protein could be observed.
- Add 100 mL of a third staining solution (0.002% Coomassie, 10% acetic acid) and again microwave to the boiling point for 1 min and 20 s.
- Discard this third stain and again rinse with distilled water and discard. At this time, bands with ~25 ng protein are visible.
- Finally, place the gels in 100 mL of destaining solution (10% acetic acid) and microwave for 1 min and 20 s. Allow to cool at room temperature for ~5 min. At this point, bands with at least 5 ng protein become visible against a light blue background. If this last step is repeated 2 or 3 times (or if the gel is left on a shaker for 15 more min), as little as 2.5 ng protein can be seen, this time against a clear background.

This method is claimed to be faster and more sensitive than any other Coomassie procedure so far reported. It appears that bringing the solution containing the gel to the boiling point is a prerequisite for the improved sensitivity and speed of this method.

5.4.4.4. Turbidimetric Protein Detection (Negative Stain) Although this staining procedure has low sensitivity, it might be useful when extracting proteins to be analyzed by MS, since often dye molecules tend to stick to the protein surface and give an erroneous signal by MS. The method is based on the observation that, when an SDS-containing gel is chilled at 0°C to 4°C, free SDS precipitates in the gel, forming an opaque background, whereas the SDS bound to protein does not precipitate, leaving a clear protein band (Wallace et al., 1974). The method presented below involves the use of high-ionic-strength solutions for precipitating SDS and producing this effect more reliably, instead of inducing SDS precipitation by chilling (Ursitti et al., 1995):

- Immediately after the SDS step, place the polyacrylamide gel containing the separated proteins into a covered basin with 10–12 gel volumes of 4 M Na acetate.
- Place on a platform shaker at low speed for 40–50 min for 1-mm-thick gels or longer for thicker gels.
- View the gel over a black, nonreflective surface using a fluorescent desk lamp to illuminate the gel from the side. If the bands are not yet visible, allow longer times for precipitation. The incubation should not proceed for too long, though, otherwise the bands will diffuse and resolution will be lost.
- For recovery of proteins by electroelution, soak gel slices in water to reduce the ionic strength.

5.4.4.5. Negative Metal Stains Similar to the method reported in Section 5.4.4.4, there exist a family of negative stains which are based on the formation of insoluble metal salts in the presence of SDS, leaving protein bands unstained when viewed against a dark background. The hypothesis is that all metal cations (e.g., Co, Cu, Zn, Ni) capable of forming sparingly soluble metal salts such as hydroxides, carbonates, and chlorides can be used to detect proteins in SDS gels. Protein visualization will occur through deposition of insoluble complexes of the type Me^{2+}–Tris–SDS to form a semiopaque background. Most of these negative stains act rapidly (within a few minutes) and do not require any protein fixation within the gel matrix, and the proteins stained in this manner are reported to be easily recovered from gels for further analysis. Some commonly used negative stains and their protein detection limits include the following:

- Copper chloride (5 ng/mm) (Lee et al., 1987)
- Zinc chloride and zinc sulfate (10–12 ng) (Dzandu et al., 1988; Fernandez-Patron et al., 1992)
- Potassium acetate (0.12–1.5 μg) (Nelles and Bamburg, 1976)
- Sodium acetate (0.1 $\mu g/mm^3$) (Higgins and Dahmus, 1979)

Detection limits for these negative stains fall between the sensitivity levels of Coomassie blue and silver stains. While these negative stains require the presence of SDS in the gel, Candiano et al. (1996) have described a negative staining technique for proteins in polyacrylamide gels operating with and without SDS based on precipitation of methyltrichloroacetate and reported to have a sensitivity of barely 0.5 ng protein. Table 5.3 gives the protocol for the negative zinc/imidazole staining.

While there are a number of advantages of this negative stain protocol, such as reversibility (Hardy et al., 1996), compatibility with Western blotting and subsequent immunodetection (Tessmer and Dernick, 1989; Wang et al., 1989), and compatibility

TABLE 5.3. Negative Zinc–Imidazole Staining for SDS–PAGE

- Prepare 100 mL of 0.5 M imidazole (solution A; 10× conc.), 100 mL of 0.5 M zinc sulfate (solution B; 10× conc.), and 100 mL of 0.5 M Tris–glycinate, pH 8.8 (destaining solution, 10x conc.).
- In separate containers, dilute one part of solution A and one part of solution B with nine parts of water. Mix the solutions thoroughly.
- Remove the SDS gel from the electrophoresis cell and place it in a small basin.
- Pour the imidazole solution (50 mL for a Bio-Rad Mini Protean II, 1-mm-thick gel will suffice) on the gel on the basin and mix gently on a platform for 10 min.
- Transfer the gel to the diluted zinc sulfate solution on a rocking platform. Allow 45 s for the gel to develop.
- Transfer the gel to a container filled with doubly distilled water and rinse for 3–5 min. Discard and replace with freshwater solution. The gel can be stored for weeks in water.
- For visualizing the protein bands, place the gel against a black background and shine light on it at a shallow angle. The protein bands will be visible as transparent bands (thus black) against an opaque, milky background.
- For recovering the protein zones, the bands can be excised and the proteins eluted by passive diffusion in 30% acetonitrile, 30% isopropanol, 30 formic acid, and 10% water.
- For destaining the gel, place it in a 50 mM (1 : 1 dilution) Tris–glycinate, pH 8.8, buffer and shake for 5 min. Replace with fresh solution and again agitate for 5 min.
- The gel is now ready for Coomassie blue or silver staining, blotting, or other manipulations.

with MS procedures (Cohen and Chait, 1997), there are also some problems, notably the difficulty of creating a photographic documentation of the gel. These gels, in fact, inherently exhibit low contrast and are usually photographed against a black background with side illumination. The light scattering in the background usually gives a poor image quality. Bricker et al. (2000) have proposed an easy documentation protocol which consists in using a low-cost flatbed digital scanner. This device utilizes reflected rather than transmitted light to acquire an image. The milky white background of negatively stained gels efficiently backscatters light, while the clear protein bands do not. When scanned with a black background, the protein bands appear dark against a white background. After scanning, the image is imported as a tagged image format (TIF) file under Corel draw and labeled as needed. The image contrast is good and such pictures are suitable for publication. More on the use of imidazole–zinc stains can be found in Castellanos-Serra et al. (1999, 2001) and in an in-depth review by Castellano-Serra and Hardy (2001).

5.4.4.6. Fluorescent Detection A large number of staining procedures are based on Coomassie stains, a family of nonpolar, sulfonated aromatic dyes typically utilized in methanol–acetic acid or methanol–trichloroacetic acid solutions, in general belonging to the category of toxic solvents. In addition, it has been demonstrated that Coomassie G solutions, containing trichloroacetic acid and alcohols, lead to irreversible acid-catalyzed esterification of Glu side-chain carboxyl groups (Haebel et al., 1998), which complicates interpretation of peptide mapping data from MS. A recent fluorescent stain of the SYPRO family described by Steinberg et al. (2000b) seems to have solved the above problems while considerably enhancing the detection sensitivity (down to about 4 ng per protein band). This novel compound, SYPRO tangerine stain, does not require solvents such as methanol or acetic/trichloracetic acids; instead, staining can be performed in a wide range of buffers or simply in 150 mM NaCl by a one-step procedure in the absence of a destaining step. Stained proteins can be excited by UV light at ~300 nm or with visible light at ~490 nm, the fluorescence emission of the dye being 640 nm. Noncovalent binding of SYPRO tangerine dye is mediated by SDS and, to a lesser extent, by hydrophobic amino acid residues in proteins. This is in contrast to acidic silver nitrate staining, which interacts primarily with Arg and Lys residues. Although SYPRO red and SYPRO orange stains (Steinberg et al., 1996a,b, 1997) have been already described for sensitive fluorescent detection of proteins in SDS gels, the novel tangerine dye appears to be an improvement in that it does not require any harsh solvent condition. All these SYPRO stains, in addition, have the advantage that, since staining appears to be due to intercalation of dye in the SDS micelle, little protein-to-protein variability is observed in SDS gels, compared with amino-directed stains such as the Coomassie stains. In case of blotting, either from 1D (SDS–PAGE) or 2D (IEF–SDS) gels, an interesting stain appears to be SYPRO ruby protein blot stain (Berggren et al., 1999, Steinberg et al., 2000a), which provides a sensitive, gentle, fluorescence-based method for detecting proteins on nitrocellulose or polyvinylidene difluoride (PVDF) membranes. SYPRO ruby is a permanent stain composed of ruthenium as part of an organic complex that interacts noncovalently with proteins. Stained proteins can be excited by UV light of ~302 nm or with visible light of ~470 nm. Fluorescence emission of the dye is approximately 618 nm. The stain can be visualized using a wide range of excitation sources utilized in image analysis systems, including a UV-B transilluminator, a 488-nm argon ion laser, a 532-nm yttrium–aluminum–garnet (YAG) laser, a blue-fluorescent light bulb, or a blue LED. The detection sensitivity of SYPRO ruby (0.25–1 ng protein/mm^2) is superior to that of amido black, Coomassie blue, and India ink staining and nearly matches colloidal gold staining. An additional advantage of SYPRO ruby blot stain is that it is fully compatible with subsequent protein analysis, such as Edman-based sequencing and MS. This type of stain is becoming increasingly popular in 2D map analysis, both on blots and directly on gel matrices (Leimgruber et al., 2002; Malone et al., 2001; Yan et al., 2000a; Patton, 2000). In addition, it would appear that quantitation with SYPRO ruby is compatible with a 1000-fold linear dynamic range (Nishihara and Champion, 2002) and that one-order-of-magnitude gain in detection limit per pixel can be attained with SYPRO red, orange, and tangerine (Woodward et al., 2001). An improved formulation of SYPRO ruby stain has been recently described that is compatible with commonly implemented protein fixation procedures (Berggren et al., 2002). As a latest addition to this family, Patton's group has described yet another fluorescent stain specific for recombinant proteins containing oligohistidine tag sequences. Two such stains have been developed: Pro-Q sapphire 365 (bright blue fluorescence at 450 nm) and Pro-Q sapphire 488 (bright green fluorescence at 515 nm). After acquiring the fluorescent signal from the tagged protein population, the entire gel can be poststained with SYPRO ruby, so as to reveal the entire protein pattern (Hart et al., 2003). Since it is becoming more and more common to subject a 2D gel matrix to multiple applications of fluorescent dyes, it has been found desirable to develop more robust and mechanically stable matrices. One such matrix, called Rhinohide liquid acrylamide, has been recently proposed by Schulenberg et al. (2003), who found it particularly resistant in serial staining of proteins for phosphorylation (with Pro-Q diamond phophoprotein stain), glycosylation (with Pro-Q emerald glycoprotein stain), followed by a general protein stain, such as SYPRO ruby.

Other stains which have recently come to the limelight are the *N*-hydroxysuccinimidyl esters of cyanine dyes, such as Cy3 and Cy5, which are used for quantitative, differential proteomic display in 2D maps, as illustrated in the first part of this book (Unlu et al., 1997; Tonge et al., 2001). These dyes react covalently with Lys residues and should be used under conditions of minimal labeling (~5%), ideally just a single Lys residue, in order to keep proteins soluble and prevent strong shifts of the labeled versus unlabeled polypeptide population in the SDS–PAGE run (each dye molecule adds ~500 Da to the labeled polypeptide!). Due to inherent problems with these labeling conditions, a novel set of these dyes aimed at saturation labeling via reaction on Cys residues was developed (Shaw et al., 2003).

Alternative dyes which might be attractive in prelabeling are those directed to Cys residues. Analysis of the SWISS-PROT database shows that 88% of all proteins, 92% of yeasts proteins, and 89% of human proteins reported there have at least a single Cys residue. Of these total Cys proteins, 98% have 20 or fewer Cys residues, with 72% of proteins containing 1–10 Cys residues, while only <2% of proteins have more than 20 Cys (Vuong et al., 2000). A set of fluorescent, thiol-reactive dyes has been recently evaluated by Tyagarajan et al. (2003). Such dyes are either of the iodoacetamide or maleimide (reactive group) dyes. The iodoacetamide derivatives (BODIPY TMR cadaverine IA and BODIPY FI C1-IA), although highly specific for Cys and able to maintain the original protein pI, are quickly destroyed by thiourea in the solubilizing cocktail (Galvani et al., 2001). The maleimide-type species (Thio-Glo I and rhodamine red C2) are insensitive to thiourea but exhibited nonspecific labeling at high dye–thiol ratios. In addition, labeling with both families of dyes was inhibited by tris(2-carboxyethyl)phosphine (TCEP), a reducing agent for –S–S– bridges. Nevertheless, proper labeling conditions (a 9 : 1 ratio of TCEP over thiols in proteins and about a 1.125 : 1 ratio of the dye over TCEP, i.e., 10 : 1 of dye over thiol) ensure proper disulfide bridge reduction and specific labeling. Covalent derivation with these dyes was shown to be compatible in gel digestion and peptide mass fingerprinting by MALDI–TOF MS, which does not seem to be the case with monobromobimane-labeled proteins (Berggren et al., 2000).

The scenario with fluorescent labels is in continuous expansion, though, and papers dealing with this topic are numerous. Among the latest stains, a new one called lightning fast is based on the fluorophore epicoccone from the fungus *Epicoccum nigrum*. Due to its very broad absorption spectrum, this dye can be excited by a range of sources, including UV-A (365 nm), UV-B (3023 nm), xenon-arc lamps, 488- and 457-nm argon ion lasers, 473- and 532-nm neodymium–yttrium aluminum garnet solid-state lasers, and 543-nm helium–neon lasers. Its maximum fluorescence emission is at 610 nm. The limit of detection in 1D SDS–PAGE is less than 100 pg of protein, at least one order of magnitude greater than SYPRO ruby and silver stains. Additionally, its quantitative linearity range spans over more than four orders of magnitude (Mackintosh et al., 2003). In yet another report (Kang et al., 2003), three hydrophobic fluorescein dyes (5-dodecanoyl amino fluorescein 5-hexadecanoyl amino fluorescein, and 5-octadecanoyl amino fluorescein) were examined as noncovalent fluorescent stains of protein bands in SDS–PAGE. Effective incorporation of the dyes to proteins in gels was accomplished either by simply adding dyes at the protein fixation step or by treating gels with a stain solution after the fixation. The sensitivity of these hydrophobic fluoresceins was ~1 ng/band for most proteins; in the best cases, as low as 0.1 ng was successfully visualized.

5.4.5. Possible Sources of Artifactual Protein Modification

If a protein extracted from an SDS gel has to be subjected to analysis by high-resolution techniques, such as MALDI–TOF, one should be aware of a number of potential modifications which proteins in SDS-mediated techniques can experience. Mass spectrometry–based methods have successfully identified a number of such modifications, including cysteine–acrylamide adducts (Bordini et al., 1999a,b), cysteine–β-mercaptoethanol adducts (Klarskov et al., 1994), methionine oxidation (Fantes and Furminger, 1967), formylation of Ser and Thr (Ehring et al., 1997; Goodlett et al., 1990), and aggregates with certain staining agents (Ogorzalek Loo et al., 1996). It is worth noting that most of these modifications are detected for molecular mass values below 30 kDa, possibly because mass resolution is lost above this size for proteins extracted from diverse matrices. These problems are aggravated in SDS gels because, typically, in most user laboratories such gels are not "washed," that is, are still full of unreacted monomers. On the contrary, today, in all isoelectric focusing techniques, the gel matrix is "clean," in that the matrix is fully washed so as to remove unreacted products and catalysts, dried, and reswollen when used. This is a normal procedure since IEF gels are usually run in horizontal chambers with an open face, whereas SDS gels, in general, are run vertically, enclosed between two glass plates, which would render such a washing step highly unpractical. As demonstrated by Caglio et al. (1994), polymerization in detergents can rarely be driven above 70% conversion (especially at high surfactant levels), which means that a huge excess of unreacted acrylamide is available in the gel for reacting with proteins. A number of reports have appeared on the use of scavengers, which include 3-mercaptopropanoic acid for Tris–Tricine gels (Ploug et al., 1989, 1992), glutathione or sodium thioglycolate for modified SDS–PAGE gels (Moos et al., 1988), or free cysteine for IPG gels (Chiari et al., 1992). These modifications, together with the removal of SDS from eluted proteins prior to their analysis by MALDI–TOF (Puchades et al., 1999), are improving the mass accuracy and diminishing the risk of protein modification. However, in untreated gels, it has recently been found that severe alkylation by free, excess acrylamide in a gel occurs for a number of proteins [e.g., ubiquitin, α-lactalbumin, β-lactoglobulin B (LG-B)] (Galvani et al., 2000). Moreover, other modifications can take place during the blotting procedure quite often performed when transferring proteins out of SDS gel matrices. It is quite typical, in this last case, to use formic acid in the transfer solution, which results in massive formylation of proteins too. A case is illustrated in Figure 5.7. With LG-B (M_r = 18,285) run in and extracted

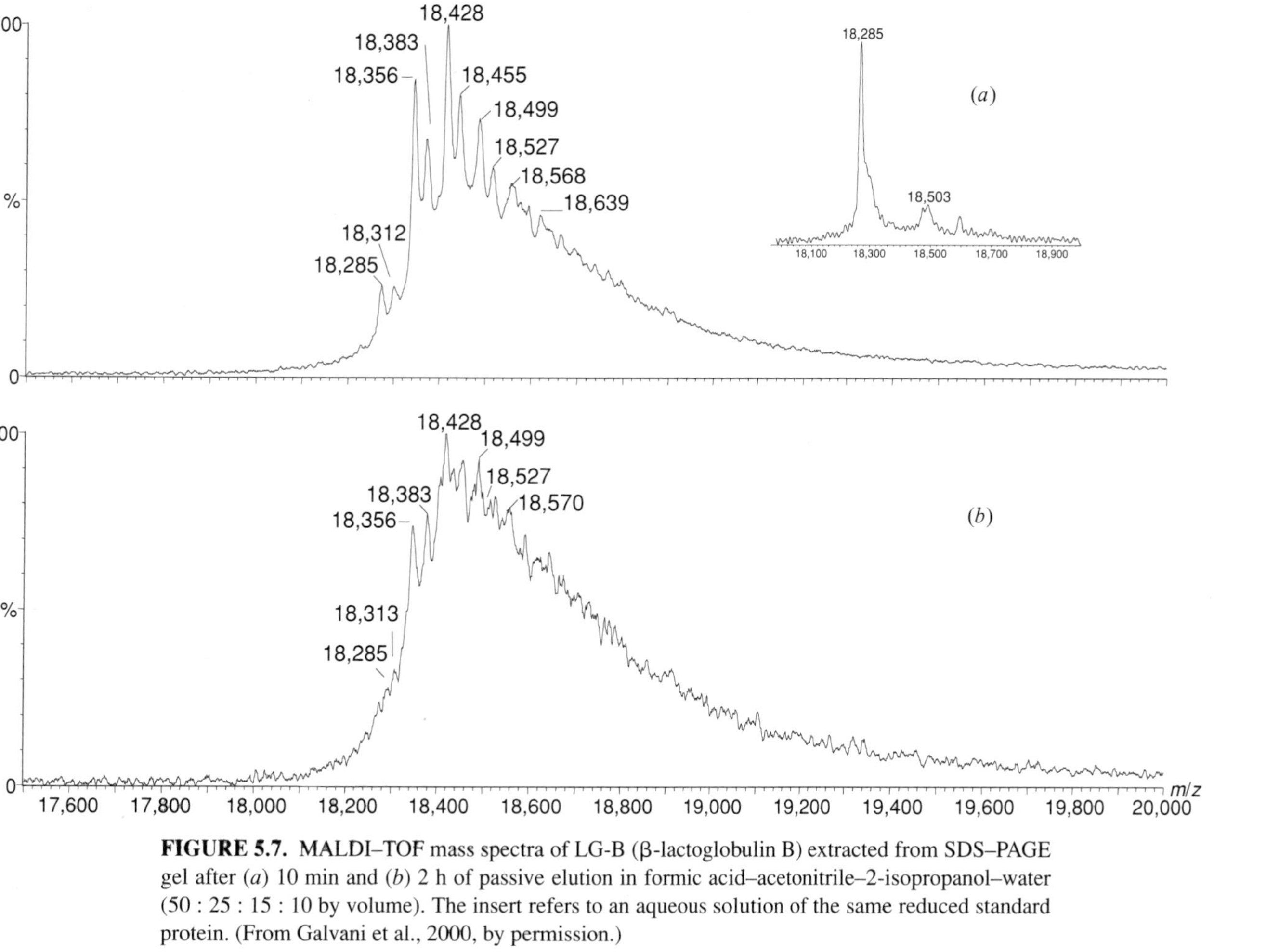

FIGURE 5.7. MALDI–TOF mass spectra of LG-B (β-lactoglobulin B) extracted from SDS–PAGE gel after (*a*) 10 min and (*b*) 2 h of passive elution in formic acid–acetonitrile–2-isopropanol–water (50 : 25 : 15 : 10 by volume). The insert refers to an aqueous solution of the same reduced standard protein. (From Galvani et al., 2000, by permission.)

from an SDS gel, MALDI–TOF revealed an impressive number of peaks: a series due to alkylation by acrylamide (spaced at intervals of 71 Da, centerd at m/z ratios of 18,356, 18,428, 18,499, and 18,568) and a second series due to formylation (on Ser and/or Thr residues; spaced at intervals of 28 Da, observed at m/z values of 18,312, 18,383, 18,455, and 18,527), for a total of more than 10 peaks extending in mass from 18,285 (control LG-B, which had become by far the least abundant species!) up to 18,639. In the case of ubiquitin ($M_r = 8564$), the unmodified protein was observed at m/z 8566, while a series of peaks (up to 10) between m/z values of 8593 up to 8835 were attributed to multiple formylation events (ubiquitin contains three Ser and seven Thr residues, which are described as potential formylation sites) (see Fig. 5.8). The fact that no acrylamide adducts can be seen here is not surprising, since ubiquitin does not contain any Cys residue. Remedies to this will be discussed below.

5.4.6. On Use and Properties of Surfactants

Four major classes of surfactants are commonly used in biological work: anionic, cationic, zwitterionic, and nonionic. The structural formulas, chemical and trade names, and general properties of a number of them have been given by Hjelmeland and Chrambach (1981), Hjelmeland (1986), Helenius et al. (1979), Neugebauer (1989), and Helenius and Simons (1975). Table 5.4 lists some of the most popular nonionic compounds, and Table 5.5 catalogs some of the zwitterionic ones. In addition to a polar group, each surfactant possesses a hydrophobic moiety, and the combination of hydrophobic and hydrophilic sections in the molecule (which is thus said to be amphiphilic) provides the basis for detergent action. Nonionic detergents are a special class since they are virtually never a single species of molecules but instead are a group of structurally related compounds. This variability is due to a statistical distribution of chain lengths formed in the manufacture of these detergents by the polymerization of ethylene oxide. The result, at best, is a single hydrophobic moiety to which polyoxyethylene chains of variable length are attached. Due to their amphiphilic properties, all detergents, when dispersed in water, tend to form aggregates in which the hydrophobic section of the molecule is protected from the solvent. These occur as monolayers at the solvent surface and as micelles in solution. Of these two, the micelle is our primary concern since this is the functional unit in protein solubilization. In the micelle (so called because the monomers enter into it at variable stoichiometries), molecules are arranged in such a way that polar groups are out and are thus exposed to the solvent while the hydrophobic segments are buried in the interior, shielded from water. According to Carey and Small (1972), clusters of SDS are actually ellipsoids, being some 2.4 nm across the minor axis and composed of approximately 100 SDS residues, depending on the ionic strength and solvent composition.

Surfactants in solution are characterized by the critical micellar concentration (CMC), which is defined as the concentration of detergent at which micelles form. Below the CMC, monomer concentration naturally increases as a function of total detergent, as no micelles exist. Above the CMC value, however, added detergent raises the concentration of micelles while leaving constant the monomer concentration, since the CMC is an effective solubility limit of detergent monomer (Tanford, 1973).

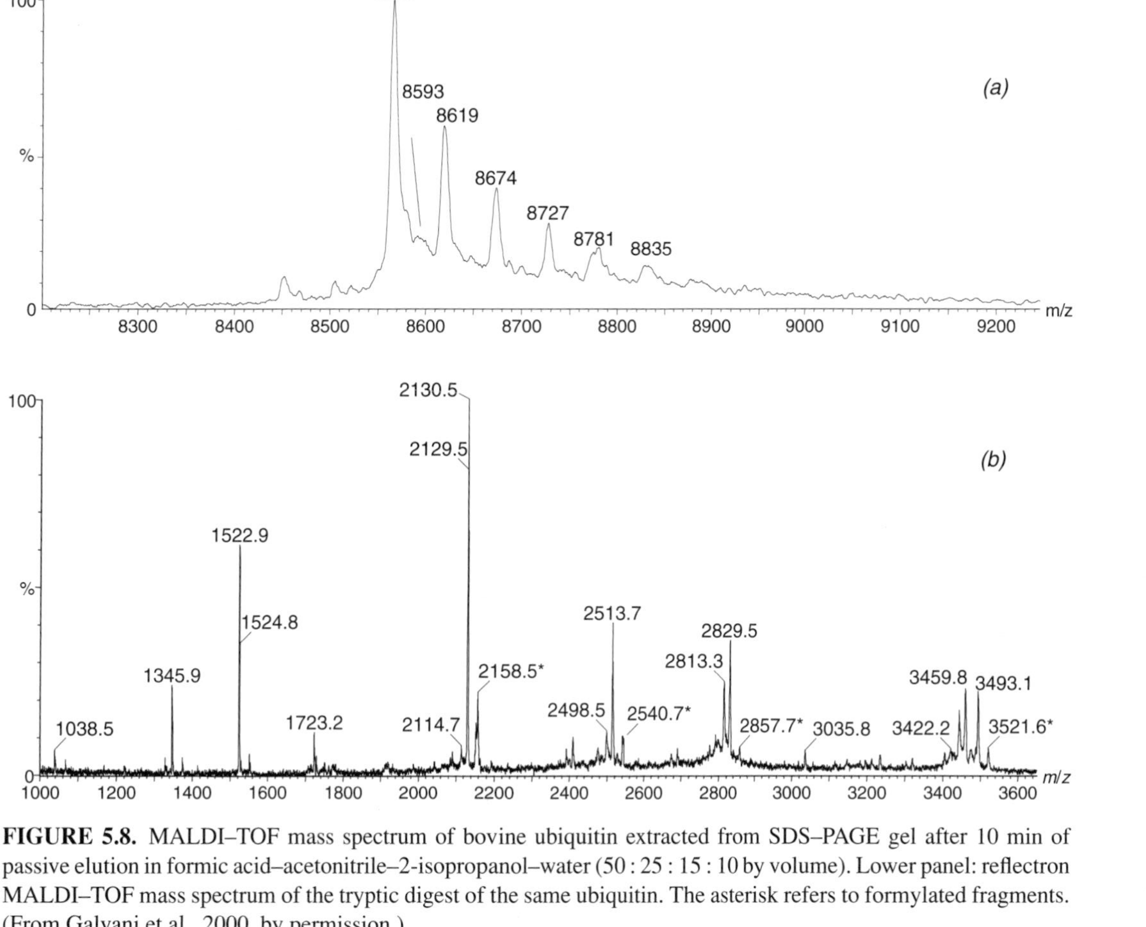

FIGURE 5.8. MALDI–TOF mass spectrum of bovine ubiquitin extracted from SDS–PAGE gel after 10 min of passive elution in formic acid–acetonitrile–2-isopropanol–water (50 : 25 : 15 : 10 by volume). Lower panel: reflectron MALDI–TOF mass spectrum of the tryptic digest of the same ubiquitin. The asterisk refers to formylated fragments. (From Galvani et al., 2000, by permission.)

TABLE 5.4. Nonionic Surfactants

Detergent	Formula	CMC (mM)	M_r
Triton X-100 (polyethylene glycol *tert*-octylphenyl ether)	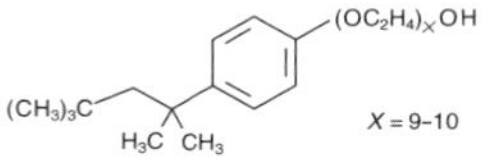 $(CH_3)_3C$; H_3C CH_3; $(OC_2H_4)_XOH$; $X = 9–10$	0.02–0.09	625
Brij 35 (polyoxyethylene lauryl ether)	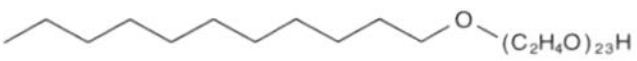 O $(C_2H_4O)_{23}H$	0.05–0.1	1225
Tween 20 (polyoxyethylene sorbitanmonolaurate)	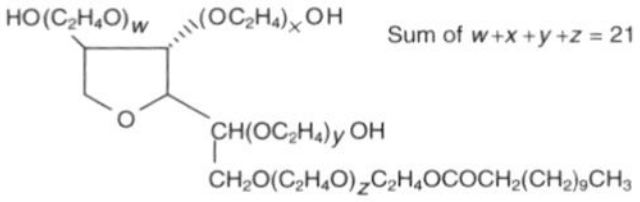 $HO(C_2H_4O)_w$; $(OC_2H_4)_XOH$; Sum of $w+x+y+z = 21$; O; $CH(OC_2H_4)_yOH$; $CH_2O(C_2H_4O)_ZC_2H_4OCOCH_2(CH_2)_9CH_3$	0.06	1228

TABLE 5.5. Zwitterionic Surfactants

Detergent	Formula	CMC (mM)	M_r
CHAPS (3-[(3-cholamidopropyl) dimethylammonio] -1-propanesulfonate)	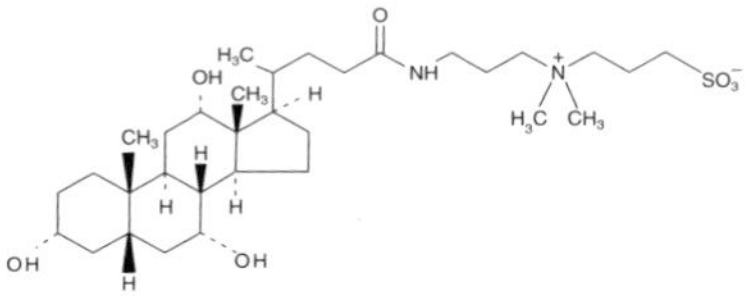 H_3C; O; NH; N^+; SO_3^-; H_3C CH_3; OH; CH_3; H; CH_3; H; H; H; OH; H; OH	6–10	614.9
Caprylyl sulfobetaine SB-10 (*N*-decyl-*N*,*N*-dimethyl- 3-ammonio-1-propanesulfonate)	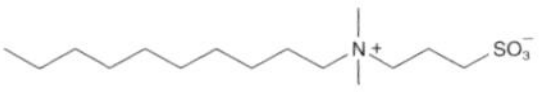 $N+$; SO_3^-	25–40	307.6
Lauryl sulfobetaine SB-12 (*N*-dodecyl-*N*,*N*-dimethyl-3-ammonio-1-propanesulfonate)	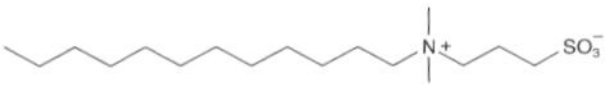 $N+$; SO_3^-	2–4	335.6
Palmityl sulfobetaine SB-16 (*N*-hexadecyl-*N*,*N*-dimethyl-3-ammonio-1-propanesulfonate)	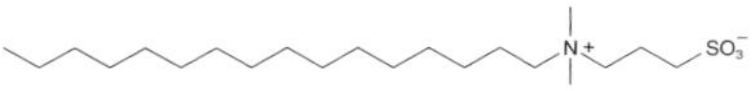 $N+$; SO_3^-	0.01–0.06	391.6

TABLE 5.6. Aggregation Number, micellar Mass, and CMC for Selected Surfactants*

Surfactant	Aggregation Number	Micellar Mass	CMC (mM)	Conditions
SDS	62	18,000	8.2	H_2O
	126	36,000	0.52	0.5 M NaCl
CTAB	169	62,000	—	13 mM KBr
Triton X-100	140	90,000	0.240	H_2O
Triton N-100	100	66,000	0.075	H_2O
Lubrol PX	106	64,000	—	H_2O
SB_{3-14}[a]	80	≈30,000	0.6	H_2O
Sodium cholate	2–4	900–1800	13–15	H_2O
Sodium deoxycholate	4–10	1700–4200	4–6	H_2O

Source: Helenius and Simons (1975).
[a] Sulfobetaine with a tetradecyl carbon tail.

Each detergent is further characterized by two other functional parameters in addition to the CMC value: an aggregation number (i.e., the number of monomers in the cluster) and the micellar molecular mass (Table 5.6). Two phenomena are readily apparent: (a) a single surfactant can exist in different micellar sizes and (b) among the different classes of detergents, the micellar mass can vary greatly. Thus, increasing ionic strengths drive the SDS micelle toward a double-mass and double-aggregation number while lowering drastically the CMC value from 8.2 to 0.52 mM. In general, ionic detergents have higher CMCs than nonionic species, with the CMC of SDS (8.2 mM in plain water) being some two orders of magnitude above the CMC of Triton N-101 (0.075 mM). On the other hand, nonionic detergents (such as Tritons) tend to have a micellar mass considerably higher that that of ionic species, and this has important practical consequences. Due to their larger micellar size, when used in electrophoresis, nonionic detergents greatly increase the solvent viscosity, so that protein migration rates will be considerably decreased. Proteins themselves will bind sizable amounts of nonionic detergents, in proportion to their number of hydrophobic groups, so that their electrophoretic mobility in sieving matrices will be a function of the new Stokes radius of the complex.

Use of neutral surfactants, such as Triton X-100 or Nonidet P-40, enables one to add another fractionation parameter to electrophoretic separations, namely the ability to resolve two macromolecular species of identical size and charge but differing in hydrophobicity. We found this out when performing IEF of human globin chains for thalassemia screening (Righetti et al., 1979; Comi et al., 1979; Saglio et al., 1979). Upon IEF in 8 M urea, the separation between β- and γ-chains was very poor, but when 1% to 3% NP-40 was added to the IEF gel, two remarkable results were obtained: Not only were β- and γ-globins amply resolved, but also γ-chains were split into two zones, which were found to be the phenotypes produced by two different genes, called Aγ and Gγ, bearing an Ala-to-Gly substitution in residue 136 of the γ-chains. It must be emphasized that, when electrophoresis is performed in 8 M urea, higher amounts of surfactants (e.g., 1% to 2%, as compared with 0.1% to 0.2% in the absence of urea) should be used, since detergent binding is reduced by these levels of urea and also the CMC value of surfactants might be markedly altered. This IEF separation of Aγ and Gγ could be reproduced also by free-solution CZE, which suggests that this is a particular case of the vast class of electrophoretic separations called by Terabe et al. (1984, 1985) "micellar electrokinetic chromatography" (MEKC), a unique mode of CZE in that it is capable of separating uncharged compounds. This is due to the fact that uncharged solutes exhibit different micelle–water partition coefficients, according to their relative hydrophobicities. As the charged surfactant micelles migrate in the electric field with a velocity proportional to their m/z ratio, uncharged solutes will migrate, too, at a velocity proportional to their residence time in the micelles. Thus, MECK can be viewed as a chromatographic technique in which the migrating charged micelles act as pseudostationary phases. The MECK can be viewed as a hybrid of reverse-phase liquid chromatography (RPLC) and CZE, as the separation process incorporates hydrophobic and polar interactions, a partitioning mechanism, and electromigration. In MEKC, anionic alkyl chain surfactants, especially SDS, have been the most widely used species.

The popularity of SDS can be attributed to its high aqueous solubility, low CMC, low Kraft point, small UV molar absorptivity, even at low wavelengths, availability, and cost. Serendipitously, SDS has provided the right type of selectivity for many solute mixtures. Sodium dodecyl sulfate is a stronger hydrogen bond donor as compared to most other surfactant systems studied so far, such as bile salts, cationic CTAB surfactants, and methacrylate-based copolymers. Consequently, SDS, should be a better surfactant type in many situations considering that the great majority of small solutes that are separated by MEKC contain a hydrogen bond acceptor group, such as nitro, carbonyl, and cyano (Khaledi, 1998; Quirino and Terabe, 1999).

5.4.7. The Use of Surfactants Other Than SDS

Williams and Gratzer (1971) reported that highly basic proteins such as protamines precipitate in SDS and that highly acidic proteins such as ferredoxins behave anomalously during SDS–PAGE, probably owing to poor SDS binding. They therefore replaced SDS with the cationic detergent cetyltrimethylammonium bromide (CTAB). In all other respects, the method is the same as with SDS, except that migration is toward the cathode, which should thus be the lower electrode in a vertical apparatus. With a number of proteins of known size on a gel at $T = 10\%$, $C = 5\%$, a linear plot of R_f versus log M_r was obtained over a range of about 10–40 kDa. Cationic surfactants are endowed with many desirable properties. They solubilize and denature proteins efficiently as long as synergistic denaturation by urea or heating is carried out, but they also precipitate large polyanions, including nucleic acids and charged polysaccharides, thereby facilitating the removal of these interfering substances and also the extraction of bound proteins (Willard et al., 1980).

However, a simple substitution of CTAB for SDS in this way results in the formation of a dense precipitate within the gel which has been attributed to cetyltrimethylammonium persulfate (Eley et al., 1979). Williams and Gratzer (1971) overcame the problem of visualizing the separated bands in opaque gels by using dansyl prelabeling and fluorescent detection rather than dye staining. This does not circumvent the objections that the formation of the precipitate results in a nonuniform distribution of ammonium persulfate, leading to nonuniform gel polymerization, and a nonuniform distribution of CTAB, both of which might affect the protein–surfactant interaction. In addition, the presence of precipitate particles themselves may influence protein migration. There might also be formation of detergent–dye precipitates during any staining/destaining process. Panyim et al. (1977) overcame this problem by staining and destaining at 80°C to 100°C.

There are two ways by which these difficulties can be circumvented. First, if ammonium persulfate is to be used as catalyst, the gels should be prepared in the absence of surfactant. The usual sample preparation and electrophoretic procedures are then followed, with surfactant included in the sample and in the upper electrode reservoir. Residual persulfate in the gel moves ahead of the CTAB during electrophoresis, so that no precipitate is formed. The second approach is to use different polymerization catalysts.

Marjanen and Ryrie (1974) used $T = 10\%$, $C = 2.67\%$ gels containing 0.1% CTAB photopolymerized with 2 mg/mL riboflavin (see also Mocz and Balint, 1984). The buffer used throughout in both gels and electrode chambers was 0.1 M Na phosphate, pH 6.0, added with 0.1% CTAB, but cacodylate (pH 6.0), citrate (pH 6.0), and succinate (pH 5.0) buffers were also satisfactory. Samples were pretreated by heating for 30 min at 70°C in 10 mM Na phosphate buffer, pH 6.0, in the presence of 1% CTAB and 1% dithiothreitol. Methylene blue was used as the tracking dye. When 18 standard proteins were examined, a linear plot of R_f versus log M_r was obtained with an accuracy very similar to that of SDS–PAGE, but a number of membrane proteins failed to migrate, probably owing to the lack of CTAB binding. Other workers used a mixture of ascorbic acid, hydrogen peroxide, and ferrous ion (Fenton's reagent) as a catalyst (MacFarlane, 1983) or even light and uranium (Deshpande et al., 1986).

In a thorough investigation on the factors affecting CTAB–PAGE, Eley et al. (1979) adopted gels of $T = 2.5\%$ to $T = 15\%$, with a low $\%C$ (1.33%) made up in 100 mM Na phosphate buffer, pH 7.0, and 0.1% CTAB. The gels were photopolymerized with flavin mononucleotide/TEMED. Samples were pretreated by heating on a boiling water bath for ca. 4 min in a buffer containing 1% CTAB and 10% β-mercaptoethanol; after cooling, a small amount of malachite green was added as tracking dye. Staining with Coomassie blue R-250 showed that, although plots of R_f versus log M_r were sigmoidal, in $T = 7.5\%$ gels linearity was ensured in the 36–96-kDa M_r range. Gels with $T < 7.5\%$ were recommended for proteins with $M_r < 70$ kDa and $T > 7.5\%$ for those with M_r below about 30 kDa. More recently, a discontinuous CTAB–PAGE system has been described by Akins et al. (1992).

N-Cetylpyridinium chloride (CPC) is another cationic surfactant that has been used for PAGE applications. Schick (1975) reported linear plots of R_f versus log M_r over the range 17–160 kDa for gels with $T = 10\%$ or $T = 12\%$ (at $C = 2.7\%$) made up in potassium acetate buffer, pH 3.7, containing 0.01% CPC in both gel and reservoir buffer, which was 0.3 M glycinium acetate, pH 3.7. A ferric sulfate/ascorbic acid catalyst system had to be employed due to the low pH in the gel phase.

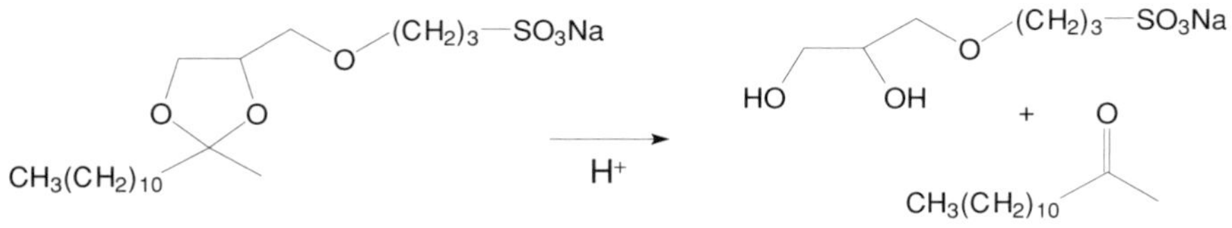

$CH_3(CH_2)_{10}$ ... O O ... O ... $(CH_2)_3$—SO_3Na $\xrightarrow{H^+}$ HO OH O $(CH_2)_3$—SO_3Na + O $CH_3(CH_2)_{10}$

FIGURE 5.9. Formula of an ALS (left side) and its degradation products after acid hydrolysis (right side). (From Zeller et al., 2002, by permission.)

It would thus appear that both CTAB and CPC can be used to solubilize proteins, including membranaceous structures. Thus, it is reasonable to assume that, in most cases, enough surfactant molecules are bound to the protein to produce complexes with a reasonably constant m/z ratio and with similar hydrodynamic shapes. Such complexes resemble SDS–protein mixed micelles in electrophoretic behavior and thus quite accurate M_r values can be gained.

5.4.7.1. Acid-Labile Surfactants Recently, a new class of compounds has emerged as a replacement for SDS: acid-labile surfactants (ALSs). The ALSs are a long-chain derivatives of 1,3-dioxolane sodium propyloxy sulfate that have similar denaturing and electrophoretic properties as SDS. Acid-labile surfactant decomposes at low pH and its use in PAGE and 2D mapping appears to be advantageous for subsequent MS analysis (Zeller et al., 2002; Ross et al., 2002; Meng et al., 2002). This author has observed that protein recovery and peptide extraction from the gel were enhanced, in turn improving protein identification from Coomassie-stained spots extracted from 2D gels. The structure and decomposition at acidic pH of an ALS are shown in Figure 5.9. Dioxolane derivatives having surfactant properties were patented by McCoy (1976) and the aspect of biodegradability was stressed for some soap and anionic polymer formulations (Hales et al., 1995). Laundry detergents containing nonionic surfactants with the 1,3-dioxane and/or 1,3-dioxolane function (Galante et al., 1998) and light-duty liquid cleaning microemulsions containing shorter alkyl chain derivatives of 1,3-dioxolane (Drapier, 1998) were patented in 1998. Also cationic micelles of (2-alkyl-1,3-dioxolane-4-yl) (methyl)trimethylammonium bromides (C_n–D-TAB; where $C_n = C_9H_{19}, C_{11}H_{23}, C_{13}H_{27}$) have been reported (Wilk et al., 1994). Although the use of ALS is still in its experimental phase (Koenig et al., 2003), the technique appears to be promising.

5.4.8. Anomalous Behavior

There are a number of exceptions to the binding rule of 1.4 g SDS/g protein, thus leading to possible anomalous migration and erroneous M_r determination (see also Page 280). First, high ionic strengths reduce the amount of SDS bound to proteins (Reynolds and Tanford, 1970a,b), thus interfering with M_r assessments by SDS–PAGE. However, it was also found (See et al., 1985) that the M_r of proteins in the M_r 20–66-kDa interval can be estimated with precision by SDS–PAGE even in the presence of NaCl in samples up to 0.8 M. Samples containing either up to 0.5 M KCl or ammonium sulfate up to 10% saturation also seem to give reliable results in SDS–PAGE (Weber and Osborn, 1975). Although in the vast majority of cases the M_r values of polypeptides, as estimated by SDS–PAGE, are reasonably accurate, there are some exceptions worth mentioning, as described below:

(a) Glycoproteins often exhibit abnormal migration during SDS–PAGE, giving different apparent M_r values when determined in different gel concentrations. This is caused by their hydrophilic glycan moiety, which reduces the hydrophobic interactions between the protein and SDS, thus preventing binding of SDS in the correct ratio. A more accurate procedure appears to be the use of an SDS gradient gel (Segrest and Jackson, 1972). As an alternative, Tris–borate EDTA buffers allow complexation to the sugars to such an extent that the increased negative charge due to the borate–diol complexes often offsets the loss of charge due to decreased SDS binding (Poduslo, 1981).
(b) Proteins with high acidic residue contents, such as caldesmon and tropomyosin, also migrate anomalously in SDS–PAGE (Bryan, 1989). This may be due to the repulsion of the negatively charged SDS by the acidic residues. Neutralization of the negative charge of the acidic groups (by, e.g., titration at appropriate pH) restores the normal electrophoretic mobility of these proteins in SDS–PAGE (Graceffa et al., 1992).
(c) Highly basic proteins, such as histones and troponin I, typically give abnormally higher M_r values by SDS–PAGE. The lower electrophoretic migration of such basic proteins is presumably due to the reduction of the m/z ratio of the SDS–polypeptide complex as a result of the high proportion of basic amino acids (Panyim and Chalkley, 1969).
(d) Proteins with a high proline content and other unusual amino acid sequences such as ventricular myosin light chain 1 and collagenous polypeptides may also have abnormally high M_r values as determined by SDS–PAGE. The anomaly might be due to the alteration of conformation of the SDS–protein complex.
(e) Finally, polypeptides with M_r below 10 kDa are not properly resolved in uniform concentration SDS gel and their mass cannot be assessed with precision. Better results can be obtained in pore gradient gels or in 8 M urea–SDS gels containing a higher percentage of crosslinker (Swank and Munkres, 1971). These aspects will be treated in more detail below.

5.5. GEL CASTING AND BUFFER SYSTEMS

The following will be a detailed treatment of various methods for performing SDS–PAGE. We will start with the standard method, that is, the original one in conventional buffers, only to proceed to multiphasic buffers and then porosity gradients, techniques which have much more to offer in terms of resolving power, although with the added burden of some extra experimental work. Prior to that, however, we will have an excursus on sample treatment in preparation for SDS–PAGE, since this is common to all procedures adopted.

5.5.1. Sample Pretreatment

Standard proteins and the unknowns must all be treated prior to electrophoresis in order to ensure reproducible binding of SDS. Pretreatment also has the scope of inactivating any proteolytic enzyme which may be present in the sample and give rise to spurious bands by generating breakdown products. It is also customary, in order to minimize proteolysis, to add cocktails of inhibitors consisting of natural and synthetic protease inactivators to the sample. The former class comprises trypsin inhibitors such as α_1-antitrypsin and Kazal- or Kunitz-type inhibitors, whereas the last class is made of irreversible inhibitors for serine, cysteine, and aspartyl proteases. Among them, we can recall phenylmethylsulfonyl fluoride (PMSF), diisopropyl fluorophosphate (mustard gas, used in Belgium during World War 1 as nerve gas to kill troops, thus a very toxic compound!), and aminoethylbenzylsulfonyl fluoride (Rabilloud, 1996). The problem of protease action is not abolished, unfortunately, when denaturing solubilization is performed. Evidence for proteolysis after solubilization in 9 M urea (Colas des Francs et al., 1985) or SDS (Granzier and Wang, 1993) has in fact been described and is also documented by the fact that peptide mapping can be carried out in dilute SDS (Cleveland et al., 1977). It is possible that the kinetics of denaturation of proteases in urea or SDS solutions are slower than those of normal proteins, so that they have some period of time for working on already fully unfolded proteins. Fortunately, this problem seems to be more important in plant tissues than in other biological samples (Colas des Francs et al., 1985) and is minimized by boiling the sample in SDS, so that thermal denaturation and SDS denaturation act synergistically.

Another important aspect in SDS–electrophoresis is the breaking of disulfide bonds, which is in general achieved by an equilibrium displacement process in a large excess of free thiols. Typically, β-mercaptoethanol, thioglycerol, or cysteamine is utilized, although rather high concentrations are needed (e.g., 0.2 M) for rupturing disulfide bonds in proteins. However, dithiothreitol (DTT) or dithioerythritol (DTE) act at much lower concentrations (e.g., 20–50 mM) because of the intramolecular, cyclic condensation process during their oxidation, which will drive the equilibrium more efficiently toward protein reduction (see Fig. 5.10). Problems with any of these reducing agents are due to the fact that dissolved oxygen can reoxidize thiols into disulfides, with the concomitant risk of reoxidation of protein thiols. This reoxidation process can also occur during SDS–electrophoresis because polyacrylamide gels can provide an oxidizing environment due to residual persulfate. A drastic way to prevent reoxidation of reduced thiols in proteins is via nucleophilic substitution with iodoacetamide (Gurd, 1967) or nucleophilic addition with maleimides (Riordan and Vallee, 1967), vinylpiridine (Griffith, 1980), or even acrylamide (Brune, 1992). These are all valid alternatives, with the proviso that utmost care should be exerted when eluted polypeptides have to be further characterized by MS. In this last case, erroneous mass determination could ensue if (a) mixed populations of reacted sulfide groups are present or (b) alkylation proceeds also onto other residues (e.g., ε-amino group of

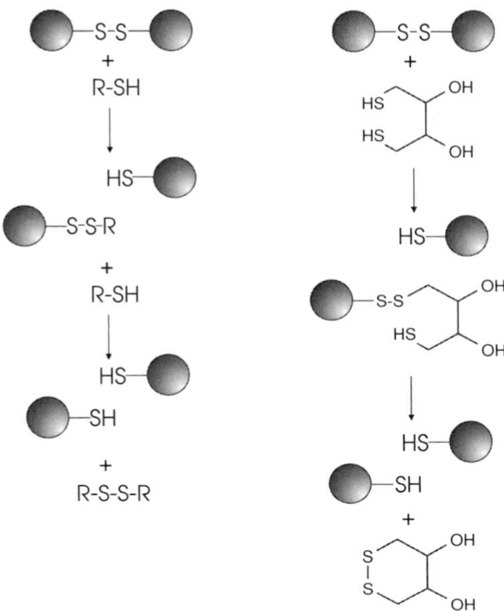

FIGURE 5.10. Mechanism of disulfide reduction with free thiols. Left column: reduction with free monothiols (e.g., β-mercaptoethanol). Right column: reduction with cyclizable dithiols (e.g., DTT). The shaded spheres represent protein surfaces. (From Rabilloud, 1996, by permission.)

Lys). In both cases, erroneous M_r values will be obtained, which would hamper the search for polypeptide identity on databases. An alternative would be to use tributyl phosphine (Bu_3P), which was the first phosphine species used for disulfide reduction in biochemistry (Ruegg and Rüdinger, 1977). This compound, which reacts with disulfides as shown in Figure 5.11, has been claimed to have a number of advantages. First, the reaction is stoichiometric, which allows the use of very low concentrations (a few mM). Second, phosphines are not as sensitive as other thiols to dissolved

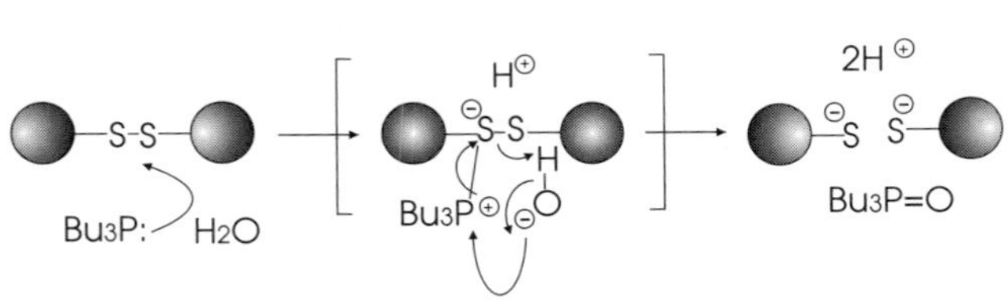

FIGURE 5.11. Mechanism of disulfide reduction with phosphines (e.g., tributylphosphine). Although the overall mechanism is known, the degree of concentration in the electron transfer process between water, the disulfide, and the phosphine is speculative (intermediate between brackets). The shaded spheres represent protein surfaces. (From Rabilloud, 1996, by permission.)

oxygen. Third, because of the limited concentration of the agent, subsequent alkylation is much easier to perform. In addition, due to the fact that phosphines do not add to double bonds, as thiols do, one-step protocols confronting the protein simultaneously with the reducing (e.g., 2 mM phosphine) and alkylating (e.g., 10 mM vinylpyridine) agents can be adopted (Kirley, 1989; Rabilloud, 1996). The main drawback of tributyl phosphine is that it is volatile and toxic and has a rather unpleasant odor. Another curious aspect of tributyl phosphine, at present not reported in the literature, is that, when this reagent is analyzed by MS, it is shown to be extensively oxidized to Bu_3PO, an event which should occur only when it reacts with disulfide bridges for reducing them. This occurs not only in old bottles of Bu_3P but also in fresh preparations and when such bottles are rigorously kept under nitrogen. It is enough to pipette Bu_3P in the typical buffer used for reduction (10 mM Tris, pH 8.3), and within a few minutes (just the time to bring the solution to the MS instrument and perform the analysis), it can be seen that ~98% of it is already present as Bu_3P0, with only ~2% remaining in the reduced form (Hamdan and Righetti, unpublished). However, curiously, such preparations are just as effective in bringing (and maintaining) reduction of –S–S– bridges in proteins, which suggests that the amount needed for reduction is not in the millimolar, but rather in the micromolar, range.

5.5.2. Standard Method Using Continuous Buffers

The simple standard method of Weber et al. (1972) and Weber and Osborn (1975) is usually adequate for a rapid assessment of a purified protein M_r. It is rapid and easy to perform and is very suitable for use with inexpensive and unsophisticated instrumentation. This is why it will be treated here, although it is clear that the use of multiphasic buffers and/or porosity gradient gels gives higher resolution. Obviously, SDS of high purity should be used, especially if subsequent removal of SDS and detection of enzyme activity is envisaged (Blank et al., 1982, 1983). It should be remembered that higher order impurities, such as C_{14} and C_{16} tails, in addition to binding tenaciously to the protein backbone, could give rise to extra bands (Margulies and Tiffany, 1984).

Although there are exceptional circumstances when undissociated protein may persist (Bryce et al., 1978), the following pretreatment procedure is generally found satisfactory (standard protocol of Weber et al., 1972): To a 10 mM phosphate buffer, pH 7.0, containing 1% SDS and 1% β-mercaptoethanol, add protein to a final concentration of 1 mg/mL. Heat to 100°C for 2 min and cool. A small amount (2% to 10%) of sucrose or glycerol is then added to increase the sample density. Bromophenol blue is a suitable tracking dye and can be added either to the sample or to the cathode (upper) buffer chamber. Another common dye for tracking is pyronin-Y, although both dyes migrate with the discontinuity front (in multiphasic buffers) only over a limited range of gel concentrations. For this reason, Neff et al. (1981) have proposed thymol blue, since this dye travels with the true discontinuity front up to $T = 20\%$ gel concentrations. Typically, apply, 20 μL sample solution to pockets precast in the gel (when staining with Coomassie blue, 2–4 μg protein/band suffices for visualization).

TABLE 5.7. Composition of Standard SDS Gels of Various Porosities

Component	$T = 5\%$	$T = 7.5\%$	$T = 10\%$	$T = 15\%$
Acrylamide, g	4.85	7.28	9.70	14.55
Bis, g	0.15	0.22	0.30	0.45
Gel buffer (0.2 M phosphate, pH 7.2), mL	50	50	50	50
SDS, g	0.1	0.1	0.1	0.1
H_2O to 95 mL, then add TEMED, mL	0.15	0.15	0.15	0.15
Ammonium persulfate (1.5%), mL	5	5	5	5

5.5.2.1. Composition of Gels and Buffers Gel concentrations of from 3% to 20% or over can be used in any type of PAGE apparatus, the choice of %T depending upon the M_r range under investigation. Dunker and Rueckert (1969) suggest a $T = 15\%$ gel for 10–60-kDa proteins, a $T = 10\%$ gel for 10–100-kDa species, and $T = 5\%$ gels for 20–350-kDa polypeptides. The compositions of solutions for making suitable gels are shown in Table 5.7 (Weber and Osborn, 1969). If gels are to be stored before use, cold storage should be avoided because of the poor solubility of SDS at 4°C. Gels can be kept at room temperature for 1–2 weeks since SDS is a powerful bacteriostatic. The apparatus buffer used in the standard method is the same as that used in the gel, that is, 0.1 M Na phosphate, pH 7.2, containing 0.1% SDS, so that the buffer system is homogeneous throughout. Depending on pocket size, volumes of ~20–30 μL of sample can be loaded, with concentrations for each protein species of ~2 μg being adequate for detection. Since SDS–polypeptide complexes have a net negative charge at pH 7.2, they migrate toward the anode, which should thus be the lower electrode in vertical gel apparatuses. The typical voltage gradients given are of the order of 30 V/cm, with running times of 2–3 h, depending on gel length.

An interesting variant of this procedure is the high-temperature SDS–PAGE of Haeberle (1997). It is known that one of the drawbacks of gel slabs run without temperature control is the "smiling effect," by which the protein bands in the center lines migrate faster than those in the outer lines, presumably because the temperature at the center of the gel is higher than at the edges, due to uneven heat dissipation and/or uneven current densities. Cooling the gels and running them at lower voltages minimize this temperature gradient, thus resulting in a more uniform migration rate. However, this remedy slows the rate of protein migration and increases the electrophoretic time to several hours. A valid alternative is to use high-temperature runs. Haeberle (1997) has proposed runs in $T = 8\%, C = 5\%$ gels at 70°C in a modified Porzio and Pearson (1977) buffer having the following composition:

- Electrode buffer: 50 mM Tris, 150 mM Gly, 0.1% SDS
- Gel buffer: 100 mM Tris, 300 mM Gly, 0.1% SDS, 5% glycerol, 0.5 mM NaN_3
- Sample buffer: 62.5 mM Tris free base, 3% SDS, 20% glycerol, 6 mg/mL DTT, traces of bromophenol blue, mixed 1:1 (v/v) with sample

Uniform heating of the gel was accomplished by completely submerging the gel in the lower (anodal) buffer preheated to 70°C. The advantages claimed are as follows:

- Greatly accelerated rate of crosslinking
- Compatibility with high-salt loads (up to 700 mM KCl)
- More uniform migration of dye and proteins
- Much reduced running times

The first point is quite important, since Righetti and Caglio (1993) have demonstrated that polyacrylamide gels contain a large number of pendant, unreacted double bonds due to the crosslinker. In turn, such unreacted bonds can covalently affix migrating polypeptides to the gel matrix (Bonaventura et al., 1994). High-temperature SDS–PAGE is stated to work equally well for 1–5- and 0.75-mm-thick gels. However, the temperature should not be increased above 70°C, since above this critical value rubber gaskets leak and bubbles begin to form between the gel and the glass plates. The present protocol is particularly well suited for continuous buffer systems such as the one reported above. In discontinuous buffers, such as Laemmli's (1970), one should remember to correct for the temperature effect by titrating the various buffers at appropriate pH values, measured at 70°C, due to the rather high temperature dependence of Tris buffers ($\Delta pK/°C = 0.031$).

5.5.3. Use of Discontinuous Buffers

As in detergent-free systems, the use of multiphasic buffers in SDS–PAGE gives much sharper bands and better resolution; thus is it almost universally used nowadays. The theory of zone stacking and the other processes involved in SDS-containing gels has been comprehensively discussed by Wyckoff et al. (1977). It is important here to recall that the nature of stacking is somewhat altered in the presence of SDS, since SDS-coated proteins have a constant m/z ratio. As a consequence, they will migrate with the same mobility and thus will automatically stack. Moreover, as the net charge of SDS–protein micelles does not vary in the pH 7–10 interval, mobility is not affected within this pH range. It is therefore not strictly necessary to have a discontinuity in pH, and unstacking of the proteins will simply occur by the change in gel concentration as these analytes leave the stacking gel to enter the running gel. Also, the observation that SDS migrates with a mobility higher than SDS–protein micelles in a restrictive gel means that SDS will overtake zones of proteins in the resolving gel provided that it is included in the sample and upper buffer reservoir. Thus, in principle, SDS could be omitted from both the stacking and separation gels (Allen et al., 1984), although few investigators seem to have taken advantage of this option. This type of electrophoresis is typically run at alkaline pH values (8.8–9.2), although an acidic system operating at about pH 4 has also been reported (Jones et al., 1981), offering comparable results (although, in this last case, lithium, rather than SDS, was required, on account of its better solubility at acidic pH values). Below we will present the two most popular discontinuous systems, those of Neville (1971) and Laemmli (1970),

TABLE 5.8. Composition of Buffers and Gels employed in Method of Neville (1971)

	pH	Composition
Upper buffer reservoir (cathode)	8.64	40 mM boric acid; 41 mM Tris; 0.1% SDS
Stacking gel buffer	6.10	26.7 mM H_2SO_4; 54.1 mM Tris
Running gel buffer	9.18	30.8 N HCl; 424.4 mM Tris
Lower buffer reservoir (anode)	9.18	Same as running gel buffer
Stacking gel ($T = 3.2\%$, $C = 6.25\%$)	6.10	3.0 g acrylamide; 0.2 g Bis; buffer to 100 mL; 0.15 mL TEMED[a]; 0.05 g APS[b]
Running gel ($T = 11.1\%$, $C = 0.9\%$)	9.18	11.0 g acrylamide; 0.1 g Bis; buffer to 100 mL; 0.15 mL TEMED; 0.05 mL APS

[a] *N*, *N*, *N*′, *N*′-Tetramethylethylene diamine.
[b] Ammonium persulfate.

which have been widely used, especially the latter, which retains its popularity in spite of the introduction of modified procedures (Baumann et al., 1984).

5.5.3.1. Method of Neville Typical compositions of buffers and gels used by Neville (1971) are shown in Table 5.8. The composition given here for the running gel buffer results in an actual running pH of 9.50, which, together with gels of $T =$ 11.1%, seems suitable for the majority of separations. Neville (1971) also describes alternative compositions for both the running gel ($T = 7.5\%$) and buffers giving running pH values of 8.64–10.0. The sample preparation is as described in Section 5.5.1, except that the stacking gel buffer is used in place of 0.1 M phosphate buffer. As in the 1970s gel slabs were not so popular, the settings given below refer to gel tubes, as commonly adopted in disc electrophoresis. Gel cylinders of 5 mm diameter were run at 1.5 mm per tube at 25°C until the bromophenol blue marker dye reached the bottom of the tube. The exact position of the dye, which migrates at the borate–sulfate ion interface, was marked by inserting a thin piece of stainless steel surgical wire into each gel before staining, so that the mobilities of protein bands relative to the dye could be calculated after staining. Using this procedure, Neville and Glossman (1971) resolved 35–40 individual protein bands from cell plasma membranes on a single, 1D gel. It should be noted that, although the %*T* compositions of both stacking and running gels in the present method and in the one of Laemmli (1970) are nearly identical, the %*C* levels in the two methods are substantially different. Whereas Laemmli uses the same amount of crosslinker in both gel phases ($C = 2.6\%$), Neville (1971) uses rather high levels in the stacking gel ($C = 6.25\%$) and very low levels in the running gel ($C = 0.9\%$). There might be good reasons for this: High %*C* in stacking gels was originally proposed by Ornstein (1964) and Davis (1964) for disc electrophoresis on the assumption that gels of high %*C* would be much more porous than analogous gels of the same %*T* but with low %*C* (an assumption fully justified; see Righetti et al., 1981). On the other hand, running gels of very low %*C* are recommended (especially in a gel slab format) since they can be dried without cracking due to their better elasticity. The discontinuous buffer system of Neville (1971), based on a borate–sulfate boundary, was selected from the series of 4269 multiphasic buffers

TABLE 5.9. Composition of Buffers and Gels Employed in Method of Laemmli (1970)

	pH	Composition
Upper buffer reservoir (cathode)	8.3	25 mM Tris; 192 mM glycine; 0.1% SDS
Stacking gel buffer	6.8	Tris adjusted to pH 6.8 with HCl and diluted to 125 mM; 0.1% SDS
Running gel buffer	8.8	Tris adjusted to pH 8.8 with HCl and diluted to 375 mM; 0.1% SDS
Lower buffer reservoir (anode)	8.3	Same as upper buffer reservoir
Stacking gel ($T = 3.1\%$, $C = 2.6\%$)	6.8	3.0 g acrylamide; 0.08 g Bis; buffer to 100 mL; 0.025 mL TEMED; 0.025 g APS
Running gel ($T = 10.3\%$, $C = 2.6\%$)	8.8	10.0 g acrylamide; 0.27 g Bis; buffer to 100 mL; 0.025 mL TEMED; 0.025 g APS

calculated by Jovin (1973), and it was designed to stack all the SDS-saturated proteins while leaving behind unstacked or partially saturated species.

5.5.3.2. Method of Laemmli Table 5.9 shows the compositions of the gels and buffers described by Laemmli (1970). In the original method, Laemmli poured 10 cm of running gel into tubes 15 cm long and 6 mm inner diameter. After polymerization, this was overlaid with 1 cm of stacking gel. Protein samples of 0.2–0.3 mL volume were made up in a solvent containing 62.5 mM Tris–HCl buffer, pH 6.8, 2% SDS, 10% glycerol, 5% β-mercaptoethanol, and 0.001% bromophenol blue tracking dye and were heated for 1.5 min in a boiling water bath to ensure complete dissociation and optimum SDS binding. Electrophoresis was run at 3 mA per gel cylinder until the tracking dye reached the bottom of the tube (about 7 h). This method was soon adapted to slab gels. For example, using 35-cm-long gel slabs, Ames (1974) separated approximately 60 protein bands in a single run from samples of bacterial cells, the limit of detection being well below 0.2 μg per band.

An interesting point to the users of the SDS–PAGE technique, in all of its possible variants, is the possibility of recovery of biological activity after the run. This would permit in situ zymogramming, that is, the detection of an enzyme-driven reaction directly in the gel slab, as originally reported by Hunter and Markert (1957) and subsequently refined by Markert and Moller (1959), who suggested the word *isozyme* to describe the enzyme bands which appeared after specific staining. This method [for reviews see Brewer and Sing (1970) and Richardson et al. (1986)] is what made disc electrophoresis extremely popular, since it allowed, for instance, population genetics and refined biochemical studies in even plain tissue homogenates. Of course, in all types of zone electrophoresis, this was quite simple, since the proteins were separated under native conditions, but it would appear to be anathema in SDS–PAGE, where the process of saturating the polypeptide chain with SDS leads to strong denaturation. A recent report by Wang (2000) suggests that this could be possible, at least for some classes of macromolecules such as protein kinases. After the SDS–PAGE step, the gel is rinsed for 1 h at room temperature in 100 mM Tris–HCl buffer, pH 8.0, containing 5 mM β-mercaptoethanol and 20% isopropanol. For renaturation, the gel is then incubated at 4°C for 16 h in the same buffer, devoid of isopropanol but with added 0.04% Tween 40. At this point, assays for kinase activity can be performed in the presence of the appropriate substrates. Presumably, isopropanol in the first step is used for disaggregating the SDS micelles; by the same token, the Tween 40 present in the second step might be needed for replacing the SDS moiety and preventing massive protein precipitates, although microaggregates might still be present, since there does not seem to be massive leach-out of proteins after such a long period of incubation.

In both systems, though, the resolution of polypeptides below 14 kDa is not sufficient, since such smaller M_r chains comigrate with the SDS front. One possibility is to use the technique of Schägger and von Jagow (1987), which adopts an additional spacer gel between stacking and resolving gels, an increase of buffer molarity to 1 mol/L Tris–HCl, a lowering of gel pH to 8.4, and a change of terminating ion from Tris to Tricine. In this way the destacking of smaller M_r polypeptides and the SDS front is much more efficient: The method claims linear resolution from as low as 1 kDa up to 100 kDa. However, this system has problems too: The running time is much longer, due to the high buffer molarity, and Tricine (as well as Bicine) is not compatible with ammonia silver stains. An improved variant has been described by Tastet et al. (2003): use of taurine as the trailing ion titrated to a pH value (8.05) coincident with the pK of Tris (the buffering counterion). This system ensures unstacking of smaller M_r chains, down to 6 kDa, while maintaining excellent resolving power up to 200 kDa. For the first time, 2D maps run in this system exhibited well-resolved chains in the M_r range 6000–140,000 Da without resorting to porosity gradient gels.

5.5.4. Porosity Gradient Gels

Since the migration of SDS–protein complexes in a gel matrix is mostly determined by its size, it is not surprising that an improved fractionation can be achieved on gels of graded porosity (gels with increasing concentration gradients). In such gradients the migration velocity of a band is inversely related to time and varies exponentially with the distance traveled. Experimentally it is found that not only does the rate of migration of components through a gel gradient vary inversely with time, but also, after a sufficient time, a stable pattern will develop in which the different components continue to move slowly but their relative positions remain constant. This has led to the concept of pore limit, which is defined by Slater (1969) as the distance migrated from the origin in a specific gradient after which further migration occurs at a slow rate directly proportional to time. Thus the technique of pore gradient electrophoresis is often referred to as pore limit electrophoresis, although this latter term is perhaps an unfortunate one, since it has led to the common misconception that once the pore limit is reached migration stops altogether. Since this is not so (see Rodbard et al., 1971), it seems preferable to refer to the method as gel gradient electrophoresis.

For a complex mixture there seems to be little doubt that under appropriate conditions a gradient gel can give a resolution superior to a gel of a single concentration and it is not necessary for this that all components should have migrated as far as their pore limits. Part of this high resolution results from the fact that, throughout the run, the leading edge of any particular band is moving through more concentrated gel than the trailing edge and hence encounters greater resistance, resulting in a band-sharpening effect (Margolis and Kenrick, 1968). There are two further practical advantages of this method. First, since after the initial "sorting-out" process a relatively stable band pattern is formed, it is not necessary to control the electrophoretic conditions so precisely as in other electrokinetic procedures, at least for qualitative work. Second, once bands have migrated well into the gel and approached their pore limits, diffusion is greatly reduced, so that the gels can be kept unfixed for long periods of time with little loss of resolution (Margolis, 1973).

Gels can be prepared with any shape of concentration gradient to suit the particular requirements of the separation. However, the bulk of published work refers to the use of linear or simple concave gradients. Since the pore size is a function of $1/\sqrt{\%T}$, in a linear gradient the pore size changes more gradually at high, rather than at low, values of $\%T$. Thus small molecules have to migrate through a considerable length of gel with a porosity close to the pore limit before finally reaching that limit. Convex concentration gradients would naturally aggravate this situation, but with concave gradients the pore limit is reached more rapidly than with linear gradients. Virtually any device for preparing solution gradients can be applied to the formation of gradient gels. The simplest apparatus for preparing linear gradients requires only two identical vessels, a stirrer and some tubing, as shown in Figure 5.12. If they are identical, the shape of the vessels is immaterial since, as long as they are at the same height, when one volume of solution is withdrawn from flask C, half a volume will flow from reservoir B into C to maintain hydrostatic equilibrium. If C (the mixing chamber) is filled with the high $\%T$ solution and B with the dilute solution and C is continuously stirred, then the $\%T$ of the solution harvested from C and filling the gel

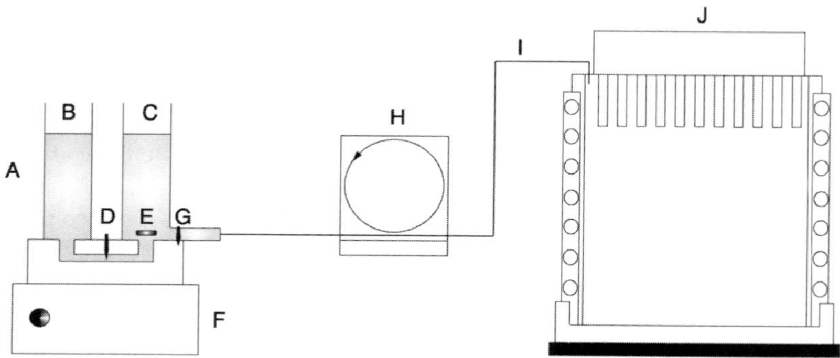

FIGURE 5.12. Apparatus for casting linear gradient polyacrylamide gels: A, two-chamber gradient mixer; B, reservoir; C, mixing chamber; D, valve; E, stirring bar; F, magnetic stirrer; G, stopcock; H, peristaltic pump; I, tubing; J, gel slab cassette. (From Dunn, 1993, by permission.)

slab cassette J will decrease linearly. Cassette J, upon polymerization, will contain a linearly decreasing porosity gradient from top to bottom. Many published methods are basically only variants of this simple concept (see, e.g., Caton and Goldstein, 1971; Arcus, 1967; Pratt and Dangerfield, 1969; and the review by Gianazza and Righetti, 1979). Small corrections to the volumes of solutions in B and C are made to allow for the different densities of the two solutions, the volume displaced by the stirring bar, and the differences in level produced by the dynamic forces resulting from the stirring action. For example, Pharmacia Biotec provides, with its gradient mixer, a compensating rod taken from the cone designed by Svensson (1967) for pouring sucrose density gradients in preparative IEF. The quality of the generated gradient can be checked by adding a dye such as *p*-nitrophenol (Lorentz, 1976) or simply bromophenol blue (Görg et al., 1980) to one of the vessels and then scanning the gel wih a densitometer.

To simultaneously cast a number of gel slabs, a widely used procedure for gradient gel preparation is that of Margolis and Kenrick (1968), as exemplified in Figure 5.13, in which the components are fed into the bottom of a rectangular mold in which the gel plates are supported on a plastic net platform. The apparatus has the advantage that a number of slabs are prepared at the same time, so the gradient reproducibility within a batch should be very high. Obviously, as the gradient solution is delivered from the bottom of the tower, the low %*T* solution should enter first and be displaced upward by the more dense liquid elements being pumped in. The same two authors stress that the most important precaution is to avoid convection currents, including those generated by the heat produced during the polymerization reaction. This is helped both by the

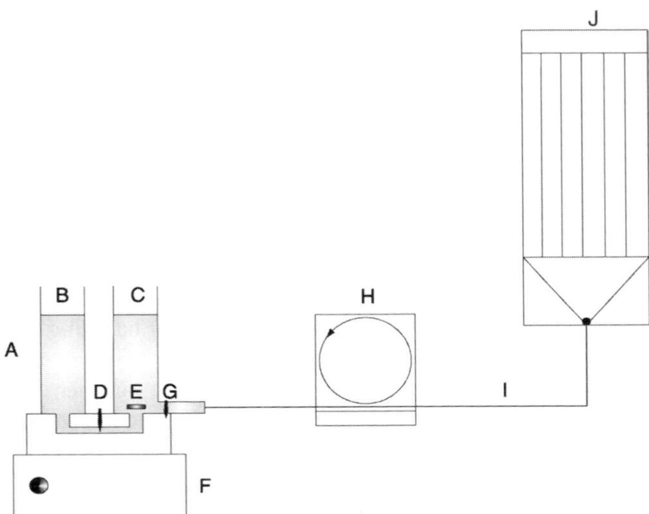

FIGURE 5.13. Apparatus for casting batches of linear gradient polyacrylamide gels: A, two-chamber gradient mixer; B, reservoir (high %*T*); C, mixing chamber (low %*T*); D, valve; E, stirring bar; F, magnetic stirrer; G, stopcock; H, peristaltic pump; I, tubing; J, stack of gel slab cassettes. (From Dunn, 1993, by permission.)

inclusion of a sucrose (or glycerol) density gradient colinear with the gradient gel mixture itself and by having a gradient in the concentration of catalysts [TEMED and APS (Ammonium Persulfate)] so that polymerization will occur first in the top (most dilute) part of the gel gradient and proceed downward. Using gradients of catalysts is too cumbersome and rarely adopted today; in addition, Righetti et al. (1994) have demonstrated that in the presence of density gradients and thin gels (as much in vogue today) such convective flows (which would render the gel highly inhomogeneous) are essentially abolished.

If desired, a stacking gel can be applied above a gradient slab by merely leaving sufficient space between the plates when adding the gradient mixture and then by pouring the stacking gel mixture in the usual way once the gradient has polymerized. In turn, on top of the stacking layer a further portion of gel mixture can be poured and a slot former inserted for seeding the samples. The concentration limits chosen for the gradient depend upon the samples to be separated. Gradients of 4% to 24, 4% to 26, 4% to 30, 2% to 30%, and so on, are generally considered to be suitable for use with most protein samples (e.g., serum, urine), although, if electrophoresis is unduly prolonged, it is possible for proteins with M_r below 30,000 Da to migrate off the end of such gels. Felgenhauer (1979) used 20% to 50% gradients for proteins below 30,000 Da. Gradients of 2.5% to 12% or 2% to 16% (Caton and Goldstein, 1971) are well suited for mixtures of proteins in the M_r range 100,000–5,000,000 Da. It is best to maximize the range of pore sizes for any particular gel concentration range, and since this is achieved with 5% crosslinking (Rodbard et al., 1971), it is common to use $C = 5\%$ in both gradient making solutions at any value of $\%T$. However, Campbell et al. (1983) found that, while pore size is minimal with about $C = 5\%$ for gels with $T < 20\%$, the proportion of crosslinker required for minimum pore size increases with higher values of T. They reported that gels with maximum sieving effect were obtained with gels of $T = 40\%$ and $C = 12.5\%$ and suggested that these should be the limiting conditions on gradient gels. Using gradient gels with $T = 3\%$, $C = 4\%$ at the top, increasing to $T = 40\%$, $C = 12.5\%$ at the bottom, these authors achieved excellent resolution of proteins covering a very broad size range from 670,000 Da down to 14,000 Da and even of peptides produced by trypsin digestion of BSA. However, the gradient of both acrylamide and crosslinker is seldom used today due to the extra burden of preparing different stock solutions of acrylamide/Bis. All the above steps can be performed with greater versatility and much higher control on the gradient reproducibility if a microcomputer is used for handling the gradient, as shown in Figure 5.14. In this case, an electromagnetic two-way valve is activated for drawing appropriate liquid amounts from vessels D or E, but then a small-volume mixing chamber (G) has to be placed between the valve and the peristaltic pump to allow for thorough mixing of the two limiting solutions.

5.5.5. Peptide Mapping by SDS–PAGE

In many cases relationships between proteins cannot be proved unequivocally by electrophoretic mobility alone, especially if they are closely related; thus further analysis

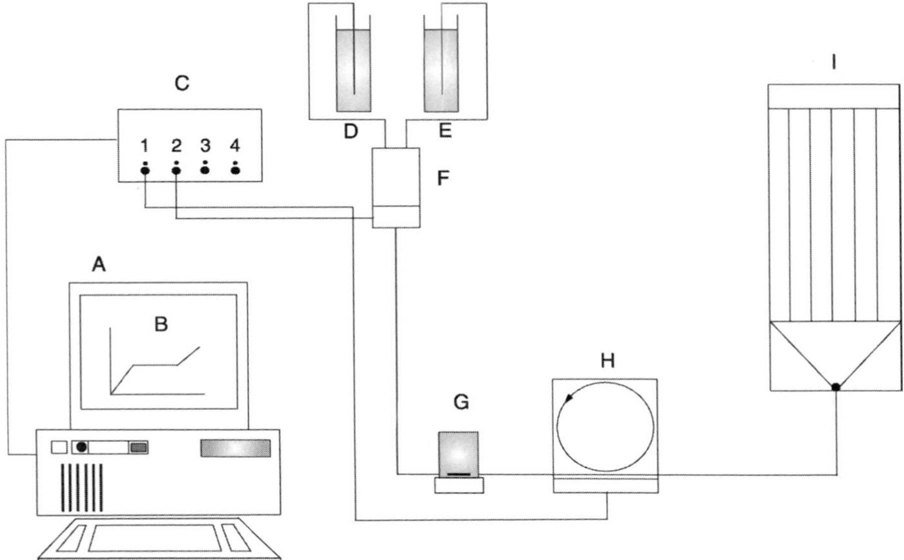

FIGURE 5.14. Microcomputer-controlled apparatus for preparation of gradients of polyacrylamide gels in slab format: A, microcomputer; B, shape of gradient to be generated; C, switch controller; D, reservoir for high %*T* acrylamide solution; E, reservoir for low %*T* acrylamide solution; F, electromagnetic two-way valve; G, mixing chamber; H, peristaltic pump; I, tower containing gel cassettes. (From Dunn, 1993, by permission.)

is required. If a protein has already been extensively purified, it can be digested in vitro and then the peptides thus produced analyzed by SDS–PAGE. This is at the heart of the method of Cleveland et al. (1977), who digested such purified proteins in a buffer already containing some SDS (0.5%) via a number of proteolytic enzymes, such as papain, chymotrypsin, or *Staphylococcus aureus* protease, and then analyzed the peptides thus produced by SDS–PAGE. However, this method is of limited applicability, since in many cases purified protein samples are not available yet the investigator requires information on the similarities between individual protein bands obtained by gel electrophoresis. To solve this problem, techniques for in situ peptide mapping have been developed and the methodology extensively described (Gooderham, 1986; Andrews, 1998). In these cases, the sample proteins are best separated by a preliminary electrophoretic step using virtually any of the usual methods (native electrophoresis, SDS–PAGE, IEF, 2D PAGE). The gels are then briefly stained and destained and placed on a transparent plastic sheet over a light box. Individual protein zones are then cut out with a scalpel of razor blade and trimmed to a size small enough to fit easily into the sample wells of an SDS–PAGE gel to be used for the peptide mapping stage. Before doing this, however, the pieces of gel are soaked for 30 min in

125 mM Tris–HCl, pH 6.8, containing 0.1% SDS. Below is a step-by-step protocol:

- Stain the polyacrylamide gel containing the separated proteins for 5–10 min in Coomassie blue staining solution (0.25% w/v Coomassie, 50% v/v methanol, and 10% v/v acetic acid) and for 10–15 min immediately destain 5% acetic acid, 10% methanol.
- Place the gel on the light box and cut out the protein zones of interest using a scalpel (or razor blade).
- Keeping each protein sample separate, trim and/or cut the recovered gel fragments such that they will fit into the sample wells of the second slab gel.
- Add 10 mL of 125 mM Tris–HCl, pH 6.8, 0.1% SDS, to each sample and leave at room temperature for 30 min.
- Discard this washing solution and place the gel fragments into separate wells of the second slab gel formed in a 5-cm-long stacking gel layer. Use a thin spatula to guide the fragments to the bottom of the wells.
- Fill the regions around the gel fragments with 125 mM Tris–HCl, pH 6.8, 0.1% SDS, 20% glycerol, delivered from a syringe.
- Overlay each sample with 10 μL of proteinase solution (1–100 μg/mL proteinase in 125 mM Tris–HCl, pH 6.8, 0.1% SDS, 10% glycerol, 0.005% bromophenol blue).
- Begin electrophoresis. When the bromophenol blue reaches near the bottom of the stacking gel, turn off the electrical current for 20–30 min to allow proteinase digestion of the sample to occur.
- Resume electrophoresis. Continue until the bromophenol blue nears the bottom of the gel slab.
- Detect the separated peptides by Coomassie blue or silver staining. Alternatively, detect radiolabeled peptides by using autoradiography or fluorography.

The method of choice in this protocol for peptide mapping is an SDS–PAGE run in a discontinuous Laemmli (1970) buffer (Table 5.9). However, since peptides are in general considerably smaller in size than proteins, it is preferable to adopt a $T = 15\%$ running gel, or better a linear porosity gradient gel, such as $T = 5\%$ to $T = 20\%$ or $T = 10\%$ to $T = 25\%$, according to the size range of the generated peptides. Figure 5.15 summarizes the various steps described and the results expected upon staining. Table 5.10 gives an extensive list of proteases used for in situ peptide mapping. Alternatively, although less frequently, chemical cleavage methods can be adopted, as listed in Table 5.11.

Cutting out a gel in which a preliminary separation has been done, a number of individual bands to be loaded in the sample slots of a second-dimension gel for peptide mapping might be too cumbersome when the sample under analysis is highly heterogeneous. At this point, one might just as well cut out the entire gel strip of the first dimension without excising any individual gel segment and subject the entire strip to a second dimension SDS–PAGE after suitable protease digestion procedures.

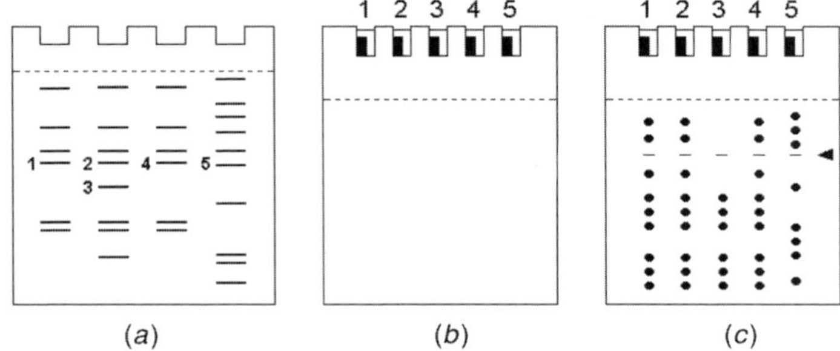

FIGURE 5.15. Representation of key steps in in situ peptide-mapping experiment. (*a*): Protein samples are separated by gel electrophoresis and stained with Coomassie brilliant blue R-250. (*b*) Bands containing proteins of interest (1–5) are cut out of the gel, equilibrated, and placed on a second gel together with a suitable protease. (*c*) Proteins and protease are then run into the gel and the resulting peptides are visualized by, e.g., silver staining. The position of the protease is indicated by the arrowhead. Analysis of the peptide map shows that proteins 1, 2, and 4 are homologous, whereas protein 3 is a specific degradation product of these proteins. Protein 5 has no significant homology to the other proteins. (From Gooderham, 1986, by permission.)

TABLE 5.10. Proteinases Used for Selective Peptide Bond Hydrolysis of Proteins for Peptide Mapping

Proteinase	pH Optimum	Bond Specificity
Staphylococcus aureus V8 protease	4–8	Glu-X, Asp-X
α-Chymotrypsin	7–9	Trp-X, Tyr-X, Phe-X, Leu-X
Trypsin	8–9	Arg-X, Lys-X
Pepsin	2–3	N- and C-sides of Leu, aromatic residues, Asp and Glu
Thermolysin	≈8	Ile-X, Leu-X, Val-X, Phe-X, Ala-X, Met-X, Tyr-X
Subtilisin	7–8	Broad specificty
Pronase (*Streptomyces griseus*) Protease	7–8	Broad specificty
Ficin	7–8	Lys-X, Arg-X, Leu-X, Gly-X, Tyr-X
Elastase	7–8	C-side of nonaromatic neutral amino acids
Papain	7–8	Lys-X, Arg-X, (Leu-X), (Gly-X), (Phe-X)
Clostripain	7–8	Arg-X
Lys-C endoproteinase (*Lysobacter enzymogenes*)		Lys-X
Asp-N endoproteinase (*Pseudomonas fragilis*)		X-Asp, X-Cys
Arg-C endoproteinase (mouse submaxillary gland)		Arg-X

TABLE 5.11. Chemical Methods for Cleavage of Peptide Bonds in Proteins

Reagent	Bonds cleaved
Cyanogen bromide	–Met–X–
Hydroxylamine	–Asn–Gly–
BNPS skatole	–Trp–X–
N-Chlorosuccinimide (*N*-bromosuccinimide)	–Trp–X–
Partial acid hydrolysis	–Asp–Pro–
Partial basic cleavage	–Ser–X–
Heat (110°C, pH 6.8, 1–2 h)	–Asp–Pro–
2-Nitro-*S*-thiocyanobenzoate	–X–Cys–

This is the method described by Bordier and Crettol-Jarvinen (1979) and Saris et al. (1983). Basically, the sequence of steps is much the same as the one outlined in the above protocol, with the use of discontinuous SDS–PAGE with Laemmli's buffer. The main exception is the fact that there will be no sample wells cast into the stacking gel, but the entire gel strip of the first dimension will be overlaid on top of this gel and cemented in situ with 1% melted agarose. Once the agarose has set, on top of it 1 mL of protease solution will be poured and electrophoresis started, as described above. Here, too, once the tracking dye has migrated the entire length of the stacking gel, the current will be turned off for 30 min in order to let enzyme digestion take place and then resumed as usual.

5.5.6. SDS–PAGE in Photopolymerized Gels

There are two main problems with the standard polymerization procedure in the presence of TEMED and APS. First, due to excess, unreacted acrylamide left behind in the gel matrix, alkylation of Cys residues in proteins is an ever-present hazard (Chiari et al., 1992). Second, due to the strong oxidation power of APS, Cys may be destroyed or Met altered (Klarskov et al., 1994). Photopolymerization systems are theoretically well suited for solving these problems, since polymerization could be driven to >95%, because radicals are continuously produced as long as light is absorbed by the dye, as opposed to a single burst of radicals in the case of chemical initiators such as persulfate. Lyubimova et al. (1993) described photopolymerization in the presence of a triad of novel catalysts: 100 μM methylene blue (MB) combined with a redox couple, 1 mM sodium toluenesulfinate (STS), a reducer, and 50 μM diphenyliodonium chloride (DPIC), an oxidizer (see Fig. 5.16 for the formulas). The results were unique: Onset of polymerization could be induced in a matter of seconds and proceeded at a very high speed until completion within, typically, 30 min. The viscoelastic properties of these gels were even better than those of persulfate-activated gels, suggesting a more homogenous gel structure. The elastic modulus exhibited a maximum in correspondence with a minimum of permeability, both situated at a 5% crosslinker. A theoretical study confirmed the very high reaction kinetics

Diphenyliodonium chloride (oxidizer) $M_r = 316.57$

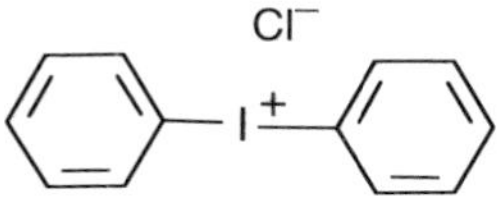

Cl^-

I^+

Sodium toluene sulfinate (reducer) $M_r = 178.18$

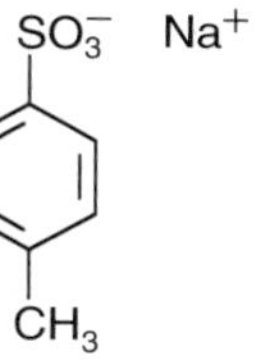

SO_3^- Na^+

CH_3

Methylene blue $M_r = 319.86$

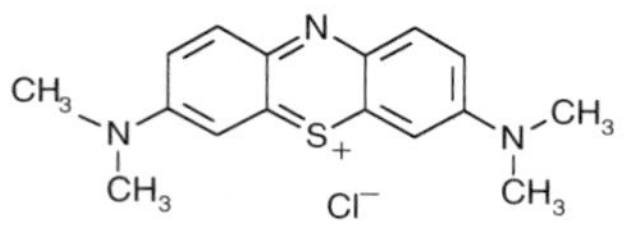

CH_3 N CH_3 N S+ Cl^- N CH_3 CH_3

Riboflavin-5′-phosphate $M_r = 514.4$

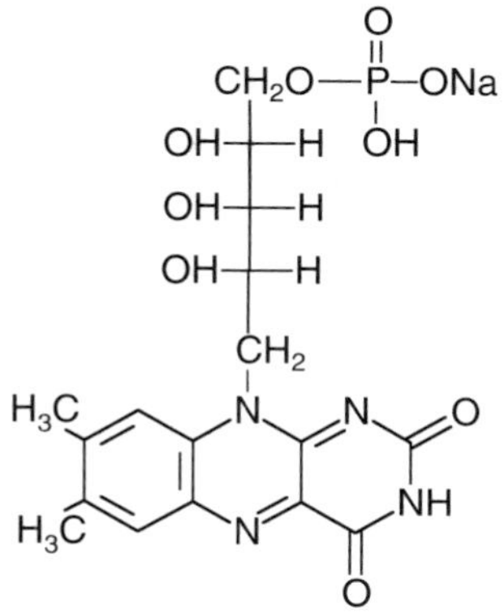

O

CH_2O—P—ONa

OH—H OH

OH—H

OH—H

CH_2

H_3C N N O

NH

H_3C N

O

FIGURE 5.16. Formulas of the triad of catalysts for photopolymerization: methylene blue (MB), sodium toluenesulfinate (STS), the reducer, and 50 μM diphenyliodonium chloride (DPIC), the oxidizer. Alternatively (or in combination with MB), riboflavin-5′-phosphate can be used (bottom formula).

of this system together with the extraordinary conversion efficiency (Lyubimova and Righetti, 1993). The "methylene blue approach" has been described in a number of additional reports. In a first study, another advantage of MB catalysis was discovered: its insensitivity to pH in the gelling solution. Persulfate–TEMED polymerization has pH optima in the pH 7–10 range; riboflavin–TEMED only in the pH 4–7 interval; ascorbic acid, ferrous sulfate, and hydrogen peroxide only at pH 4; and TEMED and hydrosulfite a reasonable conversion only in the pH 4–6 interval. However, MB-driven reactions work at full speed in the pH 4–9 range and very well up to pH 10, a real record (Caglio and Righetti, 1993a). Other interesting data on matrix structure came with further studies on the MB-driven system: Topological data could be derived by starting polymerization at 2°C and continuing, after the gel point, at 50°C. The data suggested that at 2°C the nascent chains formed clusters held together by hydrogen bonds (melting point at 28°C). Such clusters were subsequently "frozen" into the 3D space as the pendant double bonds of the crosslinkers kept reacting (Righetti and Caglio, 1993). This would lead to turbid, highly porous and unelastic gels. However, if polymerization was continued at the gel point, at 50°C, clear and elastic gels would be obtained. This was an additional proof of the widely held belief that turbid gels would indicate large-pore structures. Other remarkable features were then discovered: for example, the insensitivity of MB-driven catalysis to oxygen versus a large inhibition in persulfate reactions. Conversely, whereas 8 M urea would substantially accelerate persulfate catalysis, it left undisturbed the MB system. However, when polymerization occurred in a number of hydroorganic solvents [all in a 50 : 50 v/v ratio; dimethyl sulfoxide (DMSO), tetramethyl urea, formamide, dimethyl formamide], persulfate catalysis was severely inhibited whereas MB photopolymerization was essentially unaffected (Caglio and Righetti, 1993b). The same applied to polymerization in detergents (Caglio et al., 1994). It thus appeared that MB catalysis was a unique process proceeding at optimum rate under the most adverse conditions and able to ensure nearly complete monomer conversion into the growing polymer.

Given all the above advantages, why has MB-driven photopolymerization not become popular? A reason could be the one reported by Rabilloud et al. (1995), who claimed poor transfer of proteins in 2D maps and potential adsorption of such analytes into one (or more) of the triad of catalysts grafted at the chain termini. A partial remedy to these problems came by drastically reducing the amounts of the three initiators in the gelling solution: 500 μM STS, 20 μM DPIC, and 30 μM MB. Such lower levels were found not to affect the final properties of the gels. This new, reduced-level photopolymerization system was in fact found to give excellent results in the separation of histones at acidic pH values (pH ~4) in the classic urea–acetic acid–Triton system (Rabilloud et al., 1996). On the basis of these findings, we believe that one of the main reasons for such a poor performance of MB-catalyzed gels could be the very presence of MB, grafted covalently into part of the population of polyacrylamide chains in the matrix. Since MB is positively charged up to about pH 10, it could very well interact with the negatively charged SDS–protein micelles, producing smears during migration and poor transfer from the gel. It is a fact that this does not occur with histones, which are highly positively charged at pH 4 of this electrophoretic system and thus repelled by the like charge of MB. On these assumptions, we have made "zwitterionic" gels, that is, gels photopolymerized with 30 μM of MB (positively charged) in the presence of 30 μM of riboflavin 5′-phosphate (a negatively charged dye). On the assumption that there would be equal chances for these two catalysts to be incorporated into the gel matrix, said matrix would then be zwitterionic and have minimal tendency to adsorb protein–SDS micelles by ion pairing. Excellent gels were obtained in such a zwitterionic polymerization system, not only for marker proteins but also for total cell lysates, such *as Escherichia coli* total proteins (Olivieri, Castagna, and Righetti, unpublished). Blotting onto nitrocellulose membranes was also shown to be essentially quantitative. As an extra bonus, no oxidation of proteins by the catalysts occurred, as in the case of persulfate-driven polymerization. Notwithstanding these advantages, alkylation of proteins by free acrylamide (even when driving the reaction to 95%, high %*T* gels would still contain a large excess of unreacted monomers over the potential alkylation sites of proteins) still occurred, suggesting that there is no simple cure to this problem. Thus, if one wants to avoid protein alkylation by free acrylamide during SDS–PAGE, there appear to be only two possible remedies: (a) use washed matrices, as suggested by Westermeier (1997), or (b) use prealkylated proteins, that is, protein samples which are not only reduced but also alkylated with iodoacetamide, as suggested by any textbook on protein chemistry (see also Chapter 6). After all, the site of attack has been demonstrated to be almost exclusively the free –SH group of Cys residues (at least during the normal times of electrophoresis); thus, their alkylation efficiently suppresses further reaction with acrylamide during the SDS–PAGE step.

5.5.7. Blue Native PAGE and Other Native PAGE Protocols

Although this method, which deals with the analysis of native proteins, might not appear to belong to a discussion of SDS–PAGE, a classical denaturing system, there are in reality striking similarities with the latter, justifying its inclusion here. Blue native (BN)–PAGE is a charge-shift method originally developed by Schägger and von Jagow (1991) for running proteins under native conditions around neutral pH. A negatively charged dye (Coomassie blue G-250) is added to the native proteins, thus inducing a charge shift as it binds to the surface hydrophobic domains. The electrophoretic mobility of such proteins will then be mainly determined by the negative charges of bound Coomassie dye to the point at which even basic proteins will migrate to the cathode at pH 7.5. The drawback of this method is that not all native proteins bind to the Coomassie dye, although most of them will (Schägger et al., 1994). Blue native PAGE differs from SDS–PAGE in two respects: the Coomassie molecules bind to the protein surface (not to the interior) and the ratio of bound Coomassie to protein is variable, so that the final m/z ratio of different proteins is not constant, as it very nearly is in SDS–PAGE. There are two consequences: The dye does not denature the proteins and, as a consequence, it does not allow accurate molecular mass determination in uniform concentration gels. Nevertheless, Schagger (1995) demonstrated that, when using BN–PAGE in polyacrylamide gradient gels, one can obtain a reasonably accurate M_r determination of native proteins. This is due to the fact that the pore size gradient of such gels determines the end position of migration of proteins based on their size (electrophoresis to the pore limit). In BN–PAGE, the sample buffer is typically 750 mM aminocaproic acid, 50 mM BisTris–HCl, 1.25% dodecyl maltoside, and 0.35% Coomassie blue G.250, titrated to a final pH of 7.0. The main features of BN–PAGE include the following:

(a) Increased solubility and reduced aggregation of native proteins due to the presence of negatively charged Coomassie on the protein surface
(b) A working pH near neutrality, which minimizes protein denaturation and is specially suited for membrane proteins
(c) The visibility of the stained protein zones during migration

In addition, the oligomeric states of native proteins can be determined if one combines BN–PAGE in a first-dimension separation with SDS–PAGE in the second dimension. This method would thus appear to be highly suitable for analysis of molecular mass, oligomeric state, and homogeneity of native proteins. More recently, this method has been reproposed for analysis of the mitochondrial proteome, particularly in regard to the respiratory chain proteins (Brookes et al., 2002).

On a similar line of thinking, Bertsch and Kassner (2003) have described a method for selective staining of native proteins for hydrophobic surface patches. This would allow the identification of a subproteomic library exhibiting particular structural features related to protein functions, which could facilitate functional and structural characterization of new proteins. These authors listed quite a number of proteins known to bind hydrophobic ligands (such as lipophilic small molecules, lipids, other hydrophobic peptides and proteins) and concluded that, if specific stains for such hydrophobic surface patches could be devised, one could differentially display such species, via 1D and 2D electrophoresis, by comparison with general stains such as Coomassie blue. Two such specific stains for hydrophobic domain were identified, bromophenol blue (BPB) and 8-anilino-1-naphthalene-sulfonic acid (ANSA), the latter, in virtue of its fluorescence, permitting a sensitivity of ~65 ng/band. With a set of marker proteins, in both 1D and 2D electrophoresis, BSA monomer and dimer, phosphorylase b, and lactic acid dehydrogenase were shown to stain with both BPB and ANSA, whereas ovalbumin, carbonic anhydrase, trypsin inhibitor, and β-lactoglobulin, all well-known hydrophilic proteins, did not. Competitive displacement experiments between the two dyes as well as with hydrophobic exogenous drugs could be devised.

5.6. BLOTTING PROCEDURES

When analyzing complex samples by SDS–PAGE, it is often desirable to use a method allowing specific detection of some components in the mixture. One of the most sensitive methods is immunodetection, by which specific antibodies (either polyclonal or monoclonal) are used for precipitating components following an electrophoretic separation. This procedure, known as immunofixation, works well with cellulose acetate membranes and agarose gels [it was in fact first reported by Grabar and Williams (1953)], but it can hardly be adopted when running polyacrylamide gels, due to their restrictive nature (i.e., small pore size). This last problem was overcome by the development of blotting techniques in which the separated proteins are transferred from the polyacrylamide gel onto the surface of a thin matrix such as nitrocellulose. As a result, the proteins are immobilized onto the open face of this thin sheet (via hydrophobic bonds in the case of nitrocellulose) and are thus readily accessible to interaction with antibodies and other ligands. This procedure originated from a technique developed by Southern (1975) for transferring DNA (Southern blotting) and was subsequently adapted for RNA transfer (Northern blotting). Almost inevitably, when this procedure was subsequently adapted to proteins (Towbin et al., 1979), it became popularly known as Western blotting. The major steps in Western blotting are shown in Figure 5.17. After terminating the electrophoretic step (in general, but not exclusively, an SDS–PAGE, either 1D or 2D), the gel should be briefly incubated in the appropriate transfer buffer for removal of SDS and other gel constituents which could cause problems during transfer. This also minimizes the shrinking or swelling of the gel during transfer and allows potential protein renaturation (if possible). After transfer by a

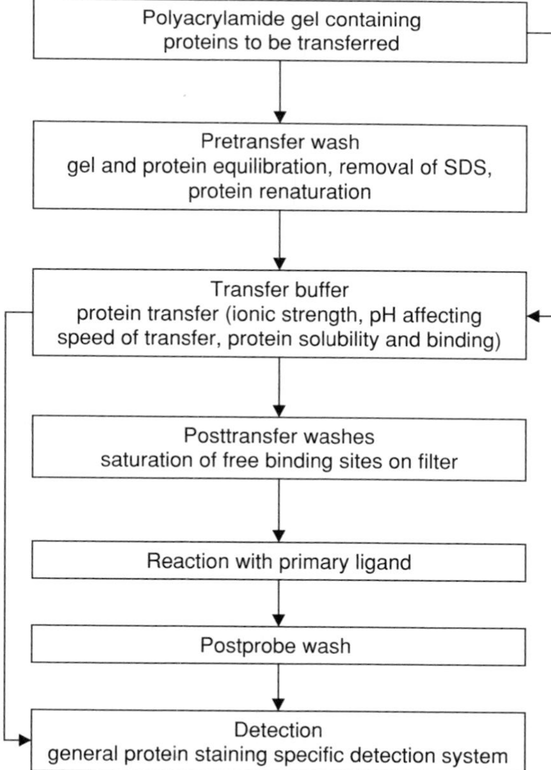

FIGURE 5.17. Flow sheet for performing protein transfer from gel and subsequent detection steps on membrane.

suitable procedure, the protein pattern can be visualized using a total protein stain. Several general detection methods for proteins on Western blots have been described by Merril and Washart (1998), to whom the readers are referred for further details [see also a review by Bini et al. (2000) related to blotting and immunoblotting in 2D map analysis]. A variety of membranes have been described for implementing protein transfer from polyacrylamide matrices. Nitrocellulose is the most commonly used blotting membrane at this time. Interaction of denatured polypeptides with nitrocellulose is probably mediated mostly by hydrophobic interaction, although the complete mechanism is not yet fully understood. However, nitrocellulose has low affinity for some proteins, particularly those of low M_r (Kakita et al., 1982). Given this problem, some authors have suggested covalent fixation of proteins to nitrocellulose via glutaraldehyde (Kay et al., 1983) or *N*-hydroxysuccinimidyl-*p*-azidobenzoate (Kakita et al., 1982). Other matrices that have been developed include diazobenzyloxymethyl (DBM) modified cellulose paper (Alwine et al., 1979) and diazophenylthioether (DPT) paper (Reiser and Wardale, 1981). Negatively charged proteins interact electrostatically with the positively charged diazonium groups of these matrices followed by irreversible covalent linkage via azo derivatives (Alwine et al., 1979). The DPT paper has been shown to be more stable than DBM paper but equally efficient. Diazo paper is reported to show less resolution of separated proteins compared to nitrocellulose or nylon membranes (Burnette, 1981). In addition, glycine may interfere with DBM protein blotting (Kakita et al., 1982). Nylon membranes are also used for protein blotting. Two such commercially manufactured membranes are GeneScreen (from NEN, Boston, MA, USA) and Zeta-bind from AMF/Cuno (Meriden, CT, USA). Zeta-bind has a very high protein-binding capacity (Gershoni and Palade, 1982) and requires extensive blocking for eliminating nonspecific binding. Nylon membranes consistently provide a higher efficiency of protein transfer from SDS gels than nitrocellulose (Gershoni and Palade, 1982, 1983). However, one disadvantage of nylon matrices is that they bind the common anionic dyes such as Coomassie blue and amido black, which may result in background staining that could obscure the detection of transferred proteins. Polyvinylidene difluoride (PVDF) membranes can also be utilized for protein blotting (Pluskal et al., 1986); they are hydrophobic in nature and proteins bound to them can be detected by most anionic dyes as well as by immunodetection protocols.

5.6.1. Capillary and Electrophoretic Transfer

When originally described (Southern, 1975), blotting was performed by capillary forces in a setup similar to that shown in Figure 5.18. The gel is overlayed with the membrane (nitrocellulose in this case) and then with a large stack of dry filter paper sheets. The assembly is then covered with a rigid plate (glass or plastic) on top of which a weight (0.5–1 kg) is placed. The dry stack of filter papers pulls liquid (with the dissolved proteins) from the gel layer, allowing the proteins to be eluted from the gel matrix and be trapped by the membrane. Although this method is not much in vogue today for SDS–PAGE, it can have remarkable efficiency when adopted for eluting proteins from agarose gels (Desvaux et al., 1990), with transfer times of less than 1 min! Pressure blotting can also be efficiently adopted for thin

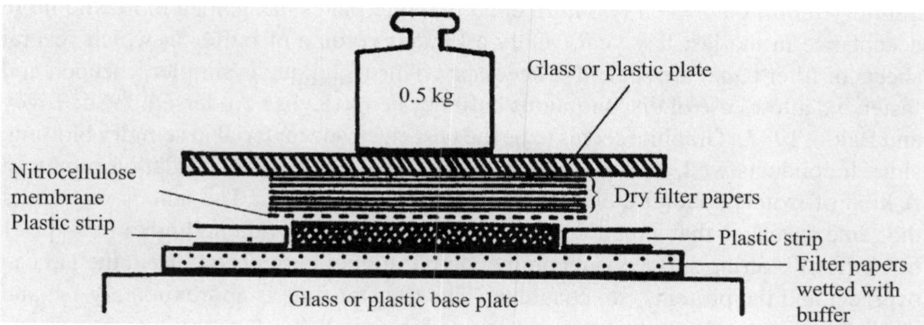

FIGURE 5.18. Setup for capillary blotting. The polyacrylamide gel is represented by the hatched area between the two plastic strips.

polyacrylamide IEF gels, with a contact time of about 1 h. An alternative procedure is vacuum blotting (Olszewska and Jones, 1988), by which efficient transfers (30–40 min, instead of overnight for a typical pressure blot) can be achieved. The other advantages of vacuum blotting are that (a) it is quantitative, there are no back transfers; (b) it leads to sharper zones and better resolution; and (c) it reduces the mechanical stress on the gel. By far, however, the preferred method today is electrophoretic blotting, performed either by tank or semidry procedures. A typical setup is shown in Figure 5.19; note that the trapping membrane is typically positioned on the anodic side of the polyacrylamide gel, since after SDS–PAGE the protein–SDS complex has a high net negative charge. Originally, vertical buffer tanks with coiled platinum wire electrodes fixed on the two sides were used. For this technique, the gel and blotting membrane are clamped in grids between filter papers and sponge pads and suspended in the tank filled with buffer. The transfers usually occur overnight.

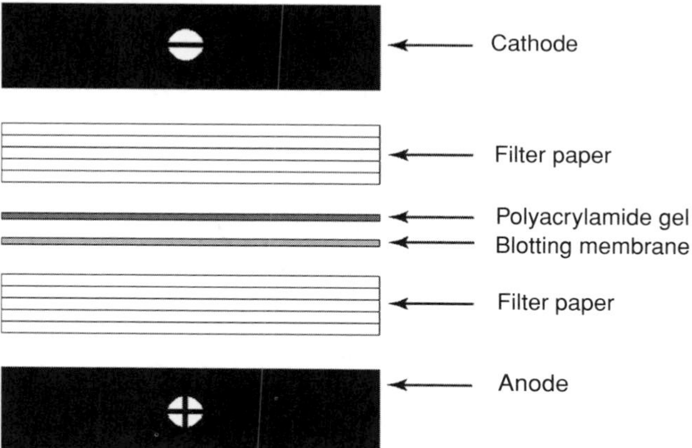

FIGURE 5.19. Experimental setup for wet or semidry blotting from SDS–PAGE gels. Note that the blotting membrane is placed on the anodic side of the gel due to the very high net negative charge of the protein–SDS complex.

Semidry blotting between two horizontal graphite plates has gained more and more acceptance in the last few years. Only a limited volume of buffer in which several sheets of filter paper are soaked is necessary. This technique is simpler, cheaper, and faster and allows use of discontinuous buffer systems (Kyhse-Andersen, 1984; Tovey and Baldo, 1987). Graphite seems to be the best electrode material in semidry blotting, since it conducts well, does not overheat, and does not catalyze oxidation products. A kind of isotachophoretic effect takes place in this system: The anions migrate at the same speed, so that a regular transfer takes place. A current no higher than 0.8–1 mA/cm^2 of blotting surface is recommended. If higher currents are used, the gel can overheat and the proteins can coagulate. The transfer time is approximately 1 h and depends on the thickness and concentration of the gel. When longer transfer times are required, such as for thick (1-mm) or highly concentrated gels, a weight is placed on the upper plate so that the electrolyte gas is expelled out of the sides. Johansson (1987) has also described a double-replica blotting: An alternating electric field is applied on a blotting sandwich with a membrane on each side of the gel with increasing pulse time, so that two symmetrical blots result. This could be quite useful if the membranes have to be probed with different detection systems, such as different antibodies.

Ready-made or home-made gels backed by support films are today quite common for the zone electrophoresis and IEF of proteins. These films are impermeable to current and buffer flow, so they must be separated from the gels prior to performing electrophoretic transfers or capillary blotting. To separate the gel from the film without damage, an apparatus exists (Fig. 5.20) with a taut thin steel wire which is pulled between them.

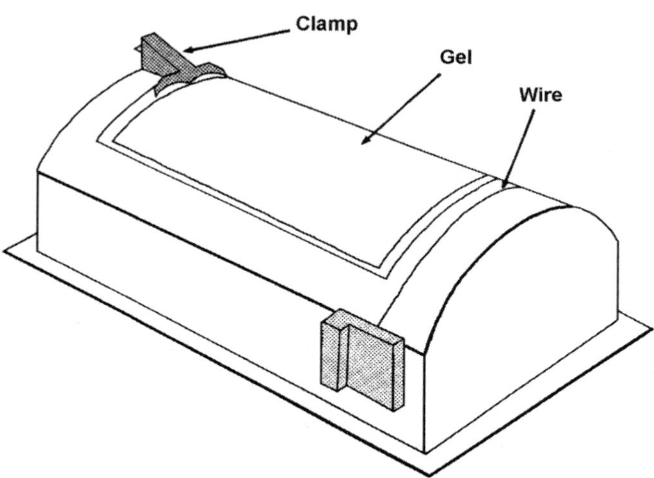

FIGURE 5.20. Guillotine-like apparatus for detaching gel matrix from backing plastic support. The gel is clamped on the curved surface of the shieldlike instrument, where the steel wire will slide at the interface between gel and plastic backing.

5.6.2. Detection Systems after Blotting

These systems comprise general staining protocols as well as specific detection methods. In the first category, besides staining with amido black or Coomassie brilliant blue, mild staining methods such as the very sensitive India ink method (Hancock and Tsang, 1983) or reversible ones such as with Ponceau S (Salinovich and Montelaro, 1986) or Fast Green FCF have been described. One can also use the following:

- A general immunostain (Kittler et al., 1984)
- Colloidal gold or colloidal iron (Moeremans et al., 1985, 1986)
- Autoradiography
- Fluorography (Burnette, 1981)

The principle of autoradiography and fluorography (coupled often to the use of intensifying screens) is illustrated in Figure 5.21, whereas Figure 5.22 shows how the

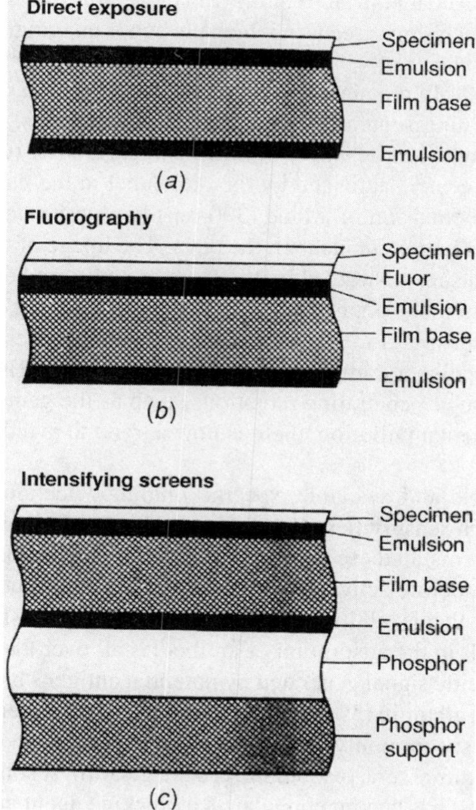

FIGURE 5.21. Three classical methods of autoradiography: (*a*) by direct exposure, (*b*) by fluorography, and (*c*) with use of intensifying screens. (Courtesy of Eastman Kodak Company.)

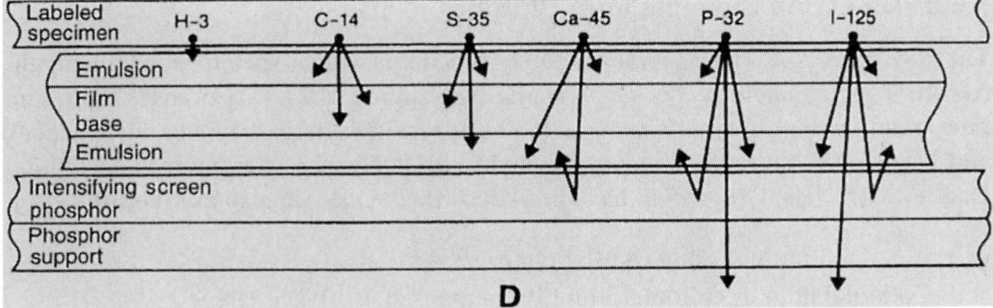

FIGURE 5.22. Penetration of different radiations from radioisotopes into X-ray film. Note that the intensifying screen phosphor method can only be used with high-energy emitters (e.g., gamma radiations) since they can penetrate through the film thickness and excite the phosphor underneath. (Courtesy of Eastman Kodak Company.)

radiation of different isotopes can penetrate to a given depth through the top and bottom emulsions, which coat both sides of the film base. More recent developments call for new types of intensifying screens in which phosphor imaging plates are composed of a thin (~500-μm) layer of very small crystals of BaFBr–Eu^{2+} in a plastic binder (Johnston et al., 1990). In practice, this phosphor imaging plate is exposed to a dried 1D or 2D gel containing separated radiolabeled protein zones. After exposure, the plate is transferred to a scanner where light from a HeNe laser (633 nm) is absorbed by the metastable F centers activated by the radiolabel in the phosphor screen. This in turn results in the emission of a blue (390-nm) luminescence proportional to the original amount of radiation incident on the plate. The image of the bands or spots in the original gel is thus reconstructed by the scanner and stored electronically. These techniques were extremely popular in the 1970s and 1980s, especially for revealing spots on 2D maps, due to their extremely high sensitivities and were typically performed on dried polyacrylamide gel slabs (Bonner, 1983). However, due to risks involved with the use of penetrating radiations, such as the gamma emitter ^{125}I, and the risk of environmental pollution, there is now a trend to avoid radioactivity in the laboratory.

Whenever possible and available, specific immunodetection methods are used, since they often reach sensitivities close to those of radiolabeling without generating environmental problems. In these procedures, after the proteins have been transferred to the membrane but before activating any immunochemical protocol, the membrane must be deactivated, that is, all its unoccupied binding sites must be blocked. Failure to do so would result in the adsorption of antibodies all over the membrane surface, obliterating any specific signal generated by potential antigens blotted onto the membrane. Bovine serum albumin (3% to 5% w/v) in phosphate-buffered saline (PBS), 1 h incubation is the most commonly used blocking agent, although other proteins could be used (e.g., other animal sera, ovalbumin, casein, gelatin). A solution of nonfat dried milk (3% w/v in PBS) has become popular as a blocking agent and usually results in very low background staining. At this point, washing the excess of blocking agent, the membrane is ready for incubation with antibodies. Although it would be possible to label the primary antibody (or other ligand) with a suitable reporter group for a direct visualization of reactive proteins on the blot, this approach is not popular, since it requires derivatization of the primary antibody, a process which could adversely affect its specificity or affinity. Therefore, it is usual to use an indirect, or sandwich, approach by utilizing labeled secondary (or tertiary) antibody reagents. Thus, following blocking, the membrane is incubated with a solution containing the appropriately diluted specific primary antibody. The membrane is then washed and reacted with a solution containing a secondary antibody specific for the species and immunoglobulin class of the primary antibody. The secondary antibody can be fluorescently labeled [e.g., with fluorescein isothiocyanate (FITC)], radiolabeled (usually with ^{125}I), or conjugated to an enzyme (e.g., horseradish peroxidase, alkaline phosphatase, β-galactosidase, or glucose oxidase). Alternatively, the secondary antibody can be replaced by appropriately labeled staphylococcal protein A or streptococcal protein G. These reagents bind specifically to the F_c region of immunoglobulins (Ig), with protein G reacting with a broader range of species and classes than does protein A. Methods employing fluorescently labeled secondary antibodies require the use of UV illumination, while radiolabeled antibody procedures depend on time-consuming autoradiographic steps. Thus, methods based on the use of enzyme-conjugated secondary antibody reagents have become most popular, and kits of these reagents are commercially available. Figure 5.23 shows how this system of antibodies builds up onto the spot

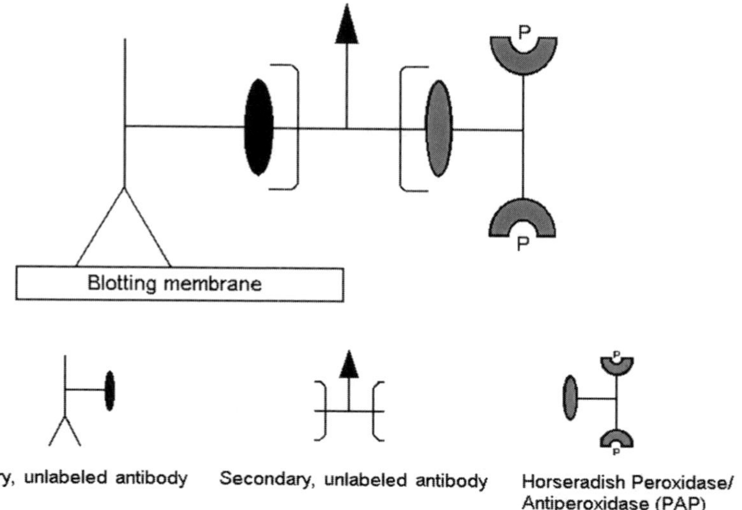

FIGURE 5.23. Scheme of system of antibodies building up onto spot containing specific antigen blotted onto membrane after 1D or 2D electrophoretic separation in polyacrylamide gel. In this example a cascade of three antibodies is used. Note that the last antibody in this cascade is covalently reacted with a reporter/amplifier, in this case the enzyme peroxidase from horseradish. Revelation will take place when the membrane will be confronted with appropriate substrates and dyes which, upon reaction, will form insoluble, stable, colored reaction products on the site.

containing the specific antigen blotted onto the membrane, although in this example a cascade of three antibodies is used. It is seen that the last antibody in this cascade carries an enzyme which will be used for detection and signal amplification. Detection is in fact achieved by activating an enzyme reaction by using appropriate substrates which form insoluble, stable colored reaction products at the sites on the blot where the enzyme-conjugated secondary (or tertiary, as in Fig. 5.23) antibody is bound. The two most popular substrates for use with peroxidase-conjugated antibodies are diamino benzidine, which gives brown bands, and 4-chloro-1-naphthol, which produces purple bands. The best substrate system for the visualization of alkaline phosphatase-conjugated antibodies is a mixture of 5-bromo-4-chloro indoxyl phosphate and nitroblue tetrazolium.

Other detection systems have been developed to increase the sensitivity of detection of proteins on blots. One approach is the use of colloidal gold-labeled antibodies (Daneels et al., 1986) or protein A. These reagents have the advantage that the stain is visible, due to its reddish-pink color, without further development; their sensitivity can be further increased by silver enhancement. Triple-antibody probing methods can also be used for increasing the sensitivity of detection (see Fig. 5.23). In this case, preformed complexes of enzyme and antibodies are linked by secondary antibodies to the primary antibodies, examples being the peroxidase–antiperoxidase (PAP) and alkaline phosphatase–anti-alkaline phosphatase (APAP) procedures. Secondary antibodies can also be conjugated with the steroid hapten, digoxigenin, which can then be detected by using an enzyme-conjugated antidigoxigenin antibody (Kessler, 1991).

Another popular method for increasing the sensitivity of detection on blots exploits the specificity of the reaction between the vitamin biotin ($M_r = 224$ Da) and the protein avidin. Antibodies can be readily conjugated with biotin and the resulting conjugates used as the secondary reagents for probing blot transfers. A third step must then be used for visualization using avidin conjugated with a suitable reporter enzyme (e.g., peroxidase, β-galactosidase, alkaline phosphatase, or glucose oxidase). Even greater sensitivity can be achieved at this stage by using preformed complexes of biotinylated enzyme with avidin, since many enzyme molecules are present in these complexes, thus generating an enhanced signal. Egg white avidin ($M_r = 68{,}000$) is often used, but streptavidin ($M_r = 60{,}000$, from *Streptomyces avidinii*) is today preferred, since it has a pI close to neutrality and is not glycosylated.

In a series of recent papers, Patton's group has outlined a number of novel, highly sensitive fluorescent staining procedures for electroblots on nitrocellulose or polyvinylidene difluoride membranes. In a first approach, Kemper et al. (2001) proposed staining with SYPRO rose plus, a europium-based metal chelate stain claimed to exhibit exceptional photostability. This stain can be excited at 254 nm as well as at 302 or 365 nm. Additionally, it has a long emission lifetime, potentially allowing time-resolved detection as it is readily removed from proteins by incubation at mildly alkaline pH values. In a second approach, Martin et al. (2002) described a simultaneous red/green dual-fluorescence detection on electroblots by using BODIPY TR-X succinimidyl ester and ELF-39 phosphate. The former would label all proteins (red signal), whereas the latter would detect specifically only target proteins in combination with alkaline phosphatase–conjugated reporter molecules. In a third approach, Martin et al. (2003) outlined a simultaneous trichromatic fluorescence detection protocol on Western blots by which all proteins would first be covalently labeled with the dye 2-methoxy-2,4-diphenyl-2(2*H*)-furanone (MDPF) (blue fluorescence) followed by detection of target proteins via alkaline-phopshatase-conjugated reporter molecules in combination with the fluorogenic substrate 9*H*-(1,3-dichloro-9,9-dimethylacridin-2-one-7-yl)phosphate (red fluorescence) and ending with horseradish peroxidase–conjugated reporter molecules in combination with Amplex gold reagent (yellow fluorescence). Fluorescent labeling of proteins with Nile red has also been suggested (Daban, 2001). Needless to say, all SYPROs (orange, red, ruby, tangerine) were found to be fully compatible with all MS protein analysis methods (Lauber et al., 2001). By the same token, it is understood that multiple replica blottings can be made from a single 2D gel (Petersen, 2003).

5.7. CONCLUSIONS

Sodium dodecyl sulfate–polyacrylamide gel electrophoresis seems to have reached a plateau in terms of method development. It is hard to conceive that the technique could be further improved, since just about all possible methodological advances known in electrokinetic methods have been applied here. Also detection techniques seem to have been pushed to the maximum, so it is hard to conceive, by present-day knowledge, what new developments could possibly be made. Sodium dodecyl sulfate–polyacrylamide gel electrophoresis is a fundamental method for 2D map analysis, since it represents the second-dimension run and the run that will finally remain in the record and it is at the end of this step that proteins are stained or blotted and extracted and further analyzed with the powerful tools today available in proteomics, such as MALDI–TOF MS of both the intact polypeptide chain and its peptides generated by proteolytic digestion. It might be asked which, of all possible techniques here described, is best suited for 2D mapping. In the early days (e.g., in the Human Anatomy Program of the Andersons) it was customary to perform the SDS–PAGE in concave porosity gradients. Today this method has been largely abandoned in favor of linear porosity gradients, which are much more user friendly and easy to be implemented. In addition, in many reports, one could find the use of a double zone-sharpening technique, namely the use of discontinuous buffers (typically the Laemmli) coupled to porosity gradients in the running gel slab. This technique is also of diminishing importance today, again due to too-cumbersome manipulations. It has been found that one of the two methods suffices, so, if one adopts discontinuous buffers, the stacking of zones is adequate to give high resolution, rendering unnecessary the simultaneous use of a porosity gradient in the running slab. Conversely, if one does (or cannot) use buffer discontinuities, then a porosity gradient is a must. With this caveat, we think it is high time that we lead the readers into the realm of 2D map analysis, as expounded in the following chapter.

REFERENCES

Akins, R. E., Levin, P. M., Tuan, R. S. (1992) *Anal. Biochem.* **202,** 172–179.

Allen, R. C., Saravis, C. A., Maurer, H. R. (1984) *Gel Electrophoresis and Isoelectric Focusing of Proteins*, deGruyter, Berlin, pp. 41–43.

Alwine, J. C., Kemp, D. J., Parker, B. A., Reiser, J., Renart, J., Stark, R. G., Whal, G. M. (1979) *Methods Enzymol.* **68,** 220–242.

Ames G. F. L. (1974) *J. Biol. Chem.* **249,** 634–639.

Andrews, A. T. (1986) *Electrophoresis: Theory, Techniques and Biochemical and Clinical Applications*, Clarendon, Oxford, pp. 118–141.

Andrews, A. T. (1998) in Hames, B. D. (Ed.), *Gel Electrophoresis of Proteins: A Practical Approach*, 3rd ed., Oxford University Press, Oxford, pp. 213–235.

Arcus, A. C. (1967) *Anal. Biochem.* **18,** 381–384.

Barger, B. O., White, F. C., Pace, J. L., Kemper, D. L., Ragland, W. L. (1976) *Anal. Biochem.* **70,** 327–335.

Baumann, H., Cao, K., Howald, H. (1984) *Anal. Biochem.* **137,** 517–525.

Berggren, K., Chernokalskaya, E., Steinberg, T. H., Kemper, C., Lopez, M. F., Diwu, Z., Huagland, R. P., Patton, W. F. (2000) *Electrophoresis* **21,** 2509–2521.

Berggren, K., Steinberg, T. H., Lauber, W. M., Carroll, J. A., Lopez, M. F., Chernokalskaya, E., Zieske, L., Diwu, Z., Huagland, R. P., Patton, W. F. (1999) *Anal. Biochem.* **276,** 129–143.

Berggren, K. N., Schulenberg, B., Lopez, M. F., Steinberg, T. H., Bogdanova, A., Smejkal, G., Wang, A., Patton, W. F. (2002) *Proteomics* **2,** 486–498.

Bertsch, M., Kassner, R. J. (2003) *J. Proteome Res.* **2,** 469–475.

Bérubé, B., Coutu, L., Lefievre, L., Begin, S., Dupont, H., Suillivan, R. (1994) *Anal. Biochem.* **217,** 331–333.

Bini, L., Liberatori, S., Magi, B., Marzocchi, B., Raggiaschi, R., Pallini, V. (2000) in Rabilloud, T. (Ed.), *Proteome Research: Two-Dimensional Gel Electrophoresis and Identification Methods*, Springer, Berlin, pp. 127–141.

Blank, A., Silber, J. R., Thelen, M. P., Dekker, C. A. (1983) *Anal. Biochem.* **135,** 423–430.

Blank, A., Sugiyma, R. H., Dekker, C. A. (1982) *Anal. Biochem.* **120,** 267–273.

Bonaventura, C., Bonaventura, J., Stevens, R., Millington, D. (1994) *Anal. Biochem.* **222,** 44–48.

Bonner, W. M. (1983) *Methods Enzymol.* **96,** 215–225.

Bordier, C., Crettol-Jarvinen, A. (1979) *J. Biol. Chem.* **254,** 2565–2570.

Bordini, E., Hamdan, M., Righetti, P. G. (1999a) *Rapid Commun. Mass Spectrom.* **13,** 1818–1827.

Bordini, E., Hamdan, M., Righetti, P. G. (1999b) *Rapid Commun. Mass Spectrom.* **13,** 2209–2215.

Bosshard, H. F., Datyner, A. (1977) *Anal. Biochem.* **82,** 327–333.

Brewer, G. J., Sing, C. F. (1970) *An Introduction to Isozyme Techniques,* Academic, New York.

Bricker, T. M., Green-Church, K. B., Limbaugh, P. A., Frankel, L. K. (2000) *Anal. Biochem.* **278,** 237–239.

Brookes, P. S., Pinner, A., Ramachandran, A., Coward, L., Barnes, S., Kim, H., Darley-Usmar, V. M. (2002) *Proteomics* **2,** 969–977.

Brune, D. R. (1992) *Anal. Biochem.* **207,** 285–290.

Bryan, J. (1989) *J. Muscle Res. Cell Motil.* **10,** 95–100.

Bryce, C. F. A., Magnusson, C. G. M., Crighton, R. R. (1978) *FEBS Lett.* **96,** 257–260.

Burlingame, A. L., Boyd, R. K., Gaskell, S. J. (1998) *Anal. Chem.* **70,** 647R–716R.

Burnette, W. N. (1981) *Anal. Biochem.* **112,** 195–203.

Caglio, S., Chiari, M., Righetti, P. G. (1994) *Electrophoresis* **15,** 209–214.

Caglio, S., Righetti, P. G. (1993a) *Electrophoresis* **14,** 554–558.

Caglio, S., Righetti, P. G. (1993b) *Electrophoresis* **14,** 997–1003.

Campbell, W. P., Wrigley, C. W., Margolis, J. (1983) *Anal. Biochem.* **129,** 31–36.

Candiano, G., Bruschi, M., Musante, L., Santucci, L., Ghiggeri, G. M., Carnemolla, B., Orecchia, P., Zardi, L., Righetti, P. G. (2004) *Electrophoresis* **25,** 1327–1333.

Candiano, G., Porotto, M., Lanciotti, M., Ghiggeri, G. M. (1996) *Anal. Biochem.* **243,** 245–250.

Carey, M. C., Small, D. M. (1972) *Arch. Int. Med.* **130,** 506–537.

Castellanos-Serra, L., Hardy, E. (2001) *Electrophoresis* **22,** 864–873.

Castellanos-Serra, L., Proenza, W., Huerta, V., Moritz, R. L., Simpson, R. J. (1999) *Electrophoresis* **20,** 732–737.

Castellanos-Serra, L., Vallin, A., Proenza, W., Le Caer, J. P., Rossier, J. (2001) *Electrophoresis* **22,** 1677–1685.

Caton, J. E., Goldstein, G. (1971) *Anal. Biochem.* **42,** 14–20.

Chen, N., Chrambach, A. (1996) *Anal. Biochem.* **242,** 64–67.

Chiari, M., Righetti, P. G., Negri, A., Ceciliani, F., Ronchi, S. (1992) *Electrophoresis* **13,** 882–884.

Chrambach, A., Rodbard, D. (1971) *Science* **172,** 440–451.

Cleveland, D. W., Fischer, S. G., Kirschner, M. W., Laemmli, U. K. (1977) *J. Biol. Chem.* **252,** 1102–1106.

Cohen, S. L., Chait, B. T. (1997) *Anal. Biochem.* **247,** 257–267.

Colas des Francs, C., Thiellement, H., De Vienne, D. (1985) *Plant Physiol.* **78,** 178–182.

Comi, P., Giglioni, B., Ottolenghi, S., Gianni, M. A., Ricco, G., Mazza, U., Saglio, G., Camaschella, C., Pich, P. G., Gianazza, E., Righetti, P. G. (1979) *Biochem. Biophys. Res. Commun.* **87,** 1–8.

Daban, J. R. (2001) *Electrophoresis* **22,** 874–880.

Daneels, G., Moeremans, M., De Raeymaeker, M., De Mey, J. (1986) *J. Immunol. Methods* **89,** 89–91.

Davis, B. J. (1964) *Ann. N.Y. Acad. Sci.* **121,** 404–427.

Deshpande, V. V., Bodhe, A. M., Pawar, H. S., Vartak, H. G. (1986) *Anal. Biochem.* **153,** 227–229.

Desvaux, F. X., David, B., Peltre, G. (1990) *Electrophoresis* **11,** 37–41.

Dohnal, J. C., Garvin, J. E. (1979) *Biochim. Biophys. Acta* **576,** 393–403.

Drapier, J. B. E. (1998) Colgate-Palmolive Co., U.S. Pat. No. 5,840,676, November 24.

Dunker, A. K., Rueckert, R. R. (1969) *J. Biol. Chem.* **244,** 5074–5080.

Dunn, M. J. (1993) *Gel Electrophoresis: Proteins*, Bios Scientific Publishers, Oxford, pp. 27–29.

Dzandu, J. K., Johnson, J. F., Wise, G. F. (1988) *Anal. Biochem.* **174,** 157–167.

Ehring, H., Strömberg, S., Tjernberg, A., Norèn, B. (1997) *Rapid Commun. Mass Spectrom.* **11,** 1867–1871.

Eley, M. H., Burns, P. C., Kannapell, C. C., Campbell, P. S. (1979) *Anal. Biochem.* **92,** 411–418.

Fairbanks, G., Steck, T. L., Wallach, D. F. H. (1971) *Biochemistry* **10,** 2606–2617.

Fantes, K. H., Furminger, I. G. S. (1967) *Nature* **215,** 750–751.

Felgenhauer, K. (1979) *J. Chromatogr.* **173,** 299–306.

Ferguson, K. A. (1964) *Metabolism* **13,** 985–995.

Fernandez-Patron, C., Castellanos-Serra, L., Rodriguez, P. (1992) *BioTechniques* **12,** 564–568.

Fish, W. W., Reynolds, J. A., Tanford, C. (1970) *J. Biol. Chem.* **245,** 5166–5168.

Galante, D. C., Hoy, R. C., Joseph, F., King. S. W., Smith, C. A. (1998) Union Carbide Corp., U.S. Pat. No. 5,744,065, April 28.

Galvani, M., Bordini, E., Piubelli, C., Hamdan, M. (2000) *Rapid Commun. Mas Spectrom.*, **14,** 18–25.

Galvani, M., Rorathi, L., Hamdan, M., Herbert, B., Righetti, P.G. (2001) *Electrophoresis* **22,** 2066–2074.

Gershoni, J. M., Palade, G. E. (1982) *Anal. Biochem.* **124,** 396–402.

Gershoni, J. M., Palade, G. E. (1983) *Anal. Biochem.* **131,** 1–10.

Gharahdaghi, F., Weinberg, C. R., Meagher, D. A., Imai, B. S., Mische, S. M. (1999) *Electrophoresis* **20,** 601–605.

Gianazza, E., Righetti, P. G. (1979) in Righetti, P. G., Van Oss, C. J., Vanderhoff, J. W. (Eds.), *Electrokinetic Separation Methods*, Elsevier, Amsterdam, pp. 293–311.

Gooderham, K. (1986) in Dunn, M. J. (Ed.), *Gel Electrophoresis of Proteins*, Wright, Bristol, pp. 312–322.

Goodlett, D. R., Armstrong, F. B., Creech, J. R., Van Breemen, R. B. (1990) *Anal. Chem.* **186,** 116–120.

Görg, A., Postel, R., Westermeier, E., Gianazza, E., Righetti, P. G. (1980) *J. Biochem. Biophys. Methods* **3,** 273–284.

Grabar, P., Williams, C. A. (1953) *Biochim. Biophys. Acta* **10,** 193–200.

Graceffa, P., Jancso, A., Mabuchi, K. (1992) *Arch. Biochem. Biophys.* **297,** 46–51.

Granzier, H. L. M., Wang, K. (1993) *Electrophoresis* **14,** 56–64.

Griffith, I. P. (1972) *Anal. Biochem.* **46,** 402–412.

Griffith, O. W. (1980) *Anal. Biochem.* **106,** 207–212.

Gurd, F. R. N. (1967) *Methods Enzymol.* **11,** 532–541.

Haebel, S., Albrecht, T., Sparbier, K., Walden, P., Korner, R., Steup, M. (1998) *Electrophoresis* **19,** 679–686.

Haeberle, J. R. (1997) *BioTechniques* **23,** 638–640.

Hales, S. G., Ezat, K., Polywka, R. (1995) Conopco Inc., U.S. Pat. No. 5,439,997, August 8.

Hancock, K., Tsang, V. C. W. (1983) *Anal. Biochem.* **133,** 157–162.

Hardy, E., Santana, H., Hernandez, L., Fernandez-Patron, C., Castellanos-Serra, L. (1996) *Anal. Biochem.* **240,** 150–152.

Hart, C., Schulenberg, B., Diwu, Z., Leung, W. Y., Patton, W. F. (2003) *Electrophoresis* **24,** 599–610.

Hashimoto, F., Horigome, T., Kanbayashi, M., Yoshida, K., Sugano, H. (1983) *Anal. Biochem.* **129,** 192–199.

Helenius, A., McCaslin, D. R., Fries, E., Tanford, C. (1979) *Methods Enzymol.* **56,** 734–749.

Helenius, A., Simons, K. (1975) *Biochim. Biophys. Acta* **415,** 29–79.

Higgins, R. C., Dahmus, M. E. (1979) *Anal. Biochem.* **93,** 257–260.

Hjelmeland, L. M. (1986) *Methods Enzymol.* **124,** 135–164.

Hjelmeland, L. M., Chrambach, A. (1981) *Electrophoresis* **2,** 1–11.

Hunter, R. L., Markert, C. L. (1957) *Science* **125,** 1294–1295.

Ibel, K., May, R. P., Kirschner, K., Szadkowski, H., Mascher, E., Lundahl, P. (1990) *Eur. J. Biochem.* **190,** 311–318.

Ibel, K., May, R. P., Sandberg, M., Mascher, E., Greijer, E., Lundahl, P. (1994) *Biophys. Chem.* **53,** 77–84.

Inouye, M. (1971) *J. Biol. Chem.* **246,** 4834–4839.

Jarush, F., Sonenberg, M. (1950) *Anal. Chem.* **22,** 175–177.

Jeannot, M. A., Zheng, J., Li, L. (1999) *J. Am. Soc. Mass Spectrom.* **10,** 512–518.

Johansson, K. E. (1987) *Electrophoresis* **8**, 379–383.

Johnston, R. F., Pickett, S. C., Barker, D. L. (1990) *Electrophoresis* **11,** 355–360.

Jones, G. D., Wilson, M. T., Darley-Usmar, V. M. (1981) *Biochem. J.* **193,** 1013–1218.

Jovin, T. M. (1973) *Biochemistry* **12,** 871–879.

Kakita, K., O'Connell, K., Permuth, M. A. (1982) *Diabetes* **31,** 648–652.

Kang, C., Kim, H. J., Kang, D., Jung, D. Y., Suh, M. (2003) *Electrophoresis* **24,** 3297–3304.

Kay, M. M. B., Goodman, S. R., Sorensen, K., Whiffield, C. F., Wong, P., Zaki, L., Rudloff, V. (1983) *Proc. Natl. Acad. Sci. USA* **80,** 1631–1635.

Kemper, C., Berggren, K., Diwu, Z., Patton, W. F. (2001) *Electrophoresis* **22,** 881–889.

Kessler, C. (1991) *Molec. Cell. Probes* **5,** 161–205.

Khaledi, M. G. (1998) in Khaledi, M. G. (Ed.), *High Performance Capillary Electrophoresis* Wiley, New York, pp. 77–140.

Kirley, T. L. (1989) *Anal. Biochem.* **180,** 231–236.

Kittler, J. M., Meisler, N. T., Viceps-Madore, D. (1984) *Anal. Biochem.* **137,** 210–216.

Klarskov, K., Roecklin, D., Bouchon, B., Sabatie, J., Van Dorssalaer, A., Bischoff, R. (1994) *Anal. Biochem.* **216,** 127–134.

Koenig, S., Schmidt, O., Rose, K., Thanos, S., Besselmann, M., Zeller, M. (2003) *Electrophoresis* **24,** 751–756.

Kurien, B. J., Scofield, R. H. (1998) *Indian J. Biochem. Biophys.* **35,** 385–389.

Kyhse-Andersen, J. (1984) *J. Biochem. Biophys. Methods* **10,** 203–209.

Laemmli, U. K. (1970) *Nature* **227,** 680–685.

Lambin, P. (1978) *Anal. Biochem.* **85,** 114–125.

Lambin, P., Fine, J. M. (1979) *Anal. Biochem.* **98,** 160–168.

Lauber, W. M., Carroll, J. A., Dufield, D. R., Kiesel, J. R., Radabaugh, M. R., Malone, J. P. (2001) *Electrophoresis* **22,** 906–918.

Lee, C., Levin, A., Branton, D. (1987) *Anal. Biochem.* **166,** 308–312.

Leimgruber, M. R., Malone, J. P., Radabaugh, M. R., LaPorte, M. L., Violand, B. N., Monahan, J. B. (2002) *Proteomics* **2,** 135–144.

Liang, X., Bai, J., Liu, Y. H., Lubman, D. M. (1996) *Anal. Chem.* **68,** 1012–1018.

Lorentz, K. (1976) *Anal. Biochem.* **76,** 214–219.

Lundahl, P., Greijer, E., Sandberg, M., Cardell, S., Eriksson, K. O. (1986) *Biochim. Biophys. Acta* **873,** 20–26.

Lundahl, P., Watanabe, Y., Takagi, T. (1992) *J. Chromatogr.* **604,** 95–102.

Lyubimova, T., Caglio, S., Gelfi, C., Righetti, P. G., Rabilloud, T. (1993) *Electrophoresis* **14,** 40–50.

Lyubimova, T., Righetti, P. G. (1993) *Electrophoresis* **14,** 191–201.

MacFarlane, D. (1983) *Anal. Biochem.* **132,** 231–235.

Mackintosh, J. A., Choi, H. Y., Bae, S. H., Veal, D. A., Bell, P. J., Ferrari, B. C., Van Dyk, D. D., Verrills, N. M., Paik, Y. K., Karuso, P. (2003) *Proteomics* **3,** 2273–2288.

Malin, M. J., and Chapoteau, E. (1981) *J. Chromatogr.* **219,** 117–122.

Malone, J. P., Radabaugh, M. R., Leimgruber, R. M., Gerstenecker, G. S. (2001) *Electrophoresis* **22,** 919–932.

Margolis, J. (1973) *Lab. Pract.* **22,** 107–111.

Margolis, J., Kenrick, K. G. (1968) *Anal. Biochem.* **25,** 347–351.

Margulies, M. M., Tiffany, H. L. (1984) *Anal. Biochem.* **136,** 309–315.

Marjanen, L. A., Ryrie, I. J. (1974) *Biochim. Biophys. Acta* **371,** 442–449.

Markert, C. L., Moller, F. (1959) *Proc. Natl. Acad. Sci. USA* **45,** 753–754.

Martin, K., Hart, C., Liu, J., Leung, W. Y., Patton, W. F. (2003) *Proteomics* **3,** 1215–1227.

Martin, K., Hart, C., Schulenberg, B., Jones, L., Patton, W. F. (2002) *Proteomics* **2,** 499–512.

Mascher, E., Lundahl, P. (1989) *J. Chromatogr.* **476,** 147–158.

Mattice, W. L., Riser, J. M., Ckark, D. S. (1976) *Biochemistry* **15,** 4264–4272.

McCoy, D. R. (1976) Texaco Inc., U.S. Pat. No. 3,948,953, April 6.

Meng, F., Cargile, B. J., Patrie, S. M., Johnson, J. R., McLoughlin, S. M., Kelleher, N. L. (2002) *Anal. Chem.* **74,** 2923–2929.

Merril, C. R., Washart, K. M. (1998) in Hames, B.D. (Ed.), *Gel Electrophoresis of Proteins: A Practical Approach* Oxford University Press, Oxford, pp. 53–91.

Mocz, J. R., Balint, J. P. (1984) *Anal. Biochem.* **143,** 283–292.

Moeremans, M., Daneels, G., De Mey, J. (1985) *Anal. Biochem.* **145,** 315–321.

Moeremans, M., De Raeymaeker, M., Daneels, G., De Mey, J. (1986) *Anal. Biochem.* **153,** 18–22.

Moos, Jr., M., Nguyen, N. Y., Liu, T. Y. (1988) *J. Biol. Chem.* **263,** 6005–6009.

Mortz, E., Krogh, T. N., Vorum, H., Goerg, A. (2001) *Proteomics* **1,** 1359–1363.

Mukerjee, P. (1956) *Anal. Chem.* **28,** 870–872.

Neff, J. L., Muniz, N., Colburn, J. L., de Castro, A. F. (1981) in Allen, R. C., Arnaud, P. (Eds.), *Electrophoresis '81: Advanced Methods, Biochemical and Chemical Applications*, de Gruyter, Berlin, pp. 49–63.

Nelles, L. P. Bamburg, J. R. (1976) *Anal. Biochem.* **73,** 522–531.

Nelson, C. A. (1971) *J. Biol. Chem.* **246,** 3895–3901.

Neugebauer, J. M. (1989) *Methods Enzymol.* **182,** 239–252.

Neuhoff, V., Arold, N., Taube, D., Ehrhardt, W. (1988) *Electrophoresis* **9,** 255–262.

Neuhoff, V., Stamm, R., Eibl, H. (1985) *Electrophoresis* **6,** 427–448.

Neville, D. M. (1971) *J. Biol. Chem.* **246,** 6328–6334.

Neville, D. M., Glossmann, H. (1971) *J. Biol. Chem.* **246,** 6335–6340.

Nishihara, J. C., Champion, K. M. (2002) *Electrophoresis* **23,** 2203–2215.

Ochs, D. (1983) *Anal. Biochem.* **135,** 470–474.

Ogorzalek Loo, R. R., Stevenson, T. I., Mitchell, C., Loo, J. A., Andrews, P. C. (1996) *Anal. Chem.* **68,** 1910–1915.

Olszewka, E., Jones, K. (1988) *Trends Gen.* **4,** 92–94.

Ornstein, L. (1964) *Ann. N.Y. Acad. Sci.* **121,** 321–349.

Panyim, S. (1971) *J. Biol. Chem.* **246,** 7557–7560.

Panyim, S., Chalkley, R (1969) *Arch. Biochem. Biophys.* **130,** 337–346.

Paniym, S., Thitipongpanich, R., Supatimusro, D. (1977) *Anal. Biochem.* **81,** 320–326.

Patton, W. F. (2000) *Electrophoresis* **21,** 1123–1144.

Payne, J. W. (1973) *Biochem. J.* **135,** 867–873.

Petersen, A. (2003) *Proteomics* **3,** 1206–1214.

Pitt-Rivers, R., Impiombato, F. S. A. (1968) *Biochem. J.* **109,** 825–833.

Ploug, M., Jensen, A. L., Barkholt, V. (1989) *Anal. Biochem.* **181,** 33–38.

Ploug, M., Stoffer, B., Jensen, L. (1992) *Electrophoresis* **13,** 148–152.

Pluskal, M. G., Przekop, B., Kavonian, M. R., Vecoli, C., Hicks, D. A. (1986) *BioTechniques* **4,** 272–283.

Poduslo, J. F. (1981) *Anal. Biochem.* **114,** 131–139.

Poduslo, J. F., Rodbard, D. (1980) *Anal. Biochem.* **101,** 394–406.

Porzio, M. A., Pearson, A. M. (1977) *Biochim. Biophys. Acta* **490,** 27–34.

Pratt, J. J., Dangerfield, W. G. (1969) *Clin. Chim. Acta* **23,** 189–201.

Puchades, M., Westman, A., Blennow, K., Davidsson, P. (1999) *Rapid. Commun. Mass Spectrom.* **13,** 344–349.

Quirino, J. P., Terabe, S. (1999) *J. Chromatogr. A* **856,** 465–482.

Rabilloud, T. (1996) *Electrophoresis* **17,** 813–829.

Rabilloud, T., Girardot, V., Lawrence, J. J. (1996) *Electrophoresis* **17,** 67–73.

Rabilloud, T., Vincon, M., Garin, J. (1995) *Electrophoresis* **16,** 1414–1422.

Ragland, W. L., Pace, J. L., Kemper, D. L. (1974) *Anal. Biochem.* **59,** 24–33.

Reiser, J., Wardale, J. (1981) *Eur. J. Biochem.* **114,** 569–574.

Reynolds, J. A., Tanford, C. (1970a) *Proc. Natl. Acad. Sci. USA* **66,** 1002–1007.

Reynolds, J. A., Tanford, C. (1970b) *J. Biol. Chem.* **245,** 5161–5165.

Richardson, B. J., Baverstock, P. R., Adams, M. (1986) *Allozyme Electrophoresis,* Academic, Sydney.

Righetti, P. G., Bossi, A., Giglio, M., Vailati, A., Lyubimova, T., Briskman, V. A. (1994) *Electrophoresis* **15,** 1005–1013.

Righetti, P. G., Brost, B. C. W., Snyder, R. S. (1981) *J. Biochem. Biophys. Methods* **4,** 347–363.

Righetti, P. G., Caglio, S. (1993) *Electrophoresis* **14,** 573–582.

Righetti, P. G., Gianazza, E., Gianni, A. M., Comi, P., Giglioni, B., Ottolenghi, S., Secchi, C., Rossi-Bernardi, L. (1979) *J. Biochem. Biophys. Methods* **1,** 47–59.

Riordan, J. F., Vallee, B. L. (1967) *Methods Enzymol.* **11,** 532–541.

Rodbard, D., Chrambach, A. (1970) *Proc. Natl. Acad. Sci. USA* **65,** 970–977.

Rodbard, D., Chrambach, A. (1971) *Anal. Biochem.* **40,** 95–134.

Rodbard, D., Kapadia, G., Chrambach, A. (1971) *Anal. Biochem.* **40,** 135–144.

Ross, R. S., Lee, P. L., Smith, D. L., Langridge, J. I., Whetton, A. D., Gaskell, S. J. (2002) *Proteomics* **2,** 928–936.

Rothe, G. M. (1982) *Electrophoresis* **3,** 255–262.

Rothe, G. M. (1991) in Chrambach, A., Dunn, M. J., Radola, B. J. (Eds.), *Advances in Electrophoresis* VCH, Weinheim, pp. 252–358.

Rothe, G. M., Maurer, W. D. (1986) in Dunn, M. J. (Ed.), *Gel Electrophoresis of Proteins,* Wright, Bristol, pp. 37–140.

Ruegg, U. T., Rüdinger, J. (1977) *Methods Enzymol.* **47,** 111–116.

Saglio, G., Ricco, G., Mazza, U., Camaschella, C., Pich, P. G., Gianni, A. M., Gianazza, E., Righetti, P. G., Giglioni, B., Comi, P., Gusmeroli, M., Ottolenghi, S. (1979) *Proc. Natl. Acad. Sci. USA* **76,** 3420–3424.

Salinovich, O., Montelaro, R. C. (1986) *Anal. Biochem.* **156,** 341–347.

Samsò, M., Daban, J. R., Hansen, S., Jones, G. R. (1995) *Eur. J. Biochem.* **232,** 818–824.

Saris, C. J. M., van Eenbergen, J., Jenks, B. G., Bloemers, H. P. J. (1983) *Anal. Biochem.* **132,** 54–60.

Schägger, H. (1995) *Electrophoresis* **16,** 763–770.

Schägger, H., Cramer, W. A., von Jagow, G. (1994) *Anal. Biochem.* **217,** 220–230.

Schägger, H., von Jagow, G. (1987) *Anal. Biochem.* **166,** 368–379.

Schägger, H., von Jagow, G. (1991) *Anal. Biochem.* **199,** 223–230.

Schick (1975) *Anal. Biochem.* **63,** 345–350.

Schulenberg, B., Arnold, B., Patton, W. F. (2003) *Proteomics* **3,** 1196–1205.

See, Y. P., Olley, M. P., Jackowski, G. (1985) *Electrophoresis* **6,** 382–388.

Segrest, J. P., Jackson, R. L. (1972) *Methods Enzymol.* **28B,** 54–60.

Segrest, J. P., Jackson, R. L., Andrews, E. P., Marchesi, V. T. (1971) *Biochem. Biophys. Res. Commun.* **44,** 390–395.

Shapiro, A. L., Vinuela, E., Maizel, J. V. (1967) *Biochem. Biophys. Res. Commun.* **28,** 815–820.

Shapiro, S. Z. (1987) *J. Immunol. Methods* **102,** 143–146.

Shaw, J., Rowlinson, R., Nickson, J., Stone, T., Sweet, A., Williams, K., Tonge, R. (2003) *Proteomics* **3,** 1181–11195.

Shi, Q., Jackowski, G. (1998) in Hames, B. D. (Ed.), *Gel Electrophoresis of Proteins: A Practical Approach*, Oxford University Press, Oxford, pp. 13–34.

Shirahama, K., Tsujii, K., Takagi, T. (1974) *J. Biochem. (Tokyo)* **75,** 309–319.

Sigrist, H. (1974) *Anal. Biochem.* **57,** 564–568.

Sinha, P., Poland, J., Schnoelzer, M., Rabilloud, T. (2001) *Proteomics* **1,** 835–840.

Slater, G. G. (1969) *Anal. Chem.* **41,** 1039–1043.

Sokoloff, R. L., Frigon, R. P. (1981) *Anal. Biochem.* **118,** 138–141.

Southern, E. M. (1975) *J. Mol. Biol.* **98,** 503–517.

Steele, Jr., J. C. H., Nielsen, T. B. (1978) *Anal. Biochem.* **84,** 218–224.

Steinberg, T. H., Chernokalskaya, E., Berggren, K., Lopez, M. F., Diwu, Z., Haugland, R. P., Patton, W. F. (2000a) *Electrophoresis* **21,** 486–496.

Steinberg, T. H., Haugland, R. P., Singer, V. L. (1996a) *Anal. Biochem.* **239,** 238–2245.

Steinberg, T. H., Jones, L. J., Haugland, R. P., Singer, V. L. (1996b) *Anal. Biochem.* **239,** 223–227.

Steinberg, T. H., Lauber, W. M., Berggren, K., Kemper, C., Yue, S., Patton, W. F. (2000b) *Electrophoresis* **21**, 497–508.

Steinberg, T. H., White, H. M., Singer, V. L. (1997) *Anal. Biochem.* **248,** 168–172.

Stoklosa, J. T., Latz, H. W. (1974) *Biochem. Biophys. Res. Commun.* **60,** 590–596.

Strupat, K., Karas, M., Hillenkamp, F. (1994) *Anal. Chem.* **66,** 471–480.

Svensson, H. (1967) *Prot. Biol. Fluids* **15,** 515–522.

Swaney, J. B., Woude, G. V. F., Bachrach, H. L. (1974) *Anal. Biochem.* **58,** 337–346.

Swank, R. W., Munkres, K. D. (1971) *Anal. Biochem.* **39,** 462–470.

Takagi, T., Tsujii, K., Shirahama, K. (1975) *J. Biochem. (Tokyo)* **77,** 939–947.

Tanford, C. (1973) *The Hydrophobic Effect*, Wiley, New York.

Tanford, C. (1980) *The Hydrophobic Effect. Formation of Micelles and Biological Membranes*, 2nd ed., Wiley, New York, pp. 159–164.

Tastet, C., Lescuyer, P., Diemer, H., Luche, S., van Dorsselaer, A., Rabilloud, T. (2003) *Electrophoresis* **24,** 1787–1794.

Terabe, S., Otsuka, K., Ando, T. (1985) *Anal. Chem.* **57,** 834–839.

Terabe, S., Otsuka, K., Ichikawa, K., Tsuchiya, A., Ando, T. (1984) *Anal. Chem.* **56,** 111–116.

Tessmer, U., Dernick, R. (1989) *Electrophoresis* **10,** 177–279.

Tonge, R., Shaw, J., Middleton, B., Rowlinson, R., Rayner, S., Young, J., Pognan, F., Hawkins, E., Currie, I., Davison, M. (2001) *Proteomics* **1,** 377–396.

Tovey, E. R., Baldo, B. A. (1987) *Electrophoresis* **8,** 384–387.

Towbin, H., Staehelin, T., Gordon, J. (1979) *Proc. Natl. Acad. Sci. USA* **76,** 4350–4354.

Tyagarjan, K., Pretzer, E., Wiktorowicz, J. E. (2003) *Electrophoresis* **24,** 2348–2358.

Unlu, M., Morgan, M. E., Minden, J. S. (1997) *Electrophoresis* **18,** 2071–2077.

Ursitti, J. A., DeSilva, T., Speicher, D. W. (1995) in Coligan, J. E., Dunn, B. M., Ploegh, H. L., Speicher, D. W., Wingfield, P. T. (Eds.), *Current Protocols in Protein Science* Wiley, New York, pp. 10.6.1–10.6.8.

Vuong, G. L., Weiss, S. M., Kammer, W., Priemer, M., Vingron, M., Nordheim, A., Cahill, M. A. (2000) *Electrophoresis* **21,** 2594–2605.

Waehneldt, T. V. (1975) *Bio-Systems* **6,** 176–187.

Wallace, R. W., Yu, P. H., Dieckert, J. P., Die kert, J. W. (1974) *Anal. Biochem.* **61,** 86–92.

Wallstén, M., Lundahl, P. (1990) *J. Chromatogr.* **512,** 3–12.

Wang, D., Dzandu, J. K., Hussain, M., Johnson, R. M. (1989) *Anal. Biochem.* **180,** 311–313.

Wang, H. C. R. (2000) *BioTechniques* **28,** 232–238.

Weber, K., Osborn, M. (1969) *J. Biol. Chem.* **244,** 4406–4412.

Weber, K., Osborn, M. (1975) in Neurath, K., Hill, R. L. (Eds.), *The Proteins*, Vol 1, Academic, New York, pp. 179–199.

Weber, K., Pringle, J. R., Osborn, M. (1972) *Methods Enzymol.* **26,** 3–13.

Weidekamm, E., Wallach, D. F. H., Flückiger, R. (1973) *Anal. Biochem.* **54,** 102–108.

Westerhuis, W. H., Sturgis, J. N., Niederman, R. A. (2000) *Anal. Biochem.* **284,** 143–152.

Westermeier, R. (1997) *Electrophoresis in Practice*, VCH, Weinheim, pp. 167–188.

Wilk, K. A., Bieniecki, A., Matuszewka, B. (1994) *J. Phys. Org. Chem.* **7,** 646–651.

Willard, K. E., Giometti, C. S., Anderson, N. L., O'Connor, T. E., Anderson, N. G. (1980) *Anal. Biochem.* **100,** 289–296.

Williams, J. G., Gratzer, W. B. (1971) *J. Chromatogr.* **57,** 121–130.

Wolf, B., Michelin Lausarot, P., Lesnaw, J. A., Reichman, M. E. (1970) *Biochim. Biophys. Acta* **200,** 180–183.

Wong, C., Sridhara, S., Bardwell, J. C. A., Jakob, U. (2000) *BioTechniques* **28,** 426–432.

Woodward, A. M., Kaderbhai, N., Kaderbhai, M., Shaw, A., Rowland, J., Kell, D. B. (2001) *Proteomics* **1,** 1351–1358.

Wu, W., Welsh, M. J. (1995) *BioTechniques* **20,** 386–388.

Wyckoff, M., Rodbard, D., Chrambach, A. (1977) *Anal. Biochem.* **78,** 459–469.

Yan, J. X., Harry, R. A., Spibey, C., Dunn, M. J. (2000a) *Electrophoresis* **21,** 3657–3665.

Yan, J. X., Wait, R., Berkelman, T., Harry, R. A., Westbrook, J. A., Wheeler, C. H., Dunn, M. J. (2000b) *Electrophoresis* **20,** 3666–3672.

Yarmola, E., Chen, N., Yi, D., Chrambach, A. (1998) *Electrophoresis* **19,** 206–211.

Yefimov, S., Yergey, A. L., Chrambach, A. (2000) *J. Biochem. Biophys. Methods* **42,** 65–79.

Yokota, H., Mori, K., Kaniwa, H., Shibanuma, T. (2000) *Anal. Biochem.* **280,** 188–189.

Zeller, M., Brown, E. K., Bouvier, E. S. P., Koenig, S. (2002) *J. Biomol. Techn.* **13,** 1–4.

6

TWO-DIMENSIONAL MAPS

6.1. INTRODUCTION

Proteomics is an emerging area of research in the postgenomic era that deals with the global analysis of gene expression via a combination of techniques for resolving, identifying, quantitating, and characterizing proteins (Banks et al., 2000). In addition, a fundamental part of proteomics is bioinformatics, for storing, communicating, and interlinking proteomic and genomic data as well as mapping information from genome projects (Wilkins et al., 1997). Although each technique can be applied independently, their impact can be maximized when used in concert for the study of complex biological problems.

For the last 25 years, 2D PAGE has been the technique of choice for analyzing the protein composition of a given cell type and for monitoring changes in gene activity through the quantitative and qualitative analysis of the thousands of proteins that orchestrate various cellular functions. Proteins are usually the functional molecules and, therefore, the most likely components to reflect qualitative (expression of new proteins, posttranslational modifications) and quantitative (up and down regulation, coordinated expression) differences in gene expression (Celis et al., 1998). Just as an example of the incredible growth in the field, in the journal *Electrophoresis* alone a large number of special issues have appeared over the years collecting several hundreds of articles dealing with 2D PAGE and its application in any possible area of biological research (Appel et al., 1997, 1999; Celis, 1989, 1990ab, 1991, 1992, 1994ab, 1995, 1996, 1999; Damerval and de Vienne, 1988; Dunn, 1991, 1995, 1997, 1999, 2000; Huber, 2000; Humphery-Smith, 1997; Lottspeich, 1996; Tümmler, 1998;

Proteomics Today. By Mahmoud Hamdan and Pier Giorgio Righetti
ISBN 0-471-64817-5

Williams, 1998). This incredible downpour of issues has subsided since, meanwhile, a number of new journals, such as *Proteomics, Journal of Proteome Research, and Molecular and Cellular Proteomics*, have appeared dealing with this topic.

Notwithstanding its extraordinary resolving power, even 2D PAGE seems to have reached a plateau in the number of proteins that can be resolved and detected in a single 2D map. The advent of IPGs (Righetti, 1990) has certainly made a big improvement, but its resolving power cannot be pushed any further. Also, solubilizing cocktails have improved due to the introduction of new surfactants (Chevallet et al., 1998) and new reducing agents (Herbert et al., 1998); however, progress seems to be leveling off. The same applies to staining protocols, which seem not to have produced any increment in detection sensitivity in the last 5 years (Rabilloud, 2000a). Yet, the enormous complexity of the proteome (it is hypothesized that up to 40,000 genes could be present in the human genome, but the proteins in the proteome could be numbered in the millions if one considers all the possible posttranslational modifications, protein splicing, and the variegated world of immunoglobulins, to name just a few causes of such polydispersity; Anderson and Anderson, 2002) calls for improvements in resolving power, possibly coupled to increments in detection sensitivity for tracking trace components. It is the aim of the present chapter not only to review the evolution of 2D PAGE up to the present but also to suggest new avenues and potential new developments in the field. As it stands today, 2D PAGE is a complex field in which, perhaps, electrophoretic knowledge has become minor: 30% of it belongs to the fine art of electrokinetic methodologies, combining with proper skills an isoelectric focusing step to an orthogonal SDS–PAGE; 30% of it is proper use and knowledge of MS, especially in the variant MALDI–TOF, for a proper assessment of the precise M_r value of polypeptide chains and their fragments; another 30% is computer science coupled to good skills in interrogating data bases and extracting pertinent data. The remaining 10%, perhaps, is fantasy and intuition, sprinkled in the above cocktail. Shake it (don't stir it) and serve it properly chilled.

6.1.1. Early Days and Evolution of 2D PAGE

Any single-dimension method (e.g., IEF, SDS–PAGE, and isotachophoresis, to name just a few high-resolution techniques) cannot resolve more than 80–100 different components, and even that only under the most favorable circumstances. Thus, when a sample is highly heterogeneous, other methods have to be sought. One such procedure is 2D PAGE, which exploits a combination of two single-dimension runs. Two-dimensional maps could be prepared by using virtually any combination of 1D methods (not necessarily electrophoretic; Giddings, 1991), but the one that has won universal recognition is that combining a charge (typically an IEF protocol) to a size (e.g., SDS–PAGE) fractionation, since this results in a more even distribution of components over the surface of the map. While this combination of separation methods was used at quite an early stage in the development of 2D macromolecular mapping, it was the elegant work of O'Farrell (1975) that really demonstrated the full capabilities of this approach [see also Klose (1975) and Scheele (1975)]. He was able to resolve and detect about 1100 different proteins from lysed *Escherichia coli* cells on a single 2D map and suggested that the maximum resolution capability might have been as high as 5000 different proteins. Apart from the meticulous attention to detail, major reasons for the advance in resolution obtained by O'Farrell, compared to earlier workers, included the use of samples labeled with ^{14}C or ^{35}S to high specific activity and the use of thin (0.8-mm) gel slabs for the second dimension, which could then be dried easily for autoradiography. This detection method was able to reveal protein zones corresponding to one part in 10^7 of the sample (usually 1–20 μg was applied initially, since higher loads caused zone spreading, although up to 100 μg could be loaded). Coomassie blue, in comparison, was about three orders of magnitude less sensitive and could reveal only about 400 spots. For the first dimension, O'Farrell adopted gel rods of 13 cm in length and 2.5 mm in diameter. The idea was to run samples fully denatured in what became known as the "O'Farrell lysis buffer" (9 M urea, 2% Nonidet P-40, 2% β-mercaptoethanol, and 2% carrier ampholytes, in any desired pH interval). For the second SDS–PAGE dimension, O'Farrell (1975) used the discontinuous buffer system of Laemmli (1970) and, for improved resolution, a concave exponential gradient of polyacrylamide gel (usually in the intervals, $T = 9\%$ to $T = 15\%$ or $T = 10\%$ to $T = 14\%$, although wide porosity gradients, e.g., $T = 5\%$ to $T = 22.5\%$, were also suggested). It is thus seen that, since its very inception, O'Farrell carefully selected all the best conditions available at the time; it is no wonder that his system was adopted as such in the avalanche of reports that soon followed. He went as far as to recognize that some protein losses could occur during the equilibration of the IEF gel prior to running of the SDS–PAGE. Depending on the identity of the protein and the duration of equilibration, losses were estimated to vary from 5% up to 25%. Even for that O'Farrell proposed a remedy: He reported that, in such cases, the equilibration step could be omitted and, for minimizing streaking in the second dimension, due to only partial SDS saturation, one could increase the depth of the stacking gel in the SDS–PAGE run from 2.5 to 5 cm. The increased length of the stacking gel coupled to lower initial voltage gradients and higher amounts of SDS in the cathodic reservoir allowed for proper saturation of the protein species by the SDS moiety. It is no surprise that, with such a thorough methodological development, there were hardly any modifications to this technique, something that in the scientific world is a very rare event (as soon as a method is published, usually a cohort of modifications is immediately reported, in the hope that the "second discoverer" will be the winner!).

Some minor modifications, over time, were adopted and they certainly helped in further improving the technique. The factors affecting the resolution of 2D maps were discussed by a number of workers (e.g., Tracy et al., 1982; Burghes et al., 1982; Duncan and Hershey, 1984), as a result of which it appeared that it was beneficial to reduce the concentration of acrylamide in the first-dimension IEF gel to as low as $T = 3.5\%$ (a smart idea, indeed, as will be seen below) and to run the IEF gel at 800–1000 V for at least 10,000 V/h (we will see that with IPGs it is not uncommon to deliver much stronger conditions than those, up to 60,000 V/h). In addition, Perdew et al. (1983) proposed to replace the nonionic detergent Nonidet P-40 with similar proportions of the zwitterionic detergent CHAPS (3-[(3-cholamidopropyl)dimethylammonium]-1-propane sulfonate), which they claimed had superior membrane protein-solubilizing properties and was particularly effective at disaggregating hydrophobic proteins. By the same token, to overcome the difficulties of separating basic proteins in the first IEF dimension, O'Farrell et al. (1977) introduced the technique of nonequilibrium pH gradient isoelectric focusing (NEPHGE), in which the sample was applied at the acidic end of the gel and the IEF terminated prior to reaching steady-state conditions. Elegant microversions of O'Farrell's method were reported by Ruechel (1977) and Poehling et al. (1980), in which the first-dimension IEF run was performed with gels in small capillary tubes followed by a second-dimension SDS–PAGE step on postage-stamp-size gel slabs with a gradient with $\%T$ of 1% to 40% or 6% to 25%.

At the opposite end of the scale, instead of miniaturization, Anderson and Anderson (1977, 1978a,b) started thinking of "large-scale biology" and building instrumentation (called the ISO-DALT system) for preparing and running a large number of O'Farrell-type gels together. This approach greatly enhanced reproducibility and comparison between the resulting protein maps while enabling a very large number of samples to be handled in a short time. A variation (termed BASO-DALT), giving enhanced resolution of basic proteins by employing NEPHGE in the first dimension, was also described (Willard et al., 1979). These approaches made practical the Molecular Anatomy Program at the Argonne National Laboratory in the United States, the object of which was to be able to fractionate human cells and tissue with the ultimate aim of being able to describe completely the products of human genes and how these vary between individuals and in disease (Anderson and Anderson, 1982, 1984). It is also due to the Herculean efforts of the two Andersons that present-day 2D maps have reached such a high stage of evolution [it is worth a perusal through two special issues of *Clinical Chemistry* (Young and Anderson, 1982; King, 1984) to see the flurry of papers published and the incredible level of development reached already in the years 1982–1984!].

6.1.2. A Glimpse at Modern Times

Although the power of 2DE as a biochemical separation technique has been well recognized since its introduction, its application, nevertheless, has become particularly significant in the past few years as a result of a number of developments, outlined below:

- The 2D technique has been tremendously improved to generate 2D maps that are superior in terms of resolution and reproducibility. This new technique utilizes a unique first dimension that replaces the carrier ampholyte-generated pH gradients with IPGs and replaces the tube gels with gel strips supported by a plastic film backing (Bjellqvist et al., 1982). Immobilized pH gradients are so precise that they allow excellent correlation between experimentally found and theoretically predicted pI values of both proteins and peptides (Bjellqvist et al., 1993a,b; Shimura et al., 2002; Cargile et al., 2004).
- Methods for the rapid analysis of proteins have been improved to the point that single spots eluted or transferred from single 2D gels can be rapidly identified.

Mass spectroscopic techniques have been developed that allow analysis of very small quantities of proteins and peptides (Aebersold and Leavitt, 1990; Patterson and Aebersold, 1995; Lahm and Langen, 2000). Chemical microsequencing and amino acid analysis can be performed on increasingly smaller samples (Lottspeich et al., 1999). Immunochemical identification is now possible with a wide assortment of available antibodies.

- More powerful, less expensive computers and software are now available, allowing routine computerized evaluations of the highly complex 2D patterns.
- Data about entire genomes (or substantial fractions thereof) for a number of organisms are now available allowing rapid identification of the genes encoding a protein separated by 2DE.
- The World Wide Web (WWW) provides simple, direct access to spot pattern databases for the comparison of electrophoretic results and to genome sequence databases for assignment of sequence information.

A large and continuously expanding application of 2DE is proteome analysis. The proteome is defined as "the PROTEin complement expressed by a genOME"; thus it is a fusion word derived from two different terms (Wilkins et al., 1996; Pennington et al., 1997). This analysis involves the systematic separation, identification, and quantitation of a large number of proteins from a single sample. Two-dimensional PAGE is unique not only for its ability of simultaneously separating thousands of proteins but also for detecting post- and cotranslational modifications, which cannot be predicted from genome sequences. Other applications of 2D PAGE include identification (e.g., taxonomy, forensic work), the study of genetic variation and relationships, the detection of stages in cellular differentiation and studies of growth cycles, the examination of pathological states and diagnosis of disease, cancer research, monitoring of drug action, and studies on vaccines. Among the books dedicated to this topic, one could recommend those of Wilkins et al. (1997), Kellner et al. (1999), Rabilloud (2000b), Righetti et al. (2001b), James (2001), Westermeier and Naven (2002), and Grandi (2004). Among book chapters, Westermeier (1997) and Hanash (1998) are recommended, to name just a few.

Perhaps a key point of the success of present 2D PAGE was the introduction of the IPG technique, as stated above, and in particular the early recognition that wide, nonlinear pH gradients could be generated covering the pH 3.5–10 interval. Such a gradient (see Fig. 4.14) was calculated for a general case involving the separation of proteins in complex mixtures such as cell lysates and applied to a number of 2D separations (Gianazza et al., 1985a–c). The idea of this nonlinear gradient came from an earlier study by Gianazza and Righetti (1980), computing the statistical distribution of the pI of water-soluble proteins and showing that as much as two-thirds of them would focus in the acidic region (taking as a discriminant pH 7.0) and only one-third in the alkaline pH scale. Given this relative abundance, it was clear that an optimally resolving pH gradient should have a gentler slope in the acidic portion and a steeper course in the alkaline region. In a separation of a crude lysate of *Klebsiella pneumoniae*, a great improvement in resolution of the acidic cluster of bands was in fact obtained, without loss of the basic portion of the pattern (Gianazza et al., 1985c). This manuscript was also a cornerstone in IPG technology since it demonstrated, for the first time, that the pH gradient and the density gradient stabilizing it did not need to be colinear because the pH could be adjusted by localized flattening while leaving the density gradient unaltered. This nonlinear pH 3.5–10 gradient formed the basis (with perhaps minor modifications) of most of the wide nonlinear gradients adopted today and sold by commercial companies. The other events that made IPGs so powerful were the recognition that a much wider portion of the pH scale could be explored; thus very acidic pH intervals (down to pH 2.5) were described for equilibrium fractionation of acidic proteins, such as pepsin (Righetti et al., 1988a), as well as very alkaline intervals (e.g., pH 10–12) for analysis of high-pI proteases (subtilisins; Bossi et al., 1994b) and even histones (Bossi et al., 1994a). Already in 1990 our group described 2D maps in what was at that time the most extended pH gradient available, spanning the pH 2.5–11 interval (Sinha et al., 1990). We went so far as to describe even sigmoidal pH gradients, which have also found applications in 2D analysis (Tonani and Righetti, 1991; Righetti and Tonani, 1991). This shows that, already starting from 1990, all the ingredients for the success of 2D maps exploiting IPGs in the first dimension were fully available in the literature.

6.2. SOME BASIC METHODOLOGY PERTAINING TO 2D PAGE

Figure 6.1 shows what is perhaps the most popular approach to 2D map analysis today. The first dimension is preferably performed in individual IPG strips laid side by side on a cooling platform, with the sample often adsorbed into the entire strip during rehydration. At the end of the IEF step, the strips have to be interfaced with the second dimension, almost exclusively performed by mass discrimination via saturation with the anionic surfactant SDS. After this equilibration step, the strip is embedded on top of an SDS–PAGE slab, where the 2D run is carried out perpendicular to the 1D migration. The 2D map displayed at the end of these steps is the stained SDS–PAGE slab, where polypeptides are seen, after staining, as (ideally) round spots each characterized by an individual set of pI/M_r coordinates.

Although most of the relevant protocols for properly performing 2D maps can be found in the respective chapters devoted to IEF (and IPG) and to SDS–PAGE, it is worth recalling some important methodologies, especially those involving sample solubilization and preparation prior to the first-dimension IEF/IPG step. Appropriate sample preparation is absolutely essential for good 2D results. Due to the great diversity of protein sample types and origins, only general guidelines for sample preparation are provided here. The optimal procedure should ideally be determined empirically for each sample type. If sample preparation is performed properly, it should result in complete solubilization, disaggregation, denaturation, and reduction of the proteins in the sample. We recall here that different treatments and conditions are required to solubilize different types of samples: Some proteins are naturally found in complexes with membranes, nucleic acids, or other proteins; some proteins form various nonspecific aggregates; and some others precipitate when removed form their physiological environment. The effectiveness of solubilization depends on the

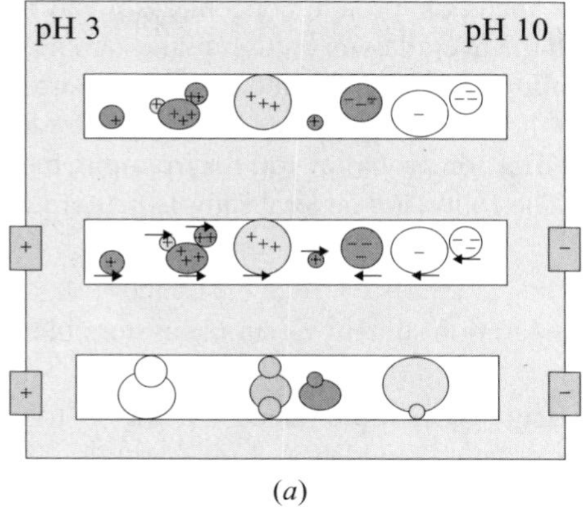

(*a*)

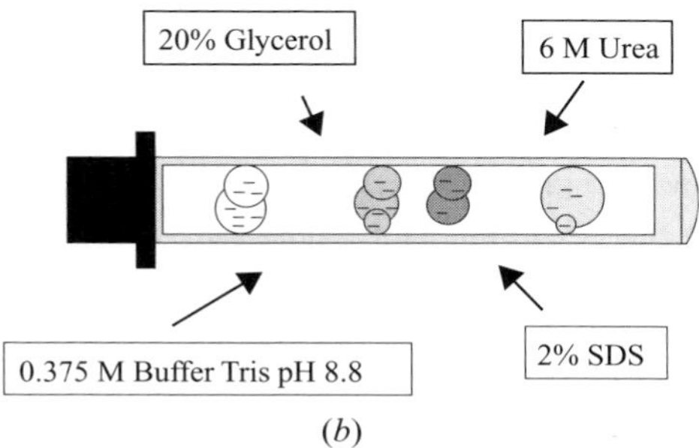

(*b*)

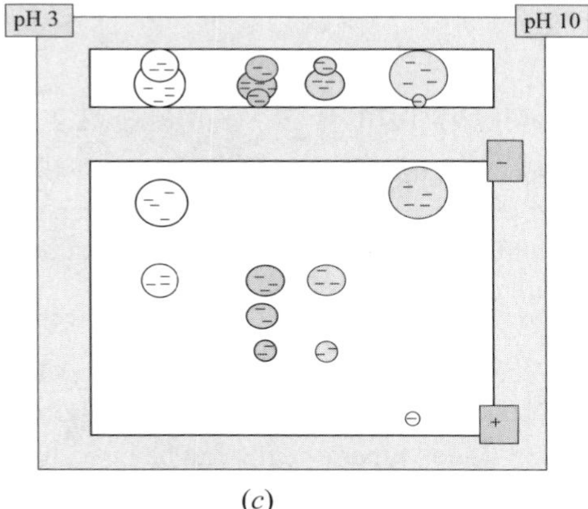

(*c*)

FIGURE 6.1. Pictorial representation of steps required for obtaining 2D map with (*a*) first dimension typically done in IPG strips followed by (*b*) equilibration of focused strips in SDS-interfacing solution and finally (*c*) second-dimension run in SDS–PAGE.

choice of cell disruption methods, protein concentration and dissociation methods, choice of detergents, and the overall composition of the sample solution. Lenstra and Bloemendal (1983), Molloy et al. (1998), and an entire issue of *Methods of Enzymology* (Deutscher, 1990) can be consulted as general guides to protein purification and/or for specific solubilization problems for, for example, membrane proteins and other difficult samples. The following general sample preparation guidelines should always be kept in mind:

- Keep the sample preparation strategy as simple as possible so as to avoid protein losses.
- Cells and tissues should be disrupted in such a way as to minimize proteolysis and other modes of protein degradation. Cell disruption should be done at low temperatures and should ideally be carried out directly in strongly denaturing solutions containing protease inhibitors.
- Sample preparation solutions should be freshly made or stored frozen as aliquots. High-purity, deionized urea should always be used.
- Preserve sample quality by preparing it just prior to IEF or by storing samples in aliquots at -80°C. Samples should not be repeatedly thawed.
- Remove all particulate material by appropriate centrifugation steps. Solid particles and lipids should be eliminated because they will block the gel pores.
- In the presence of urea, samples should never be heated. Elevated temperatures produce higher levels of cyanate from urea, which in turn can carbamylate proteins. A specific technique has been described for producing "carbamylation trains" as pI markers by boiling specific proteins in urea (Anderson and Hickman, 1979; Hickman et al., 1980; Tollaksen et al., 1981).

Below we will briefly describe some general methods pertaining to sample preparation.

6.2.1. Methods of Cell Disruption

These methods can be divided into two categories: "soft" and harsher methods. Soft methods are in general employed when the sample under investigation consists of cells that can be lysed under mild conditions [e.g., tissue culture cells, red blood cells (RBCs). Briefly, they are as follows:

- *Osmotic Lysis.* When isolated cells are suspended in hypotonic solution, they swell and burst, releasing all the cellular content (Dignam, 1990).
- *Freeze–Thaw Lysis.* Many types of cells can be lysed by subjecting them to one or more cycles of quick freezing and subsequent thawing. (Bollag and Edelstein, 1991; Toda et al., 1994).
- *Detergent Lysis.* Most detergents solubilize cellular membranes, thereby lysing cells and liberating their content.

Harsh methods are those employing, in general, mechanical means for cell or tissue disruption. They can be summarized as follows:

- *Sonication.* Ultrasonic waves generated by a sonicator lyse cells through shear forces. Care should be exerted for minimizing heating and foaming (Kawaguchi and Karamitsu, 1995; Teixera-Gomes et al., 1997).
- *French Pressure Cell.* This cell can be pressurized well above 1000 atm. Cells are lysed by shear forces resulting from forcing a cell suspension through a small orifice at high pressure. This is particularly useful for microorganisms with a cell wall (e.g., bacteria, algae).
- *Grinding.* Some types of cells can be broken by hand grinding with a mortar and pestle, often admixed with quarzite or alumina powders. This is well suited for solid tissues and microorganisms.
- *Mechanical Homogenization.* This is one of the most popular methods for soft, solid tissues. Classical devices are the Dounce and the Potter-Elvehjem homogenizers. Blenders and other electrically driven devices are often used for large samples (Gengenheimer, 1990; Theillet et al., 1982; Wolpert and Dunkle, 1983).
- *Glass-Bead Homogenizers.* The abrasive force of the vortexed beads is able to break cell walls, liberating the cell contents. This is useful for cell suspensions and microorganisms (Jawinski, 1990; Cull and McHenry, 1990).

6.2.2. Proteolytic Attack during Cell Disruption

When cells are lysed, hydrolases (phosphatases, glycosidases, and especially proteases) are in general liberated or activated. In the presence of chaotropes, glycosidases and phosphatases are quickly denatured, but the denaturation of proteases might have slower kinetics. For instance, Cleveland et al. (1977) performed peptide mapping in dilute SDS solutions, suggesting that proteases had the ability to work in such denaturing media. This often results in many altered spots in the subsequent 2D analysis, to the point that band recognition is no longer possible. Proteases are in general inhibited when the tissue is disrupted directly in strong denaturants, such as 8 M urea, 10% TCA, or 2% SDS (Damerval et al., 1986; Wu and Wang, 1984; Harrison and Black, 1982; Granzier and Wang, 1993; Colas de Francs et al., 1985). But there are still two major problems: (a) some proteases might be more resistant to denaturation than the majority of the other cellular proteins, so that, while the former have been unfolded, proteases might have some time for attack before being denatured too; (b) in other cases (e.g., RBCs) the cells might have to be lysed under native conditions in order to eliminate some cellular components, so that there is ample time for strong proteolytic aggression. In any event, proteases are less active at lower temperature, which is why, during cell disruption, low temperatures (2–4°C) are in general recommended. In addition, most tissue proteases are inactive above pH 9.5, so that proteolysis can often be inhibited by preparing the sample in Tris-free base, sodium bicarbonate, or basic carrier ampholytes at pH values close to 10. Some proteases, however, might retain activity even when all the above precautions are taken. Thus, as a safety precaution, it is advisable to use a cocktail of protease inhibitors. Such combinations are available from a number of commercial sources and in general comprise both chemical and proteinaceous inhibitors. Below, a list of the most common inhibitors is given:

- *Phenylmethylsulfonyl Fluoride (PMSF).* One of the most common inhibitors, it is used at concentrations up to 1 mM. It is an irreversible inhibitor that binds covalently into the active site of serine proteases and some cysteine proteases. Since PMSF is quickly inactivated in aqueous solutions, it should be prepared just prior to use. Since it is also less active in the presence of DTT or thiol reagents, the latter species should preferably be added at a later stage.
- *Aminoethyl Benzylsulfonyl Fluoride or Pefabloc SC (AEBSF).* It is similar to PMSF but is more soluble and less toxic. It is used at concentrations up to 4 mM.
- *EDTA or EGTA.* Generally used at 1 mM concentrations, they inhibit metalloproteases by chelating the metal ions required for activity.
- *Peptide Protease Inhibitors.* They are (a) reversible inhibitors, (b) active in the presence of DTT, and (c) active at low concentrations under widely different conditions. Examples are leupeptin, active against serine and cysteine proteases; pepstatin, inhibiting aspartyl proteases, such as pepsin; aprotinin, active against many serine proteases; and bestatin, an inhibitor of aminopeptidase. In general, they are used at a level of 2–20 μg/mL.
- *Tosyl Lysine Chloromethyl Ketone (TLCK) and Tosyl Phenyl Chloromethyl Ketone (TPCK).* These rather similar compounds irreversibly inhibit many serine and cysteine proteases. Both are used at levels of 0.1–0.5 mM.
- *Benzamidine.* It inhibits serine proteases and is used at concentrations of 1–3 mM.

The importance of safeguarding your sample against accidental proteolytic attack during preparation should never be underestimated. A case in point is the dramatic example of 2D maps of RBC membranes, given below, when processed in the absence or presence of proper inhibitor cocktails (Figs. 6.2*a*,*b*). The primary concern when preparing RBC membranes for electrophoresis is to remove hemoglobin, which is present at extremely high levels in RBCs and can cause severe streaking in 2D maps. Most of the procedures in the literature involve washing the RBCs in physiological buffered saline followed by lysis and washing in hypotonic buffer. However, there has been a lack of attention paid to the activation of proteases upon RBC lysis. Indeed, Heegaard and Poglod (1991) stated that no differences in the protein patterns were observed between protease-treated and standard, non-protease-treated RBC membranes. Interestingly, the same authors also reported that there was a large difference in the number of RBC membrane spots and in their distribution in the 2D maps shown by seven previous papers on RBC membranes. The number of RBC membrane 2D spots reported in the seven papers quoted by Heegaard and Poglod (1991) varied from less than 100 to more than 600, which should have been a strong indication that there were major discrepancies in the sample preparation methodologies presented. Rabilloud

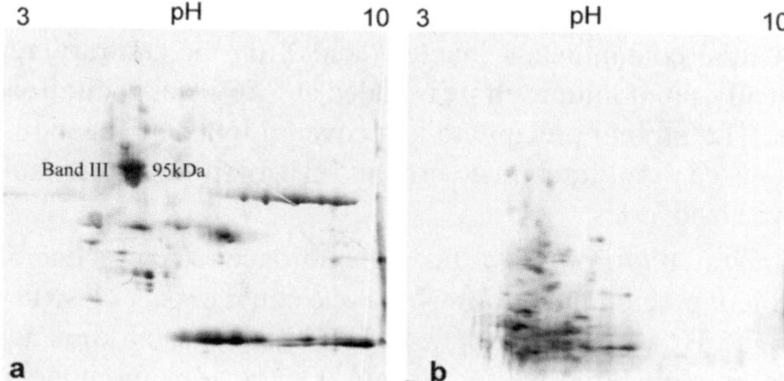

FIGURE 6.2. Two-dimensional maps of RBC membranes: (*a*) treated with protease inhibitors during lysis and washing steps; (*b*) control, untreated with protease inhibitors. Note the massive proteolytic action in (*b*). The numbers at the gel top indicate the pH interval (pH 3–10, nonlinear). (From Olivieri et al., 2001, by permission.)

et al. (1999) have shown 2D maps of RBC membranes using a conventional protocol, protease inhibitors, and a highly solubilizing cocktail containing urea, thiourea, and amidosulfobetaine 14, a novel zwitterionic surfactant which has improved the solubility of membrane proteins from a variety of sources (Chevallet et al., 1998). The 2D map shown in Figure 6.2*a* is a silver-stained $T = 10\%$ to $T = 20\%$ gel of RBC membranes prepared using protease inhibitors (Olivieri et al., 2001). The pattern is quite similar to the RBC profiles obtained by Rabilloud et al. (1999) and major landmark membrane proteins such as Band III (95 kDa) are visible. In contrast, the 2D map in Figure 6.2*b* is a silver-stained $T = 10\%$ to $T = 20\ \%$ gel of RBC membranes prepared without protease inhibitors. Clearly, in the absence of inhibitors there has been massive degradation of the high-molecular-mass proteins and the resulting mixture of peptides barely extends above a M_r of 50 kDa.

6.2.3. Precipitation Procedures

Precipitation protocols are optional steps in sample preparation for 2D maps. It should be borne in mind that precipitation followed by resolubilization in sample solution is typically employed for selectively separating proteins from other contaminants, such as salts, detergents, nucleic acids, and lipids, but that rarely are the recoveries 100% (in fact, often it could be much lower than that). Thus, employing a precipitation step prior to a 2D map may alter the protein profile in the final 2D image. Therefore such protocols should not be adopted if one aims at obtaining a complete and accurate profiling of all proteins in the sample under analysis. Precipitation followed by resuspension can also be utilized as a sample concentration step from dilute sources (e.g., urines, plant tissues), although in this latter case Centricon tubings might be preferred, since during the centrifugation step both concentration and desalting occur simultaneously. Below is a list of the most common precipitation methods:

- *Ammonium Sulfate Precipitation (Salting Out).* This is one of the oldest methods known in biochemistry. At high salt concentrations, proteins lose water in their hydration shell and thus tend to aggregate and precipitate out of solution. Many potential contaminants (nucleic acids), on the contrary, remain in solution. Typically, ammonium sulfate is added at >50% concentration and up to full saturation. The protein precipitate is recovered by centrifugation. This method typically is used to subfractionate proteins in a mixture, so one should not expect high sample recoveries.

- *Trichloroacetic Acid Precipitation.* Trichloroacetic acid is one of the most effective protein precipitants, from which one could expect close to 100% sample recovery. The TCA is added to 10% to 20% final concentration and proteins are allowed to precipitate on ice for 30 min. Direct tissue homogenization in this medium could also be adopted. The protein pellet recovered by centrifugation should be washed with acetone or ethanol for proper TCA removal. Since TCA is a strong denaturant/precipitant, not all proteins might resolubilize in the typical cocktails used for 2D maps [e.g., the O'Farrell (1975) concoction].
- *Acetone Precipitation.* This solvent is commonly used for precipitating proteins, since it leaves in solution many organic-solvent soluble contaminants, such as detergents and lipids. In general, an excess of at least three volumes of ice-cold acetone is added to the extract and proteins are allowed o precipitate at −20°C for a few hours. Proteins are pelletted by centrifugation and acetone is removed in a gentle nitrogen stream or by lyophilization (Holloway and Arundel, 1988; Flengsrud and Kobro, 1989; Matsui et al., 1997).
- *Precipitation with TCA–Acetone.* This combination is often used for precipitating proteins in preparation for a 2D map analysis, and it is more effective than either TCA or acetone alone. The sample can be directly lysed or disrupted in 10% TCA in acetone in the presence of 20 mM DTT. Proteins are allowed to precipitate for 1 h at −20°C. After centrifugation, the pellet is washed with cold acetone in the presence of 20 mM DTT (Tsugita et al., 1996; Granier, 1988). This method is, in reality, a modification of an old procedure utilizing acetone/HCl for precipitation of heme-free globin chains from hemoglobins (Clegg et al., 1966). In fact, colorless globins would be recovered, with the red heme group left behind in the acetone solvent.
- *Precipitation with Ammonium Acetate in Methanol after Phenol Extraction.* This technique is much in vogue for plant samples, due to their content of high levels of substances interfering with 2D maps (e.g., polyphenols). The proteins are first extracted in water- or buffer-saturated phenol and subsequently precipitated by adding 0.1 M acetate in methanol. The recovered pellet is washed a few times with the same solvent and finally with acetone (Usada and Shimogawara, 1995).

The above methods have been recently evaluated by Jiang et al. (2004), who have reported that TCA and acetone precipitation as well as ultrafiltration delivered a higher protein recovery compared to ammonium sulfate and chloroform/methanol steps. More tips on sample preparation can be found in the review by Shaw and Riederer (2003).

6.2.4. Removal of Interfering Substances

The following section is taken in large part from the review of Rabilloud (1996), to which the readers are referred for further details. Nonprotein impurities in the sample can interfere with separation and subsequent detection of spots in 2D maps, so sample preparation should preferably include steps for eliminating such substances from the analyte. Below is a list of the major contaminants and of removal techniques:

- *Salts, Residual Buffers.* In general, high amounts of salts are present in halophilic organisms or in some biological fluids (urine, sweat, and, to a lesser extent, plasma and spinal fluid). Removal techniques are dialysis, spin dialysis, gel filtration, and precipitation/resolubilization. Dialysis is one of the most popular methods (e.g., Manabe et al., 1982), but it is time consuming. Spin dialysis is quicker, but protein adsorption onto the dialysis membrane might be a problem. Other methods for salt removal are based on precipitation of proteins with dyes (Marshall and Williams, 1992; Marshall et al., 1995).
- *Nucleic Acids (DNA, RNA).* These macromolecules increase sample viscosity and cause background smears. In addition, high-M_r nucleic acids can clog gel pores. They are also able to bind to many proteins through electrostatic interactions, preventing proper focusing and producing severe streaking (Heizmann et al., 1980). It has also been demonstrated that they form complexes with carrier ampholytes (Galante et al., 1976). For removal, the sample can be treated with protease-free DNase/RNase mixtures. This is often accomplished by adding one-tenth of the sample volume of a solution of 1 mg/mL DNase, 0.25 mg/mL RNase, and 50 mM $MgCl_2$ and incubating in ice (O'Farrell, 1975). In SDS–PAGE, though, DNA and RNA are only a problem because they clog the gel pores (Glass et al., 1981). With proteins that bind tenaciously to nucleic acids, even in presence of 8 M urea, solubilization can be achieved by adding competing cations, such as protamine (Sanders et al., 1980), or lecithins at acidic pH values (Willard et al., 1979). Other methods are based on the selective precipitation of nucleic acids by metallic cations such as calcium (Mohberg and Rusch, 1969) or lanthanum (Yoshida and Shimura, 1972).
- *Polysaccharides.* The problems due to polysaccharides are similar to those due to nucleic acids, although they are fortunately less severe. Uncharged polysaccharides (starch, glycogen, etc.) pose problems only because they are huge molecules which could clog the pores of polyacrylamide matrices. The problems are more severe for complex polysaccharides such as mucins, hyaluronic acids, dextrans, and so on, since they contain negative charges favoring protein binding. In more severe cases, precipitation could occur (Gianazza and Righetti, 1978). Some of them, like heparin, have been shown to give a multitude of bands due to complexation with acidic carrier ampholytes (Righetti et al., 1978). For their removal, TCA, ammonium sulfate, or phenol/ammonium acetate precipitation of sample proteins could be used.

- *Lipids.* These give two kinds of problems, depending on whether they are present as monomers or as assemblies (e.g., in membranes). As monomers, they typically bind to specific proteins, lipid carriers, and could thus give rise to artifactual heterogeneity. The problem is more severe with membranous material. In general, the presence of detergents in the solubilization solution should disaggregate lipids and delipidate and solubilize the proteins. However, if large amounts of lipids are present, chemical delipidization prior to sample resolubilization might be necessary. Delipidation is achieved by extraction with organic solvents (Van Renswoude and Kemps, 1984) or of mixtures thereof (Radin, 1981), often containing chlorinated solvents (Wessel and Flugge, 1984). Partial but useful delipidation with ethanol or acetone can also be achieved (Menke and Koenig, 1980; Penefsky and Tzagoloff, 1971).
- *Ionic Detergents.* The most popular ionic detergent is SDS, as adopted for solubilizing difficult samples. Sodium dodecy/sulfate forms strong complexes with proteins, and the resulting negatively charged complex will not focus unless SDS is removed or sequestered. In general, for its removal, the SDS-solubilized sample is diluted into a solution containing high levels of nonionic or zwitterionic detergents, such as CHAPS, Triton X-100, and Nonidet P-40, so that the final SDS concentration is no greater than 0.25% and the ratio of the other detergent to SDS is at least 8 : 1 (Ames and Nikaido, 1976). A note of caution: Strongly hydrophobic proteins, which absolutely need SDS for their solubilization, will in general precipitate out of solution when SDS is exchanged with any other detergent. Thus, the idea of solubilizing them with SDS is fallacious if a 2D separation process is sought.
- *Other Compounds.* A number of additional, interfering compounds can be found, mainly in extracts from plants. These include lignins, polyphenols, tannins, alkaloids, and pigments (Gengenheimer, 1990). Polyphenols bind proteins via hydrogen bonds when they are in a reduced state, but, when oxidized, they form covalent bonds with them. A method for eliminating polyphenols is to use polyvinylpyrrolidone as trapping agent (Cremer and Van de Walle, 1985). Addition of reducing agents during extraction (ascorbate, DTT, sulfite) is also helpful, since it prevents phenolic oxidation. Polyphenol oxidase can also be inhibited with thiourea or diethyldithiocarbamic acid.
- *Insoluble Material.* Particulate material in the sample resulting from incomplete solubilization can obstruct gel pores and result in poor focusing. Its presence could cause severe streaking when the sample is applied from a cup, rather than being uniformly adsorbed into the entire gel strip. Care should be taken that the sample is clarified by centrifugation prior to application to the IEF/IPG strip.

The importance of safeguarding your sample against any of the interfering substances, as listed above, should never be underestimated if top performance in 2D maps is sought. To our reckoning, though, the biggest offender could be the type and amount of salts present in the sample just prior to application to the IEF/IPG strip. Since its inception, the IPG technique was recognized to be quite tolerant to salt levels present in the sample. This was publicized as one of the greatest advantages of the IPG technique as opposed to CA–IEF, known to be quite sensitive even to low salt levels in the sample. Thus biological samples (containing high salt and dilute proteins) were thought as being amenable to IPG runs without prior dialysis or concentration. This statement turned out to be fallacious, as demonstrated by Righetti et al. (1988a): It is true when referring to the IPG matrix, which in principle can stand any amount of salt, but it is not true when referring to the protein sample. Salts formed from strong acids and bases (e.g., NaCl, Na_2SO_4, Na_2HPO_4) that are present in a protein sample applied to an IPG gel induce protein modification (e.g., oxidation of iron moiety in Hb) already at low levels (5 mM) and irreversible denaturation (precipitation) at higher levels (>50 mM). This effect is due to production of strongly alkaline cationic and strongly acidic anionic boundaries formed by the splitting of the salt's ion constituents, as the protein zone *is not and cannot be buffered by the surrounding gel until it physically migrates into the IPG matrix.* To explain the phenomenon in more detail, Figure 6.3*a* shows what happens in the sample liquid droplet, containing high salt levels (in this case, 100 mM NaCl), as soon as the voltage is applied. Within a few minutes, at an applied voltage drop of 200 V/cm, the anodic end of the sample layer reaches a pH as low as 1, with an apparently more modest pH increment in the rear (cathodic) boundary. These extreme pH values generated in the two boundaries are also a function of the initial applied voltage (Fig. 6.3*b*): At moderate applied voltages (e.g., 500 V) no adverse pH boundaries are generated, while at progressively higher voltages strongly adverse pH zones are generated that are able to denature and precipitate the protein macroions present in the sample layer. Substitution of "strong" salts in the sample zone with salts formed by weak acids and bases (e.g., Tris–acetate, Tris–glycinate, Good's buffers) essentially abolishes both phenomena, oxidation and irreversible denaturation. Suppression of strong salt effects is also achieved by adding CAs to the sample zone, in amounts proportional to the salt present (e.g., by maintaining a salt–CA molar ratio of ~1 : 1). Low-voltage runs for extended initial periods (e.g., 4 h at 500 V) are also beneficial (see Fig. 6.3*b*). Although the data here presented refer to native protein runs, it must be emphasized that they apply to denatured samples too, especially if these samples are run side by side in a continuous gel slab rather than in individual strips. In the former case, variable salt levels in the various samples would cause fanning out of the proteins fronts, which would thus invade laterally all adjacent sample tracks and completely ruin the IPG pattern. Even in individual strips, excess salt in the sample can considerably quench migration of proteins to the pI value; the salts will have to migrate out and collect in the electrode reservoirs before any appreciable protein migration will ensue. As a result, the run might take an incredible length of time, to the point at which proteins (and the gel matrix too) could hydrolize, especially in runs performed in rather alkaline (or acidic) pH ranges. We would thus like to offer the following "pentalogue":

- Avoid high salt levels in your sample (>40 mM).
- Avoid salts formed by strong acids and bases (e.g., NaCl, Na_2SO_4, Na_2HPO_4).

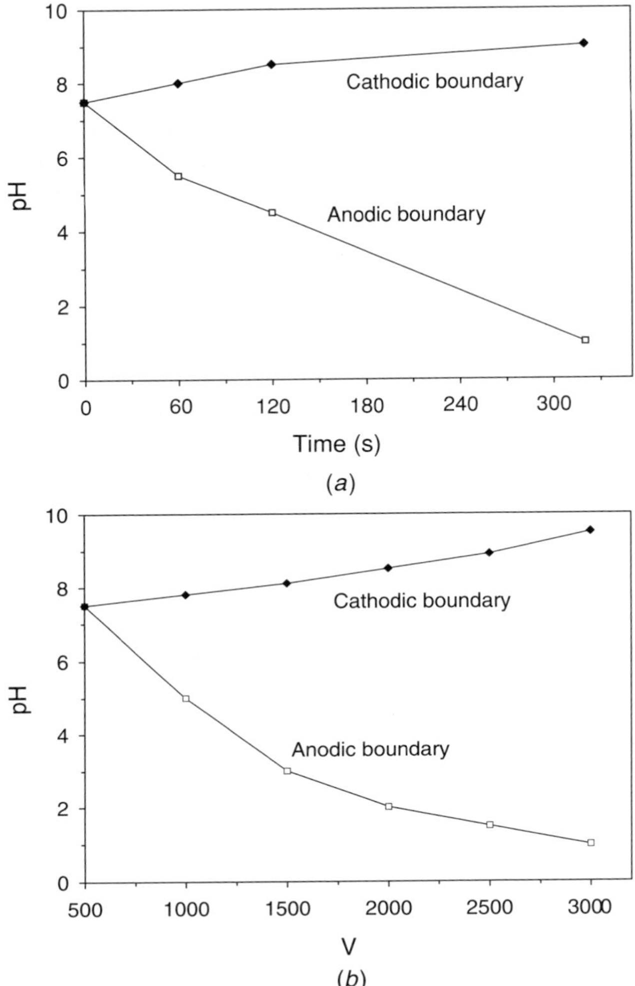

FIGURE 6.3. (*a*) Assessment of pH of salt ion boundaries in sample zone as function of time. Pockets cast in the middle of the gel (pH 7.5) were filled with 20 mL of 100 mM NaCl. The cathodic and anodic hedges of the pocket were covered with thin strips of alkaline and acidic pH indicators, respectively. The pH in the two boundaries was assessed by visual inspection of the color changes at the given time intervals at constant 2000 V. (*b*) Assessment of pH of salt ion boundaries in sample zone as function of applied voltage. The experiment of part (*a*) was repeated, except that pH estimations were made as a function of different voltage gradients applied (from 500 to 3000 V) after 10 min from the application of the electric field. Here, as well as in (*a*), the alkaline pH estimates must be regarded as approximate, since the pH indicators, being negatively charged, move away from the Na^+ boundary toward lower pH values. Conversely, the pH of the anodic boundary is a much better estimate since the pH indicator, when it starts leaching out of the filter paper strip, moves with the Cl^- boundary. (From Righetti et al., 1988a, by permission.)

- In the presence of high salt levels, when running native proteins, add high levels of carrier ampholytes (e.g., 3% to 4% to 50 mM salt).
- If salt is needed for, for example, sample solubility, use only salts formed from weak acids and bases (e.g., Tris–acetate, Tris–glycinate) or any of the Good's buffers (e.g., ACES, BES, MOPS, see p. 202) titrated around the p*K* of their amino group (with an appropriate weak counterion!).
- In the presence of high salt levels, run your sample at low voltages for several hours (possibly no greater than 2–300 V) so as to prevent formation of strongly acidic and alkaline boundaries.

6.2.5. Solubilization Cocktail

We have seen in Section 6.1.1 that for decades the most popular lysis solution has been the O'Farrell cocktail (9 M urea, 2% Nonidet P-40, 2% β-mercaptoethanol, and 2% carrier ampholytes, in any desired pH interval). Although much in vogue also in present times, over the years new, even more powerful solubilizing mixtures have been devised. We will discuss below the progress made in sample solubilization, since this is perhaps the most important step for success in 2D map analysis. Great efforts were dedicated to such developments, especially in view of the fact that many authors noted that hydrophobic proteins were largely absent from 2D maps (see, e.g., Wilkins et al., 1998). These authors noted that, quite strikingly, in three different species analyzed (*E. coli, Bacillus subtilis, Saccharomyces cerevisiae*), all proteins above a given hydrophobicity value were completely missing, independent of the mode of IEF (soluble CAs or IPG). This suggested that the initial sample solubilization was the primary cause for loss of such hydrophobic proteins. The progress made in solubilizing cocktails can be summarized as follows (see also the reviews by Molloy, 2000; Rabilloud and Chevallet, 2000; Santoni et al., 2000b):

1. *Chaotropes.* Although urea (up to 9.5 M) has been for decades the only chaotrope used in IEF, recently thiourea has been found to further improve solubilization, especially of membrane proteins (Molloy et al., 1998; Rabilloud et al., 1997; Musante et al., 1997; Pasquali et al., 1997; Rabilloud, 1998). The inclusion of thiourea is recommended for use with IPGs, which are prone to adsorptive losses of hydrophobic and isoelectrically neutral proteins. Typically, thiourea is added at concentrations of 2 M in conjunction with 5–7 M urea. The high concentration of urea is essential for solvating thiourea, which is poorly water soluble (only ~1 M in plain water; Gordon and Jencks, 1963). Among all substituted ureas (alkyl ureas, both symmetric and asymmetric), Rabilloud et al. (1997) found thiourea to the best additive. However, it would appear that not much higher amounts of thiourea can be added to the IPG gel strip, since it seems that at >2 M concentrations this chaotrope inhibits binding of SDS in the equilibration step between the first and second dimensions, thus leading to poor transfer of proteins into the 2D gel. It should also be remembered that urea in water exists in equilibrium with ammonium cyanate, whose level increases with increasing pH and temperature (Hagel et al., 1971). Since

cyanate can react with amino groups in proteins, such as the N-terminus α-amino or the ε-amino groups of Lys, these reactions should be avoided since they will produce a false sample heterogeneity and give wrong M_r values upon peptide analysis by MALDI–TOF MS. Thus, fresh solutions of pure-grade urea should be used, in concomitance with low temperatures and with the use of scavengers of cyanate (such as the primary amines of carrier ampholytes or other suitable amines). In addition, protein mixtures solubilized in high levels of urea should be subjected to separation in the electric field as soon as possible: In the presence of the high voltage gradients typical of the IEF protocol, cyanate ions are quickly removed and no carbamylation can possibly take place (McCarthy et al., 2003), whereas it will if the protein–urea solution is left standing on the bench.

2. *Surfactants.* These molecules are always included in solubilizing cocktails to act synergistically with chaotropes. Surfactants are important in preventing hydrophobic interactions due to exposure of protein hydrophobic domains induced by chaotropes. Both the hydrophobic tails and the polar head groups of detergents play an important role in protein solubilization (Jones, 1999). The surfactant tail binds to hydrophobic residues, allowing dispersal of these domains in an aqueous medium, while the polar head groups of detergents can disrupt ionic and hydrogen bonds, aiding in dispersion. Detergents typically used in the past included Nonidet P-40 or Triton X-100, in concentrations ranging from 0.5% to 4%. More and more, zwitterionic surfactants, such as CHAPS, are replacing those neutral detergents (Hochstrasser et al., 1988; Holloway and Arundel, 1988), often in combination with small amounts (0.5%) of Triton X-100. In addition, low levels of CAs (<1%) are added, since they appear to reduce protein–matrix hydrophobic interactions and overcome detrimental effects caused by salt boundaries (Rimpilainen and Righetti, 1985). Linear sulfobetaines are now emerging as perhaps the most powerful surfactants, especially those with at least a 12-carbon tail (SB 3-12). They were in fact already demonstrated to be potent solubilizers of hydrophobic proteins (e.g., plasma membranes), except that they had the serious drawback of being precipitated out of solution due to low urea compatibility (only 4 M urea for SB 3-12) (Satta et al., 1984; Dunn and Burghes, 1983; Willard et al., 1979). This has prompted the synthesis of more soluble variants (Gianazza et al., 1987). The inclusion of an amido group along the hydrophobic tail greatly improved urea tolerance, up to 8.5 M, and ameliorated separation of some proteins of the erythrocyte membranes (Rabilloud et al., 1990). Recently, Chevallet et al. (1998), Tastet et al. (2003), and Luche et al. (2003) embarked on a project for determining the structural requirements for the best possible sulfobetaine. Three major features were found to be fundamental: (a) an alkyl or aryl tail of 14–16 carbons, (b) a sulfobetaine head of 3 carbons, and (c) a 3-carbon spacer between the quaternary ammonium and the amido group along the alkyl chain. In particular, when analyzing the three basic parts of a surfactant molecule, the following requirements were found to be of major importance: (a) in the hydrophilic, zwitterionic head, the best distance between the two charged groups would be a propyl chain (butyls would not improve the molecules, ethyls would render it much less soluble); (b) the presence of a linker region (usually an amidopropyl), that is, the bridge linking the hydrophilic head to the hydrophobic tail, bridgeless detergents being almost inefficient; and (c) the

absence, in the hydrophobic part, of any polar atoms. Moreover, such a hydrophobic tail should be composed of linear alkyl (no longer that a C16) or alkylaryl, the phenyl ring being close to the hydrophilic head. The most promising surfactants were found to be ASB 14 (amidosulfobetaine) containing a 14-C linear alkyl tail and C8Ø (see page 360), which possesses a *p*-phenyloctyl tail (see Fig. 6.4). Both of these reagents have since been used successfully in combination with urea and thiourea to solubilize integral membrane proteins of both *E. coli* (Molloy et al., 2000) and *Arabidopsis thaliana* (Santoni et al., 1999, 2000a,b). Recently, Malone and Kramer (2000) have reevaluated the use of different detergents and proposed the following choices: (a) when working with mixtures of hydrophobic and hydrophilic proteins, like those seen in whole-cell samples, combinations of nonionic/zwitterionic surfactants with hydrophobicity/hydrophilicity indexes at the low and high end of the scale offer best solubilization results; (b) conversely, when analyzing water-soluble samples, such as serum or aqueous extracts, best performances are obtained with a single zwitterionic detergent. In all cases, they claim that, for most samples, best results are obtained when less than 2% total detergent is used for solubilization, although this seems to apply not to the first but to the second dimension, since such low levels of surfactants would allow better SDS equilibration of the IPG strip with shorter incubation times.

3. *Reducing Agents.* Thiol agents are typically used to break intramolecular and intermolecular disulfide bridges. Cyclic reducing agents such as DTT or dithioerythritol (DTE) are the most commonly reagents admixed to solubilizing cocktails. These chemicals are used in large excess (e.g., 20–40 mM) so as to shift the equilibrium toward oxidation of the reducing agent with concomitant reduction of the protein disulfides. Because this is an equilibrium reaction, loss of the reducing agent through migration of proteins away from the sample application zone can permit reoxidation of free Cys to disulfides in proteins, which would result not only in horizontal streaking but also, possibly, in formation of spurious extra bands due to scrambled –S–S– bridges and their crosslinking different polypeptide chains. Even if sample is directly reswollen in the dried IPG strip, as customary today, the excess DTT or DTE will not remain in the gel at a constant concentration, since, due to their weakly acidic character, both compounds will migrate above pH 7 and be depleted from the alkaline gel region. Thus, this will aggravate focusing of alkaline proteins and be one of the many factors responsible for poor focusing in the alkaline pH scale. This problem is not trivial at all and deserves further comments. For example, Righetti et al. (1989b), when reporting the focusing of recombinant pro-urokinase and urinary urokinase, two proteins with rather alkaline pI values, in IPG gel strips, detected a continuum of bands focusing in the pH 8–10 region, even for the recombinant protein, which exhibited a single, homogeneous band by SDS–PAGE. Since this protein has an incredible number of Cys residues (no less than 24!), this extraordinary heterogeneity was attributed to formation of scrambled disulfide bridges not only within a single polypeptide chain but also among different chains in solution. Curiously this happened even if the protein was not subjected to reduction of –S–S– bridges prior to the IPG fractionation, but this could also have a logical explanation. According to Bordini et al. (1999a,b), when probing the alkylation by acrylamide of –SH groups in proteins by MALDI–TOF, it was found that the primary site of attack, even in

Name	Formula	M_r
CHAPS 3-[(3-Cholamidopropyl) dimethylammonio] -1-propanesulfonate	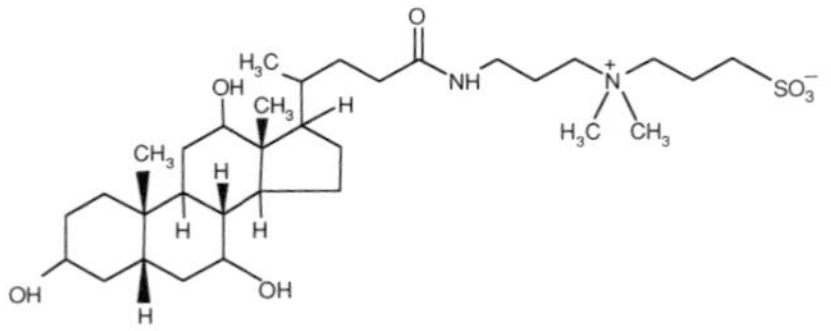	614.9
ASB14 Tetradecanoylamido propyl dimethyl ammonio propane sulfonate	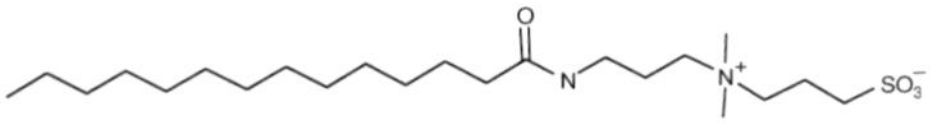	434
ASB16 Exadecanoylamido propyl dimethyl ammonio propane sulfonate	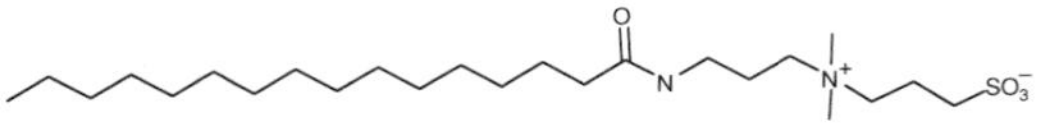	462
C8Ø (4-Octyl benzoylamido propyl dimethyl ammonio propane sulfonate)	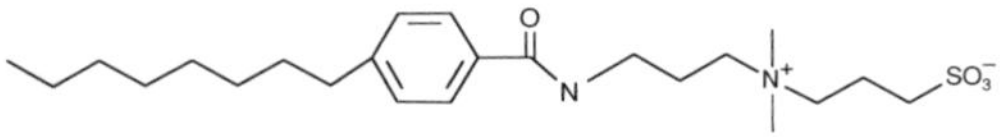	440
C6Bz (4-Hexyl)benzyl dimethyl ammonio propane sulfonate	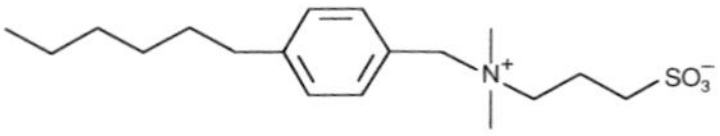	341
ØC6 6-Phenylhexyl dimethyl ammonio propane sulfonate	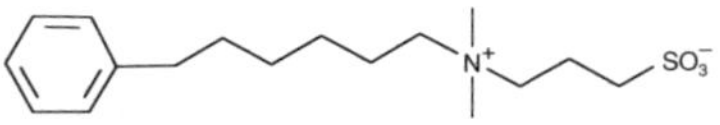	326
C7BzO 3-(4-Heptyl) phenyl 3-hydroxy propyl dimethyl ammonio propane sulfonate	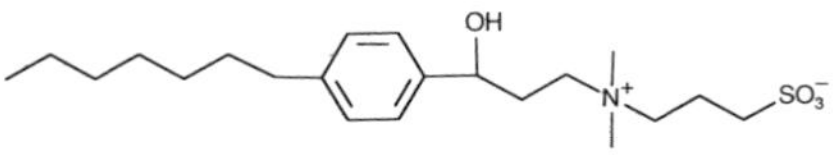	399

FIGURE 6.4. Structural formulas of some of the most powerful surfactants used in 2DE. (From Rabilloud et al., 1999, by permission.)

proteins having both disulfide bridges and free –SH groups, was not the free –SH residues, as it should, but it was systematically one of the –SH engaged in disulphide bridges! This could only be explained by assuming that, at alkaline pH values (the incubation was carried out at pH $\sim$ 10), disulfide bridges are weakened and probably constantly broken and reformed. The situation would be aggravated when using β-mercaptoethanol, since the latter compound has an even lower pK value; thus it is more depleted in the alkaline region and will form a concentration gradient toward pH 7, with a distribution in the gel following its degree of ionization at any given pH value along the IEF strip (Righetti et al., 1982). This is probably the reason for the dramatic loss of any pH gradient above pH 7.5, lamented by most users of conventional IEF in CAs, when generating 2D maps. The most modern solution to all the above problems appears to be the use of phosphines as alternative reducing agents. Phosphines (Ruegg and Ruedinger, 1977) operate in a stoichiometric reaction, thus allowing the use of low concentrations (barely 2 mM). The use of tributyl phosphine (TBP) was recently proposed by Herbert et al. (1998), who reported much improved protein solubility for both Chinese hamster ovary cell lysates and intractable, highly disulfide crosslinked wool keratins. Tributyl phosphine thus offers two main advantages: It can be used at much reduced levels as compared to other thiolic reagents (at least one order of magnitude lower concentration) and it can be uniformly distributed in the IPG gel strip (when rehydrated with the entire sample solution) since, being uncharged, it will not migrate in the electric field. Major drawbacks of TBP are that it is volatile, toxic, rather flammable in concentrated stocks, and easily oxidized when in contact with air. As a summary of the above, it would appear that one of the most powerful solubilization cocktails is the one shown in Figure 6.5.

4. *Reduction and Alkylation Prior to First Dimension or between First and Second Dimensions?* This is also a very important aspect of sample solubilization and pretreatment in 2D map analysis. Alkylation will prevent all the noxious phenomena reported above, such as re-formation of disulfide bridges, producing smears and even spurious bands due to interchain crosslinking, even among unrelated polypeptide chains. However, although reduction and alkylation are performed by anyone working with 2D maps, this last step is done not at the beginning of the sample treatment, just prior to the first-dimension IEF/IPG run, but between the first and second dimensions,

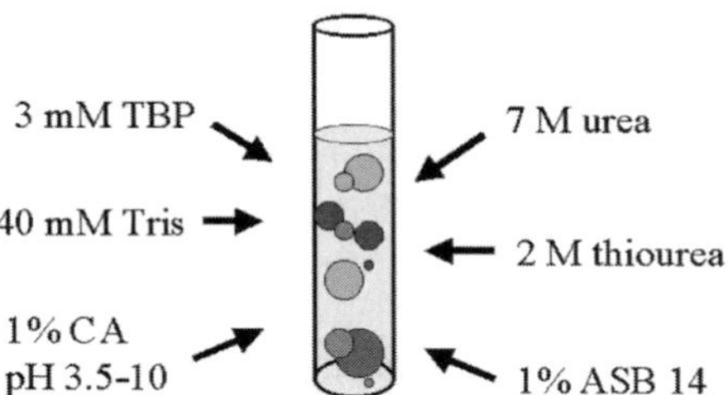

FIGURE 6.5. Composition of cocktail currently preferred for best sample solubilization in preparation of first-dimension run. In some cases, the level of urea is reduced from 7 to 5 M. The wide-range carrier ampholyte solution is added to facilitate sample entry in the IPG strip.

during the interfacing of the IEF/IPG strip with SDS-denaturing solution, in preparation for the SDS–PAGE final step. This probably stems from earlier reports by Goerg et al. (1987, 1988), who recommended alkylation of proteins with iodoacetamide during the interfacing between the first and second dimensions on the grounds that this treatment would prevent point streaking and other silver-staining artifacts. Clearly, in light of the above discussion, it appears to be a much smarter move to reduce and alkylate the sample first that is, prior to the first IEF/IPG dimension. The drawbacks for alkylation with iodoacetamide between the 2D runs have also been highlighted, recently, by Yan et al. (1999), although on different grounds than discussed above. These authors noted that in most 2D protocols the discontinuous Laemmli (1970) buffer is used, which calls for a stacking and sample gels to be equilibrated in a pH 6.8 buffer. To prevent pH alterations, most people use a modified stacking buffer with a reducing agent (DTT or DTE) and alkylating agent (iodoacetamide) at pH 6.8, so that the strips can be loaded as such after these two treatments, avoiding any further pH manipulations (Dunn, 1993). However, at this low pH both reduction and alkylation are not as efficient, since the optimal pH for these reactions is usually at pH 8.5–8.9 (Yan et al., 1998). As a result of this poor protocol, Yan et al. (1999) have reported additional alkylation by free acrylamide during the SDS–PAGE run, with the same protein exhibiting a major peak of Cys-carboxyamidomethyl and a minor one of Cys-propionamide. It should be borne in mind, in fact, that, whereas the risk of acrylamide adduct formation is much reduced in IPG gels (but not in unwashed IEF gels!), it is quite real in SDS–PAGE gels, due to the fact that these gels are not washed and that surfactants, in general, hamper incorporation of monomers into the growing polymer chain (Caglio et al., 1994). Different alkylating residues on a protein will complicate their recognition by MALDI–TOF analysis, a tool much in use today in proteomics. Here, too, however, although Yan et al. (1999) clearly identified the problem, they did not give the solution we are proposing here, namely to reduce and alkylate the sample prior to any electrophoretic step. They suggest retaining the original protocol of alkylation between the two dimensions but increasing the pH of the equilibration to pH 8.0, adding a much large amount of alkylating agent (125 mM iodoacetamide), and increasing the time of incubation to 15 min. We again stress that, although this new protocol is an improvement over previous ones, it still does not cure the problems of the first IEF/IPG dimension, namely smears and formation of spurious bands, both due to reoxidation of reduced but nonalkylated Cys residues.

5. *Alkylation with Iodoacetamide in Presence of Thiourea: A Conflicting Situation.* We have seen above that reduction and alkylation prior to any electrophoretic step are a must if one wants to avoid a number of artifacts typically encountered in the alkaline pH region. For example, Herbert et al. (2001) have demonstrated, by MALDI–TOF MS, that failure to alkylate prior to the IEF/IPG step would result in a large number of spurious spots in the alkaline pH region due to “scrambled” disulfide bridges among like and unlike polypeptide chains. This series of artifactual spots comprises not only dimers but also an impressive series of oligomers (up to nonamers) in the case of simple polypeptides such as the human α- and β-globin chains, which

possess only one (α-) or two (β-) –SH groups. As a result, misplaced spots are to be found in the resulting 2D map if performed with the wrong protocol. Subsequently, the same group (Galvani et al., 2001) has monitored the kinetics of protein alkylation by iodoacetamide over the period 0–24 h at pH 9. Alkylation reached ~70% in the first 2 min, yet the remaining 30% required up to 6 h for further reaction (which, however, never quite reached 100%). The use of SDS during the alkylation step resulted in a strong quenching of this reaction (thus further corroborating the notion that alkylation between the first and second dimensions is a useless procedure), whereas 2% CHAPS exerted a much reduced inhibition. The same authors reported another disturbing phenomenon: During alkylation of α- and β-globin chains, substantially different results were obtained when the sample was dissolved in plain 8 M urea or in the mixture 5 M urea + 2 M thiourea. In the former case, almost complete disappearance of all homo- and hetero-oligomers was achieved, whereas in the latter case appreciable amounts of dimers and trimers were still left over, even upon prolonged incubation. A thorough investigation by ESI–MS demonstrated that thiourea was competing with the free –SH groups of proteins for the reaction and was scavenging iodoacetamide at a fast rate (in ca. 5 min of incubation all iodoacetamide present in solution had fully disappeared). Iodoacetamide was found to add to the sulfur atom of thiourea; the reaction was driven in this direction also by the fact that, once this adduct was formed, thiourea was deamidated and the reaction product generated a cyclic compound (a thiazolinone derivative). Thus, those using urea/thiourea solutions should be aware of this side reaction and of the potential risk of not achieving full alkylation of the free, reduced –SH groups in Cys residues. Fortunately, alkylation of proteins is still substantial if iodoacetamide is added as a powder to the protein solution treated with the reducing agent, instead of as a solution. In addition, due to the fact that all iodoacetamide is scavenged by thiourea, the final concentration of thiourea in the solubilizing solution will be reduced by the same molar amounts of iodoacetamide added. As an alternative to alkylating with iodoacetamide, alkylation with acrylamide (producing a thioether derivative, Cys-S-β-propionamide) has been recommended (Mineki et al., 2002). More recently, another unique advantage was found in this protocol of immediate sample reduction and alkylation: protection of the protein from degradation. Herbert et al. (2003) have just reported an unexpected artifact occurring during the focusing step of alkaline proteins: β-elimination. Upon prolonged focusing, the Cys residues start losing the –SH group (desulfurization). The resulting β-alanine residue will be dehydrated with formation of a double bond and subsequent rupture of the bond in the polypeptide chains. This process produces a number of large peptides from intact proteins. However, if all Cys residues are alkylated, this degradation pathway is almost abolished.

As a conclusion, reduction and alkylation at the very start of any 2D fractionation would have the following immediate advantages (Herbert et al., 2001; Galvani et al., 2001):

1. Considerably reduce smears in the alkaline region, above pH 8.
2. Prevent formation of spurious bands due to mixed disulfide bridges.

3. Abolish formation of a mixed population of Cys-propionamide and Cys-carboxyamidomethyl species due to alkylation with acrylamide and iodoacetamide, respectively.
4. Prevent degradation of alkaline proteins into large peptides via a desulfurization event.

There remains the aspects of how to perform it. If reduction is done with DTT or DTE, it must be a two-step operation. The sample, brought to pH 8.5–8.9 (with a weak base such as free Tris and/or basic CAs!), is first incubated for ca. 1 h to allow for full reduction of –S–S– bridges. After that, a twofold molar excess of alkylating agent (iodoacetamide) is added and the sample left to incubate for an additional hour (or more) to allow for full alkylation (of both free Cys groups in proteins and free thiols in the DTT or DTE additives). Now, due to the fact that DTT or DTE is typically used at ~40 mM level, this means that 80–100 mM iodoacetamide will have to be added. This per se does not pose any risk in the subsequent IEF/IPG fractionation since this molecule is neutral, so large amounts can be tolerated in the focusing step. However, upon prolonged focusing (1–2 days, as often applied in IPG runs), there could be the risk of partial hydrolysis of the amido bond in iodoacetamide, thus provoking undue currents and destabilizing the focusing process (with the proviso, though, that in the presence of thiourea no iodoacetamide will be left over and the alkylation will be poor anyhow; since thiourea is universally used today, we recommend abandoning iodoacetamide in favor of acrylamide or other alkylating agents, as discussed below). Here is where the use of TBP might be preferred. Since the levels used of TBP as a reducing agent are minute (2 mM), this means that the levels of alkylating agent (iodoacetamide) to be added will also have to be small, typically of the order of 5–10 mM, that is, at least one order of magnitude less than in the case of DTT or DTE reduction. But there is more to it. Due to the fact that TBP does not react with some alkylating agents, such as acrylamide and 4-vinylpyridine (4VP) (Ruegg and Ruedinger, 1977), it could offer the opportunity of a simplified, one-step reduction and alkylation procedure using TBP and acrylamide (as alkylating agent) simultaneously, as proposed by Molloy (2000). However, since both acrylamide and iodoacetamide rarely reach an alkylation efficiency better than 80%, the best protocol would be the use of 2VP or 4VP: recent data suggest that, indeed, vinylpyridines offer two unique advantages: (a) 100% reactivity and (b) 100% specificity. This is due to the fact that the reaction can be conducted at pH 6.8, that is, halfway between the pK values of the –SH group and the weakly basic group of 2VP and 4VP. Moreover, since 2VP can be prepared in a deuterated form, it can be efficiently used in quantitave proteomics (Sebastiano et al., 2003). As an alternative to alkylation, Olsson et al. (2002) have proposed oxidation of thiol groups in proteins to mixed disulfides by using, in the sample and in the gel rehydration solution, an excess of hydroethyldisulfide (HO–CH_2–CH_2–S–S–CH_2–CH_2–OH) (HED). Note that this is an equilibrium reaction; thus an excess HED has to be present at all times during the IEF/IPG run. Other alkylating agents described are β-(4-hydroxyphenyl)ethyliodoacetamide and 1,5-I-AEDANS (Hudson–Weber reagent) (Vuong et al., 2000), although for the purpose of enhancing sensitivity via radioactive labeling.

6.2.6. Sample Application

This ritual is just as important for successful 2D fractionation as all the steps described above. There are different modes of sample application to the first-dimension IEF/IPG strip:

1. *Pulse Loading.* This is the method typically adopted in all fractionation techniques, including chromatography, since due to entropic forces tending to dissipate the sample zone, the latter has in general to be applied as a very thin zone. This, of course, does not hold true for any focusing procedure which reaches steady-state conditions. Nevertheless, it has been customary for years to apply the sample (especially in the old procedure of CA–IEF) in pockets precast in the gel or in surface basins or adsorbed onto granulated materials (e.g., Sephadex, Paratex films; see, as an example of surface basins, Fig. 6.6*a*). Sample cups did seem to offer a valid alternative to loading, since they permitted higher volume loadings (up to 100 μL) than with any other conventional means. To this purpose, Amersham Pharmacia Biotech devised a bar holder able to position battery of cups on a linear array of prehydrated IPG strips. This holder would gently press the cups against the gel surface so as to prevent sample leakage; each cup would then be filled with the desired volume of any sample (see Fig. 6.6*b*). This method too is being used less and less, since it was found

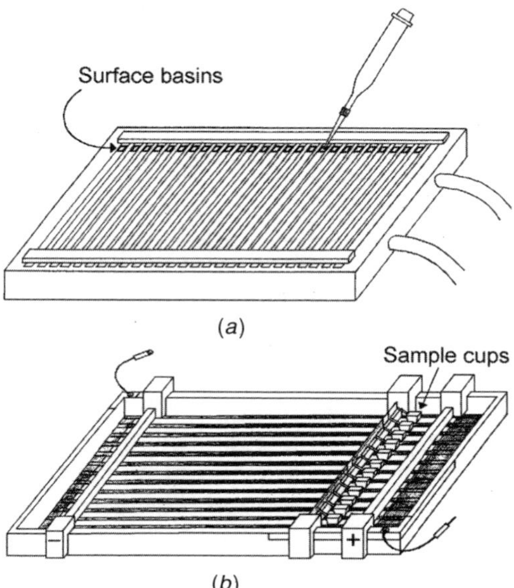

FIGURE 6.6. Scheme of different sample application procedures to rehydrated IPG strips: (*a*) surface basins (long strips at two gel extremities are electrode reservoirs, i.e., filter paper strips soaked in anolyte and catholyte, respectively); (*b*) battery of cups supported on holding bar for sample application to rehydrated IPG gel strips.

that significant amounts of proteins were lost at the application point, due to protein aggregation and precipitation as they migrated from the free liquid phase into the gel, with the ultimate result being poor resolution and "underloading" (Bjellqvist et al., 1993a). This effect has been explained as a concentration phenomenon occurring at the gel–sample interface, where proteins massively accumulate as they are dramatically slowed down in their transit from a liquid to a gel phase. At this interface also anomalous ionic boundaries will form and the two concomitant phenomena would favor protein aggregation and precipitation (Righetti et al., 1988a; Rabilloud et al., 1994).

2. *In-Gel Rehydration Loading.* This is currently the preferred method (Schupbach et al., 1991; Rabilloud et al., 1994; Sanchez et al., 1997; Pasquali et al., 1997) since it facilitates higher protein loads and reduces focusing times. With this technique, the dehydrated IPG strip is directly reswollen with the protein sample dissolved in the rehydration solution (notice that, in this last case, the same solution is used to solubilize the protein and rehydrate the IPG strip) (see Fig. 6.7). After a suitable rehydration time (approximately 5–6 h), the IPG strip is ready for the first-dimension IEF, with the proteins already uniformly distributed within the gel matrix. The clear advantage of in-gel rehydration is the large volume of sample that can be applied (up to 450 μL for a 18-cm-long, 4-mm-wide, 0.5-mm-thick IPG strip) compared to 100 μL for conventional cup loading. The other major advantage is minimization of sample aggregation and precipitation: Since the sample is diluted through the entire gel strip, no local, accidental buildup of proteins can occur (except, of course, in the focusing zone, but here, even if large sample amounts are collected and precipitate

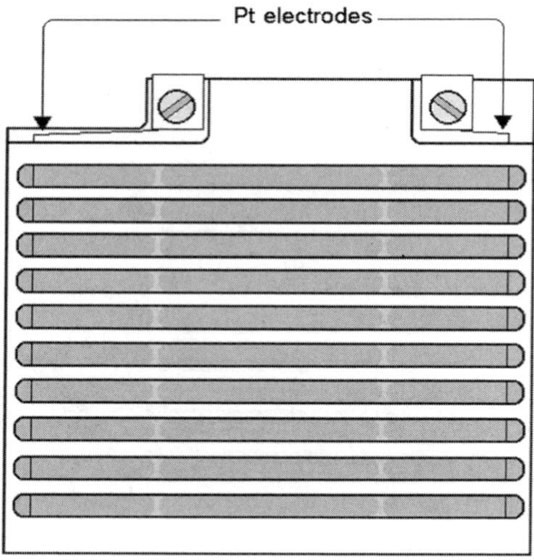

FIGURE 6.7. Pictorial representation of IPG gel strips rehydration tray for sample passive or active diffusion into IPG gel. Since the tray is provided with electrode contacts at the two ends, active sample loading can be implemented by applying a gentle voltage gradient during rehydration. (Courtesy of Bio-Rad and Proteome Systems Ltd.)

TABLE 6.1. Typical IEF Regime Using Multiphor II (Amersham-Pharmacia) with In-Gel Rehydration Loading for Samples Containing Minimal Salt (<50 mM)

Phase Number	Voltage (V)	Maximum Current[a] (mA)	Time (h)
1	150	5	0.1
2	300	3	0.1[b]
3	1000	3	0.5[b]
4	2500	1	0.5–1
5	5000	1	3–5

Source: Molloy, 2000.

[a]If the maximum current exceeds these limits, there is a danger of the IPG strip burning. As a precaution, the voltage should be decreased and the IEF time extended at the lower voltage.

[b]The time may be lengthened depending on the protein load and salt contamination.

locally, it would not be a problem since separation has already occurred; it must be borne in mind that the real danger of protein aggregation and precipitation, as typical of pulse loading, is the fact that these events occur among unlike proteins, so that ultimately highly heterogeneous proteins collect into a single precipitin zone, a process lethal to any separation operation). The third advantage of this protocol is that, due to dilution of proteins throughout the gel, many accelerated focusing protocols can be adopted, utilizing higher voltage gradients at the very start of the IEF/IPG run. Thus, with in-gel sample rehydration, it is rare for IEF to exceed a total of 30,000 V/h, even with milligram protein loads (Molloy, 2000). Table 6.1 gives the protocol adopted by Molloy (2000) for in-gel hydration loading procedures. The in-gel rehydration, as compared to sample cup loading, has been recently quantitatively evaluated by Zuo and Speicher (2000), who have reported that, at high sample loads, cup loading can result in as much as 50% sample loss, whereas in the case of in-gel rehydration, even when applying up to 0.5 mg protein, rarely sample losses of >15% to 20% are observed. In turn, these results have been challenged by Barry et al. (2003), who claim that, at least in the pH 6–11 range, cup loading is by far superior than active or passive rehydration methods, always producing the greatest number of detectable spots. Instrumental to good protein recoveries seems to be also the addition of 2% carrier ampholytes in the sample buffer, whereas use of thiourea did not significantly affect protein recoveries, although it did improve resolution in the final 2D gel. Interestingly, the same authors have demonstrated that another 10% sample loss is experienced during the two-step equilibration procedure prior to the second-dimension SDS–PAGE, as recommended by Goerg et al. (1988). Also, Kramer and Malone (2000) underline the importance of adding CAs to the IPG strip for improved protein entrance and recoveries. They suggest that best results are obtained with blends of CAs from different manufacturers.

3. *In-Gel Hydration Loading under Low Voltage.* This technique, which we call active sample loading, is distinguished from the previous one (called passive) in that during IPG strip rehydration a gentle voltage gradient (typically 50 V) is applied, typically for the entire duration of the reswelling (overnight). It is believed that this procedure would further facilitate sample entry. Olivieri et al. (2001) have found an

interesting aspect of such active sample loading: its ability to facilitate entry of high-M_r proteins. In passive rehydration, spectrins, large (280-kDa for α-chains) filamentous proteins, which represent the major constituents of the cytoskeletal network underlying the RBC plasma membrane and are typically associated with band 4.1 and actin to form the cytoskeletal superstructure of the RBC plasma membrane, are absent from the 2D maps but are present when adopting active rehydration. When Heegaard and Poglod (1991) reported 2D maps of RBCs, spectrins were highly visible as a string of bands penetrating for a short distance into the top (the most porous) part of the SDS–PAGE slab. Their work, though, was on CA–IEF, whereas the reports of Olivieri et al. (2001) and that of Rabilloud et al. (1999) are the only ones in which RBCs are analyzed by IPG–DALT, that is, by adopting immobilized pH gradients in the first dimension. In the late 1980s, Rabilloud et al. (1987b) reported the massive loss of all large proteins in Immobiline gels and found that even the free monomers contained in the bottles would precipitate large macromolecules out of solution. Later, the problem was found to be the autopolymerization of basic Immobilines fired by the radioactive cloud of Chernobyl to oligomers and *n-mers* able to crosslink and aggregate all large proteins (a ferritin precipitation test had been devised for detecting the presence of such homo-oligomers; Rabilloud et al., 1987a). The problem was finally solved by the introduction of the Immobiline II generation, in which *n*-propanol (an inhibitor of radical polymerization) was used as a solvent for alkaline Immobilines (Gåveby et al., 1988). Even though today IPGs should be trouble free, often large, filamentous proteins have difficulties in penetrating and migrating through the Immobiline strip for reasons not yet completely understood. At least the active rehydration method adopted by Olivieri et al. (2001) seems to partially alleviate this problem. With active loading, at least three major spectrin bands entered the gel, whereas nothing was visible in the case of passive sample entry.

4. *Soft and Thicker Gels for High-Capacity Loading.* With regular IPG strips (0.5 mm thick, 0.3 mm wide, different lengths from 7 to 18 cm), it is difficult to load more than 1–2 mg total protein. Although for an analytical gel this amount is considered to be a high load, it is often not enough if one aims at detecting and eluting, for further chemical characterization, a number of minor spots. In this last case, considerably higher sample loads are required, often greater by at least one order of magnitude. It appears to be possible to increase the sample load to 10 mg (and even higher) total protein per IPG strip provided the following modifications are followed. First the polyacrylamide matrix can be made "softer," that is, it can be cast more diluted. Typically, all commercial, ready-made IPG strips are made to contain $T = 4\%$ acrylamide (with various levels of crosslinker, usually $C = 2\%$ to $C = 5\%$) and are only 0.5 mm thick. Diluting the gel matrix greatly enhances the gel load ability, as demonstrated in the 1980s by Gelfi and Righetti (1983) and Righetti and Gelfi (1984). For example, when decreasing the $\%T$ from 6% to 3%, the protein load per unit of gel volume can be more than doubled; below $T = 3\%$, the sample load ability increases exponentially. It was in fact on the basis of these observations that preparative IPGs in gel matrices were abandoned and evolved into multicompartment electrolyzers with Immobiline membranes, where preparative separations of proteins occur in a liquid vein and the gel phase is reduced to an isoelectric membrane functioning as a titrating unit (Righetti et al., 1989c, 1990). Candiano et al. (2002) and Bruschi

et al. (2003) have in fact demonstrated that, by casting gels with a matrix content as low as $T = 3\%$, considerably higher protein loads and faster focusing times are ensured. The problem of entry of large proteins (>150 kDa) can also be alleviated by abandoning IPG gels in favor of agarose CA–IEF gels, as proposed by Oh-Ishi et al. (2000) and Hirabayashi (2000), with all the problems connected with carrier ampholyte focusing, of course. As a second modification, one can increase the gel thickness from 0.5 to 1 mm: This higher thickness also allows higher sample loads and is still compatible with second-dimension SDS–PAGE, which is routinely run in gels 1–1.5 mm thick. As a third modification, the width of the IPG strip can be increased from 3 mm (as available from commercial sources) to 6 mm, this also allowing doubling of the sample load. Note that this width is also compatible with proper stacking when using discontinuous SDS–PAGE, such as the Laemmli (1970) buffer. With these three simple modifications, it is common to be able to load 10 mg total proteins and even higher levels.

Some final comments are due in this section about two other aspects: IPG strip pretreatment before the run and a comparative study on the different instruments used for the first-dimension run.

1. *IPG Strip Prewashing.* A number of reports have lamented, in the past, horizontal and point streaking of proteins due to the first-dimension run. In a recent report, Chan et al. (1999) have claimed complete elimination of horizontal streaking of basic proteins if the commercial, dried IPG strips were washed in 100 mM ascorbic acid, pH 4.5, for 24 h before use and if the subsequent IEF run were performed at 40°C, instead of 15°C, as customary. Interestingly, this procedure of reduction in ascorbic acid had been proposed in 1989b by Righetti et al., who observed that, during gel polymerization in the presence of persulfate, the basic Immobilines would be oxidized to different extents, forming *N*-oxides. These oxides, in turn, would oxidize proteins migrating through the relevant gel zones (i.e., in the alkaline milieu). These authors had recommended reduction for only 45 min, whereas Chan et al. (1999) now propose as much as 24 h washing. Perhaps this elimination of horizontal streaking might be due to prevention of oxidation of reduced (but not blocked) –SH groups in proteins by the *N*-oxides in the Immobilines. Since re-formation of disulfide bridges is a time-dependent phenomenon, its avoidance might indeed reduce or abolish horizontal streaking. If this is the case, probably the novel procedure we have recommended (reduction and alkylation directly during the sample solubilization step, prior to any electrophoretic run) should take care of that, thus avoiding such a lengthy washing procedure.

2. *Use of Different IEF Chambers for IPG Run.* The performance of the three most popular chambers for the first-dimension IEF/IPG run (Multiphor II and IPGphor from Amersham-Pharmacia and the Protean IEF cell from Bio-Rad) has been recently evaluated by Choe and Lee (2000) using 18-cm-long IPG strips run under various conditions. The Multiphor II consistently resulted in the highest number of spots detected per gel, independent of sample type, width of the IPG interval, and method for calculating the number of spots. The Protean IEF cell had the next highest number of spots detected per gel. The IPGphor afforded good reproducibility in the total number of computer-detected spots, whereas the Protean IEF cell offered better reproducibility

in the total number of manually detected spots from gel to gel. As a final conclusion, Choe and Lee (2000), as a measure of quantitative reproducibility, suggested that the Protean IEF cell, which was the easiest instrument to use, performed better than the other instruments, although all three demonstrated good reproducibility in the experiments performed.

6.2.7. Sequential Sample Extraction

When a sample is highly complex or when extracting under widely different conditions, it might pay to apply a protocol of sequential extraction, as described by Molloy et al. (1998) in the case of total extracts of *E. coli*. This sequence is composed of three steps, as follows:

1. The cells are first ruptured in an Aminco French press with two presses at 14,000 psi. After removal of unbroken cells, the total cell lysate is diluted with ice-cold sodium carbonate, pH 11, and stirred slowly on ice for 1 h. This step brings only the most soluble proteins into the solution.
2. The precipitated, insoluble material after the carbonate treatment is recovered by centrifugation. The pellet is then resuspended in the classical reagent for 2D maps (8 M urea, 4% CHAPS, 100 mM DTT, and 0.5% alkaline carrier ampholytes). This step solubilizes more hydrophobic proteins.
3. After centrifugation, the remaining, insoluble pellet is further treated with the enhanced solubilizing solution, incorporating 5 M urea, 2 M thiourea, 2% ASB 14, and 2 mM TBP.

When the three resulting 2D maps were compared, it was seen that, except for spots appearing in more than one gel, the protein patterns were quite specific; in addition, the final gel displayed several abundant protein spots that were missing from the two previous ones. Subsequent identification confirmed these as integral outer membrane proteins which had not been detected previously in any 2D maps reported for *E. coli*, owing to their greater hydrophobicity and thus to their poor solubility with classical sample solubilization cocktails. This was despite the high copy number of some of the identified proteins (e.g., the *E. coli* integral outer membrane protein, OmpA, was present at 10^5 copies per cell) (Nikaido, 1996; Molloy et al., 1999). From this point of view, it is of interest to have an estimate of the copy number needed before a protein can be detected on a gel. This problem has been addressed by Herbert et al. (1997), who gave the following guidelines, as also visualized in Figure 6.8, which represents the relationship between the quantity of protein loaded onto a 2D gel and the final concentration of a 20-kDa protein present at 10, 10^3, and 10^5 copies per cell. On a double-log plot, these three protein levels are indicated by three diagonal lines correlating the two ordinates, both representing the on-gel concentration of said protein, the one to the left expressed in grams, the one to the right expressed in moles. Assuming that no prefractionation is performed, with a loading capacity of 10 mg per gel, only the proteins present at more than 10^5 copies per cell (17 pmol) can be easily identified using analytical procedures. With a highly sensitive silver stain, proteins

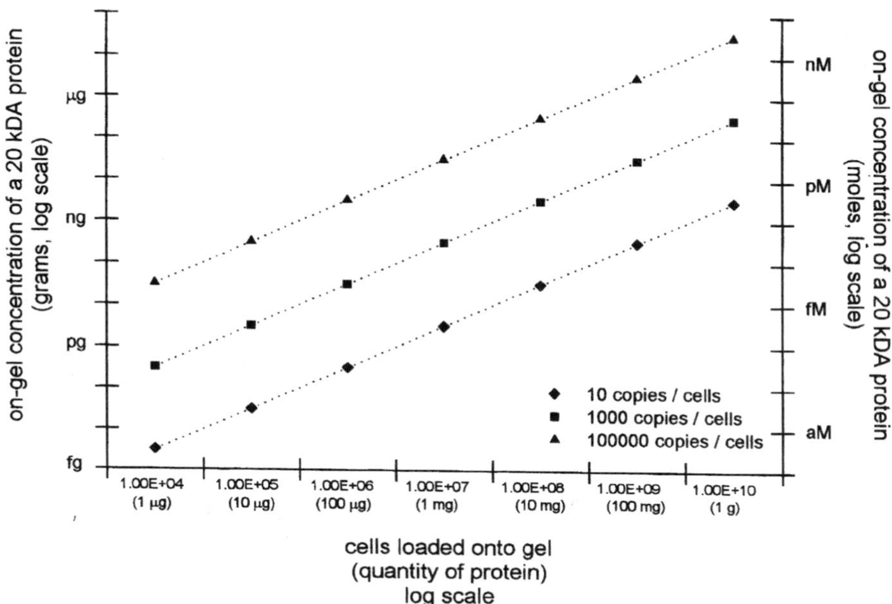

FIGURE 6.8. Final concentration of 20-kDa protein after 2D gel electrophoresis and how it varies with quantity of cells of proteins loaded onto gel and the copy number of protein per cell. This is based on 10^9 cells being 1 g fresh weight, 200 mg dry weight and containing 100 mg protein. (From Herbert et al., 1997, by permission.)

present at $>10^3$ copies per cell (17 pmol) will be detectable but will be difficult to analyze with analytical techniques. Immunoblotting, using a combination of high-affinity monoclonal antibodies and enhanced chemiluminescence, will allow detection of proteins present at >10 copies per cell (1.7 fmol). Therefore, the proportion of total protein complement that can be seen on a gel for any tissue will depend the copy number, the quantity of protein loaded on the gel, and the method of detection. For example, on a typical $160 \times 180 \times 1.5$-mm 2D PAGE gel, not more than 20% of the expressed genes of mammalian cells (assuming 5000 proteins corresponding probably to 10,000–15,000 polypeptides or isoforms) can be detected, representing only the tip of the proteome "iceberg" in this case (Celis et al., 1992). Since many of the low-copy-number proteins are likely to have important regulatory functions in cells, it is clear that many of them will go undetected and that the major challenge of today proteomics will be to find ways to make such rare proteins visible. Prefractionation seems to be definitely a major solution to this problem.

6.2.8. True Artifacts and Fata Morganas

In closing this section on 2D map analysis, we will briefly summarize the obscure field of artifacts, often invoked since the inception of 2D map analysis but never quite demonstrated until recent findings.

- *Carbamylation.* Carbamylation is widely quoted as being a problem in 2D gel analysis and the associated sample preparation steps. This modification occurs when isocyanate, a urea breakdown product, covalently modifies lysine residues, thus inducing a change in the pI. Urea is used at up to 9 M concentrations in sample preparation and 2D gels because of its ability to disrupt protein structure and effect denaturation without the need for ionic surfactants such as SDS. McCarthy et al. (2003) have studied carbamylation using 7 M urea and 2 M thiourea under a range of experimental temperatures to establish when and if it occurs and what can be done to minimize the modification. The actual time required for protein extraction from a tissue is usually short compared to the time required for procedures such as reduction and alkylation and IPG rehydration and focusing. Therefore, the temperature during these postextraction procedures is the most critical factor. Their experiments have shown that carbamylation simply does not occur during electrophoresis in the presence of urea, even with prolonged run times. However, under poorly controlled sample preparation and storage conditions it can become a major event.
- *Deamidation.* Hydrolysis of asparagines and glutamine residues during the IEF/IPG process has often been claimed as one of the major sources for the strings of spots often seen in acidic proteins, especially sera. The only report hinting at such deamidation events came from Sariouglu et al. (2000). These authors generated 2D maps of human plasma and, upon analysis of tryptic fragments of trains of spots, found the 1-Da difference typical of Asn → Asp and Gln → Glu transitions. Yet, their conclusions were that such processes occurred in vivo, during the lifetime of plasma proteins, not in vitro, during sample analysis. In addition, such extensive trains of spots could only be found in sera, not in other human tissues (e.g., liver). There is ample evidence that their conclusions could be right: In an extensive investigation on the heterogeneity of maize zeins (storage proteins with a very high content of Gln and Asn residues, up to 25% of total amino acid residues), in order to extensively deamidate such residues, Righetti et al. (1977) reported incubation times of 3 weeks in borate buffer, pH 9.5, at 55°C.
- *Oligomerization.* This problem has recently been found to generate true artifacts, arising from regeneration of disulfide bridges between reduced but not alkylated polypeptide chains during IEF at alkaline pH values (Herbert et al., 2001). As stated above, this is amply cured by properly alkylating with iodoacetamide, acrylamide, or, better, 2VP and 4VP. Although these procedures have been recently challenged by Luche et al. (2004), these authors did not seem to be aware of the work of Sebastiano et al. (2003) on VPs or the work of Mineki et al. (2002), who reported 97% alkylation of SH groups in BSA with high levels of acrylamide. With the latter compound, Luche et al. (2004) report spurious alkylation on Lys residues, but they make the fundamental mistake of not removing the excess alkylant during the IPG run (where alkylation will continue undisturbed!).
- β-*Elimination (or Desulfuration).* This is the only true noxious and unexpected artifact in proteome analysis (Herbert et al., 2003). It results on the loss of an

H_2S group (34 Da) from Cys residues for proteins focusing in the alkaline pH region. Upon such an elimination event, a dehydro alanine residue is generated at the Cys site. In turn, the presence of a double bond in this position elicits lysis of the peptide bond, generating a number of peptides of fairly large size from an intact protein. The first process seems to be favored by the electric field, probably due to the continuous harvesting of the SH^- anion produced. The only remedy found to this noxious degradation pathway is the reduction and alkylation of all Cys residues prior to their exposure to the electric field. Alkylation appears to substantially reduce both β-elimination and the subsequent amido bond lysis.

6.3. PREFRACTIONATION TOOLS IN PROTEOME ANALYSIS

Although both the first (IEF/IPG) and second (SDS–PAGE) separation dimensions appear to be performing at their best, there remains the fact that, due to the vast number of polypeptide chains present in a tissue lysate, especially at acidic pH values (pH 4–6), resolution is still not enough. An illuminating example comes from some recent articles on the plasma proteome (Anderson and Anderson, 2002; Pieper et al., 2003). If one assumes that there are just 500 true "plasma proteins", each present in 20 variously glycosylated forms and in five different sizes, one would end up with 50,000 molecular forms. If one further hypothesizes that there exist 50,000 gene products in the human proteome, each having on average 10 splice variants, cleavage products, and posttraslational modifications, this would yield a further 500,000 protein forms. Finally, one might consider that the immunoglobulin class might contain perhaps 10,000,000 different sequences. Adding all of this up, one can imagine the immense complexity of just the human plasma proteome, further complicated by the fact that its dynamic range might be more than 10 orders of magnitude (Anderson and Anderson, 2002). This view seems to be shared by Rammensee (2004), simply based on the idea of protein splicing. By slicing a protein into pieces, stitching different portions together, and then cutting out strings of nine sequential amino acids from the melded pieces, cells can manufacture a tremendous variety of new species from the original proteins coded for by the some 30,000 genes present in humans.

Some recent papers suggest that the problem of this vast complexity could be overcome by running a series of narrow-range IPG strips (covering no more than 1 pH unit, "zoom gels"; Hoving et al., 2000; Westbrook et al., 2001; note that "zoom" and "ultrazoom" gels are quite important for avoiding or at least minimizing the problem of spot overlap, an ever-present hazard in 2D maps; see Pietrogrande et al., 2002, 2003) on large-size gels (18 cm or more in the first dimension, large-format slabs in the second dimension), which would dramatically increase the resolution (Corthals et al., 2000; Wildgruber et al., 2000). The entire, wide-range 2D map would then be electronically reconstructed by stitching together the narrow-range maps. This might turn out to be a Fata Morgana, though: There remains the fact that, even when using very narrow IPG strips, they have to be loaded with the entire cell lysate, containing proteins focusing all along the pH scale. Massive precipitation will then ensue due to aggregation among unlike proteins, with the additional drawback that the proteins which should focus in the chosen narrow-range IPG interval will be strongly underrepresented, since they will be only a small fraction of the entire sample loaded. That this is a serious problem has been debated in a recent work by Gygi et al. (2000). These authors analyzed a yeast lysate by loading 0.5 mg total protein on a narrow-range (nr) IPG, pH 4.9–5.7. Although they could visualize by silvering approximately 1500 spots, they lamented that a large number of polypeptides simply did not appear in such a 2D map. In particular, proteins from genes with codon bias values of <0.1 (low-abundance proteins) were not found, even though fully one-half of all yeast genes fall into that range. The codon bias value for a gene is its propensity to use only one of several codons to incorporate a specific amino acid into the polypeptide chain. It is known that highly expressed proteins have large codon bias values (>0.2). Thus, these authors conclude that, when analyzing protein spots from 2D maps by MS, only generally abundant proteins (codon bias >0.2) can be properly identified. Thus, the number of spots on a 2D gel is not representative of the overall number or classes of expressed genes that can be analyzed. Gygi et al. (2000) have calculated that, when loading only 40 μg total yeast lysate, as done in the early days of 2D mapping, only polypeptides with an abundance of at least 51,000 copies/cell could be detected. With 0.5 mg of starting protein, proteins present at 1000 copies/cell could now be visualized by silvering, but those present at 100 and 10 copies/cell could not be detected. These authors thus concluded that the large range of protein expression levels limits the ability of the 2D MS approach to analyze proteins of medium to low abundance, and thus the potential of this technique for proteome analysis is likewise limited. As a corollary, the construction of complete, quantitative protein maps based on this approach will be very challenging, even for relatively simple, unicellular organisms.

Due to such major drawbacks, we believe that the only way out of this impasse will be prefractionation. At present, two approaches have been described: chromatographic and electrophoretic. They will be discussed below.

6.3.1. Sample Prefractionation via Different Chromatographic Approaches

Fountoulakis's group has extensively developed this approach. In a first procedure, Fountoulakis et al. (1997) and Fountoulakis and Takàcs (1998) adopted affinity chromatography on heparin gels as a prefractionation step for enriching certain protein fractions in the bacterium *Haemophilus influenzae*. This gram-negative bacterium is of pharmaceutical interest and has been recently sequenced; its complete genome has been found to comprise about 1740 open reading frames, although not more than 100 proteins had been characterized by 2D map analysis. Heparin is a highly sulfated glucosaminoglycan with affinity for a broad range of proteins, such as nucleic acid–binding proteins and growth and protein synthesis factors. On account of its sulfate groups, heparin also functions as a high-capacity cation exchanger. Thus, prefractionation on a heparin–Actigel was deemed suitable for enriching low-copy-number gene products. In fact, about 160 cytosolic proteins bound with different affinities to the heparin matrix and were thus highly enriched prior to 2D PAGE separation.

As a result, more than 110 new protein spots detected in the heparin fraction were identified, thus increasing the total identified proteins of *H. influenzae* to more than 230. In a second approach (Fountoulakis et al., 1998), the same lysate of *H. influenzae* was prefractionated by chromatofocusing on Polybuffer Exchanger. Approximately 125 proteins were identified in the eluate collected from the column. Seventy of these were for the first time identified after chromatography on the Polybuffer Exchanger, the majority of them being low-abundance enzymes with various functions. Thus, with this additional step, a total of 300 proteins could be identified in *H. influenzae* by 2D map analysis, out of a total of about 600 spots visualized on such maps from the soluble fraction of this microorganism. In yet another approach, the cytosolic soluble proteins of *H. influenzae* were prefractionated by Fountoulakis et al. (1999) by hydrophobic interaction chromatography (HIC) on a TSK phenyl column. The eluate was subsequently analyzed by 2D mapping followed by spot characterization by MALDI–TOF MS. Approximately 150 proteins bound to the column were identified, but only 30 for the first time. In addition, most of the proteins enriched by HIC were represented by major spots, so that enrichment of low-copy-number gene products was only modest. In total, with all the various chromatographic steps adopted, the number of proteins so far identified could be increased to 350.

The same heparin chromatography procedure was subsequently applied by Karlsson et al. (1999) to the prefractionation of human fetal brain soluble proteins. Approximately 300 proteins were analyzed, representing 70 different polypeptides, 50 of which were bound to the heparin matrix. Eighteen brain proteins were identified for the first time. The polypeptides enriched by heparin chromatography included both minor and major components of the brain extract. The enriched proteins belonged to several classes, including proteosome components, dihydropyrimidinase-related proteins, T-complex protein 1 components, and enzymes with various catalytic activities.

In yet another variant, Fountoulakis et al. (2000) reported enrichment of low-abundance proteins of *E. coli* by hydroxyapatite chromatography. The complete genome of *E. coli* has now been sequenced and its proteome analyzed by 2D mapping. To date, 223 unique loci have been identified and 201 protein entries were found in the SWISS-PROT 2D PAGE on ExPASy server using the Sequence Retrieval System query tool (http://www.expasy.ch/www/sitemap.html). Of the 4289 possible gene products of *E. coli*, about 1200 spots could be counted on a typical 2D map when $\sim$2 mg total protein was applied. Possibly, most of the remaining proteins were not expressed in sufficient amounts to be visualized following staining with Coomassie blue. Thus, it was felt necessary to perform a prefractionation step, this time on hydroxyapatite beads. By this procedure, approximately 800 spots corresponding to 296 different proteins were identified in the hydroxyapatite eluate. About 130 new proteins that had not been detected in 2D gels of the total extract were identified for the first time. This chromatographic step, though, enriched both low-abundance as well as major components of the *E. coli* extract. In particular, it enriched many low-M_r proteins, such as cold-shock proteins. Figure 6.9 displays 2D maps from pools 19 and 21 of fractions collected from the hydroxyapatite prefractionation of an *E. coli* total lysate. On a similar line of thinking, Harrington et al. (1992) reported the use of cation exchange chromatography for enriching DNA-binding proteins on tissue

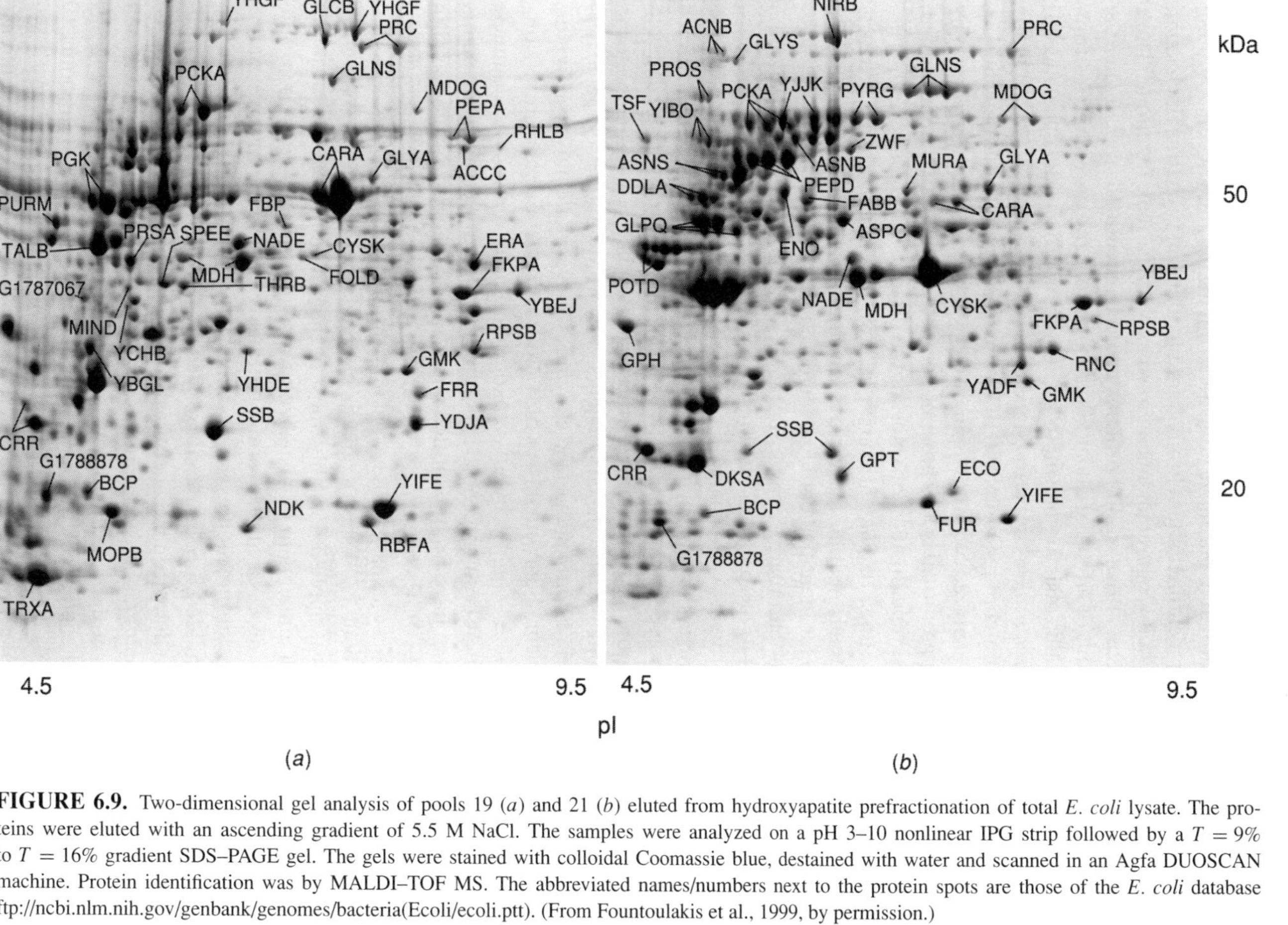

FIGURE 6.9. Two-dimensional gel analysis of pools 19 (*a*) and 21 (*b*) eluted from hydroxyapatite prefractionation of total *E. coli* lysate. The proteins were eluted with an ascending gradient of 5.5 M NaCl. The samples were analyzed on a pH 3–10 nonlinear IPG strip followed by a $T = 9\%$ to $T = 16\%$ gradient SDS–PAGE gel. The gels were stained with colloidal Coomassie blue, destained with water and scanned in an Agfa DUOSCAN machine. Protein identification was by MALDI–TOF MS. The abbreviated names/numbers next to the protein spots are those of the *E. coli* database ftp://ncbi.nlm.nih.gov/genbank/genomes/bacteria(Ecoli/ecoli.ptt). (From Fountoulakis et al., 1999, by permission.)

extracts prior to 2D analysis. Here, too, since the basic domains do not seem to influence the overall pI of such proteins, one sees a general spread of the prefractionated sample across the complete pI range of a 2D map.

More recently, Krapfenbauer et al. (2003) have adopted a double-prefractionation scheme: ultracentrifugation for collection of separate cytosolic, mitochondrial, and microsomal fractions followed by additional fractionation of the cytosolic fraction via ion exchange chromatography. They could thus detect and identify some 437 rat brain proteins, about double what had been previously cataloged.

Although the work presented by Fountoulakis's group is impressive and truly innovative in proteome analysis, we believe that, nevertheless, there are some inherent drawbacks to this approach. We list some of them here:

- In general, huge amounts of salts are needed for elution from chromatographic columns, up to 2.5 M NaCl.
- As a consequence, loss of entire groups of proteins could ensue during the various manipulations (dialysis of large salt amounts, concentration of highly diluted pools of eluted fractions).
- The eluted fractions do not represent narrow pI cuts but in general are constituted by proteins having pI values in the pH 3–10 range. Thus, one should still use wide pH gradients in analyses.

In particular, it is known that high salt amounts could induce irreversible adsorption of proteins to the resin used for fractionation, resulting in irreversible loss of spots during the subsequent 2D mapping. When eluting from ion exchange columns, moreover, not only will the collected fractions be quite dilute, necessitating a concentration step, but also they will contain huge amounts of salt, rendering them incompatible with the IEF/IPG first-dimension step. Thus, although the approach by Fountoulakis's group could be quite attractive for analysis of some protein fractions in a total cell lysate, it would clearly not work for exploring the entire proteome due to all the potential losses described above. We discuss below a number of electrophoretic prefractionation steps.

6.3.2. Sample Prefractionation via Different Electrophoretic Techniques

Unlike its chromatographic counterpart, preparative electrophoretic techniques are scarcely used today (Righetti et al., 1992), due to the cumbersome instrumental setup and the limited load ability. This limitation stems from the very principle of electrophoresis: Electrokinetic methodologies are quite powerful, in fact much too powerful, and this comes with a steep price. This extraordinary power (call it wattage) has to be dissipated from the separation chamber, thus seriously limiting any attempt at scaling-up, such as increasing the thickness of a gel slab or the diameter of a column. However, all focusing techniques do not suffer from these limitations: in conventional IEF in soluble CAs, the typical conductivity at the steady state is about two orders of magnitude lower than in ordinary buffers (Righetti, 1980); in turn, immobilized pH gradients offer a further decrement of conductivity of at least another order of magnitude as compared to CA–IEF (Mosher et al., 1986). Considering, in addition, that proteins fractionated by any IEF/IPG preparative protocol are both isoelectric and isoionic (in fact, fully desalted), their interfacing with the first dimension of a 2D map is a straightforward procedure.

6.3.2.1. Rotationally Stabilized Focusing Apparatus: Rotofor There is a long history behind this remarkable invention by M. Bier going back to the doctoral thesis of Hjertén (1967): As he was trained in astrophysics, Hjertén's apparatus was a "Copernican revolution" in electrokinetic methodologies. In fact, Hjertén was, the first to propose electrophoretic separations in the free zone (i.e., in the absence of anticonvective, capillary media such as polyacrylamide and agarose gels), but he had to fight the noxious phenomenon of electrodecantation induced by gravity. He thus devised rotation of the narrow-bore tubes used as electrophoretic chambers around a horizontal axis, mimicking celestial planetary motion! This must have spurred the fantasy of Svensson-Rilbe, in those days a colleague at Uppsala University, who finally described a mammoth multicompartment electrolyzer capable of fractionating proteins in the gram range (Jonsson and Rilbe, 1980). The cell was assembled from 46 compartments, accommodating a total sample volume of 7.6 L, having a total length of 1 m—hardly user friendly! Cooling and stirring were affected by slow rotation of the whole apparatus in a tank filled with cold water. Bier's 50-year-long love affair with preparative electrophoresis in free solution produced, as a last evolutionary step, a remarkable gadget, dubbed the Rotofor (Bier, 1998): It was happiness for all parties involved on both sides of the counter, users and producers alike. Users found it quite simple to operate, and the producer, Bio-Rad, sold over $10 million worth of instruments. A preparative-scale Rotofor is capable of being loaded with up to 1 g of protein in a total volume of up to 55 mL. A mini-Rotofor with a reduced volume of about 18 mL is also available. A typical instrumental setup is shown in Figure 6.10. It is assembled from 20 sample chambers separated by liquid-permeable nylon screens, except at the extremities, where cation and anion exchange membranes are placed against the anodic and cathodic compartments, respectively, so as to prevent diffusion within the sample chambers of noxious electrode products. At the end of the preparative run, the 20 focused fractions are collected simultaneously by piercing a septum at the chamber's bottom via 20 needles connected to a vacuum source. The narrow-pI-range fractions can then be used to generated conventional 2D maps. This is the original approach described by Hochstrasser et al. (1991). However, in recent times, this methodology has taken another, unexpected turn: The Rotofor is used directly as the first dimension of a peculiar 2D methodology in which each fraction is further analyzed by hydrophobic interaction chromatography using nonporous reverse-phase high-performance liquid chromatography (HPLC) (Zhu et al., 2003). Each peak collected from the HPLC column is then digested with trypsin and subjected to MALDI–TOF MS analysis and MSFit database searching. By this approach, Wall et al. (2000) have been able to resolve a total of about 700 bands from a human erythroleukemia cell line. It should be stated, though, that the pI accuracy of this methodology (which is still based on conventional CA–IEF) is quite poor: It ranges from ±0.65 to ±1.73 pI units, a large error indeed. In a similar line of thinking, Davidsson et al. (2001) have subfractionated human cerebrospinal fluid and brain

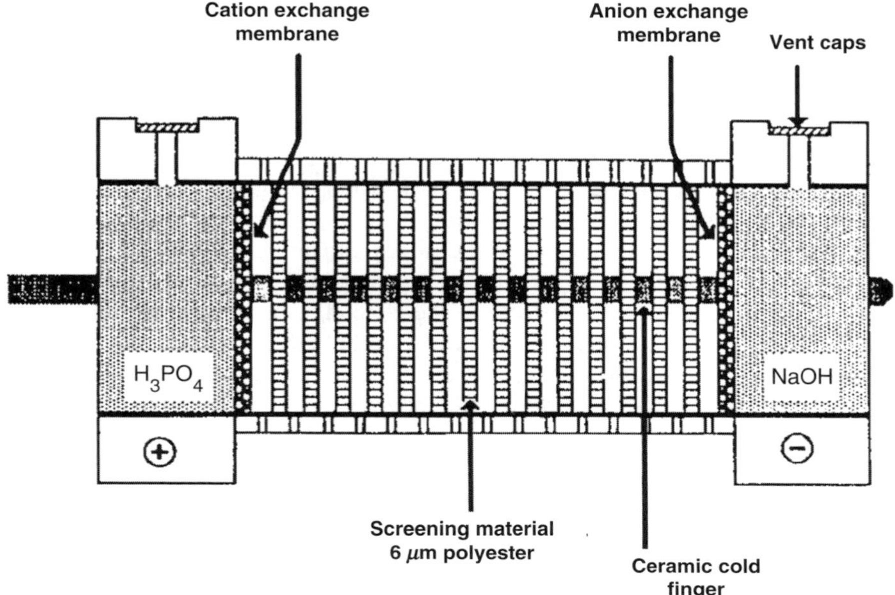

FIGURE 6.10. Schematic presentation of Rotofor instrument. Rotation and the screen partitioning are essential for good separations. (From Bier, 1998, by permission.)

tissue, whereas Wang et al. (2002) have mapped the proteome of ovarian carcinoma cells. More recently, Xiao et al. (2004) have reported an unexepected application of the Rotofor, not only for the fractionation of intact proteins in the presence of carrier ampholytes but also for the fractionation of peptide digests of an entire proteome (in this case, human serum) in an ampholyte-free environment. The peptides themselves would act as CA buffers and create a pH gradient via an "autofocusing" process (with a caveat, though: The pH gradient will be quite poor, since only a few peptides have good buffering power and conductivity in the pH 5–8 range).

6.3.2.2. Continuous Free-Flow Isoelectric Focusing This liquid-based IEF technique (FF-IEF) relies on the principle of free-flow electrophoresis (FFE) described by Hannig (1982). In FF-IEF, samples are continuously injected into a carrier ampholyte solution flowing as a thin film (0.4 mm thick) between two parallel plates, and by introducing an electric field perpendicular to the flow direction, proteins are separated by IEF according to their different pI values and finally collected into up to 96 fractions. Key advantages of the method are improved sample recovery (due to absence of gel media or membranous material) and sample loading capacity (due to continuous sample feeding; this step is not rate limiting). The fundamental principles of FFE, including commercial instrumentation, have been described in a recent review (Krivankova and Bocek, 1998). It is of interest to recall here that perhaps one of the first applications of continuous-flow (CF) IEF for preparative protein fractionation came from Fawcett (1973), who, however, devised a vertical, CF chamber filled with a Sephadex bed as an anticonvective medium. In one version, Gianazza et al. (1975) fractionated synthetic, wide-range carrier ampholytes in a chamber with just 12 outlets at the bottom as sample collection ports. The system could be run for weeks almost unattended by arranging a continuous sample input fed by constant hydrostatic pressure via a Mariotte flask. The Hannig apparatus went through successive designs and improvements, from an original liquid descending curtain to the present commercial version, dubbed Octopus, exploiting an upward liquid stream (Kuhn and Wagner, 1989). An example of this modern FFE version is shown in Figure 6.11. Instrumental to the success of the method, especially when run for long periods of time, is the constancy of the elution profile at the collection port, so that the same protein species is always collected into the same test tube. Thus, electroendoosmotic flow should be suppressed via a number of ways, including glass-wall silanol deactivation and addition of polymers (e.g., 0.1% hydroxypropylmethyl cellulose),

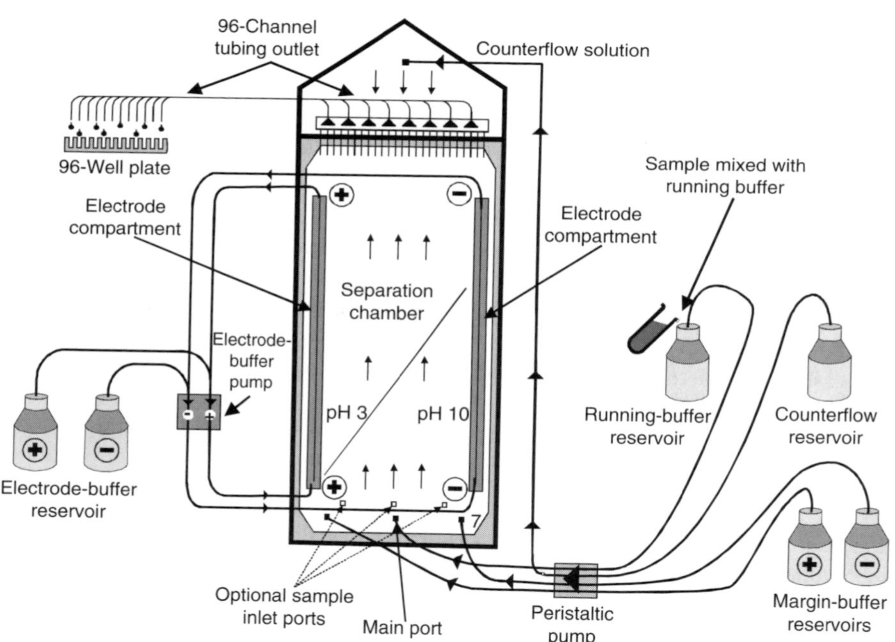

FIGURE 6.11. Schematic representation of continuous FFE apparatus (Octopus). Separation chamber dimensions: 500 × 100 × 0.4 mm); electrode length: 500 mm (electrical contact with separation area is made via a polyacetate PP60 membrane); distance between electrodes: 100 mm; chamber depth: 0.4 mm, created by insertion of polyvinyl chloride papers between separation plates. Focused protein samples are collected into 96-well plates via an in-line multichannel outlet. The volume of each fraction is typically ~2 mL. (Courtesy of Dr. H. Wagner.)

providing dynamic wall coating and proper liquid viscosity. The first report on the use of CF-IEF for prefractionation of total cell lysates (HeLa and Ht1080 cell lines), in view of a subsequent 2D map, is perhaps the one of Burggraf et al. (1995), who collected individual or pooled fractions for further 2D analysis. Theoretical modeling of the process was later presented by Soulet et al. (1998). In a variant of the above approach, Hoffmann et al. (2001) have proposed CF-IEF as the first dimension of a 2D map, the eluted fractions being directly analyzed by orthogonal SDS–PAGE. In turn, individual bands in the second SDS dimension were eluted and analyzed by ESI–IT–MS. By this approach, they could identify a number of cytosolic proteins of a human colon carcinoma cell line. One advantage of CF-IEF is immediately evident from Hoffman et al.'s data: Large proteins (e.g, vinculin, $M_r = 116.6$ kDa) could be well recovered and easily identified; on the contrary, recovery of large-M_r species has always been problematic in IPG gels.

6.3.2.3. Continuous Free-Flow Electrophoresis The original Hanning (1982) apparatus was, in fact, designed for electrophoresis in the zonal (ZE) mode and for separation of intact cells and organelles, which would have very low diffusion coefficients in the ZE mode. This approach has been recently applied to improved proteome analysis via continuous free-flow purification of mitochondria from *S. cerevisiae* (Zischka et al., 2003). These authors reported a considerable increment in purity of the mitochondria via ZE-FFE as compared to their isolation via differential centrifugation. Moreover, there was an extra benefit: 2D maps of proteins extracted from ZE-FFE purified mitochondria exhibited a much higher level of undegraded proteins. On the contrary, about 49% of the proteins from crude mitochondria, as isolated via centrifugation, were found to be degraded or truncated. In recent times, the ZE-FFE instrument has been miniaturized by Kobayashi et al. (2003). Their separation chamber is barely 66×70 mm in size, with a gap between the two Pyrex glass plates of only 30 μm. The liquid curtain and the sample are continuously injected from five and one holes, respectivley, at the top (see Fig. 6.12*a*). At the bottom of the chamber, a micromodule fraction collector consisting of 19 stainless steel tubes is conneted perpendicular to the chamber and liquid stream (see Fig. 6.12*b*). It is anticipated that this apparatus could be quite valuable for prefractionation in proteome analysis.

6.3.2.4. Gradiflow The Gradiflow is a multifunctional electrokinetic membrane apparatus that can process and purify protein solutions based on differences of mobility, pI, and size (Margolis et al., 1995; Horvàth et al., 1996). The general operating scheme of the Gradiflow is shown in Figure 6.13*a*, whereas Figure 6.13*b* shows how to operate it for collecting protein fractions within an approximate pI range (Locke et al., 2002). Its interfacing with 2D map analysis was demonstrated by Corthals et al. (1997), who adapted this instrument for prefractionation of native human serum and enrichment of protein fractions. In a more recent report (Locke et al., 2002), this device was also shown to be compatible, in the prefractionation of baker's yeast and Chinese snow pea seed total cellular extracts, with the classical denaturing/solubilizing

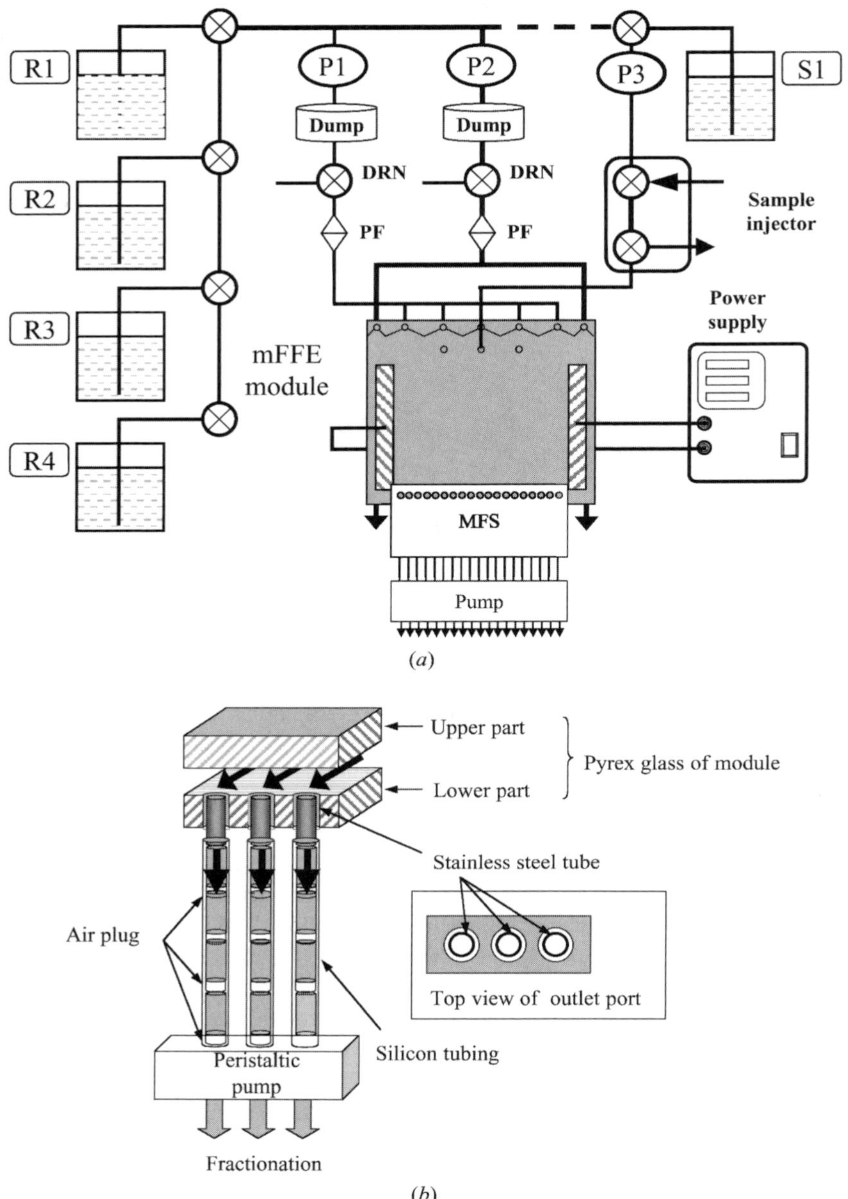

FIGURE 6.12. (*a*) Scheme of miniaturized ZE-FFE apparatus. Reservoir R1 is used for the electrophoresis buffer, whereas R2 (0.1 M aqueous NaOH–ethanol, 50 : 50 v/v), R3 (0.01 M HCl), and R4 (80% aqueous ethanol) are used for washing cycles. S1 is the sample reservoir, P1–P3 are peristaltic pumps for pouring the solutions into the separation chamber, the electrode reservoirs, and the sample port, respectively. D is a dumper, DRN a drain duct, PF a fuse unit for excessive pressure. (*b*) Scheme of micromodule fraction separator (MFS). A cross section of the separation chamber's bottom with upper and lower parts of Pyrex glass is shown. (From Kobayashi et al., 2003, by permission.)

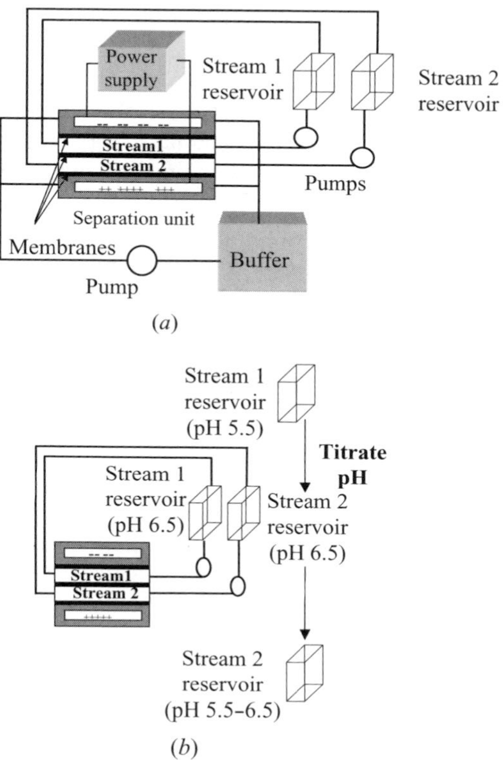

FIGURE 6.13. (*a*) Schematic diagram of Gradiflow. Ice-cold buffer circulates around streams 1 and 2 to cool the sample as it passes through the separation unit. The streams are separated from the buffer and from each other by the cartridge configuration of three membranes. The middle membrane has a larger pore size to permit protein permeation between streams 1 and 2. (*b*) Example of isolation of 1-pH-unit fraction. The stream 1 reservoir from the initial separation at pH 5.5 contains proteins with pI >5.5. This sample is collected, and the pH is adjusted to 6.5 and loaded into stream 2 for the second fractionation. The second step separates proteins at pH 6.5; proteins with pI >6.5 will move into stream 1 and proteins with pI <6.5 remain in stream 2. The final sample in the stream 2 reservoir contains proteins with pI values approximating the interval pH 5.5–6.5. (From Locke et al., 2002, by permission.)

solutions of 2D maps, comprising urea/thiourea and surfactants. Whereas in the case of size fractionation this can be achieved with polyacrylamide-coated membranes at different %*T* and %*C* for sieving of macromolecules in given M_r ranges, its use for separating approximate pI fractions is more complex and can be better visualized by referring to Figure 6.13*b*. In this example, the goal is to collect proteins within a pH 5.5–6.5 range; this would need two sequential Gradiflow manipulations. In the first separation, the sample is buffered at pH 5.5 and the resulting stream 1 fraction collected. In turn, the pH of this faction is adjusted to 6.5 and subjected to a second processing in stream 2. The fraction collected from this last stream (thus processed a second time at altered pH) will represent proteins with approximate pI values between pH 5.5 and 6.5. Instrumental to this setup is the use of low-conductivity buffers, such as those devised by Bier et al. (1984), so as to allow reasonably high voltage gradients and relatively short separation times. This technique, though, has been described only at the methodological level and has not proven its worth, as yet, in "mining below the tip of the iceberg" for finding low-abundance and membrane proteins.

6.3.2.5. Sample Prefractionation via Multicompartment Electrolyzers with Immobiline Membranes This kind of equipment represents, perhaps, the most advanced evolutionary step in all preparative electrokinetic fractionation processes (Righetti et al., 1989c, 1990, 1992). This method relies on isoelectric membranes fabricated with the same Immobiline chemicals adopted in IPG fractionations (Righetti, 1990). The advantages of such a procedure are immediately apparent:

- It offers a method that is fully compatible with the subsequent first-dimension separation, a focusing step based on the Immobiline technology. Such a prefractionation protocol is precisely based on the same concept of immobilized pH gradients, and thus protein mixtures harvested from the various chambers of this apparatus can be loaded onto IPG strips without any need for further treatment, in that they are isoelectric and devoid of any nonamphoteric ionic contaminant.
- It permits harvesting a population of proteins having pI values precisely matching the pH gradient of any narrow (or wider) IPG strip.
- As a corollary of the above point, much reduced chances of protein precipitation will occur. In fact, when an entire cell lysate is analyzed in a wide gradient, there are fewer risks of protein precipitation; on the contrary, when the same mixture is analyzed in a narrow gradient, massive precipitation of all non-isoelectric proteins could occur, with a strong risk of coprecipitation of proteins that would otherwise focus in the narrow-pH interval.
- Due to the fact that only the proteins cofocusing in the same IPG interval will be present, much higher sample loads can be operative, permitting detection of low-abundance proteins.
- Finally, in samples containing extreme ranges in protein concentrations (such as human serum, where a single protein, albumin, represents >60% of the total species), one could assemble an isoelectric trap narrow enough to just eliminate the unwanted protein from the entire complex, this too permitting much higher sample loads without interference from the most abundant species.

The original apparatus, as miniaturized by Herbert and Righetti (2000), Righetti et al. (2001a), and Herbert et al. (2004), is shown schematically in Figure 6.14*a*. In this exploded view, two terminal electrode chambers are used to block, in between, three sample chambers. Figure 6.14*b* shows the missing elements between each chamber: the isoelectric, buffering membranes used to define, in each chamber, a given pI

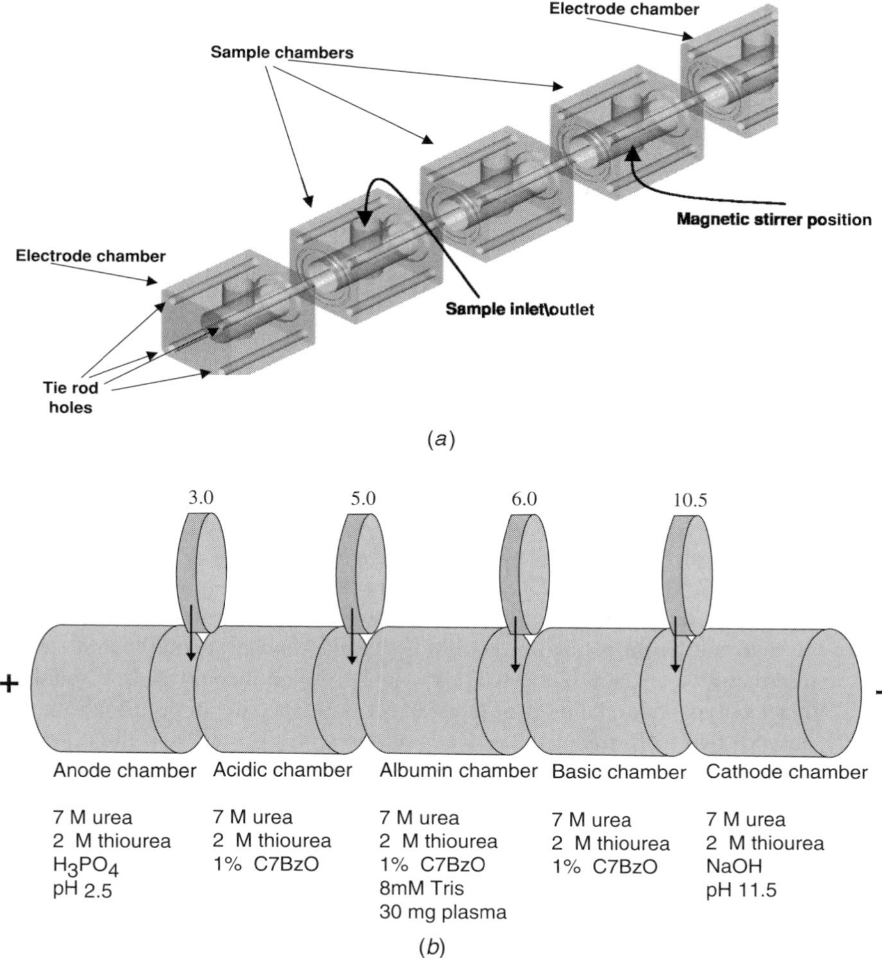

FIGURE 6.14. (*a*) Exploded view of miniaturized multicompartment electrolyzer operating with isoelectric membranes. An assembly with only five chambers is shown. (*b*) Assembly of MCE with five chambers and four membranes having pI selected for purification of human serum. Note that the pI 5.0 and 6.0 membranes have been chosen for capturing albumin within this chamber. (Courtesy of B. Herbert and P. G. Righetti, unpublished.)

interval into which proteins of intermediate pI values will be trapped by a titration process. In this case, a setup made of five chambers is divided by four membranes, having pI values of 3.0, 5.0, 6.0, and 10.5, selected so as to trap albumin, the most abundant protein in human serum, in a central, narrow-pI chamber. This permits concentration of other fractions and detection of many more dilute species in other regions of the pH scale. By properly exploiting this prefractionation device, Pedersen et al. (2003) have been able to capture and detect much more of the "unseen" yeast membrane proteome. Here is a summary of their results:

- 780 protein isoforms identified,
- 323 unique proteins (genes),
- 105 integral membrane proteins (33%),
- 54 membrane-associated proteins (17%),
- 159 total membrane-associated proteins (50%), and
- 90 proteins with codon bias index (CBI) <0.2 (27%).

The importance of some of these findings is here highlighted: integral membrane proteins are rarely seen in 2D maps; proteins with CBI < 0.2 represent low-abundance proteins and are scarcely detected in 2D maps unless enriched by some prefractionation protocol. The power of this methodology, which not only relies on prefractionation steps but also deals with the analysis of intact proteins rather than proteolytic digests, as customary in most protocols exploiting coupled chromatographic processes (see below), is illustrated in Figure 6.15. In yeast, there exist two forms of NADH–cytochrome *b*5 reductase, one called p34 (pI $= 8.7$, $M_r = 34$ kDa) and the other p32 (pI $= 7.8$, $M_r = 30$ kDa). The first species is an integral outer membrane protein which mediates the reduction of cytochrome *b*5; the second one is derived from the first one by an in vivo proteolytic cleavage, and it is soluble and resides in the intermembrane space. The MCE enrichment protocol coupled to 2D analysis of all intact polypeptide chains is able to detect both species. In addition, one can easily observe two isoforms of the p34 chain, one with pI $= 8.7$ and the other one more alkaline, pI $= 9.1$ (Pedersen et al. 2003; Righetti et al., 2003). Chromatographic techniques which rely on the presence of just one (or a few) peptides in a total cell digest would not have been able to detect any of these biologically relevant forms!

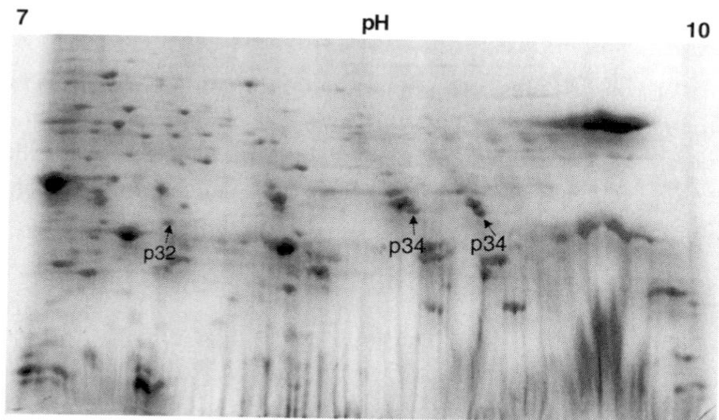

FIGURE 6.15. Example of prefractionation/enrichment of yeast membrane proteins and detection of isoforms and truncated forms of p34 NADH–cytochrome *b*5 reductase. (From Pedersen et al., 2003, by permission.)

6.3.2.6. Off-Gel IEF in Multicompartment Devices This is the latest evolution in preparative techniques based on contact with IPG matrices (or membranes), as described in Section 6.3.4. Just like the MCE technique, off-gel electrophoresis has been devised for separation of proteins according to their pI and for their direct recovery in solution without adding buffers or ampholytes (Ros et al., 2002). The principle is to place a sample in a liquid chamber which is positioned on top of an IPG gel. The gel buffers a thin layer of the solution in the liquid chamber and the proteins are charged according to their pI values and to the pH imposed by the gel. Theoretical calculations and modeling have shown that the protonation of an ampholyte occurs in the thin layer of solvation closed to the IPG gel–solution interface (Arnaud et al., 2002). Upon application of a voltage gradient, perpendicular to the liquid chamber, the electric field penetrates into the channel and extracts all charged species (those having pI values above and below the pH of the IPG gel), thus vacating them from the sample cup. After separation, only the globally neutral species (pI = pH of the IPG gel) remain in solution. This technique offers a high separation efficiency and allows easy recovery of the purified compounds directly in the liquid phase. In a further extension of this initial work, the off-gel electrophoresis format was improved and adapted to a multiwell device composed of a series of compartments of small volume (100–300 μL) and compatible with the instruments that are used in modern separation science (Michel et al., 2003). Examples of such fractionation cells are shown in Figure 6.16, which displays two such assemblies, one with 8 chambers, the other with 20 (the remaining 2 being for anolyte and catholyte, respectively). It can be seen that underlying such chambers there is an IPG gel spanning any desired pH gradient, rather than a fixed-pH gel (an isoelectric membrane, in fact), as in previous, single-cup equipment. Thus, at the end of the electrophoretic run, each chamber should, in principle, contain only that peculiar pI cut delimited by the pH value of the gel segment underlying a given cup. As the underlying gel contains a continuous pH gradient, each cup will not contain a single population of pI values, but rather a narrow (or wider) pI interval according to the pH span of the gel segment forming the floor of the given chamber. The same authors have demonstrated that this fractionation device, when operated in rather narrow ranges, can ensure a resolution of ~0.1 pH units between two neighboring protein species, since it can resolve β-lactoglobulins A and B, known to have pI values differing, precisely, by this ΔpH value.

In a revolutionary line of thinking, Zilberstein et al. (2003) have proposed a novel concept for true miniaturization of the MCE device. Whereas all instruments here shown are based on a continuous, serial pH gradient (in that standard IEF demands colinearity of the electric field and pH gradient directions), for true miniaturization, these authors have suggested and realized parallel IEF chips. The main separation tool is a dielectric membrane (chip) with conducting channels that are filled with Immobiline gels of varying pH. The membrane is held perpendicular to the applied electric field and proteins are collected in the channels whose pH values are equal to the pI of the proteins. With this chip device, these authors claim fractionation of proteins to their pI values down to several minutes, as opposed to several hours (or days) in conventional instruments.

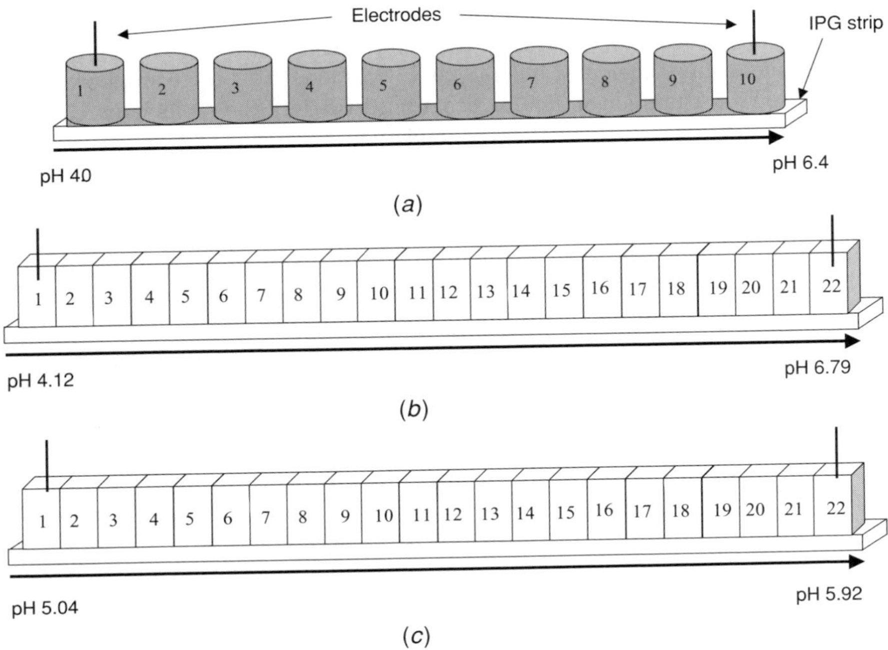

FIGURE 6.16. Experimental setup used to perform off-gel separations with multicup devices composed of (*a*) 10 or (*b*, *c*) 22 wells. The pH gradients used in each case are presented under the gels. (From Michel et al., 2003, by permission.)

6.3.3. Prefractionation via Subcellular Organelle Purification

Of course, one of the oldest and still most valid methods for prefractionation is the centrifugal cell fractionation scheme, well ingrained in classical biochemical analysis, as developed in the late 1950s and early 1960s by de Duve and other reseach groups (de Duve et al., 1955, 1959; de Duve, 1965). By this procedure, via a series of runs at different centrifugal forces, one can isolate in a reasonably pure form subcellular organelles such as nuclei, mitochondria, lysosomes, peroxisomes, synaptosomes, microbodies, and the like. Clearly, if one is investigating the proteome of such organelles, this is the most direct and proficient method for enrichment of the desired protein fraction. Such a method has been recently rediscovered and applied, in a number of reports, in proteome analysis (for reviews, see Huber et al., 2003; Dreger, 2003; Stannard et al., 2004; Jung et al., 2000a; Cordwell et al., 2000; Corthals et al., 2000). In particular, this centrifugal organelle fractionation scheme has been applied to the isolation of nuclei, since their separation is relatively simple as compared to the capture of many other subcellular organelles and the methods are very well established (Turck et al., 2004; Willsie and Clegg, 2002). For instance, in the case of human liver nuclei, Jung et al. (2000b) established a 2D reference map displaying as many as 1497 spots. Even nucleoli have been further subfractionated and their reference maps reported (Andersen et al., 2002). In this last case, in regard to Hela cells, Scherl et al. (2002) constructed a 2D nucleolar protein map displaying approximatley 350 spots. In particular, nuclear fractionation appeared to be one of the simplest processes, since nuclei are the first particles to be collected at a rather low centrifugal speed, due to their rather higher densities (the density of DNA at 4° equals ~1.75, whereas that of most proteins is only 1.188).

However, just as there are no typical cells, so there are no standard methods of cell fractionation that are equally applicable to all other cell types. The standard methodology was originally developed for just one tissue of one animal, the liver of the rat. As pointed out by Dreger (2003), conditions applying to one tissue or cell type might not necessarily apply to other tissues as well. This might be particularly true for tissue culture cells, presumably because of differences in the cytoskeletal organization. Thus, anybody attempting organelle fractionation in view of proteome analysis should first work out proper conditions and establish purity checks for such subcellular fractions.

6.3.4. Prefractionation of Membrane Proteins

It is well known that one of the major drawbacks of 2D mapping is the detection of membrane proteins, since they are sparingly soluble even in the most powerful solubilizing cocktails in vogue today. Even though membrane proteins with up to 12 transmembrane α-helices have been successfully resolved and identified by 2D MS merthodologies (Rabilloud et al., 1999), most membrane proteins have been recalcitrant to this approach. The situation is aggravated by the lack of tryptic cleavage sites in the membrane spanning domains of these proteins; this results in quite large, hydrophobic peptides that are not readily detected by MS. Several procedures have thus been described for prefractionating these complex samples. One approach is the use of organic solvents. Thus Ferro et al. (2000) have used a combination of chloroform and methanol for differential extraction of membrane proteins from chloroplasts (this procedure is also used for a delipidation step; see Simões-Barbosa et al., 2000). Blonder et al. (2002) have proposed a combination of washing isolated membranes with carbonate buffer and organic solvent extraction for increasing the number of identified hydrophobic proteins. Centrifugal sucrose density gradient fractionation for preparation of membranes is another alternative prefractionation method that was applied by Taylor et al. (2002) for isolating human mitochondria. Cloud-point extraction (or temperature-induced separation) of the nonionic detergent Triton X-114 has also been proposed by Sanchez-Ferrer et al. (1994). It is based on the fact that, when temperature is increased, a homogeneous solution of Triton X-114 separates into one detergent-enriched phase and one aqueous phase. Wissing et al. (2000) have successfully applied cloud-point extraction in Triton X-114 in combination with hydroxyapatite column chromatography for fractionation of membrane proteins prior to 2D MS analysis. More recently, Everberg et al. (2004) have proposed a novel method for prefractionation of membrane proteins via detergent-based aqueous two-phase partitioning. This two-phase system consists of the polymer poly(ethylene glycol) (average M_r is 40 kDa) and either of the two commonly used nonionic detergents

Triton X-114 or dodecyl maltoside (DDM). When mitochondria from the yeast *S. cerevisiae* were subjected to this treatment, soluble proteins partitioned mainly to the polymer phase, whereas membrane proteins were enriched in the detergent phase, as identified by 1D or 2D mapping followed by MS analysis. Prefractionation was further enhanced by addition of an anionic detergent, SDS, or a chaotropic salt, $NaClO_4$, and by raising the pH in the system. It is thus hoped that some of these methods will help to disclose larger proportions of membrane proteins in different proteomes.

6.4. MULTIDIMENSIONAL CHROMATOGRAPHY COUPLED TO MS

In spite of the contributions of 2D PAGE to proteomics, there are shortcomings to this technology. High-throughput analysis of proteomes is challenging because each spot from 2D PAGE must be individually extracted, digested, and analyzed, a time-consuming process. In addition, owing to the limited loading capacity of 2D PAGE gels and the detection limit of staining method, 2D PAGE presently has an insufficient dynamic range for complete proteome analysis. These shortcomings have encouraged development of alternative methods which have been reviewed by Washburn and Yates (2000). Such methodologies could be either purely chromatographic or hybrid chromatographic/electrophoretic methods, such as HPLC, cIEF, CZE, microcapillary chromatography, or even preparative IEF in the Rotofor coupled to nonporous, reverse-phase (RP) HPLC (Zhu et al., 2003). Just as with 2D PAGE, MS would be the method of choice for identifying proteins resolved by liquid separation protocols. This would have the advantage that the coupling with MS would be online, thus eliminating the lengthy steps for transferring proteins to MS equipment in 2D PAGE. A clarification is here due, though: The term "multidimensional" is a misnomer; these chromatographic techniques are just as two dimensional as 2D PAGE.

In chromatographic approaches the proteins, in most cases, are digested into peptides prior to separation. The advantage is that peptides (especially from membrane proteins) are more soluble in a wide variety of solvents and hence easier to separate that the parent proteins. The disadvantage is the tremendous increment in the number of species to be resolved. Since after fractionation in coupled columns the eluate is sent directly to MS instrumentation, the sample complexity might still be too high for proper analysis. While MS technology is rapidly improving, its dynamic range measurement capability is still two to three orders of magnitude less than the range of protein expression found within mammalian cells. In one of the earliest reports, Opiteck et al. (1997, 1998) described a 2D HPLC system that used size exclusion chromatography (SEC) followed by RP HPLC for mapping *E. coli* proteins. Perhaps one of the most successful approaches, though, is the one by Yates and co-workers (1997; Link et al., 1997; Washburn et al., 2001), who developed an online 2D ion exchange column coupled to RP-HPLC, termed MudPit, for separating tryptic digests of 80S ribosomes from yeast. The acidified peptide mixture was loaded onto a strong cation exchanger (SCX) column; discrete eluate fractions were then fed onto a RP column, whose effluent was fed directly into a mass spectrometer. This iterative process was repeated 12 times using increasing salt gradient elution from the SCX bed and an increasing organic solvent concentration from the RP beads (typically a

C_{18} phase). In a total yeast lysate, the MudPit strategy allowed the identification of almost 1500 proteins (Washburn et al., 2001). A similar procedure was used for separation of proteins and peptides in human plasma filtrates (Raida et al., 1999), plasma (Richter et al., 1999), blood ultrafiltrates (Schrader et al., 1997), and human urines (Heine et al., 1997). The same setup (SCX followed by C18-RP) was followed by Davis et al. (2001) for resolving a protein digest derived from a conditioned medium from human lung fibroblasts. Perhaps the most sophisticated instrumentation is the one devised by Unger and co-workers (Wagner et al., 2002) for processing proteins of $M_r < 20$ kDa. The setup consists of two gradient HPLC instruments, two UV detectors, an isocratic pump, four RP columns, an ion exchange column, four 10-port valves, an injection valve, two fraction collector stations, and a work station to control such a fully automated system. This system was applied to mapping of human hemofiltrates as well as lysates from human fetal fibroblasts. Other, hybrid systems consist in coupling, for example, an HPLC column to electrophoretic instrumentation, notably capillary electrophoresis or isoelectric focusing or even isotachophoresis (Wang et al., 2002; Mohan and Lee, 2002; Chen et al., 2002). For those, we refer the reader to some recent reviews (Issaq et al., 2002; Issaq, 2001; Shen and Smith, 2002). At times, bidimensionality can simply be achieved by coupling a single-dimension separation to powerful MS probes. A nice example of this comes from the work of Shen et al. (2001). The first-dimension separation is RPLC (reverse-phase liquid chromatography) (which discriminates on the basis of a hydrophobic parameter) coupled directly to FTICR–MS (Fourier transform ion cyclotron resonance mass spectrometry). With this coupled system, in a single run of ca. 3 h, >100,000 components could be resolved, a truly impressive achievement (see Fig. 6.17). More than 1000 soluble yeast proteins could be identified from >9000 peptide database search "hits" in a single, high-efficiency capillary RPLC–FTICR–MS run of 3 h from a total of only 40 μg sample (Shen et al., 2001).

A number of recent reviews have appeared evaluating the various 2D liquid-phase-based separations in proteomics (Romijn et al., 2003; Peng et al., 2003; Wang and Hanash, 2003; Lubman et al., 2002; Guttman et al., 2004), to which the reader is referred for further insight on this topic.

6.4.1. Eavesdropping on Thy Neighbor

Although this exhortation can hardly be found in the Gospels, it was precisely via this process that chromatographers, unhappy with the resolution of their 1D techniques, started copying electrophoreticists (who had been doing that at least ever since O'Farrell, 1975, and even before) and discovered 2D chromatography (although they pompously called it "multidimensional") in the early 1990s. The latest comers into this arena were scientists working in gas chromatography (GC): Although this topic is not strictly related to proteome analysis, it is worth a diversion. Gas chromatography (for which Archer J. P. Martin and Richard L. M. Synge shared the Nobel Prize in Chemistry in 1952) had been thought to be a technique of such exquisite resolution and peak capacity that for decades nobody bothered to improve it, until Phillips and Xu, in 1995, launched multidimensional (here, too, read two-dimensional!) chromatography

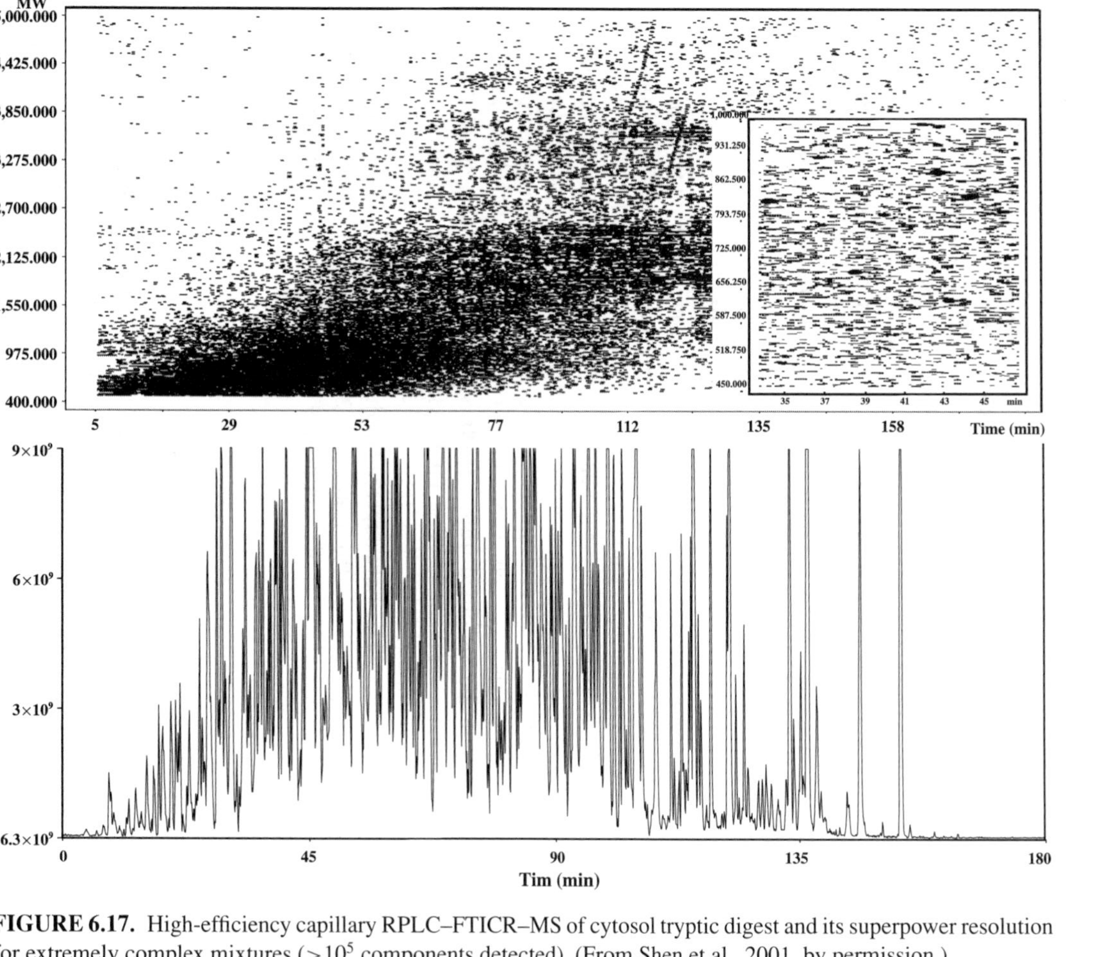

FIGURE 6.17. High-efficiency capillary RPLC–FTICR–MS of cytosol tryptic digest and its superpower resolution for extremely complex mixtures ($>10^5$ components detected). (From Shen et al., 2001, by permission.)

when they realized, already in 1991 (Liu and Phillips, 1991), that for complex mixtures the resolution of 1D GC was not adequate. Comprehensive 2D gas chromatography (GC × GC) is now rapidly gaining importance for analysis of complex samples in the modern pharmaceutical industry, in flavor and petroleum analysis, and in the wine and liquor industries and in general in many areas of food analysis, where complex, volatile components have to be separated, detected, and identified. In GC × GC, the first dimension is usually a 15–30-m-long, high-resolution capillary column (ID of 250–320 μm) with a nonpolar stationary phase, supporting mainly boiling point–based separations. The modulator is the interfacing device between the first- and second-dimension columns. Here, fractions from the first column are refocused and injected in the second column. The second GC capillary is significantly shorter than the first one (1–2 m, with ID of 50–100 μm), with a polar phase so as to provide true orthogonal separation in Giddings's (1991) terms. Separation time in the first dimension is 1–2 h, but separation in the second columns takes only seconds under essentially isothermal conditions. There are various types of modulators, including dual-stage thermal desorption modulators, cryogenic trapping and focusing, as well as the cryogenic jet system (Harynuk and Gorecki, 2003) (for a review and evaluation of modulators, see Kristenson et al., 2003). For maintaining resolution in the first dimension and avoiding the loss of ordered structures, four or more modulations are necessary for each band that is separated in the first column, and separation in the second capillary should be finished before the next pulse is injected. Such fast separations require rapid detection systems with small dead volumes and high data acquisition rates, such as the fast flame ionization detector or the micro electrocapture detector. Recently, GC × GC–TOF–MS has also been introduced (Adahchour et al., 2003), thus allowing positive identification of eluted peaks. The field recently reached a critical mass to the point that an international meeting was called by Brinkman and Vreuls (2003). It is worh a perusal through this issue, with papers ranging from analysis of petrochemical compounds (van Stee et al., 2003) to determination of atropoisomeric and planar polychlorinated biphenyls (Harju et al., 2003), natural fats and oils (Mondello et al., 2003), polynuclear aromatic hydrocarbons (Cavagnino et al., 2003), and even urban aerosols (Welthagen et al., 2003; Kallio et al., 2003). Figure 6.18 is an example of a GC × GC separation of a standard mixture of PAHs (polyaromatic hydrocarbons) in composite diesel fuel diluted 1 : 10,000 in *n*-hexane. It should be noted that the peak of acenaphthene, as resolved in the second dimension, represents barely 2 ppb (parts per billion) at a signal/noise (S/N) ratio of 27!

6.5. PROTEIN CHIPS AND MICROARRAYS

Since this technique has already been dealt with in Part I (Chapter 2), it will only be briefly dealt with here. One of the latest technology platforms that have been developed for proteomics includes chip-based arrays. Such arrays have been elaborated for separating proteins based on surface chemistries or known protein ligands (e.g., antibodies), with subsequent identification by MS (Lueking et al., 1999;

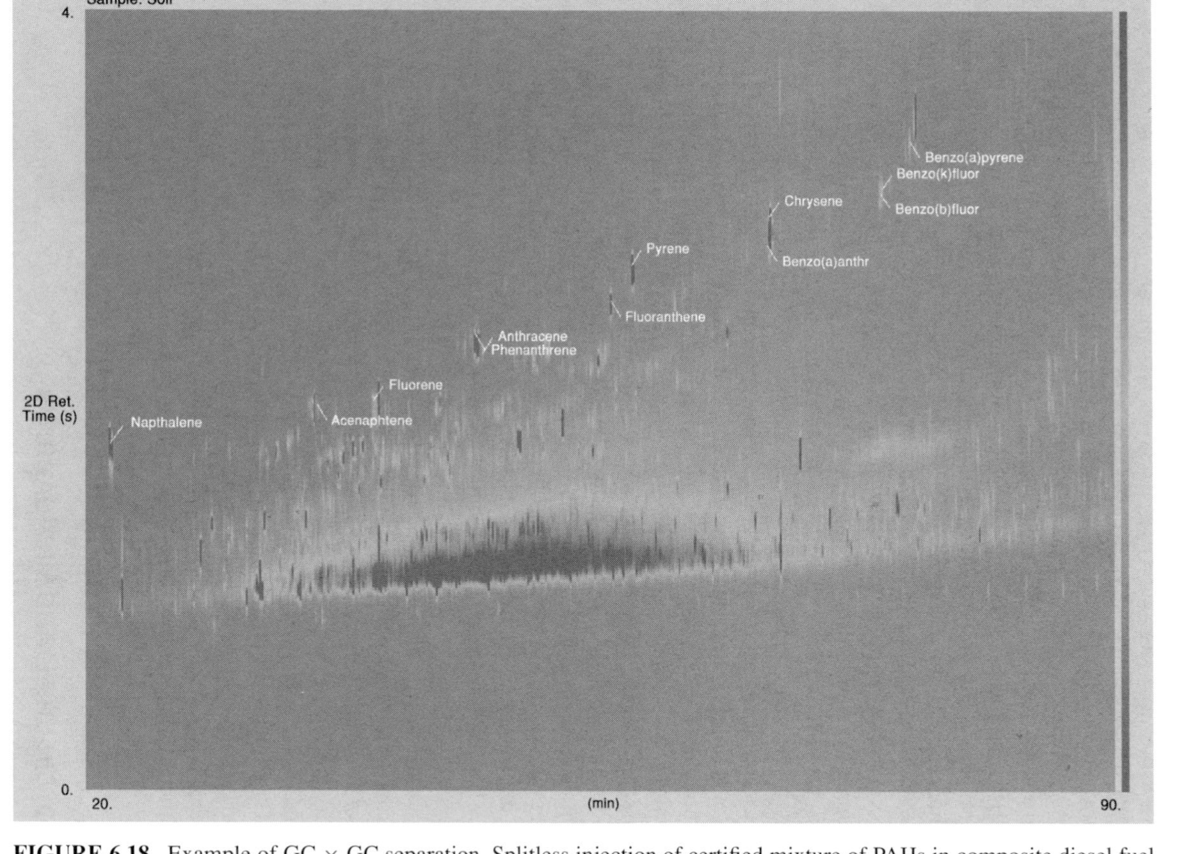

FIGURE 6.18. Example of GC × GC separation. Splitless injection of certified mixture of PAHs in composite diesel fuel diluted 1 : 10,000 in *n*-hexane. It should be noted that the peak of naphthalene, as resolved in the second dimension, is just 9 ppb (parts per billion), acenaphthene represents barely 2 ppb at a S/N ratio of 27, and fluorene is only 3.2 bbp at S/N = 55. (Modified from Cavagnino et al., 2003, by permission.)

Borrebaeck, 2000, Borrebaeck et al., 2001). Other promising applications of protein chip microarrays have been differential profiling (MacBeath and Schreiber, 2000) and high-throughput functional analysis (Fung et al., 2001; Sawyer, 2001). An example of such protein chips was shown in Fig. 2.2. Such arrays can be divided into chemical and bioaffinity surfaces. In the first case (see upper row in Fig. 2.2) such surfaces function essentially like minichromatographic columns in that they capture a given protein population by, for example, hydrophobic interaction (reverse phase), metal chelation, and different types of ion exchangers. Such chemical surfaces are not highly selective; nevertheless, they can be used in a cascade fashion; for example, proteins captured on ion exchangers can be eluted and readsorbed onto a reverse phase or on metal chelators so as to further subfractionate a most heterogeneous protein mixture as a total cell lysate. The bioaffinity surfaces (see lower row in Fig. 2.2) clearly work on a much higher selectivity principle and allow capture of a narrow and well-defined protein population immediately ready for MS analysis. How the latter will be performed is shown schematically in Figure 6.19: Once a selected protein has been captured (hopefully in a highly purified form), the surface of the chip is bombarded with a laser beam which will desorb and ionize the different proteins, which are ultimately identified via their precise molecular mass. The main difference between this kind of structural analysis and MALDI–TOF desorption ionization is that, in the latter case, the sample has to be manually transferred to a microwell plate, admixed with a special matrix (such as sinapinic acid), and then desorbed/ionized via pulse laser shots. Conversely, in the protein chips illustrated here (Fig. 6.19) a stick containing eight (or multiples of) such affinity surfaces, with the captured protein, is directly inserted in the MS instrument and desorbed by surface-enhanced laser desorption/ionization (SELDI) in the presence of added matrix (although, recently, the surface of such

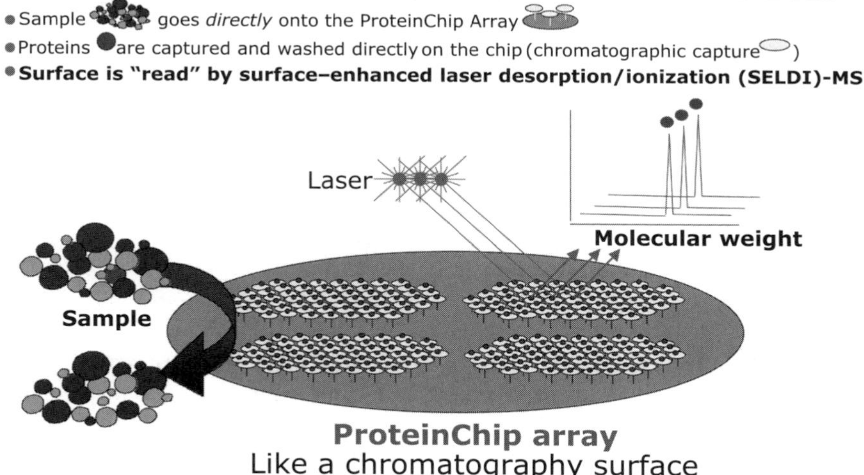

FIGURE 6.19. Scheme of processing/detection of protein species captured by different chemical surfaces (chromatographic adsorption). (Courtesy of Ciphergen.)

arrays has been already coated with energy-absorbing polymers; Voivodov et al., 1996). Such instrumentation is rapidly been adopted in many clinical chemistry laboratories around the world and might soon become part of the standard instrumentation of such laboratories. The following is a (nonexhaustive) list of papers recently published on the field of arrays exploiting SELDI technology: Lin et al. (2001); Paweletz et al. (2001); Boyle et al. (2001); Wang et al. (2001); Thulasiraman et al. (2000); Austen et al. (2000); Von Eggeling et al. (2000); Merchant and Weinberger (2000); Weinberger et al. (2002).

In terms of genuine protein microarrays, at a level comparable to DNA microarrays, though, it must be stated that the former technique is still in its infancy. DNA microarrays have already emerged as one of the most powerful methodologies for analyzing gene function (Blohm and Giuseppi-Ellie, 2001). This flourishing technology exploits certain key properties of DNA. Fluorescently labeled cDNA generated from the mRNA via reverse transcription hybridizes to a complementary single-stranded DNA sequence immobilized in an ordered array on a chip surface. In this way, thousands of genes can be probed in a single assay in a microchip. The ladscape for protein arrays is much less developed. Unlike DNA, proteins do not react by simple linear hydridization, and the properties of proteins are often altered markedly by the ambient conditions. One must guard against possible drastic changes in protein conformation, and hence of activity or affinity, as induced by immobilization onto a solid surface, derivatization with a fluorescent probe, heating, and drying. Thus, it is unlikely that the first generation of protein arrays will be able to recognize all the proteins expressed from higher eukaryote genomes. At present, the most direct approaches to such arrays appear to be the following:

1. Phage libraries (i.e., combinatorial libraries obtained via amplification of selected clones of immunoglobulin short-chain variable regions) (Armstutz et al., 2001)
2. Aptamers (i.e., single-stranded oligonucleotides with affinity for individual protein molecules) (Rhodes et al., 2001).

More on these topics can be found in recent reviews (Jenkins and Pennington, 2001; Figeys and Pinto, 2001; Cutler, 2003; Stoll, 2003).

6.6. NONDENATURING PROTEIN MAPS

Before closing this chapter on 2D maps, a digression is due to native (nondenaturing) protein maps, as opposed to polypeptide maps, that is, the standard maps obtained under fully denaturing conditions in both dimensions. Manabe's group has extensively developed this approach, which might have interesting practical implications, as outlined below. Starting in 1979, Manabe et al. (1979, 1981) proposed 2D maps for analysis of proteins in various body fluids, including human plasma, under nondenaturing conditions, so as to acquire information on proteins under physiological conditions. Major plasma proteins could thus be identified by electrophoretic blotting followed by immunochemical staining, enabling them to construct a "protein map" of human sera (Manabe et al., 1985). Enzyme activities could also be detected on such native 2D maps (Kadofuku et al., 1983); additionally, the dissociation process of lipoproteins into apolipoproteins could be analyzed by modifying the process of IEF gel equilibration (Manabe et al., 1987a,b). More recently, Manabe et al. (1999) and Manabe (2000) have proposed four types of 2D maps for obtaining systematic information on proteins and their constituent polypeptides. To facilitate the analyses, a microgel system was adopted for creating such a set of 2D maps. In type I 2D map analysis, a plasma sample was first analyzed under nondenaturing conditions in both dimensions, so as to characterize the properties of constituent proteins under physiological conditions. In type II 2D maps, the sample was analyzed by employing nondenaturing IEF in the first dimension followed by SDS–PAGE in the second dimension. In this second approach, the dissociation of noncovalently bound protein subunits could be revealed. In type III 2D gel slabs, proteins were separated again by nondenaturing IEF in the first dimension, just as performed in type II gels; however, in the second-dimension SDS–PAGE, the denaturing solution of urea–SDS was added with mercaptoethanol, so as to induce the dissociation of disulfide-bonded polypeptides. In type IV 2D mapping, full denaturing conditions were adopted in both dimensions, as routinely done in conventional proteome analysis. Figures 6.20*a*, *b* give examples of such 2D maps: the fully native (Fig. 6.20*a*) versus the fully denatured (Fig. 6.20*b*) maps of human plasma proteins. These four types of 2D maps were useful for starting the construction of a comprehensive database of plasma proteins combining the "nondenaturing protein map" with the "denaturing polypeptide map." Such an approach should be applicable to the comprehensive analysis of other complex systems of soluble proteins, such as cytosol proteins extracted from various cells.

6.7. SPOT MATCHING IN 2D GELS VIA COMMERCIAL SOFTWARE

The power of the 2D gel technique lies in its capacity to simultaneously separate thousands of proteins for their subsequent identification and quantitative comparison. Such a proteomics approach typically requires the quantitative analysis of numerous sets of gels for revealing differential protein expression across multiple experiments. As the experiments result in large amounts of data, efficient use of the 2D techniques relies on powerful and user-friendly data analysis by means of computer algorithms. A number of software packages have appeared in the last decade, as listed in Table 6.2. A few papers have been published comparing some of these packages: PDQuest and Progenesis (Rosengren et al., 2003) and Z3 and Melanie 3.0 (Raman et al., 2002). Other reports assess just a single product: Phoretix 2D full (Mahon and Dupree, 2001) and Compugen's Z4000 (Rubinfeld et al., 2003). Other articles deal with general aspects of such algorithms, such as point pattern matching, reproducibility, and matching efficiency (Panek and Vohradsky, 1999; Pleissner et al., 1999; Cutler et al., 2003; Voss and Haberl, 2000; Molloy et al., 2003; Moritz and Meyer, 2003). Although some principles have been given in the first part of this book, here we will review in

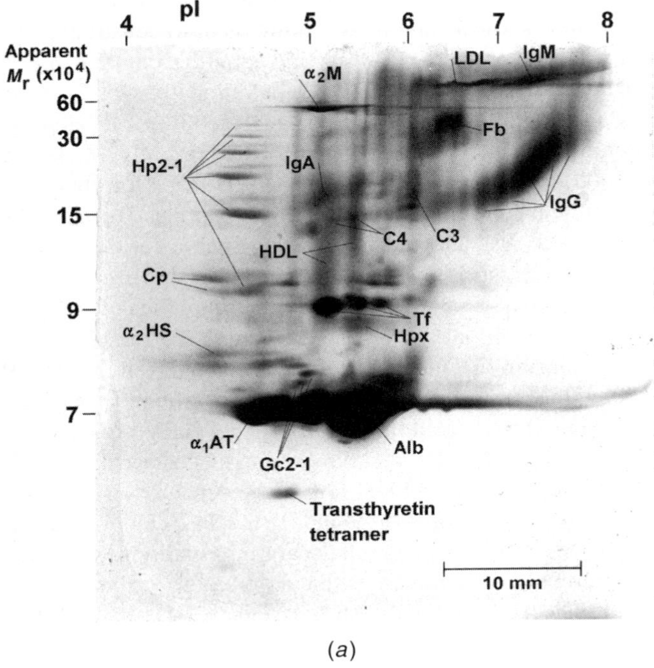

(*a*)

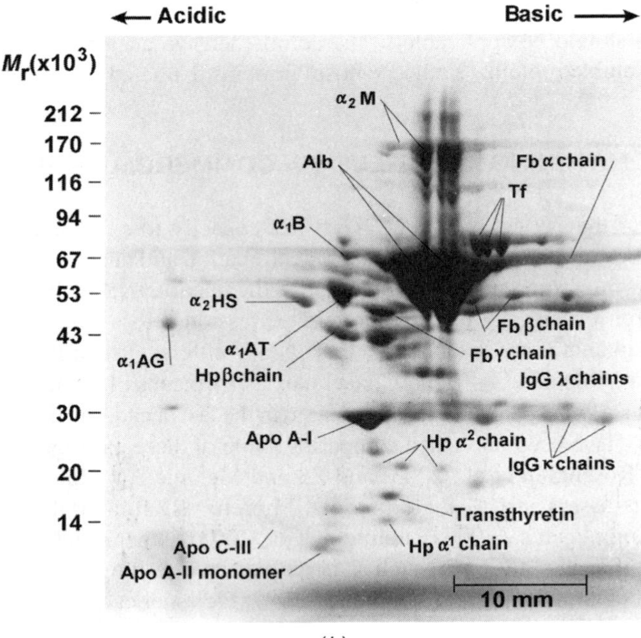

(*b*)

FIGURE 6.20.

more detail the steps required for analyzing and matching data obtained from control and treated samples run in separate gels and subsequently subjected to matching via different software packages.

Accurate quantitation relies upon computer analysis of a digitized representation of the stained gel. In the majority of cases, images are captured by laser densitometry, by phosphor imagery, or via a CCD camera. Once the image has been digitized, a standard computer-assisted analysis of 2D gels includes at least the following three basic steps: (i) protein spot detection, (ii) spot quantitation, and (iii) gel-to-gel matching of spot patterns.

Since in our laboratory we have gathered extensive experience with the PDQuest system, we will deal in depth with it (while, of course, not endorsing it). PDQuest (just as all other packages) is a software system for imaging, analyzing, and databasing 2D E gels. Once a gel has been scanned, advanced algorithms are available for removing background noise, gel artifacts, and horizontal or vertical streaking from the image. PDQuest then uses a spot segmentation facility to detect and quantify protein spots. Protein spots that change over time can be traced, quantified, displayed on-screen, and exported to other applications for statistical analysis. The spot-matching and databasing facilities make it possible to objectively compare hundreds of different gels. In more detail, the series of steps necessary for proper gel evaluation can be summarized as follows:

1. Scanning
2. Filtering images
3. Automated spot detection
4. Matching of protein profiles
5. Normalization
6. Differential analysis
7. Statistical analysis

FIGURE 6.20. (*a*) Nondenaturing (type I) 2D map of human plasma proteins. A sample of normal human plasma was subjected to IEF in the absence of denaturants and then set on a microslab gel of polyacrylamide gradient $T = 4.2\%$ to $T = 17.85\%$ linear gradient at $C = 5\%$ and the 2D run also in the absence of denaturants. Abbreviations: α_2M, α_2-macroglobulin; LDL, low-density lipoprotein; IgM, immunoglobulin M; HP 2-1, haptoglobin phenotype 2-1, polymer series; IgA, immunoglobulin A; Fb, fibrinogen; IgG, immunoglobulin G; HDL, high-density lipoprotein; C4, complement factor C4; C3, complement factor C3; Cp, ceruloplasmin; Tf, transferrin; Hpx, hemopexin; α_2HS, α_2-HS glycoprotein; α_1AT, α_1-antitrypsin; Gc2-1, Gc-globulin phenotype 2-1; Alb, albumin. (*b*) Type IV (denaturing in both dimensions) 2D map of human plasma proteins. The plasma sample was treated with 8 M urea, 1% Nonidet P-40 (NP-40), and 10% β-mercaptoethanol. After the IEF step in urea/NP-40, the strip was equilibrated with 2% SDS, 8 M urea, and 5% β-mercaptoethanol and then the second dimension was run in 0.1% SDS. The size and the acrylamide gradient of the slab gel were the same as in (*a*). All abbreviations as (*a*) (except for α_1B, α_1-glycoprotein B; α_1AG, α_1-acidic glycoprotein; Apo C-III, apolipoprotein C-III.) (From Manabe et al., 1999, by permission.)

TABLE 6.2. Commercial Software Packages Currently Available for 2D Gel Image Analysis

	Software	Company	Year of Arrival	Comments	Platforms	Images Supported
1	Delta 2D	DECODON GmbH, http://www.decodon.com	2000	Save-disabled evaluation version available	PC (Windows 98, ME, 2000, NT), Linux, Sun Solaris, Mac OS X	TIFF (8, 12, and 16 bit), JPEG, BMP, GIF, PNG.
2	GELLAB II+	Scanalytics, http://www.scanalytics.com/	1999	Trial version available	PC (Windows 95, NT)	TIFF (8 bit)
3	Melanie	Geneva Bioinformatics S.A., http://www.genebio.com	N/A[b]	30-day fully functional trial version available	PC (Windows 95, 98, 2000, NT)	TIFF (8, 16 bit), GIF, Biorad Scan
4	PD Quest	Bio-Rad Laboratories, Inc., http://www.biorad.com	1998	30-day fully functional trial version available	PC (Windows 95, 98, NT), Macintosh Power PC	TIFF (8, 16 bit)
5[a]	Phoretix 2D Advanced	Nonlinear Dynamics Ltd., http://www.nonlinear.com, http://www.phoretix.com	1991	Trial version available through sales agent	PC (Windows 95, 98, 2000, NT)	TIFF (8, 12, and 16 bit)
5.1	AlphaMatch 2D	Alpha Innotech Corporation, http://alphainnotech.com	1999	Trial version available through sales agent	PC (Windows 95, 98, 2000, ME, NT)	TIFF (8,12, and 16 bit)
5.2	Image Master 2D Elite	Amersham Pharmacia Biotech, http://www.apbiotech.com	2001	Trial version available through sales agent	PC (Windows 95, 98, 2000, ME, NT)	TIFF (8,12, and 16 bit)
5.3	Investigator HT Analyzer	Genomic Solutions, Inc., http://www.genomicsolutions.com	2000	Trial version available through sales agent	PC (Windows 98, 2000, NT)	TIFF (8, 12, and 16 bit)
6	Progenesis	Nonlinear Dynamics Ltd., http://www.nonlinear.com, http://www.phoretix.com	2001	Special hardware and software requirements	PC (Windows 2000)	TIFF (8, 12, and 16 bit)
7	Z3	Compugen http://www.2dgels.com	2000	21-day fully functional trial version available	PC (Windows 98, 2000, NT)	TIFF (8, 12, and 16 bit), JPEG, BMP, GIF, PNG, GEL, FLT
8	Proteomeweaver	Definiens (Munich, Germany)	2002	21-day fully functional trial version available	PC (Windows 2000, XP)	TIFF (8, 12, and 16 bit), JPEG, BMP, GIF, PNG, GEL, FLT

Source: Raman et al., 2002.

Note: The software packages listed in the table are comprehensive off-the-shelf-commercial software packages available for 2D gel image analysis. The information listed has been obtained from various sources, including the Internet, literature, and sales agents. Misinformation, if any, is unintentional.

[a] These software packages are essentially the same as Phoretix 2D Advanced™ and marketed under different brandnames. Contact individual companies to know about any differences.

[b] Not available.

We will now briefly describe these steps.

1. *Scanning.* This process converts signals from biological samples to digital data. A data object as displayed on the computer is composed of tiny individual screen pixels. Each pixel has an X and Y coordinate, which are the pixel's horizontal and vertical positions on the image, and a Z value, which is the signal intensity of the pixel. The total intensity of a data object is the sum of the intensities of all the pixels that make up the object. The mean intensity of a data object is the total intensity divided by the number of pixels in the object.

2. *Image Filtering.* This is a fundamental process in that it removes small noise features on an image while leaving larger features (like spots) unaffected. A *Filter Wizard* helps the user through this selection process, removing, for example, specks and other imperfections of the imaged gel.

3. *Automated Spot Detection.* A *Spot Detection Wizard* is designed to guide users through this process. One first selects the *faintest spot* in the scan (this will set the sensitivity and minimum peak-value parameters) and then the *smallest spot* (this will set the size scale parameter); then one selects the *largest spot* on the image that one wants to detect. Once the gel image is created (i.e., each pixel of a 2D scan is originally assigned an OD), the program helps the user through a number of steps, comprising *Initial Smoothing, Background Subtraction, Final Smoothing* (necessary for removing extraneous spots at or near the background level), *Locating Spots in Gel Image* (i.e., locating the center and position of each recognizable spot), *Fitting and Quantifying Spots* (fits ideal Gaussian distributions to spot centres). As a final product, one will obtain three separate images: the original *raw 2D scan*, which remains unchanged; the *filtered image*, which is a copy of the original scan that has been filtered and processed; and the *Gaussian image*, which is a synthetic image containing the Gaussian spots.

4. *Matching of Protein Profiles.* Groups of gels can be edited and matched to one another in a "match set." Match sets consist of "gel spots files" and gel images. In a match set, protein spots are matched to each other, enabling the user to compare their quantities. For matching the same protein spots between different gels, one will need to put some landmarks. Landmarks are reference spots used by PDQuest to align and position match set members for matching. They are used to compensate for slight differences and distortions in the member gels.

5. *Normalization.* When comparing gels in a Match Set, there is often some variation in spot size and intensity between gels that is not due to differential protein expression. This variation can be caused by a number of factors, including pipetting errors during sample preparation and loading, variations in sample density, and inconsistencies in staining. To accurately compare spot quantities between gels, one must compensate for these nonexpression-related variations in spot intensity. This is the process of normalization.

6. *Differential Analysis.* For quantitative analysis, at least five replicate gels should be run per sample. A replicate group is the name given to a set of gels that are duplicates of each other. For example, five separate gels of a control cell lysate theoretically should produce the same quantitation for all spots on a 2D gel. In practice,

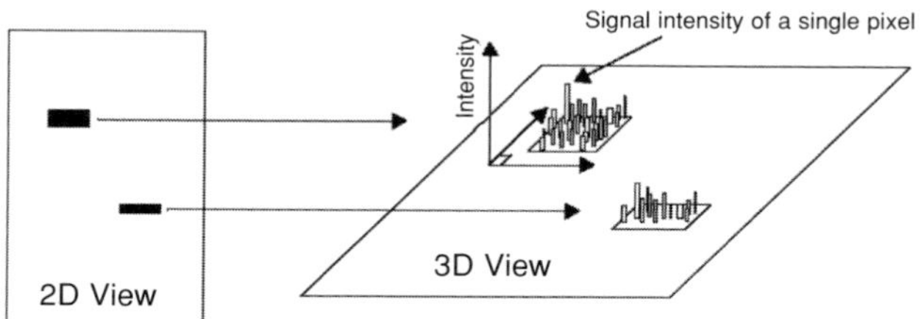

FIGURE 6.21. Representation of pixels in two digitally imaged bands in gel as 2D and 3D image. (Reprinted by permission of Bio-Rad.).

however, slight variations in the quantitation will probably be seen. Instead of choosing the one that the user thinks is best, one can take the average and use those values for spot quantitation. Once the sample groups have been created (e.g., control and drug-treated cell line), it is possible to perform the comparison between the protein profiles to find differentially expressed proteins (down regulated, up regulated, silenced, etc.). It is recalled that, due to the high variability of comparing samples run in different gels, the threshold for accepting a meaningful variation is set at a factor of 2.0, that is, only spots whose quantity in gel B is at least twice that of the corresponding spot in gel A are accepted as significantly changed (100% variation). Conversely, in the DIGE technique, due to the fact that both treated and control samples are run within the same gel, the error seems to be much reduced and spot variations of only 20% are accepted.

7. *Statistical Analysis.* Once differentially expressed proteins have been detected, it is necessary to perform a statistical analysis to find significant differences between

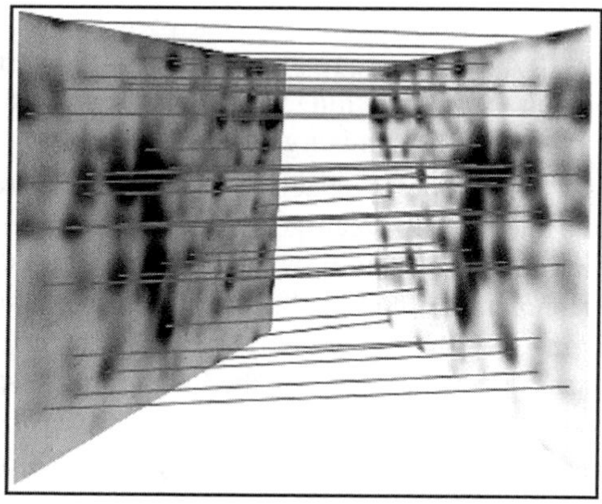

FIGURE 6.22. How protein spots of two gels are matched with PDQuest program. (Reprinted by permission of Bio-Rad.)

the two compared samples. The statistical considerations are performed usually with Student's t test ($p < 0.05$).

Efficient analysis of protein expression by using 2D data relies on the use of automated image-processing techniques. The overall success of this research depends critically on the accuracy and reliability of the analysis software. In addition, the software has a profound effect on the interpretation of the results obtained and the amount of user intervention demanded during the analysis. The choice of analysis software that best meets specific needs is therefore of interest to the research laboratory. Different packages show different strengths and weaknesses. *ImageMaster* (Amersham Biosciences) is quoted among the most accurate packages, *Z3* (Compugen Limited) appears to be the most robust to poor S/N ratio and *PDQuest* (Bio-Rad Laboratories) the most robust to spot overlap, *Melanie III* (GeneBio) performs well in all evaluations, and *Progenesis* (Nonlinear Dynamics) has the advantage of a parameter-free spot detection while also performing well in most evaluations. One should not forget, however, what is stated in Table 6.2: According to Raman et al. (2002), all the packages listed under number 5 to 5.3 appear to be essentially the same as Phoretix 2D Advanced and are marketed under different trade names!

In concluding this chapter as well as the book, it is of interest to recall some other procedures for 2D map comparisons that have recently come to the limelight. Although not as yet elaborated into commercially available software, these methods appear to be just as powerful and user friendly. In one approach, statistical treatment of 2D maps is performed via fuzzy logic (Marengo et al., 2003b,c). In this system, the signal corresponding to the presence of proteins in 2D maps is substituted with probability functions centerd on the signal itself. The standard deviation of the bidimensional Gaussian probability function employed to blur the signal allows assigning different uncertainties to the two electrophoretic dimensions. This method coupled with multidimensional scaling allows a meaningful separation of experimental 2D maps into control and treated classes with the same accuracy as commercial packages. Additionally, differences in protein expression, as visualized in spots in 2D maps, can be assessed, as shown in Figure 6.23. In another approach, three-way principal-component analysis alone is applied to classification of 2D maps, also with the same good performance (Marengo et al., 2003a). It is thus seen that, presently, for 2D map analysis, users have quite a wide range of programs, software, and approaches for solving their problems.

6.8. CONCLUSIONS

Hopefully, this book has given the reader an appreciation of all tools involved in proteome analysis, ranging from MS in all its varied aspects to classical 2DE approaches as well as to 2D chromatographic techniques. We even ventured into GC $\times$ GC, which, although not strictly pertinent in proteome analysis, shows clearly how ideas propagate from one field to another, a contagion that, unlike the pandemic diseases of the past, has had a strongly cooperative effect in enhancing progress in science.

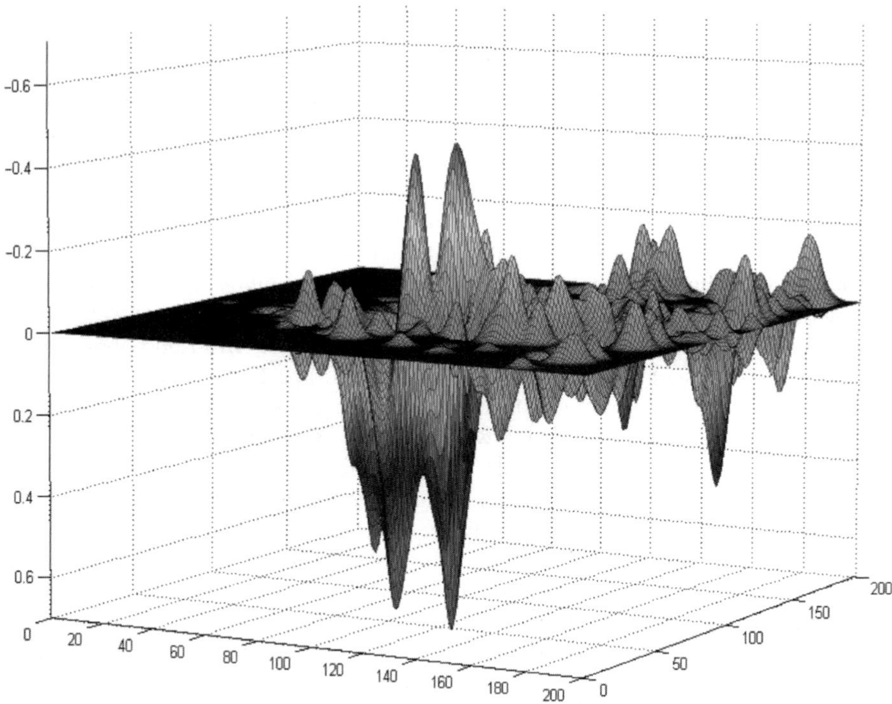

FIGURE 6.23. Fuzzy map obtained as difference of mean healthy and pathological fuzzy maps in analysis of healthy human lymph nodes and non-Hodgkin's lymphomas. In this difference map, the negative values represent the spots which characterize pathological subjects; the positive values represent the spots characterizing healthy individuals. (From Marengo et al., 2003c, by permission.)

What we can envision for the future is hard to say. If we could express a wish, it would be that there would be a mass spectrometer at every bench, not just for protein and other macromolecular analysis but also for the determination of molecular binding constants, flux rates, and all sorts of diagnostic applications. Whether the MS instruments of the future will be the size of a matchbox remains to be seen. Also, the field of classical 2D techniques has witnessed dramatic changes, from the O'Farrell (1975) to modern IPG methodologies, allowing full flexibility and control of pH gradients. There appear to be two opposite trends in this field: on the one hand, groups of scientists are pushing toward more and more miniaturization and chips and, on the other, groups are pushing toward larger and larger 2D gels (napkin size). Clearly, 2D chromatography is growing at a brisk pace, although we believe that neither technique will replace the other. Both techniques will go hand in hand for a long time in the future, since they seem to be rather complementary and to allow exploring different facets of the proteome. In the long run, proteome science will surely benefit from this panoply of tools so highly refined and developed to a level unthinkable only a few years ago. Thus, it is hoped that we will accelerate the process of breaking the proteome code, although that might still be a few decades away. Proteome science is turning out to be extremely more complex than nucleic acid science, after all!

REFERENCES

Adahchour, M., van Stee, L. L. P., Beens, J., Vreuls, R. J. J., Batenburg, M. A., Brinkman, U. A. Th. (2003) *J. Chromatogr A* **1019,** 157–172.

Aebersold, R., Leavitt, J. (1990) *Electrophoresis* **11,** 517–527.

Ames, G. F. L., Nikaido, K. (1976) *Biochemistry* **15,** 616–623.

Andersen, J. S., Lyon, C. E., Fox, A. H., Leung, A. K., Lam, Y. W., Steen, H., Mann, M., Lamond, A. L. (2002) *Curr. Biol.* **12,** 1–11.

Anderson, L. N., Anderson, N. G. (2002) *Mol. Cell. Proteomics* **1,** 845–867.

Anderson, N. L., Anderson, N. G. (1977) *Proc. Natl. Acad. Sci. USA* **74,** 5421–5426.

Anderson, N. L., Anderson, N. G. (1978a) *Anal. Biochem.* **85,** 331–340.

Anderson, N. L., Anderson, N. G. (1978b) *Anal. Biochem.* **85,** 341–354.

Anderson, N. L., Anderson, N. G. (1982) *Clin. Chem.* **28,** 739–748.

Anderson, N. L., Anderson, N. G. (1984) *Clin. Chem.* **30,** 1898–1905.

Anderson, N. L., Hickman, B. J. (1979) *Anal. Biochem.* **93,** 312–320.

Appel, R. D., Dunn, M. J., Hochstrasser, D. F. (Guest Eds.) (1997) Paper Symposium: Biomedicine and Bioinformatics, *Electrophoresis* **18,** 2703–2842.

Appel, R. D., Dunn, M. J., Hochstrasser, D. F. (Guest Eds.) (1999) Paper Symposium: Biomedicine and Bioinformatics, *Electrophoresis* **20,** 3481–3686.

Armstutz, P., Forrer, P., Zahnd, C., Pluckthun, A. (2001) *Curr. Opin. Biotechnol.* **12,** 400–405.

Arnaud, I. L., Josserand, J., Rossier, J. S., Girault, H. H. (2002) *Electrophoresis* **23,** 3253–3261.

Austen, B. M., Frears, E. R., Davies, H. (2000) *J. Pept. Sep.* **6,** 459–469.

Banks, R. E., Dunn, M. J., Hochstrasser, D. F., Sanchez, J. C., Blackstock, W., Pappin, D. J. (2000) *Lancet* **356,** 1749–1756.

Barry, R. C., Alsaker, B. L., Robinson-Cox, J. F., Dratz, E. Z. (2003) *Electrophoresis* **24,** 3390–3404.

Bier, M. (1998) *Electrophoresis* **19,** 1057–1063.

Bier, M., Mosher, R. A., Thormann, W., Graham, A. (1984) in Hirai, H., (Ed.), *Electrophoresis '83: Advanced Methods, Biochemical and Chemical Applications*, de Gruyter, Berlin, pp. 99–107.

Bjellqvist, B., Ek, K, Righetti, P. G., Gianazza, E., Görg, A., Postel, W., Westermeier, R. (1982) *J. Biochem. Biophys. Methods*, **6,** 317–339.

Bjellqvist, B., Hughes, G. J., Pasquali, C., Paquet, N., Ravier, F., Sanchez, J. C., Frutiger, S., Hochstrasser, D. (1993a) *Electrophoresis* **14,** 1023–1031.

Bjellqvist, B., Sanchez, J. C., Pasquali, C., Ravier, F., Paquet, N., Frutiger, S., Hughes, G. J., Hochstrasser, D. (1993b) *Electrophoresis* **14,** 1375–1378.

Blohm, D. H., Giuseppi-Elie, A. (2001) *Curr. Opin. Microbiol.* **12,** 41–47.

Blonder, J., Goshe, M. B., Moore, R. J., Pasa-Tolic, L., Masselon, C. D., Lipton, M. S., Smith, R. D. (2002) *J. Proteome Res.* **1,** 351–369.

Bollag, D. M., Edelstein, S. J. (1991b) *Protein Methods*, Wiley-Liss, New York.

Bordini, E., Hamdan, M., Righetti, P. G. (1999a) *Rapid Commun. Mass Spectrom.* **13,** 1818–1827.

Bordini, E., Hamdan, M., Righetti, P. G. (1999b) *Rapid Commun. Mass Spectrom.* **13,** 2209–2215.

Borrebaeck, C. A. K. (2000) *Immunol. Today* **21,** 379–382.

Borrebaeck, C. A. K., Ekstrom, S., Molmborg Hager, A. C., Nilson, J., Laurell, T., Marko-Varga, G. (2001) *BioTechniques* **30,** 1126–1132.

Bossi, A., Gelfi, C., Orsi, A., Righetti, P. G. (1994a) *J. Chromatogr. A* **686,** 121–128.

Bossi, A., Righetti, P. G., Vecchio, G., Severinsen, S. (1994b) *Electrophoresis* **15,** 1535–1540.

Boyle, M. D., Romer, T. G., Meeker, A. K., Sledjeski, D. D. (2001) *J. Microbiol. Methods* **46,** 87–97.

Brinkman, U. A. Th., Vreuls, R. J. J. (Eds.) (2003) First International Symposium on Comprehensive Multidimensional Gas Chromatography, *J. Chromatogr. A* **1019,** 1–284.

Bruschi, M., Musante, L., Candiano, G., Herbert, B., Antonucci, F., Righetti, P. G. (2003) *Proteomics* **3,** 821–825.

Burggraf, D., Weber, G., Lottspeich, F. (1995) *Electrophoresis* **16,** 1010–1015.

Burghes, A. H. M., Dunn, M. J., Dubowitz, V. (1982) *Electrophoresis* **3,** 354–360.

Caglio, S., Chiari, M., Righetti, P. G. (1994) *Electrophoresis* **15,** 209–214.

Candiano, G., Musante, L., Bruschi, M., Ghiggeri, G.M., Herbert, B., Antonucci, F., Righetti, P.G. (2002) *Electrophoresis* **23,** 292–297.

Cargile, B. J., Talley, D. L., Stephenson, Jr., J. L. (2004) *Electrophoresis* **25,** 936–945.

Cavagnino, D., Magni, P., Zilioli, G., Trestianu, S. (2003) *J. Chromatogr A.* **1019,** 211–220.

Celis, J. E. (Guest Ed.) (1989) Paper Symposium: Protein Databases in Two Dimensional Electrophoresis, *Electrophoresis* **10,** 71–164.

Celis, J. E. (Guest Ed.) (1990a) Paper Symposium: Cell Biology, *Electrophoresis* **11,** 189–280.

Celis, J. E. (Guest Ed.) (1990b) Paper Symposium: Two Dimensional Gel Protein Databases, *Electrophoresis* **11,** 987–1168.

Celis, J. E. (Guest Ed.) (1991) Paper Symposium: Two Dimensional Gel Protein Databases, *Electrophoresis* **12,** 763–996.

Celis, J. E. (Guest Ed.) (1992) Paper Symposium: Two Dimensional Gel Protein Databases, *Electrophoresis* **13,** 891–1062.

Celis, J. E. (Guest Ed.) (1994a) Paper Symposium: Electrophoresis in Cancer Research, *Electrophoresis* **15,** 307–556.

Celis, J. E. (Guest Ed.) (1994b) Paper Symposium: Two Dimensional Gel Protein Databases, *Electrophoresis* **15,** 1347–1492.

Celis, J. E. (Guest Ed.) (1995) Paper Symposium: Two Dimensional Gel Protein Databases, *Electrophoresis* **16,** 2175–2264.

Celis, J. E. (Guest Ed.) (1996) Paper Symposium: Two Dimensional Gel Protein Databases, *Electrophoresis* **17,** 1653–1798.

Celis, J. E. (Guest Ed.) (1999) Genomics and Proteomics of Cancer. *Electrophoresis* **20,** 223–428.

Celis, J. E., Ostergaard, M., Jensen, N. A., Gromova, I., Rasmussen, H. H., Gromov, P. (1998) *FEBS Lett.* **430,** 64–72.

Celis, J. E., Rasmussen, H. H., Madsen, P., Leffers, H., Honorè, B., Dejgaard, K., Gesser, B., Olsen, E., Gromov, P., Hoffman, H. J., Nielsen, M., Celis, A., Basse, B., Lauridsen, J. B., Ratz, G. P., Nielsen, H., Andersen, A. H., Walbrum, E., Kjaegaard, I., Puype, M., Van Damme, J., Vandekerckhove, J. (1992) *Electrophoresis* **13,** 893–959.

Chan, C., Warlow, R. S., Chapuis, P. H., Newland, R. C., Bokey, E. L. (1999) *Electrophoresis* **20,** 3467–3471.

Chen, J., Lee, C. S., Shen, Y., Smith, R. D., Baehrecke, E. H. (2002) *Electrophoresis* **23,** 3143–3148.

Chevallet, M., Santoni, V., Poinas, A., Rouquie, D., Fuchs, A., Kieffer, S., Rossignol, M., Lunardi, J., Garin, J., Rabilloud, T. (1998) *Electrophoresis* **19,** 1901–1909.

Choe, L. H., Lee, K. H. (2000) *Electrophoresis* **21,** 993–1000.

Clegg, J. B., Naughton, M. A., Weatherall, D. (1966) *J. Mol. Biol.* **19,** 91–101.

Cleveland, D. W., Fischer, S. G., Kirscner, M. W., Laemmli, U. K. (1977) *J. Biol. Chem.* **252,** 1102–1106.

Colas de Francs, C., Thiellement, H., de Vienne, D. (1985) *Plant Physiol.* **78,** 178–182.

Cordwell, S. J., Nouwens, A. S., Verrills, N. M., Basseal, D. J., Walsh, B. J. (2000) *Electrophoresis* **21,** 1094–1103.

Corthals, G. L., Molloy, M. P., Herbert, B. R., Williams, K. L., Gooley, A. A. (1997) *Electrophoresis* **18,** 317–323.

Corthals, G. L., Wasinger, C. V., Hochstrasser, D. F., Sanchez, J. C. (2000) *Electrophoresis* **21,** 1104–1115.

Cremer, F., Van de Walle, C. (1985) *Anal. Biochem.* **147,** 22–26.

Cull, M., McHenry, C. S. (1990) *Methods Enzymol.* **182,** 147–153.

Cutler, P. (2003) *Proteomics* **3,** 3–18.

Cutler, P., Heald, G., White, I. R., Ruan, J. (2003) *Proteomics* **3,** 392–401.

Damerval, C., de Vienne, D. (Guest Eds.) (1988) Paper Symposium: Two Dimensional Electrophoresis of Plant Proteins, *Electrophoresis* **9,** 679–796.

Damerval, C., de Vienne, D., Zivy, M., Thiellement, H. (1986) *Electrophoresis* **7,** 52–54.

Davidsson, P., Paulson, L., Hesse, C., Blennow, K., Nilsson, C. L. (2001) *Proteomics* **1,** 444–452.

Davis, M. T., Beierle, J., Bures, E. T., McGinley, M. D., Mort, J., Robinson, J. H., Spahr, C. S., Yu, W., Luethy, R., Pattersson, S. D. (2001) *J. Chromatogr. B* **752,** 281–291.

de Duve, C. (1965) *Harvey Lectures* **59,** 49–59.

de Duve, C., Berthet, J., Beaufay, H. (1959) *Progr. Biophys. Biophys. Chem.* **9,** 325–350.

de Duve, C., Pressman, B. C., Gianetto, R., Wattiaux, R., Appelmans, F. (1955) *Biochem. J.* **60,** 604–612.

Deutscher, M. P. (Ed.) (1990) *Methods Enzymol.* **182,** 1–894.

Dignam, J. D. (1990) *Methods Enzymol.* **182,** 194–203.

Dreger, M. (2003) *Eur. J. Biochem.* **270,** 589–599.

Duncan, R. Hershey, J. W. B. (1984) *Anal. Biochem.* **11,** 342–349.

Dunn, M. J. (1993) *Gel Electrophoresis: Proteins*, Bios Scientific Publisher, Oxford.

Dunn, M. J. (Guest Ed.) (1991) Paper Symposium: Biomedical Applications of Two-Dimensional Gel Electrophoresis, *Electrophoresis* **12,** 459–606.

Dunn, M. J. (Guest Ed.) (1995) 2D Electrophoresis: From Protein Maps to Genomes, *Electrophoresis* **16,** 1077–1326.

Dunn, M. J. (Guest Ed.) (1997) From Protein Maps to Genomes, Proceedings of the Second Siena Two-Dimensional Electrophoresis Meeting, *Electrophoresis* **18,** 305–662.

Dunn, M. J. (Guest Ed.) (1999) From Genome to Proteome: Proceedings of the Third Siena Two-Dimensional Electrophoresis Meeting, *Electrophoresis* **20,** 643–1122.

Dunn, M. J. (Guest Ed.) (2000) Proteomic Reviews, *Electrophoresis* **21,** 1037–1234.

Dunn, M. J., Burghes, A. H. M. (1983) *Electrophoresis* **4,** 97–116.

Everberg, H., Sivars, U., Emanuelsson, C., Persson, C., Englund, A. K., Haneskog, L., Lipniunas, P., Jornten-Karlsson, M., Tjerneld, F. (2004) *J. Chromatogr. A* **1029,** 113–124.

Fawcett, J. S. (1973) *Ann. N. Y. Acad. Sci.* **209,** 112–126.

Ferro, M., Seigneurin-Berny, D., Rolland, N., Chapel, A., Salvi, D., Garin, J., Joyard, J. (2000) *Electrophoresis* **21,** 3517–3526.

Figeys, D., Pinto, D. (2001) *Electrophoresis* **22,** 208–216.

Flengsrud, R., Kobro, G. A. (1989) *Anal. Biochem.* **177,** 33–36.

Fountoulakis, M., Langen, H., Evers, S., Gray, C., Takacs, B. (1997) *Electrophoresis* **18,** 1193–1202.

Fountoulakis, M., Langen, H., Gray, C., Takacs, B. (1998) *J. Chromatogr. A* **806,** 279–291.

Fountoulakis, M., Takacs, B. (1998) *Protein Expr. Purif.* **14,** 113–119.

Fountoulakis, M., Takacs, M. F., Takacs, B. (1999) *J. Chromatogr. A* **833,** 157–168.

Fountoulakis, M., Takacs, M. F., Berndt, P., Langen, H., Takacs, B. (2000) *Electrophoresis* **20,** 2181–2195.

Fung, E. T., Thulasiraman, V., Weinberger, S. R., Dalmasso, E. A. (2001) *Curr. Opin. Biotechnol.* **12,** 65–69.

Galante, E., Caravaggio, T., Righetti, P. G. (1976) *Biochim. Biophys. Acta* **442,** 309–315.

Galvani, M., Hamdan, M., Herbert, B., Righetti, P. G. (2001) *Electrophoresis* **22,** 2058–2065.

Gåveby, B. M., Pettersson, P., Andrasko, J., Ineva-Flygare, L., Johannesson, U., Görg, A., Postel, W., Domscheit, A., Mauri, P. L., Pietta, P., Gianazza, E., Righetti, P. G. (1988) *J. Biochem. Biophys. Methods* **16,** 141–164.

Gelfi, C., Righetti, P. G. (1983) *J. Biochem. Biophys. Methods* **8,** 156–171.

Gengenheimer, P. (1990) *Methods Enzymol.* **182,** 174–193.

Gianazza, E., Astrua-Testori, S., Righetti, P. G. (1985a) *Electrophoresis* **6,** 113–117.

Gianazza, E., Giacon, P., Astrua-Testori, S., Righetti, P. G. (1985b) *Electrophoresis* **6,** 326–331.

Gianazza, E., Giacon, P., Sahlin, B., Righetti, P. G. (1985c) *Electrophoresis* **6,** 53–56.

Gianazza, E., Pagani, M., Luzzana, M., Righetti, P. G. (1975) *J. Chromatogr.* **109,** 357–364.

Gianazza, E., Rabilloud, T., Quaglia, L., Caccia, P., Astrua-Testori, S., Osio, L., Grazioli, G., Righetti, P. G. (1987) *Anal. Biochem.* **165,** 247–257.

Gianazza, E., Righetti, P. G. (1978) *Biochim. Biophys. Acta* **540,** 357–364.

Gianazza, E., Righetti, P. G. (1980) *J. Chromatogr.* **193,** 1–8.

Giddings, J. C. (1991) *Unified Separation Science*, Wiley, New York, pp. 122–138.

Glass, W. F., Briggs, R. C., Hnilica, L. S. (1981) *Science* **211,** 70–72.

Goerg, A., Postel, W., Gunther, S. (1988) *Electrophoresis* **9,** 531–546.

Goerg, A., Postel, W., Weser, J., Gunther, S., Strahler, J. R., Hanash, S. M., Somerlot, L. (1987) *Electrophoresis* **8,** 122–124.

Gordon, J. A., Jencks, W. P. (1963) *Biochemistry* **2,** 47–57.

Grandi, G. (2004) *Genomics, Proteomics and Vaccines*, Wiley, Chichester.

Granier, F. (1988) *Electrophoresis* **9,** 712–806.

Granzier, H. L. M., Wang, K. (1993) *Electrophoresis* **14,** 56–64.

Guttman, A., Varoglou, M., Khandurina, J. (2004) *Drug Discovery Today* **9,** 136–144.

Gygi, S. P., Corthals, G. L., Zhang, Y., Rochon, Y., Aebersold, R. (2000) *Proc. Natl. Acad. Sci. USA* **97,** 9390–9395.

Hagel, P., Gerding, J. J. T., Fieggen, W., Bloemendal, H. (1971) *Biochim. Biophys. Acta* **243,** 366–373.

Hanash, S. (1998) in Hames, B. D. (Ed.), *Gel Electrophoresis of Proteins*, Oxford University Press, Oxford, pp. 189–212.

Hannig, K. (1982) *Electrophoresis* **3,** 235–243.

Harju, M., Bergman, A., Olsson, M., Roos, A., Haglund, P. (2003) *J. Chromatogr A* **1019,** 127–142.

Harrington, M. G., Coffman, J. A., Calzone, F. J., Hood, L. E., Britten, R. J., Davidson, E. H. (1992) *Proc. Natl. Acad. Sci. USA* **89,** 6252–6256.

Harrison, P. A., Black, C. C. (1982) *Plant Physiol.* **70,** 1359–1366.

Harynuk, J., Gorecki, T. (2003) *J. Chromatogr A* **1019,** 53–64.

Heegaard, N. H. H., Poglod, R. (1991) *Appl. Theor. Electrophoresis* **2,** 109–127.

Heine, G., Raida, M., Forssmann, W. G. (1997) *J. Chromatogr. A* **766,** 117–124.

Heizmann, C. W., Arnold, E. M., Kuenzle, C. C. (1980) *J. Biol. Chem.* **255,** 11504–11511.

Herbert, B., Galvani, M., Hamdan, M., Olivieri, E., McCarthy, J., Pedersen, S., Righetti, P. G. (2001) *Electrophoresis* **22,** 2046–2057.

Herbert, B., Hopwood, F., Oxley, D., McCarthy, J., Laver, M., Grinyer, J., Goodall, A., Williams, K., Castagna, A., Righetti, P. G. (2003) *Proteomics* **3,** 826–831.

Herbert, B., Righetti, P. G., McCarthy, J., Grinyer, J., Castagna, A., Laver, M., Durack, M., Rummery, G., Harcourt, R., Williams, K. L. (2004) in Simpson, R. J. (Ed.), *Purifying Proteins for Proteomics*, Cold Spring Harbor Laboratory, Cold Spring Harbor, NY, pp. 431–442.

Herbert, B. R., Molloy, M. P., Gooley, B. A. A., Walsh, J., Bryson, W. G., Williams, K. L. (1998) *Electrophoresis* **19,** 845–851.

Herbert, B. R., Righetti, P. G. (2000) *Electrophoresis* **21,** 3639–3648.

Herbert, B. R., Sanchez, J. C., Bini, L. (1997) in Wilkins, M. R., Williams, K. L., Appel, R. D., Hochstrasser, D. F. (Eds.), *Proteome Research: New Frontiers in Functional Genomics*, Springer, Berlin, pp. 13–33.

Hickman, B. J., Anderson, N. L., Willard, K. E., Anderson, N. G. (1980) in Radola, B. J. (Ed.), *Electrophoresis '79: Advanced Methods, Biochemical and Chemical Applications*, de Gruyter, Berlin, pp. 341–360.

Hirabayashi, T. (2000) *Electrophoresis* **21,** 446–451.

Hjertén, S. (1967) *Chromatogr. Rev.* **9,** 122–219.

Hochstrasser, A. C., James, R. W., Pometta, D., Hochstrasser, D. (1991) *Appl. Theor. Electrophor.* **1,** 333–337.

Hochstrasser, D. F., Harrington, M. G., Hochstrasser, A. C., Miller, M. J., Merill, C. R. (1988) *Anal. Biochem.* **173,** 424–435.

Hoffmann, P., Ji, H., Moritz, R. L., Connolly, L. M., Frecklington, D. F., Layton, M. J., Eddes, J. S., Simpson, R. J. (2001) *Proteomics* **1,** 807–818.

Holloway, P., Arundel, P. (1988) *Anal. Biochem.* **172,** 8–15.

Horvàth, Z. S., Gooley, A. A., Wrigley, C. W., Margolis, J., Williams, K. L. (1996) *Electrophoresis* **17,** 224–226.

Hoving, S., Voshol, H., van Oostrum, J. (2000) *Electrophoresis* **21,** 2617–2621.

Huber, L. A. (Guest Ed.) (2000) Paper Symposium: Proteomics of Cell Organelles, *Electrophoresis* **21,** 3329–3528.

Huber, L. A., Pfaller, K., Vietor, I. (2003) *Circulation Res.* **92,** 962–968.

Humphery-Smith (Guest Ed.) (1997) Paper Symposium: Microbial Proteomes, *Electrophoresis* **18,** 1207–1497.

Issaq, H. J. (2001) *Electrophoresis* **22,** 3629–3638.

Issaq, H. J., Conrads, T. P., Janini, G. M., Veenstra, T. D. (2002) *Electrophoresis* **23,** 3048–3061.

James, P. (2001) *Proteome Research: Mass Spectrometry*, Springer, Berlin.

Jawinski, S. M. (1990) *Methods Enzymol.* **182,** 154–174.

Jenkins, R. E., Pennington, S. R. (2001) *Proteomics* **1,** 13–29.

Jiang, L., He, L., Fountoulakis, M. (2004) *J. Chromatogr. A* **1023,** 317–320.

Jones, M. N. (1999) *Int. J. Pharm.* **177,** 137–139.

Jonsson, M., Rilbe, H. (1980) *Electrophoresis* **1,** 3–14.

Jung, E., Heller, M., Sanchez, J. C., Hochstrasser, D. F. (2000a) *Electrophoresis* **21,** 3369–3377.

Jung, E., Hoogland, C., Chiappe, D., Sanchez, J. C., Hochstrasser, D. F. (2000b) *Electrophoresis* **21,** 3483–3487.

Kadofuku, T., Sato, T., Manabe, T., Okuyama, T. (1983) *Electrophoresis* **4,** 427–431.

Kallio, M., Hyotylainen, T., Lehtonen, M., Jussila, M., Hartonen, K., Shimmo, M., Riekkola, M. L. (2003) *J. Chromatogr A* **1019,** 251–260.

Karlsson, K., Cairns, N., Lubec, G., Fountoulakis, M. (1999) *Electrophoresis* **20,** 2970–2976.

Kawaguchi, S. I., Karamitsu, S. (1995) *Electrophoresis* **16,** 1060–1066.

Kellner, R., Lottspeich, F., Meyer, H. E. (1999) *Microcharacterization of Proteins*, Wiley-VCH, Weinheim.

King, J. S. (Guest Ed.) (1984) Special Issue: Two-Dimensional Gel Electrophoresis, *Protein Mapping, Clin. Chem.* **30,** 1897–2108.

Klose, J. (1975) *Humangenetik* **26,** 231–243.

Kobayashi, H., Shimamura, K., Akaida, T., Sakano, E., Tajima, N., Funazaki, J., Suzuki, H., Shinohara, E. (2003) *J. Chromatgr. A* **990,** 169–178.

Kramer, M. R., Malone, J. P. (2000) in Pollini, V., Bini, L. (Eds.) *Abstract Book of the Fourth Siena Meeting: From Genome to Proteome: Knowledge, Acquisition and Representation*, Siena, September 4–7, p. 269.

Krapfenbauer, K., Fountoulakis, M., Lubec, G. (2003) *Electrophoresis* **24,** 1847–1870.

Kristenson, E. M., Korytar, P., Danielsson, C., Kallio, M., Brandt, M., Makela, J., Vreuls, R. J. J., Beens, J., Brinkman, U. A. Th. (2003) *J. Chromatogr A* **1019,** 65–78.

Krivankova, L., Bocek, P. (1998) *Electrophoresis* **19,** 1064–1074.

Kuhn, R., Wagner, H. (1989) *J. Chromatogr.* **481,** 343–350.

Laemmli, U. K. (1970) *Nature* **227,** 680–685.

Lahm, H. W., Langen, H. (2000) *Electrophoresis* **21,** 2105–2114.

Lenstra, J. A., Bloemendal, H. (1983) *Eur. J. Biochem.* **135,** 413–423.

Lin, S., Tornatore, P., King, D., Orlando, R., Weinberger, S. R. (2001) *Proteomics* **1,** 1172–1184.

Link, A. J., Eng, J., Schieltz, D. M., Carmack, E., Mize, G. J., Morris, D. R., Garvik, B. M., Yates III, J. R. (1997) *Nature Biotech.* **17,** 676–682.

Liu, Z., Phillips, J. B. (1991) *J. Chromatogr. Sci.* **29,** 227–237.

Locke, V. L., Gibson, T. S., Thomas, T. M., Corthals, G. L., Rylatt, D. B. (2002) *Proteomics* **2,** 1254–1260.

Lottspeich, F. (Guest Ed.) (1996) Paper Symposium: Electrophoresis and Amino Acid Sequencing, *Electrophoresis* **17,** 811–966.

Lottspeich, F., Houthaeve, T., Kellner, R. (1999) in Kellner, R., Lottspeich, F., Meyer, H. E. (Eds.), *Microcharacterization of Proteins*, Wiley-VCH, Weinheim, pp. 141–158.

Lubman, D. M., Kachman, M. T., Wang, H., Gong, S., Yan, F., Hamler, R. L., O'Neil, K. A., Zhu, K., Buchanan, N. S., Barder, T. J. (2002) *J. Chromatogr. B* **782,** 183–196.

Luche, S., Diemer, H., Tastet, C., Chevallet, M., van Dorsselaer, A., Leize-Wagner, E., Rabilloud, T. (2004) *Proteomics* **4,** 551–561.

Luche, S., Santoni, V., Rabilloud, T. (2003) *Proteomics* **3,** 249–253.

Lueking, A. M., Horn, H., Eickhoff, H., Lehrqach, H., Walter, G. (1999) *Anal. Biochem.* **270,** 103–111.

MacBeath, G., Schreiber, S. L. (2000) *Science* **289,** 1760–1763.

Mahon, P., Dupree, P. (2001) *Electrophoresis* **22,** 2075–2085.

Malone, J. P., Kramer, M. R. (2000) in Pollini, V., Bini, L. (Eds.) *Abstract Book of the Fourth Siena Meeting: From Genome to Proteome: Knowledge, Acquisition and Representation*, Siena, September 4–7, p. 270.

Manabe, T. (2000) *Electrophoresis* **21,** 1112–1116.

Manabe, T., Hayama, E., Okuyama, T. (1982) *Clin. Chem.* **28,** 824–827.

Manabe, T., Kojima, K., Jitzukawa, S., Hoshino, T., Okuyama, T. (1981) *J. Biochem.* **89,** 1317–1323.

Manabe, T., Mizuma, H., Watanabe, K. (1999) *Electrophoresis* **20,** 830–835.

Manabe, T., Tachi, K., Kojima, K., Okuyama, T. (1979) *J. Biochem.* **85,** 649–659.

Manabe, T., Takahashi, Y., Higuchi, N., Okuyama, T. (1985) *Electrophoresis* **6,** 462–467.

Manabe, T., Visvikis, S., Dumon, M.F., Clerc, M., Siest, G. (1987a) *Electrophoresis* **8,** 468–472.

Manabe, T., Visvikis, S., Steinmetz, J., Galteau, M. M., Okuyama, T. Siest, G. (1987b) *Electrophoresis* **8,** 325–330.

Marengo, E., Leardi, R., Robotti, E., Righetti, P. G., Antonucci, F. Cecconi, D. (2003a) *J. Proteome Res.* **2,** 351–360.

Marengo, E., Robotti, E., Gianotti, V., Righetti, P. G. (2003b) *An. Chim.* **93,** 105–116.

Marengo, E., Robotti, E., Righetti, P. G., Antonucci, F. (2003c) *J. Chromatogr. A* **1004,** 13–28.

Margolis, J., Corthals, G., Horvàth, Z. S. (1995) *Electrophoresis* **16,** 98–100.

Marshall, T., Abbott, N. J., Fox, P., Williams, K. M. (1995) *Electrophoresis* **16,** 28–31.

Marshall, T., Williams, K. M. (1992) *Electrophoresis* **13,** 887–888.

Matsui, N. M., Smith, D. M., Clauser, K. R., Fichmann, J., Andrews, L. E., Sullivan, C. M., Burlingame, A. L., Epstein, L. B. (1997) *Electrophoresis* **18,** 409–417.

McCarthy, J., Hopwood, F., Oxley, D., Laver, M., Castagna, A., Righetti, P. G., Williams, K., Herbert, B. (2003) *J. Proteome Res.* **2,** 239–242.

Menke, W., Koenig, F. (1980) *Methods Enzymol.* **69,** 446–452.

Merchant, M., Weinberger, S. R. (2000) *Electrophoresis* **21,** 1164–1167.

Michel, P. E., Reymond, P., Arnaud, I. L., Josserand, J., Girault, H. H., Rossier, J. S. (2003) *Electrophoresis* **24,** 3–11.

Mineki, R., Taka, H., Fujimura, T., Kikkawa, M., Shindo, N., Murayama, K. (2002) *Proteomics* **2,** 1672–1681.

Mohan, D., Lee, C. S. (2002) *Electrophoresis* **23,** 3160–3167.

Mohberg, J., Rusch, H.P. (1969) *Arch. Biochem. Biophys.* **134,** 577–589.

Molloy, M. P. (2000) *Anal. Biochem.* **280,** 1–10.

Molloy, M. P., Brzezinski, E. E., Hang, J., McDowell, M. T., VanBogelen, R. A. (2003) *Proteomics* **3,** 1912–1919.

Molloy, M. P., Herbert, B. R., Slade, M. B., Rabilloud, T., Nouwens, A. S., Williams, K. L., Gooley, A. A. (2000) *Eur. J. Biochem.* **267,** 1–12.

Molloy, M. P., Herbert, B. R., Walsh, B. J., Tyler, M. I., Traini, M., Sanchez, J. C., Hochstrasser, D. F., Williams, K. L., Gooley, A. A. (1998) *Electrophoresis* **19,** 837–844.

Molloy, M. P., Herbert, B. R., Williams, K. L., Gooley, A. A. (1999) *Electrophoresis* **20,** 701–704.

Mondello, L., Casilli, A., Tranchida, P. Q., Dugo, P., Dugo, G. (2003) *J. Chromatogr A* **1019,** 187–196.

Moritz, B., Meyer, H. E. (2003) *Proteomics* **3,** 2208–2220.

Mosher, R. A., Bier, M., Righetti, P. G. (1986) *Electrophoresis* **7,** 59–66.

Musante, L., Candiano, G., Ghiggeri, G. M. (1997) *J. Chromatogr. A* **705,** 351–356.

Nikaido, H. (1996) in Neidhardt, F. C. (Ed.), *Escherichia coli and Salmonella Cellular and Molecular Biology*, ASM Press, Washington DC, pp. 29–47.

Oda, Y., Huang, K., Cross, F. R., Cowburn, D., Chait, B. T. (1999) *Proc. Natl. Acad. Sci. USA* **96,** 6591–6596.

O'Farrell, P. H. (1975) *J. Biol. Chem.* **250,** 4007–4021.

O'Farrell, P. Z., Goodman, H. M., O'Farrell, P. H. (1977) *Cell* **12,** 1133–1140.

Oh-Ishi, M., Satoh, M., Maeda, T. (2000) *Electrophoresis* **21,** 1653.

Olivieri, E., Herbert, B., Righetti, P. G. (2001) *Electrophoresis* **22,** 560–565.

Olsson, I., Larsson, K., Palmgren, R., Bjellqvist, B. (2002) *Proteomics* **2,** 1630–1632.

Opiteck, G. J., Lewis, K. C., Jorgenson, J. W., Anderegg, R. J. (1997) *Anal. Chem.* **69,** 1518–1524.

Opiteck, G. J., Ramirez, S. M., Jorgenson, J. W., Moseley III, M. A. (1998) *Anal. Biochem.* **258,** 349–361.

Panek, J., Vohradsky, J. (1999) *Electrophoresis* **20,** 3483–3491.

Pasquali, C., Fialka, I., Huber, L. A. (1997) *Electrophoresis* **18,** 2574–2581.

Patterson, S. D., Aebersold, R. (1995) *Electrophoresis* **16,** 1791–1814.

Paweletz, C. P., Liotta, L. A., Petricoin III, E. F. (2001) *Urology* **57,** 160–164.

Pedersen, S. K., Harry, J. L., Sebastian, L., Baker, J., Traini, M. D., McCarthy, J. T., Manoharan, A., Wilkins, M. R., Gooley, A. A., Righetti, P. G., Packer, N. H., Williams, K. L., Herbert, B. (2003) *J. Proteome Res.* **2,** 303–312.

Penefsky, H. S., Tzagoloff, A. (1971) *Methods Enzymol.* **22,** 204–219.

Peng, J., Elias, J. E., Thoreen, C. C., Licklider, L. J. I. Gygi, S. P. (2003) *J. Proteome Res.* **2,** 43–50.

Pennington, S. R., Wilkins, M. R., Hochstrasser, D. F., Dunn, M. J. (1997) *Trends Cell Biol.* **7,** 168–173.

Perdew, G. H., Schaup, H. W., Selivonchick, D. P. (1983) *Anal. Biochem.* **135,** 453–459.

Phillips, J. B., Xu., J. (1995) *J. Chromatogr. A* **703,** 327–337.

Pieper, R., Gatlin, C. L., Makusky, A. J., Russo, P. S., Schatz, C. R., Miller, S. S., Su, Q., McGrath, A. M., Estock, M. A., Parmar, P. P., Zhao, M., Huang, S. T., Zhou, J., Wang, F., Esquer-Blasco, R., Anderson, N. L., Taylor, J., Steiner, S. (2003) *Proteomics* **3,** 1345–1364.

Pietrogrande, M. C., Marchetti, N., Dondi, F., Righetti, P. G. (2002) *Electrophoresis* **23,** 283–291.

Pietrogrande, M.C., Marchetti, N., Dondi, F., Righetti, P. G. (2003) *Electrophoresis* **24,** 217–224.

Pleissner, K. P., Hoffman, F., Kriegel, K., Wenk, C., Wegner, S., Sahlstrom, A., Oswald, H., Alt, H., Fleck, E. (1999) *Electrophoresis* **20,** 755–765.

Poehling, H. M., Wyss, U., Neuhoff, V. (1980) *Electrophoresis* **1,** 198–204.

Rabilloud, T. (1996) *Electrophoresis* **17,** 813–829.

Rabilloud, T. (1998) *Electrophoresis* **19,** 758–760.

Rabilloud, T. (2000a) *Anal. Chem.* **72,** 48A–55A.

Rabilloud, T. (2000b) *Proteome Research: Two-Dimensional Gel Electrophoresis and Identification Methods*, Springer, Heidelberg.

Rabilloud, T., Adessi, C., Giraudel, A., Lunardi, J. (1997) *Electrophoresis* **18,** 307–316.

Rabilloud, T., Blisnick, T., Heller, M., Luche, S., Aebersold, R., Lunardi, J., Braun-Breton, C. (1999) *Electrophoresis* **20**, 3603–3610.

Rabilloud, T., Chevallet, M. (2000) in Rabilloud, T. (Ed.), *Proteome Research: Two-Dimensional Gel Electrophoresis and Identification Methods*, Springer, Heidelberg, pp. 9–29.

Rabilloud, T., Gelfi, C., Bossi, M. L., Righetti, P. G. (1987a) *Electrophoresis* **8,** 305–312.

Rabilloud, T., Gianazza, E., Cattò, N., Righetti, P. G. (1990) *Anal. Biochem.* **185,** 94–102.

Rabilloud, T., Pernelle, J. J., Wahrmann, P., Gelfi, C., Righetti, P. G. (1987b) *J. Chromatogr.* **402,** 105–113.

Rabilloud, T., Valette, C., Lawrence, J. J. (1994) *Electrophoresis* **15,** 1552–1558.

Radin, N. S. (1981) *Methods Enzymol.* **72,** 5–7.

Raida, M., Schultz-Knape, P., Heine, G., Forssmann, W. G. (1999) *J. Am. Mass Spectrom.* **10,** 45–54.

Raman, B., Cheung, A., Marten, M. R. (2002) *Electrophoresis* **23,** 2194–2202.

Rammensee, H. G. (2004) *Nature* **427,** 203–204.

Rhodes, A., Smithers, N., Chapman, T., Parsons, S., Rees, S. (2001) *FEBS Lett.* **506,** 85–90.

Richter, R., Schultz-Knape, P., Schrader, M., Standkerm L., Jurgens, M., Tammen, H., Forssmann, W.G. (1999) *J. Chromatogr. B* **726,** 25–35.

Righetti, P. G. (1980) *J. Chromatogr.* **190,** 275–282.

Righetti, P. G. (1990) *Immobilized pH Gradients: Theory and Methodology*, Elsevier, Amsterdam, 1990.

Righetti, P. G., Barzaghi, B., Sarubbi, E., Soffientini, A., Cassani, G. (1989a) *J. Chromatogr.* **470,** 337–350.

Righetti, P. G., Brown, R., Stone, A. L. (1978) *Biochim. Biophys. Acta* **542,** 222–231.

Righetti, P. G., Castagna, A., Herbert, B. (2001a) *Anal. Chem.* **73,** 320A–326A.

Righetti, P. G., Castagna, A., Herbert, B., Reymond, F., Rossier, J. S. (2003) *Proteomics* **3,** 1397–1407.

Righetti, P. G., Chiari, Casale, E., Chiesa, C. (1989b) *Appl. Theor. Electrophor.* **1,** 115–121.

Righetti, P. G., Chiari, M., Gelfi, C. (1988a) *Electrophoresis* **9,** 65–73.

Righetti, P. G., Chiari, M., Sinha, P. K., Santaniello, E. (1988b) *J. Biochem. Biophys. Methods* **16,** 185–192.

Righetti, P. G., Faupel, M., Wenisch, E. (1992) in Chrambach, A., Dunn, M., Radola, B. J., (Eds.), *Advances in Electrophoresis*, Vol. 5, VCH, Weiheim, pp. 159–200.

Righetti, P. G., Gelfi, C. (1984) *J. Biochem. Biophys. Methods* **9,** 103–119.

Righetti, P. G., Gianazza, E., Salamini, F. Galante, E., Viotti, A., Soave, C. (1977) in Radola, B.J., Graesslin, D. (Eds.), *Electrofocusing and Isotachophoresis* W. de Gruyter, Berlin, pp. 199–211.

Righetti, P. G., Stoyanov, A. L., Zhukov, M. Y. (2001b) *The Proteome Revisited*, Elsevier, Amsterdam.

Righetti, P. G., Tonani, C. (1991) *Electrophoresis* **12,** 1021–1027.

Righetti, P. G., Tudor, G., Gianazza, E. (1982) *J. Biochem. Biophys. Methods* **6,** 219–227.

Righetti, P. G., Wenisch, E., Faupel, M. (1989c) *J. Chromatogr.* **475,** 293–309.

Righetti, P. G., Wenisch, E., Jungbauer, A., Katinger, H., Faupel. M. (1990) *J. Chromatogr.* **500,** 681–696.

Rimpilainen, M. A., Righetti, P. G. (1985) *Electrophoresis* **6,** 419–422.

Romijn, E. P., Krijgsveld, J., Heck, A. J. R. (2003) *J. Chromatogr. A* **1000,** 589–608.

Ros, A., Faupel, M., Mees, H., Oostrum, J. V., Ferrigno, R., Reymond, F., Michel, P., Rossier, J. S., Girault, H. H. (2002) *Proteomics* **2,** 151–156.

Rosengren, A. T., Salmi, J. M., Aittokallio, T., Westerholm, J., Lahesmaa, R., Nyman, T. A., Nevalainen, O. S. (2003) *Proteomics* **3,** 1936–1946.

Rubinfeld, A., Keren-Lehrer, T., Hadas, G., Smilansky, Z. (2003) *Proteomics* **3,** 1930–1935.

Ruechel, R. (1977) *J. Chromatogr.* **132,** 451–459.

Ruegg, U. T., Ruedinger, J. (1977) *Methods Enzymol.* **47,** 111–116.

Sanchez, J. C., Rouge, V., Pisteur, M., Ravier, F., Tonella, L., Moosmayer, M., Wilkins, M. R., Hochstrasser, D. (1997) *Electrophoresis* **18,** 324–327.

Sanchez-Ferrer, A., Bru, R., Garcia-Carmona, F. (1994) *Crit. Rev. Biochem. Mol. Biol.* **29,** 275–313.

Sanders, M. M., Groppi, V. E., Browning, E. T. (1980) *Anal. Biochem.* **103,** 157–165.

Santoni, V., Kieffer, S., Desclaux, D., Masson, F., Rabilloud, T. (2000a) *Electrophoresis* **21,** 3329–3344.

Santoni, V., Molloy, M., Rabilloud, T. (2000b) *Electrophoresis* **21,** 1054–1070.

Santoni, V., Rabilloud, T., Doumas, P., Rouquie, D., Fuchs, A., Kieffer, S., Garin, J., Rossignol, M. (1999) *Electrophoresis* **20,** 705–711.

Sariouglu, H., Lottspeich, F., Walk, T., Jung, G., Eckerskorn, C. (2000) *Electrophoresis* **21,** 2209–2218.

Satta, D., Schapira, G., Chafey, P., Righetti, P. G., Wahrmann, J. P. (1984) *J. Chromatogr.* **299,** 57–72.

Sawyer, T. K. (2001) *BioTechniques* **31,** 156–160.

Scheele, G. A. (1975) *J. Biol. Chem.* **250,** 5375–5385.

Scherl, A., Coute, Y., Deon, C., Calle, A., Kindbeiter, K., Sanchez, J. C., Greco, A., Hochstrasser, D. F., Diaz, J. J. (2002) *Mol. Biol. Cell.* **13,** 4100–4109.

Schrader, M., Jurgens, M., Hess, R., Schultz-Knape, P., Raida, M., Forssmann, W. G. (1997) *J. Chromatogr. A* **766,** 139–145.

Schupbach, J., Ammann, R. W., Freiburghaus, A. U. (1991) *Anal. Biochem.* **196,** 337–343.

Sebastiano, R., Citterio, A., Lapadula, M., Righetti, P. G. (2003) *Rapid Commun. Mass Spetrom.* **17,** 2380–2386.

Shaw, M. M., Riederer, B. M. (2003) *Proteomics* **3,** 1408–1417.

Shen, Y., Smith, R. D. (2002) *Electrophoresis* **23,** 3106–3124.

Shen, Y., Tlic, N., Zhao, R., Pasa-Tolic, L., Li, L., Berger, S. J., Harkewicz, R., Anderson, G. A., Belov, M. E., Smith, R. D. (2001) *Anal. Chem.* **73,** 3011–3021.

Shimura, K., Kamiya, K., Matsumoto, H., Kasai, K. (2002) *Anal. Chem.* **74,** 1046–1053.

Simões-Barbosa, A., Santana, J. M., Teixeira, A. R. L. (2000) *Electrophoresis* **21,** 641–644.

Sinha, P. K., Praus, M., Köttgen, E., Gianazza, E., Righetti, P. G. (1990) *J. Biochem. Biophys. Methods* **21,** 173–179.

Soulet, N., Roux-de-Balmann, H., Sanchez, V. (1998) *Electrophoresis* **19,** 1294–1299.

Stannard, C., Brown, L. R., Godovac-Zimmermann, J. (2004) *Current Proteomics* **1,** 13–25.

Stoll, D. (2003) *Drug Plus Int.* **2,** 12–16.

Tastet, C., Charmont, S., Chevallet, M., Luche, S., Rabilloud, T. (2003) *Proteomics* **3,** 111–121.

Taylor, S. W., Warnock, D. E., Glenn, G. M., Zhang, B., Fahy, E, Gaucher, S. P., Capaldi, R. A., Gibson, B. W., Ghosh, S. S. (2002) *J. Proteome Res.* **1,** 451–458.

Teixera-Gomes, A. P., Cloeckaert, A., Bezard, G., Dubray, G., Zygmunt, M. S. (1997) *Electrophoresis* **18,** 156–162.

Theillet, C., Delpeyroux, F., Fiszman, M., Reigner, P., Esnault, R. (1982) *Planta* **155,** 478–485.

Thulasiraman, V., McCutchens-Maloney, S. L., Motin, V. L., Garcia, E. (2000) *BioTechniques* **30,** 428–432.

Toda, T., Ishijima, Y., Matsushita, H., Yoshida, M., Kimura, N. (1994) *Electrophoresis* **15,** 984–987.

Tollaksen, S. L., Edwards, J. J., Anderson, N. G. (1981) *Electrophoresis* **2,** 155–160.

Tonani, C., Righetti, P. G. (1991) *Electrophoresis* **12,** 1011–1021.

Tracy, R. P., Currie, R. M., Young, D. S. (1982) *Clin. Chem.* **28,** 908–913.

Tsugita, A., Kamo, M., Kawakami, T., Ohki, Y. (1996) *Electrophoresis* **17,** 855–865.

Tümmler, B. (Guest Ed.) (1998) Microbial Genomes: Biology and Technology, *Electrophoresis* **19,** 467–624.

Turck, N., Richert, S., Gendry, P., Stutzmann, J., Kedinger, M., Leize, E., Simon-Assmann, P., Van Dorsselaer, A., Launay, J. F. (2004) *Proteomics* **4,** 93–105.

Usada, H., Shimogawara, K. (1995) *Plant Cell Physiol.* **36,** 1149–1155.

Van Renswoude, J., Kemps, C. (1984) *Methods Enzymol.* **104,** 329–339.

Van Stee, L. L. P., Beens, J., Vreuls, R. J. J., Brinkman, U. A. Th (2003) *J. Chromatogr. A* **1019,** 89–100.

Voivodov, K. I., Ching, J., Hutchens, T. W. (1996) *Tetrahedron Lett.* **37,** 5669–5672.

Von Eggeling, F., Davies, H., Lomas, L., Fiedler, W., Junker, K., Claussen, U., Ernst G. (2000) *BioTechiques* **29,** 1066–1070.

Voss, T., Haberl, P. (2000) *Electrophoresis* **21,** 3345–3350.

Vuong, G. L., Weiss, S. M., Kammer, W., Priemer, M., Vingron, M., Nordheim, A., Cahill, M. A. (2000) *Electrophoresis* **21,** 2594–2605.

Wagner, K., Miliotis, T., Marko-Varga, G., Bischoff, R., Unger, K. K. (2002) *Anal. Chem.* **74,** 809–820.

Wall, D. B., Kachman, M. T., Gong, S., Hinderer, R., Parus, S., Misek, D. E., Hanash, S. M., Lubman, D. M. (2000) *Anal. Chem.* **72,** 1099–1111.

Wang, H., Hanash, S. (2003) *J. Chromatogr. B* **787,** 11–18.

Wang, H., Kachman, M. T., Schwartz, D. R., Cho, K. R., Lubman, D. M. (2002) *Electrophoresis* **23,** 3168–3181.

Wang, S., Diamond, D. L., Hass, G. M., Sokoloff, R., Vessella, R. L. (2001) *Int. J. Cancer* **92,** 871–876.

Washburn, M. P., Wolters, D., Yates, J. R. (2001) *Nature Biotech.* **19,** 242–247.

Washburn, M. P., Yates III, J. R. (2000) in Blackstock, W., Mann, M. (Eds.), *Proteomics: A Trend Guide*, Elsevier, London, pp. 27–30.

Weinberger, S., Viner, R. I., Ho, P. (2002) *Electrophoresis* **23,** 3182–3192.

Welthagen, W., Schnelle-Kreis, J., Zimmermann, R. (2003) *J. Chromatogr A* **1019,** 233–250.

Wessel, D., Flugge, U. I. (1984) *Anal. Biochem.* **138,** 141–143.

Westbrook, J. A., Yan, J. X., Wait, R., Welson, S. Y., Dunn, M. J. (2001) *Electrophoresis* **22,** 2865–2871.

Westermeier, R. (1997) *Electrophoresis in Practice*, VCH, Weinheim, pp. 213–228.

Westermeier, R., Naven, T. (2002) *Proteomics in Practice*, Wiley-VCH, Weinheim.

Wildgruber, R., Harder, A., Obermeier, C., Boguth, G., Weiss, W., Fey, S. J., Larsen, P. M., Goerg, A. (2000) *Electrophoresis* **21,** 2610–2616.

Wilkins, M. R., Gasteiger, E., Sanchez, J. C., Bairoch, A., Hochstrasser, D. F. (1998) *Electrophoresis* **19,** 1501–1505.

Wilkins, M. R., Pasquali, C., Appel, R. D., Ou, K., Golaz, O., Sanchez, J. C., Yan, J. X., Gooley, A. A., Hughes, G., Humphrey-Smith, I., Williams, K. L., Hochstrasser, D. F. (1996) *BioTechnology* **14,** 61–65.

Wilkins, M. R., Williams, K. L., Appel, R. D., Hochstrasser, D. F. (Eds.) (1997) *Proteome Research: New Frontiers in Functional Genomics*, Springer, Berlin.

Willard, K. E., Giometti, C., Anderson, N. L., O'Connor, T. E., Anderson, N. G. (1979) *Anal. Biochem.* **100,** 289–298.

Williams, K. L. (Guest Ed.) (1998) Strategies in Proteome Research, *Electrophoresis* **19,** 1853–2050.

Willsie, J. K., Clegg, J. S. (2002) *J. Cell Biochem.* **84,** 601–614.

Wissing, J., Heim, S., Flohé, L., Bilitewski, U., Frank, R. (2000) *Electrophoresis* **21,** 2589–2593.

Wolpert, T. J., Dunkle, L. D. (1983) *Proc. Natl. Acad. Sci. USA* **80,** 6576–6580.

Wu, F. S., Wang, M. Y. (1984) *Anal. Biochem.* **139,** 100–103.

Xiao, Z., Conrads, T. P., Lucas, D. A., Janini, G. M., Schaefer, C. F., Buetow, K. H., Issaq, H. J., Veenstra, T. D. (2004) *Electrophoresis* **25,** 128–133.

Yan, J. X., Kett, W. C., Herbert, B. R., Gooley, A. A., Packer, N. H., Williams, K. L. (1998) *J. Chromatogr. A* **813,** 187–200.

Yan, J. X., Sanchez, J. C., Rouge, V., Williams, K., Hochstrasser, D. F. (1999) *Electrophoresis* **20,** 723–726.

Yates III, J. R., McCormack, A. L., Schieltz, D., Carmack, E., Link, A. (1997) *J. Protein Chem.* **16,** 495–497.

Yoshida, M., Shimura, K. (1972) *Biochim. Biophys. Acta* **263,** 690–695.

Young, D. S., Anderson. N. G. (Guest Eds.) (1982) Special Issue: Two-Dimensional Gel Electrophoresis, *Clin. Chem.* **28,** 737–1092.

Zhang, W., Czernik, A. J., Yungwirth, T., Aebersold, R., Chait, B. T. (1994) *Protein Sci.* **3,** 677–685.

Zhu, K., Yan, F., O'Neil, K. A., Hamler, R., Lin, L., Barder, T. J., Lubman, D. M. (2003) in Coligan, J.E., Dunn, B.M., Hidde, L.P., Speicher, D.W., Wingfield, P.T. (Eds.) *Current Protocols in Protein Science,* Vol. 3, Wiley, Hoboken, pp. 23.3.1–23.3.28.

Zilberstein, G. V., Baskin, E. M., Bukshpan, S. (2003) *Electrophoresis* **24,** 3735–3744.

Zischka, H., Weber, G., Weber, P. J. A., Posch, A., Braun, R. J., Buehringer, D., Schneider, U., Nissum, M., Meitinger, T., Ueffing, M., Eckerskorn, C. (2003) *Proteomics* **3,** 906–916.

Zuo, X., Speicher, D. W. (2000) *Electrophoresis* **21,** 3035–3047.

INDEX

Proteomics Today. By Mahmoud Hamdan and Pier Giorgio Righetti
ISBN 0-471-64817-5